D1156169

ENCYCLOPEDIA OF

PHYSICS

ENCYCLOPEDIA OF PHYSICS

JOE ROSEN, PH.D.

Facts On File, Inc.

Encyclopedia of Physics

Facts On File, Inc.
132 West 31st Street
New York NY 10001

Library of Congress Cataloging-in-Publication Data

Rosen, Joe, 1937–
Encyclopedia of physics / Joseph Rosen.
p. cm.
Includes bibliographical references and index.
ISBN 0-8160-4974-2 (hardcover)
1. Physics—Encyclopedias. I. Title.
QC5.R596 2004
530′.03—dc22
2003014963

Facts On File books are available at special discounts when purchased in bulk quantities for businesses, associations, institutions, or sales promotions. Please call our Special Sales Department in New York at 212/967-8800 or 800/322-8755.

You can find Facts On File on the World Wide Web at http://www.factsonfile.com.

Text design by Joan M. Toro
Cover design by Cathy Rincon
Illustrations by Sholto Ainslie

Printed in the United States of America

VB Hermitage 10 9 8 7 6 5 4 3 2

This book is printed on acid-free paper.

To Mira

CONTENTS

ACKNOWLEDGMENTS

The following institutions and organizations are gratefully acknowledged for supplying information, text, and images and allowing their use in this work: American Institute of Physics, American Physical Society, Annenberg Rare Book and Manuscript Library of the University of Pennsylvania, Bachrach Portrait Photography, Brookhaven National Laboratory, Burndy Library of the Massachusetts Institute of Technology, Deutsches Museum, European Organization for Nuclear Research (Organisation Européenne de la Recherche Nucléaire, CERN), Fermi National Accelerator Laboratory (Fermilab), International Bureau of Weights and Measures (Bureau International des Poids et Mesures, BIPM), International Union of Pure and Applied Physics, Los Alamos National Laboratory, Lucent Technologies Inc./Bell Labs, NASA, National Institute of Standards and Technology (NIST), Nimitz Library of the U.S. Naval Academy, Oak Ridge National Laboratory, and Stanford Linear Accelerator Center (SLAC).

In particular, thanks to the Friends of the Center for History of Physics of the American Institute of Physics for their generous support of the AIP Emilio Segrè Visual Archives, from which most of the photographs in this work were obtained, and to Heather Lindsay at the American Institute of Physics, who helped me obtain them.

I am grateful to Frank K. Darmstadt, my editor at Facts On File, for guiding me through the process of encyclopedia writing and to the other people there who contributed to producing this work.

Especially, I thank my wife, Mira, for her wonderful support and for so gracefully putting up with my many and lengthy periods of disappearance into my study during the writing of this book.

INTRODUCTION

Encyclopedia of Physics is intended to reflect today's physics as it really is. As for *today,* very little of the history and development of physics was incorporated, and only where it enhances the appreciation of some issue. The idea is to present physics, as grasped by modern physicists, to students and general readers.

What physics *really* is can be quite different from the impression one might obtain from many high school, and even college, textbooks. All too many students come away convinced that physics consists solely of a disparate and all-too-large collection of facts, definitions, and formulas. It is true that facts, definitions, and formulas have their places, but physics is much more than those. Physics is, really, a way of understanding the most fundamental aspects of nature. This understanding is based on facts, of course. Definitions are needed to ensure a common language. And formulas offer important expressions of relationships among physical quantities. But it is the unifying understanding underlying it all that is the *real* physics. The *Encyclopedia* attempts to move that aspect of physics to the forefront and highlight it. To this end, 11 essays, scattered throughout the *Encyclopedia,* discuss various aspects of what physics really is about.

Among the entries, a sampling of physics-related laboratories, institutions, and organizations is included to give the reader an inkling of what is going on in this regard. The choices were quite arbitrary, and there are many more such establishments, of similar importance, that are not included.

More than 50 physicists were picked for brief biographical entries. Here, too, the choice was rather arbitrary, and many, many more physicists deserve to be included but had to be left out due to space limitations.

Numerous figures, including drawings and photographs, accompany and illustrate the text. Some of the concepts discussed, especially those involving spatial relations in three dimensions, almost demand to be illustrated in order to make them more readily understandable.

Vector notation is used throughout the *Encyclopedia* wherever it is appropriate, while component notation is usually included as well, for those readers not yet comfortable with vector notation. The entry VECTOR should be consulted for clarification, if necessary.

An appendix to the *Encyclopedia* presents a list of all the Nobel Prize laureates in physics since the inception of the prize in 1901, including the Nobel Committee's descriptions of the accomplishments for which the prize was awarded.

The importance of the index to *Encyclopedia of Physics* cannot be overemphasized. If the reader is looking for a term or concept that is not included among the entries, the index should be consulted. And for compound terms, such as *magnetic dipole,* in order not to miss anything it is best to look for each word composing the term, both among the entries and in the index.

Encyclopedia of Physics was written especially with high-school and college students in mind, although the general reader, too, should find it useful. This is an *introduction* to today's physics—a gateway, so to speak. It makes no pretense to substitute for deeper and broader sources. The bibliography offers suggestions for additional reading and study for those interested in further depth or breadth.

Entries A–Z

aberration In OPTICS, aberration refers to any deviation from the formation of an optimal image by a LENS or MIRROR. Of course, a low-quality lens or mirror will produce a distorted or blurry image. But even a well-manufactured lens or mirror possesses inherent limitations on the quality of its images. One of the most commonly considered types of aberration is spherical aberration. This aberration is due to the fact that most mirror and lens surfaces have the shape of a part of a spherical surface, and this shape simply does not precisely produce optimal images. What that shape does afford, however, is lower cost, since spherical surfaces are much easier, and thus cheaper, to fabricate than are surfaces that do form optimal images. (For mirrors, such surfaces have the form of a paraboloid.) Spherical lenses and mirrors are nonetheless effective because they can form good approximations to optimal images as long as the LIGHT rays involved are at sufficiently small angles from the optical axis. The condition for this to be valid is that the focal length be considerably larger than the diameter of the lens or mirror. Spherical aberration can be reduced by using lenses and mirrors of sufficiently small area or by masking (i.e., "stopping down") the outer part of the lens or mirror, thus confining the light rays to the central part.

Another commonly considered type of aberration is chromatic aberration. It comes about due to the dependence of the focal length of a lens on the color of the light. That, in turn, is a result of the dependence of

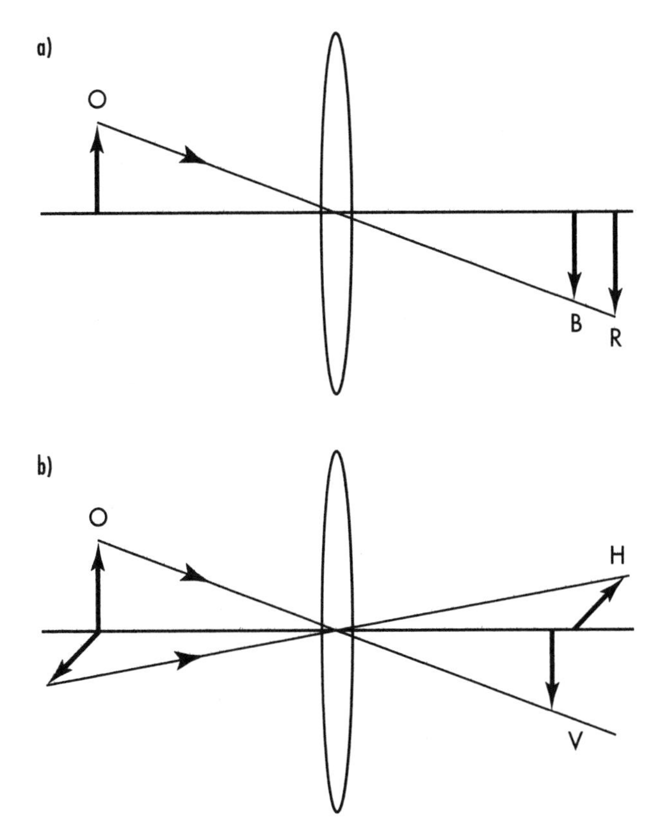

Two kinds of aberration in a lens. (a) Chromatic aberration causes light of different colors to be focused at different locations. In the figure, the red and blue components of white light from the object O each form a separate image, R and B, respectively. (b) In astigmatism, parts of the object O having different orientations, vertical and horizontal in the figure, are imaged at different locations and produce images V and H, respectively.

the INDEX OF REFRACTION of the lens material on the frequency of the light, an effect known as DISPERSION. What happens is that the same object is imaged by a lens at different locations, depending on the color of light used. With white light, as in ordinary photography, the result is an image that might be sharp for the red component of white light, say, while blurred for the blue component, which would degrade the total quality of the image. (Note that mirrors are not subject to chromatic aberration, since light is reflected off their surface and is not affected by the nature of what they are made of.) In order to reduce chromatic aberration, designers of optical systems combine lenses in such a way as to have the aberrations of the various components cancel each other out to the extent possible within the constraints of the project.

A third type of commonly considered aberration is astigmatism. This is normally a manufacturing fault. It occurs when the focal length depends on the orientation of the object (or of the lens), so that the image of the trunk of a tree, say, and that of the tree's horizontal branch are formed at different locations. In a photograph taken with an astigmatic lens, the trunk might appear sharp, while the branch is blurred. Astigmatism can be reduced, if not by replacing the lens or mirror with a better one, then by canceling it with a compensating optical component, much as astigmatism of the eye is corrected with an appropriate eyeglass lens. An additional meaning of the term *astigmatism* refers to the inherent inability of even well-made spherical lenses and mirrors to sharply focus rays from objects that are off the optical axis. There are additional kinds of optical aberration, including comma. In the latter, off-axis images of round objects have a comma-like appearance. (*See* RESOLVING POWER.)

In astronomy, aberration is the effect that the observed direction of a ray of light depends on the VELOCITY of the observer. In particular, the positions of stars in the sky appear to undergo annual variation as the Earth makes its revolution about the Sun. Astronomical aberration is caused by the finite SPEED OF LIGHT; for hypothetical infinite speed of light propagation, there would be no aberration.

absorption As ENERGY propagates through a material medium, absorption is the process in which some (or all) of the propagating energy is converted to some other form and is taken up by the medium, thus becoming lost to propagation. As a result, the INTENSITY of the propagation is reduced. A sound WAVE propagating through a wall undergoes partial or total absorption as its intensity is attenuated by the conversion of some of the acoustic energy (orderly OSCILLATIONS of MOLECULES in the wall) to thermal energy (HEAT, disordered molecular oscillations). Similarly, LIGHT passing through glass is partially, if only slightly, absorbed and loses intensity. The intensity of energy propagation through an absorbing homogeneous medium is described by Lambert's law, which states that the intensity falls off exponentially with distance:

$$I = I_0 e^{-\alpha x}$$

where I_0 is the intensity at any point in the medium, I is the intensity at distance x deeper into the medium in the direction of propagation, and α denotes the absorption coefficient. (I and I_0 are expressed in the same, appropriate UNITS. Distance x can be in any unit of length, while the unit of α is the inverse of the unit used for x. In SI, for instance, the unit of x is the meter (m), and that of α the inverse meter (1/m).) It follows that equal thicknesses of a homogeneous medium absorb equal proportions of propagating energy.

See also ACOUSTICS; SOUND.

acceleration Instantaneous acceleration, which is what is usually meant by acceleration, is the TIME rate of change of VELOCITY, the change of velocity per unit time. Like velocity, acceleration is a VECTOR, having both magnitude and direction. In a formula:

$$\mathbf{a} = d\mathbf{v}/dt$$

where $\mathbf{a}$ denotes the acceleration, $\mathbf{v}$ the velocity, and t the time. The SI UNIT of acceleration is meter per second per second (m/s^2). Velocity is in meters per second (m/s), and time is in seconds (s).

The average acceleration, $\mathbf{a}_{av}$, over the time interval between times t_1 and t_2 equals the change in velocity between the two times, divided by the elapsed time interval:

$$\mathbf{a}_{av} = [\mathbf{v}(t_2) - \mathbf{v}(t_1)]/(t_2 - t_1)$$

Here $\mathbf{v}(t_1)$ and $\mathbf{v}(t_2)$ denote the velocities at times t_1 and t_2, respectively. The instantaneous acceleration is obtained from the average acceleration by taking the limit of t_2 approaching t_1.

In a simple situation, the motion of a body might be constrained to one direction. In that case, we can drop the vector notation and use the formula:

$$a = dv/dt$$

where a and v denote, respectively, the acceleration and velocity along the line of motion. If, moreover, the acceleration is constant (i.e., the velocity changes [increases or decreases] by equal increments during equal time intervals), the formula can be further simplified to:

$$a = \Delta v/\Delta t$$

where Δv is the (positive or negative) change in velocity during time interval Δt.

Alternatively, the velocity might change in its direction but not in its magnitude, which is motion at constant SPEED. A simple case of that is constant-speed CIRCULAR MOTION. In this case, the direction of the acceleration is toward the center of the circular path—that is called centripetal acceleration—and the constant magnitude of acceleration, a, is given by:

$$a = v^2/r$$

where v denotes the speed in meters per second (m/s), and r is the radius of the circular path in meters (m).

In general, though, acceleration can involve change in both the magnitude and the direction of the velocity. An important acceleration is the acceleration of a body in FREE FALL near the surface of the Earth, also called the acceleration due to gravity, or gravitational acceleration. All bodies dropped from rest near the surface of the Earth are found to fall with the same constant acceleration (ignoring the resistance of the air). The magnitude of this acceleration is conventionally denoted by g and has the approximate value of 9.8 meters per second per second (m/s^2). Its precise value depends on a number of factors, such as geographical latitude and altitude, as well as the DENSITY of the Earth's surface in the vicinity of the falling body. GALILEO GALILEI was a pioneer in the investigation of free-fall motion. (*See* GRAVITATION.)

A more restricted use of acceleration is to indicate an increase of speed, as opposed to deceleration, which is a decrease of speed. In this usage, deceleration can be thought of as negative acceleration. However, both are included in the general meaning of acceleration that is explained above.

See also KINEMATICS.

accelerator, particle A particle accelerator is a device for giving electrically charged particles considerable KINETIC ENERGY by accelerating them to high SPEEDs. The accelerated particles can be either subatomic particles or charged ATOMs (i.e., IONs). Acceleration is achieved electromagnetically by means of static or time-varying electric or magnetic FIELDs. The particles are accelerated in a VACUUM to avoid their colliding with air MOLECULEs and thus deviating from the desired direction of motion and dissipating their kinetic energy. The uses of particle accelerators are many and varied, including research, industrial, and medical applications. In research, for example, the properties of the ELEMENTARY PARTICLEs and their INTERACTIONs, as well as the properties of NUCLEI, are studied. Industrial applications include, for instance, the production of various kinds of RADIATION for such purposes as sterilization, imaging, and the manufacture of computer chips. Medical uses involve mainly the treatment of cancer. (*See* ACCELERATION; CHARGE; ELECTRICITY; ELECTROMAGNETISM; ELEMENTARY PARTICLE; MAGNETISM.)

There are two types of particle accelerators: electrostatic and nonelectrostatic. Electrostatic accelerators accelerate charged particles by means of a FORCE acting on them due to a constant electric field, which is produced by generating and maintaining a constant high VOLTAGE. This type of accelerator accelerates particles in a continuous beam. The particle energies it can reach are limited by the maximal voltage that can be achieved. Van de Graaff accelerators belong to this category.

Alternatively, charged particles can be accelerated by time-varying electric and magnetic fields in various ways. Accelerators of this type are either linear accelerators or circular accelerators. Linear accelerators, or linacs, use only electric fields. With many electrodes arrayed along their length, these accelerators accelerate particles in bunches from one electrode to the next through the accelerator tube by means of precisely timed varying electric fields. The voltages between adjacent electrodes do not have to be especially high, so the limitation of the electrostatic accelerator is avoided, but the cumulative effect of all the electrodes is to boost the particles to high energy. There is no limit in principle to the energy that can be achieved; the longer the accelerator, the higher the energy. However, there are practical limitations to the accelerator's length.

In order to achieve even higher energies than the length limitation of linacs allows, the limitation is avoided by accelerating the particles in a circular, rather than linear, path. A circular path is maintained through the use of magnetic fields, which might be constant or might vary in time. The acceleration itself is accomplished by time-varying voltages between adjacent pairs of the electrodes that are positioned around the path. Such accelerators go by the names of cyclotron and synchrotron. The maximal achievable magnetic field is a major limiting factor in achieving high energies in such devices. The use of superconducting MAGNETs allows higher fields and consequently higher energies than are attainable by conventional magnets. (*See* SUPERCONDUCTIVITY.)

Just to give an example of what is being accomplished, the Tevatron accelerator at the FERMI NATIONAL ACCELERATOR LABORATORY (FERMILAB) can accelerate PROTONs to an energy of about 1 tera-electron-volt (TeV) = 1×10^{12} electron volts (eV) and has a diameter of approximately 2 kilometers. The LEP accelerator at the EUROPEAN ORGANIZATION FOR NUCLEAR RESEARCH (CERN) can achieve around 50 giga-electron-volts (GeV) = 50×10^9 eV for electrons and its diameter is about 8.5 kilometers.

There are many variations and combinations of particle accelerators that optimize the acceleration of particular particles and adapt them to specific applications. Particles might first be accelerated to some energy in a linear accelerator and then be injected into a circular accelerator for further acceleration. A related device is the storage ring, in which previously accelerated particles are kept in circular orbits. Counter-rotating beams of oppositely charged particles, such as of ELECTRONs and POSITRONs, can be maintained in the same storage ring and made to collide with each other again and again.

See also BROOKHAVEN NATIONAL LABORATORY; STANFORD LINEAR ACCELERATOR CENTER (SLAC).

acoustics The branch of physics that deals with SOUND and its production, propagation, and detection is called *acoustics*. Some applications of this field are: musical acoustics, understanding how musical instruments generate the sounds they do; architectural acoustics, predicting and controlling the sound in structures such as concert halls; noise abatement, controlling the levels of unwanted noise; sonar, the use of underwater sound for the detection and identification of submerged objects; and medicine, applying high-frequency sound (ultrasound) for imaging and for treatment.

action This is a SCALAR quantity characterizing a physical process, such as the motion of a body. In simple cases, it is the KINETIC ENERGY multiplied by the time duration of the process, or LINEAR MOMENTUM times spatial DISPLACEMENT, or ANGULAR MOMENTUM times angular displacement. In more general cases, the changes of kinetic energy, momentum, etc. during the process must be taken into account by integration. Its SI UNIT is joule·second (J·s), or equivalently, meter²·kilogram per second (m^2·kg/s). In CLASSICAL PHYSICS there does not seem to be much use for action, although the principle of stationary action can be used to derive the EQUATIONS OF MOTION of a system. The idea is that of all possible motions that might carry a system from a fixed initial state to a fixed final state, the motion that will actually occur is the one for which the action of the process does not change when that motion is hypothetically varied to any very similar motion between those initial and final states. In QUANTUM PHYSICS, on the other hand, action plays a fundamental role, since the PLANCK CONSTANT, which characterizes the quantum character of nature, has the DIMENSION of action. Nature seems to allow only actions that are multiples of the fundamental unit of action given by the Planck constant, $h = 6.62606876 \times 10^{-34}$ J·s.

adhesion The attraction between different materials is called adhesion. Its origin is in the electromagnetic FORCEs between the ATOMs or MOLECULEs of the two materials. The CAPILLARY FLOW of water, say, in a narrow glass tube is due to adhesion between water and glass.

See also COHESION; ELECTROMAGNETISM.

adiabatic process A process in which a system does not exchange HEAT with its surroundings is called an adiabatic process. That might be accomplished in practice by thermally insulating the system or by having the process take place so rapidly that there is no time for significant heat exchange between the system and its surroundings. Since no thermal insulation is absolute and because every process takes *some* time, these methods afford only approximate elimination of heat exchange. Thus the concept of an adiabatic process is an idealization, although a very useful one. For example, the CARNOT CYCLE, which is of great importance in THERMODYNAMICS and gives the theoretical upper limit on the efficiency of real heat engines, involves two adiabatic processes. A graphic curve showing the relation between two variables of a system during an adiabatic process, such as between PRESSURE and VOLUME, is called an adiabat.

aerosol A dispersion of droplets of liquid or small particles of solid in a gas is an aerosol. The particles or droplets are larger than molecular size but small enough that they are kept in dispersion by collisions with the molecules of the gas (BROWNIAN MOTION) and do not settle out. For example, spray cans and atomizers often produce aerosols in air. Also, smoke and smog are aerosols. And in injection-type internal-combustion engines, the fuel is introduced into the cylinders as an aerosol.

alloy A blend of a metal with one or more metallic or nonmetallic materials is an alloy. The components of an alloy do not combine chemically but, rather, are very finely mixed. An alloy might be homogeneous or might contain small particles of components that can be viewed with a microscope. Brass is an example of an alloy, being a homogeneous mixture of copper and zinc. Another example is steel, which is an alloy of iron with carbon and possibly other metals. The purpose of alloying is to produce desired properties in a metal that naturally lacks them. Brass, for example, is harder than copper and has a more goldlike color. Steel is harder than iron and can even be made rust proof (stainless steel).

alpha decay This is a form of RADIOACTIVITY, whereby an atomic NUCLEUS spontaneously emits a helium nucleus, also called an ALPHA PARTICLE, and thus transmutes into a daughter nucleus having ATOMIC NUMBER less by two and ATOMIC MASS less by four than those of the parent nucleus. Among the various types of radiation from radioactivity, the helium nuclei produced by alpha decay are relatively weakly penetrating due to their large mass while relatively highly ionizing due to their electric charge. An example of alpha decay is the spontaneous decay of radium $^{226}_{88}\text{Ra}$ (atomic number 88, atomic mass 226) to radon $^{222}_{86}\text{Rn}$ (atomic number 86, atomic mass 222):

$$^{226}_{88}\text{Ra} \rightarrow {}^{222}_{86}\text{Rn} + {}^{4}_{2}\text{He}$$

where $^{4}_{2}\text{He}$ denotes the emitted alpha particle (with atomic number 2 and atomic mass 4). Alpha decay is governed by the STRONG INTERACTION. The process of alpha decay involves the TUNNELING of an alpha particle from inside the nucleus to outside.

alpha particle This is another name for a helium NUCLEUS, consisting of two PROTONs and two NEUTRONs, and thus possessing atomic number 2 and atomic mass 4. This combination of NUCLEONs is especially stable. The name is a relic from the early years of the study of RADIOACTIVITY.

See also ALPHA DECAY.

American Institute of Physics (AIP) Founded in 1931, the American Institute of Physics is a not-for-profit membership corporation established for the purpose of promoting the advancement and diffusion of the knowledge of physics and its application to human welfare. Its headquarters are located in the American Center for Physics building in College Park, Maryland, in suburban Washington, D.C. The mission of the institute is to support physics, astronomy, and related fields of science and technology by serving its member societies, individual scientists, educators, students, research and development leaders, and the general public with programs, services, and publications.

The member societies of the AIP are: AMERICAN PHYSICAL SOCIETY, Optical Society of America, Acoustical Society of America, The Society of Rheology, American Association of Physics Teachers, American Crystallographic Association, American Astronomical Society, American Association of Physicists in Medicine, American Vacuum Society, and American Geophysical Union. (*See* ACOUSTICS; CRYSTALLOGRAPHY; GEOPHYSICS; OPTICS; VACUUM.)

The institute lists some 25 affiliated societies, including, for example, the American Institute of Aeronautics and Astronautics and the Society for Applied Spectroscopy. (*See* SPECTROSCOPY.)

In additional to the journals published by its member societies, the AIP publishes a number of its own journals, in paper or electronic formats. These include, for example, *Journal of Applied Physics, Journal of Mathematical Physics, Review of Scientific Instruments,* and *Physics Today.*

The institute maintains the Center for History of Physics, which includes the Niels Bohr Library,

The American Center for Physics, at One Physics Ellipse in College Park, Maryland, is home to the American Institute of Physics and some of the institute's member societies. Washington, D.C., is conveniently accessible from the Center for Physics. ***(Courtesy of the American Center for Physics)***

dedicated to the history of physics and allied fields. (*See* BOHR, NIELS.)

The URL of the American Institute of Physics website is http://www.aip.org/. The organization's mailing address is One Physics Ellipse, College Park, MD, 20740-3843.

American Physical Society (APS) Founded in New York City in 1899 with the mission "to advance and diffuse the knowledge of physics," the American Physical Society has grown to be the preeminent professional society of physicists in the world. Its members currently number over 42,000, most of whom (about 80 percent) are in the United States. The society's headquarters are located in the American Center for Physics building in College Park, Maryland, in suburban Washington, D.C.

In fulfillment of its mission, the APS publishes some of the world's most widely read peer-reviewed journals and holds over 20 meetings a year. The journals include the five sections of the *Physical Review,* which cover all fields of physics. It also publishes *Physical Review Letters,* generally thought to be the most

A scene from the centennial meeting of the American Physical Society, held in March 1999 in Atlanta, Georgia. ***(Courtesy of the American Physical Society)***

prestigious venue to publish important advances in physics research, *Reviews of Modern Physics,* and the all-electronic journal *Special Topics: Accelerators and Beams (STAB).* Except for the latter, all the journals appear in both on-line and print versions, with the on-line version being the official one. The APS maintains an on-line archive, called PROLA, which contains every article published in all the APS journals since their inception.

The APS organizes two general annual meetings, one of which is the largest meeting devoted to physics anywhere in the world, regularly attracting more than 5,000 attendees. The other APS meetings are on more specialized research topics and are typically organized by the individual divisions and topical groups of the APS.

In addition to those core activities, the APS serves the physics community and the general public in a variety of ways. Physics education is a major concern, and the APS runs a number of education-oriented programs, including a major one aimed at improving the preparation of science teachers. The society is also active in informing the public about physics, which it does through the media and through its website. The APS plays an important role in the international physics community, maintaining contacts and reciprocal relations with physics societies throughout the world and monitoring human-rights problems affecting physicists and other scientists. The organization bestows about 40 prizes and awards each year for achievements in physics research and for service to the physics community.

The URL of the American Physical Society website is http://www.aps.org/. APS's website for bringing the importance and excitement of physics to the general public has the URL http://PhysicsCentral.com/. The society's mailing address is One Physics Ellipse, College Park, MD, 20740-3844.

Ampère, André-Marie (1775–1836) French *Physicist, Chemist, Mathematician* Among the pioneers in laying the foundations of ELECTROMAGNETISM, André-Marie Ampère investigated, during 1820–27, the relationship between electric CURRENTs and the magnetic FIELDs they produce, as well as other magnetic effects. The SI UNIT of electric current, the ampere (A), is named after him. Largely self-educated by copious reading, Ampère survived the French Revolution, became a science teacher, and then a professor of math-

André-Marie Ampère, active in the 18th and 19th centuries, was among the pioneers in laying the foundations of electromagnetism. ***(Engraved by Ambroise Tardieu, 1825, courtesy AIP Emilio Segrè Visual Archives)***

ematics in Paris. In addition to his investigations in electromagnetism, he studied LIGHT and carried out considerable research in mathematics and chemistry.

See also AMPÈRE'S LAW; ELECTRICITY; MAGNETISM.

Ampère's law This law, named for the French physicist ANDRÉ-MARIE AMPÈRE, relates the magnetic FIELD to the electric CURRENTs producing it. More specifically, Ampère's law connects the magnetic field along a closed path, or loop, to the electric currents flowing through the surface enclosed by the loop. The mathematical form of the law is:

$$\oint_{\text{loop}} \mathbf{B} \cdot d\mathbf{s} = \mu_0 i$$

where $\oint_{\text{loop}}$ denotes integration around the loop, **B** is the magnetic field VECTOR in teslas (T), $d\mathbf{s}$ is the differential vector element of directed path length around the loop in meters (m), i is the net electric current through the surface enclosed by the loop in amperes (A), and μ_0 is the PERMEABILITY of the VACUUM, whose value is $4\pi \times 10^{-7}$ tesla·meter per ampere (T·m/A). What the law says is this: choose an arbitrary closed path—the loop—in space, and choose one of the two possible senses of traversing the loop. At each point on the loop, consider the magnitude and direction of the magnetic field at that point and the direction of the tangent line to the loop in the chosen sense of traversal. Now multiply the magnitude of the magnetic field at that point, the cosine of the angle between the direction of the magnetic field there and the direction of the tangent line to the loop, and a differential element of path length. Then sum up, or integrate, all those products for all points on the loop. That is the meaning of the left-hand side of the equation. (*See* ELECTRICITY; MAGNETISM.)

For the right-hand side, consider any surface in space having the chosen loop as its boundary. (For illustration, think of the loop as being made of wire and having been dipped in a soap solution, with a resulting soap-film surface formed inside the loop.) Now, imagine your right hand placed at that surface with your fingers curled and your thumb stretched away from your fingers and piercing the surface. Make your curled fingers indicate the chosen sense of traversal of the loop. (If they do not indicate the chosen sense of traversal, flip your hand so your thumb pierces the surface in the opposite direction). From the vantage point of the tip of your thumb and looking back toward the surface and your curled fingers, the loop is traversed in the counterclockwise sense. After that preparation, consider all the electric currents flowing through the loop (i.e., piercing the surface). Those flowing in the general direction of your thumb are considered positive, while those flowing in opposition to your thumb are negative. Form the algebraic sum—taking the just-defined signs into account—of all the electric currents through the loop, which is the net current through the loop. Multiply this by the constant μ_0, and you have the right-hand side of Ampère's law.

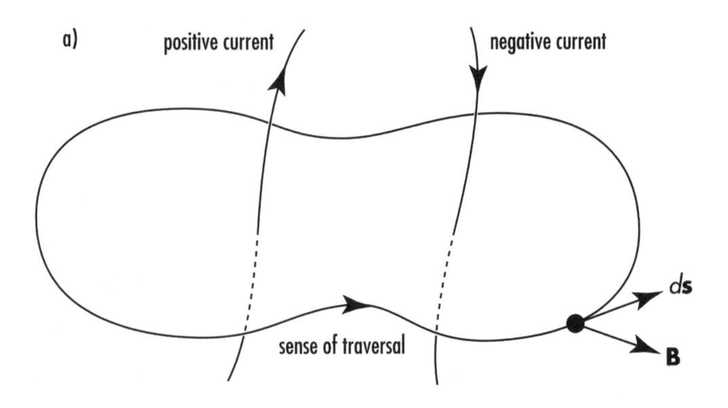

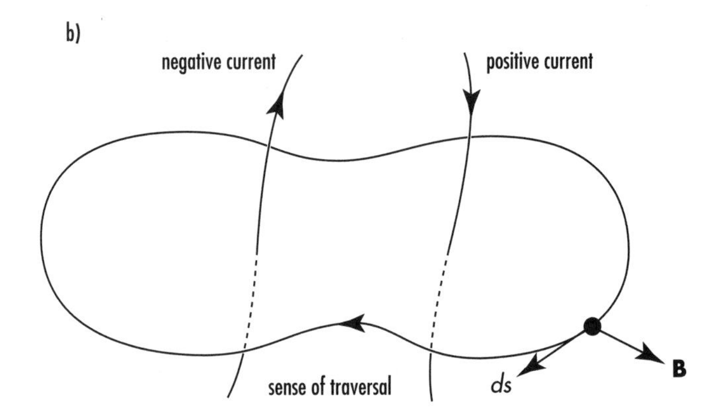

Ampère's law describes the relation of the sense of traversal of a loop in space to the signs of the electric currents flowing through the surface enclosed by the loop. B and *ds* denote the magnetic-field vector and the differential vector element of directed path length, respectively, at an arbitrary point on the loop. Both senses of traversal are shown.

Ampère's law describes a relation between the magnetic field at the points along the loop, on the one hand, and the currents flowing through the loop, on the other. Note that, although all electric currents contribute to the magnetic field at every point in space, it is only the *currents through the loop* that are involved in Ampère's law. In certain, especially symmetric, situations the law can be used to calculate the magnetic field produced by a given electric current. For a notable example, the magnitude of the magnetic field, B, in teslas (T), inside a tightly wound, long helical coil of current-carrying wire (called a solenoid)—and not too close to the ends of the coil—can be shown to be given by:

$$B = \mu_0 n i$$

where n is the number of turns of the coil per unit length and i is the current flowing through the solenoid in amperes (A).

See also BIOT-SAVART LAW; MAXWELL'S EQUATIONS.

Anderson, Philip Warren (1923–) American *Physicist* Known for his work in CONDENSED-MATTER PHYSICS, Philip Anderson's research has included the quantum theory of condensed matter, MAGNETISM, SUPERCONDUCTIVITY, and SUPERFLUIDITY. After working at the U.S. Naval Research Laboratories during World War II, he studied THEORETICAL PHYSICS at Harvard University, Cambridge, Massachusetts, and received his Ph.D. there in 1949. During 1949–84, Anderson worked at, or was associated with, BELL LABS, where he carried out many of his investigations. He has served as professor of physics at Princeton University, New Jersey, since 1975. Anderson shared the 1977 Nobel Prize in physics with Nevill Francis Mott and John H. Van Vleck "for their fundamental theoretical investigations of the electronic structure of magnetic and disordered systems."

See also QUANTUM PHYSICS.

Philip Warren Anderson is a condensed-matter theoretical physicist who has investigated magnetism, superconductivity, and superfluidity, among other areas, and shared the 1977 Nobel Prize in physics. ***(AIP Emilio Segrè Visual Archives, Physics Today Collection)***

angular acceleration *See* KINEMATICS.

angular momentum In the physics of rotational MOTION, angular momentum plays a role that is analogous to the role played by LINEAR MOMENTUM in translational (i.e., linear) motion. The angular momentum of a point particle is:

$$\mathbf{L} = \mathbf{r} \times \mathbf{p}$$

where the VECTOR **L** denotes the angular momentum of the particle in kilogram·meter^2 per second (kg·m^2/s) and **r** is the particle's position vector, the vector whose magnitude equals the particle's distance from the origin of the COORDINATE SYSTEM and whose direction is the direction of the particle as viewed from the origin, in meters (m). The vector **p** denotes the particle's linear momentum in kilogram·meters per second (kg·m/s):

$$\mathbf{p} = m\mathbf{v}$$

where m is the particle's MASS in kilograms (kg) and **v** its VELOCITY in meters per second (m/s). Note that angular momentum depends on the origin chosen, while linear momentum does not. Thus angular momentum is always with respect to some point. The angular momentum vector **L** is perpendicular to the plane in which the two vectors **r** and **p** lie. Its sense is such that if the base of **p** is attached to the base of **r** and the base of **L** joined to those bases, then from the vantage point of the tip of **L**, looking back at the plane of **r** and **p**, the rotation from **r** to **p** through the smaller angle between them (the angle less than 180°) is counterclockwise. The magnitude of the angular momentum, L, is:

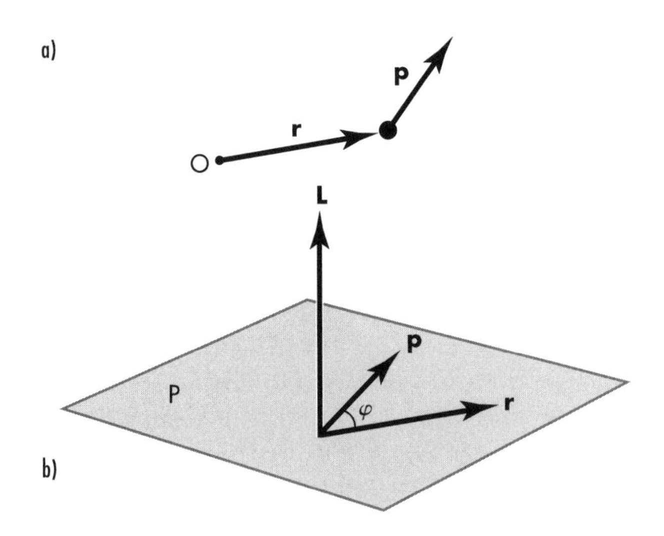

Angular momentum of a particle. (a) The angular momentum of a particle that is moving with momentum p and whose position vector with respect to the origin O is r, is L = r × p. (b) The relation among vectors r, p, and L. Plane P contains r and p. Vector L is perpendicular to P in the sense shown. The smaller angle (less than 180°) between r and p is denoted *φ*. The magnitude of the angular momentum equals *rp* sin *φ*, where *r* and *p* denote the magnitudes of r and p, respectively.

$$L = rp \sin \varphi$$

where r and p are the magnitudes of vectors **r** and **p**, respectively, and φ is the smaller angle (less than 180°) between **r** and **p**. Since $r \sin \varphi$ is the perpendicular distance of the origin from the line of the linear momentum vector, L can be thought of as the product of the magnitude of the linear momentum and the perpendicular distance of the origin from the line of the linear momentum. Alternatively, since $p \sin \varphi$ is the component of the linear momentum perpendicular to the position vector, L can be thought of as the product of the distance of the particle from the origin and the component of its linear momentum perpendicular to its position vector. (*See* ROTATION.)

Another way of viewing angular momentum is through the relation:

$$\mathbf{L} = mr^2\boldsymbol{\omega}$$

where:

$$\boldsymbol{\omega} = \frac{\mathbf{r} \times \mathbf{v}}{r^2}$$

is the angular velocity vector in radians per second (rad/s). (This follows directly from the above definition of **L**.) Then the angular momentum of a point particle is seen to be mr^2 times the angular velocity, in analogy to linear momentum as mass times linear velocity, with mr^2 serving as the analog of mass.

The total angular momentum of a system of point particles with respect to some point is the vector sum of the individual angular momenta with respect to the same point. It then follows that the magnitude of the total angular momentum of a rigid body consisting of a number of particles, with respect to any axis of rotation, is:

$$L = I\omega$$

where ω is the magnitude of the angular velocity of the body, the body's angular speed, about the axis, and I, called the MOMENT OF INERTIA of the body, is given by:

$$I = \sum m_i r_i^2$$

In this formula m_i denotes the mass of the ith particle, r_i is the particle's perpendicular distance from the axis, and the summation is taken over all the particles constituting the body. I is in units of kilogram·meter² (kg·m²). For continuous bodies, the summation is replaced by integration. For a single point particle, the moment of inertia is simply:

$$I = mr^2$$

Note that I is analogous to mass in the relation:

$$L = I\omega$$

which is the angular analog of the linear relation:

$$p = mv$$

(*See* SPEED.)

The angular analog of ISAAC NEWTON's second law of motion is:

$$\tau = I\, d\omega/dt$$

or more generally:

$$\tau = dL/dt$$

where τ denotes the magnitude of the TORQUE acting on a rigid body in newton·meter (N·m), $d\omega/dt$ is the magnitude of the body's resulting angular acceleration in radians per second per second (rad/s²), and t denotes TIME in seconds (s). (*See* ACCELERATION; NEWTON'S LAWS.)

Angular momentum obeys a CONSERVATION LAW, which states that in the absence of a net external torque acting on a system, the total angular momentum of the system remains constant in time. At the quantum level, angular momentum, including SPIN, is quantized, i.e., it takes values that are only an integer or half-integer multiple of $h/(2\pi)$, where h denotes the PLANCK CONSTANT, whose value is $6.62606876 \times 10^{-34}$ joule·second (J·s). (*See* QUANTUM PHYSICS.)

angular speed *See* KINEMATICS; SPEED.

angular velocity *See* ANGULAR MOMENTUM; VELOCITY.

antimatter This is the name given to a class of ELEMENTARY PARTICLES, each of which is closely related to a member of another class of elementary particles, called MATTER. The latter class includes the particles that make up ordinary matter, such as the ELECTRON, PROTON, and NEUTRON, as well as others. For every matter particle there exists an antimatter particle, called its antiparticle, that has some properties in common with it and other properties opposed. A particle and its antiparticle both possess the same MASS, SPIN, and magnitude of electric CHARGE, for example. On the other hand, they have opposite signs of electric charge, as well as opposite signs of charges of other kinds, such as LEPTON number or BARYON number. The antiproton has the same mass as the proton, a half unit of spin, and one elementary unit of electric charge, as does the proton. But its electric charge is negative, whereas that of the proton is positive. Also, its baryon number is –1, while that of the proton is +1. The situation for the electron is similar, but its antiparticle has its own name, the POSITRON. Certain particles, whose various charges are zero, are their own antiparticles. Such are the neutral PION and the PHOTON, for instance.

When a particle and its antiparticle collide with each other, they can undergo mutual annihilation, with the emission of a pair of photons. In this process, the energy of the pair is totally converted to electromagnetic energy. On the other hand, a sufficiently energetic photon passing through matter can convert to a particle-antiparticle pair, such as an electron and a positron, a process called PAIR PRODUCTION. Particle-antiparticle pairs feature in the picture of the VACUUM that QUANTUM PHYSICS describes. The vacuum is in a constant turmoil of activity, with particle-antiparticle pairs of all kinds ceaselessly and spontaneously materializing briefly and then dematerializing. The energy for such fleeting pair production and annihilation is "on loan" for extremely small times according to the quantum HEISENBERG UNCERTAINTY PRINCIPLE of WERNER KARL HEISENBERG.

As far as is presently known, the objects of the universe consist entirely of matter, with antimatter only being produced in various processes and annihilated soon after. The reason for that is still a mystery, especially in light of cosmological theories that have matter and antimatter being created in nearly equal amounts at an early stage in the evolution of the universe. The matter and antimatter are theorized to have almost completely annihilated each other, leaving a very slight surplus of matter (rather than of antimatter), which is what we observe today. Why there is a surplus of matter rather than of antimatter, indeed, why there is any surplus at all, are questions that are presently unanswered.

See also COSMOLOGY.

Archimedes' principle Named for the ancient Greek scientist, Archimedes' principle states that a body immersed in a FLUID, whether wholly or partially immersed, is buoyed up by a FORCE whose magnitude equals the magnitude of the WEIGHT of the fluid that the body displaces. This force is called the buoyant force. If volume V of the body (which might be the whole volume of the body or only part of it) is submerged, then V is also the volume of displaced fluid. The magnitude of the weight of that volume of displaced fluid is then $V\rho g$ in newtons (N), where ρ is the DENSITY of the fluid in kilograms per cubic meter (kg/m^3) and g denotes the ACCELERATION due to gravity with nominal value 9.8 meters per second per second (m/s^2). Thus, according to Archimedes' principle, the magnitude of the buoyant force on the body equals $V\rho g$.

See also BUOYANCY; HYDROSTATICS; SPECIFIC GRAVITY.

area The amount of a surface is its area. The SI UNIT of area is the square meter (m^2). For example, the area of a flat rectangle whose sides have lengths a and b is ab. Or, the area of the surface of a sphere of radius r is $4\pi r^2$. If all lengths characterizing a system are changed by multiplication by the same factor, then all areas of the system will be correspondingly multiplied by the square of that factor. For instance, if the sides of a rectangle are each doubled, then the area grows by a factor of four. Or, if the radius of a sphere is halved, then its surface area is reduced to a quarter of its former value.

astrophysics As the word implies, astrophysics deals with the physics of the stars. In more detail, astrophysics is the study of the physical processes involved in the formation and subsequent evolution of stars and groups of stars, such as GALAXIES. Astrophysicists combine the results of various kinds of astronomical observations with general physics understanding. Physics normally advances its grasp of nature by means of carefully controlled experiments. However, stars and galaxies cannot be experimented with. Instead, astrophysicists take advantage of the vast variety and diversity of stars and galaxies, of different kinds and at different stages of evolution, to let nature do the experimenting for them, so to speak.

Although there is still much to be learned, astrophysics has achieved a very detailed understanding of the birth, life, and death of stars. A typical star is formed when a cloud of gas and dust contracts, due to gravitational attraction, and heats up as a result of the contraction. It becomes hot enough to initiate nuclear reactions, which raise its TEMPERATURE much higher. This produces internal PRESSURE that opposes the gravitational FORCEs and halts the contraction. Thus a state of EQUILIBRIUM is reached and maintained for some time. At the beginning, the nuclear processes within the star are mostly the conversion of hydrogen to helium. As the star ages and its hydrogen supply dwindles, heavier chemical ELEMENTS are converted into even heavier ones. The star continues in equilibrium as long as there is sufficient nuclear fuel. Eventually the nuclear reactions die down, and the star can no longer maintain equilibrium. What happens then depends on the star's MASS. (*See* ATOMIC MASS; ATOMIC NUMBER; GRAVITATION; NUCLEAR PHYSICS.)

A typical low-mass star, meaning a star whose mass is less than around eight times the mass of the Sun, passes through a red-giant stage, in which it is cooler, and thus redder in color, than the Sun and about 100 times as large. It continues to evolve and becomes a red supergiant, with a small, inert carbon core surrounded by concentric shells reaching a radius as large as 500 times the radius of the Sun, in which nuclear reactions continue. The shells expand and cool and disperse into space, while the core becomes a white dwarf, shining by its stored heat. The white dwarf is about the size of the Earth and has a mass of about half that of the Sun. Eventually it cools down and darkens. During its lifetime, such a star converts hydrogen into heavier chemical elements up to, and including, carbon. By means of the dispersion of its shells, it adds those elements to the contents of interstellar space.

A typical high-mass star evolves less eventfully than does a low-mass star, but it achieves much higher temperatures, allowing it to produce chemical elements up to, and including, iron. Eventually it develops an iron core, which cannot produce heat through nuclear reactions. With no further heat coming from the core, the star collapses within seconds and then rebounds in a stupendous explosion the likes of which there is little else in the universe. This is a supernova. It is observed as a brilliant flash of light, billions of times brighter than the Sun, which dies down within a few months. In the extremely high temperatures of the explosion, all the chemical elements heavier than iron are produced. Thus a supernova ejects much material, containing all the chemical elements, into interstellar space. Some supernovae do not completely destroy the star they evolve from and leave behind a residual core, either a NEUTRON STAR (if the mass of the original star is less than some 25 solar masses) or a BLACK HOLE. The former is an extremely dense object containing approximately the mass of the Sun, but not more than three solar masses, within a diameter of about 10 kilometers. A black hole is an object whose gravity is so strong that nothing can escape from it, not even light. Black holes produced in this way have masses greater than three solar masses. (*See* DENSITY.)

It is now understood that all the heavier chemical elements, such as those we humans are made of, were "cooked up" in stars and spewed out by supernovae and by expanding shells. Thus it is literally true to say we are made of stardust.

Recent developments in astrophysics include increased understanding of the formation of galaxies and galaxy clusters. It is now understood that a super-massive black hole likely resides at the centers of galaxies.

atom The smallest entity that can be identified with a chemical ELEMENT is an atom. Thus we refer to a helium atom, for instance, or a carbon atom. Atoms are approximately spherical, with sizes on the order of 10^{-10} meter (m). Since that is considerably less than the WAVELENGTH of visible light, atoms cannot be seen, not even with the aid of the most powerful optical microscopes. Nonoptical microscopes, such as electron microscopes, can produce images of single atoms, though not of their structure. Their structure is most correctly described by QUANTUM PHYSICS. But a rough picture is that of a relatively tiny central component, called the NUCLEUS, whose size is on the order of 10^{-15} m, surrounded by a cloud of ELECTRONs. Since electrons are, at least as far as is presently known, point particles, it is not incorrect to say that the volume of atoms is mostly empty. This structure was first discovered by ERNEST RUTHERFORD. (*See* ELECTRON MICROSCOPY; RESOLVING POWER.)

The negatively charged electrons are bound in the atom by the electric attraction between them and the positively charged nucleus. They are ordered around the nucleus into shells and subshells, which are characterized by various QUANTUM NUMBERs determining the ENERGY, the magnitude of ANGULAR MOMENTUM, the direction of angular momentum, and the direction of SPIN of the electrons, as well as the sizes of the shells and subshells. It is the electron configuration that determines the chemical properties of the atom and thus of the corresponding chemical element, since chemical reactions involve only the electron component of the atom, specifically the electrons in the outermost shell. The understanding of the electron structure of atoms provided by QUANTUM MECHANICS explains the PERIODIC TABLE of the chemical elements. (*See* ELECTRICITY.)

The nucleus comprises a number of positively charged PROTONs, which give the nucleus its electric charge, and a number of electrically neutral NEUTRONs. (Protons and neutrons, the constituents of the nucleus, are collectively called NUCLEONs.) The number of protons, called the ATOMIC NUMBER, equals the number of electrons in an electrically neutral atom and thus determines the chemical identity of the atom: for hydrogen the atomic number is 1, for helium 2, for lithium 3, . . ., for uranium 92, and so on. Uranium has the highest atomic number among the naturally occurring elements. Elements with higher atomic numbers have been produced in the laboratory by causing the nuclei of lighter elements to collide and coalesce. Particle accelerators are used for this purpose. (*See* ACCELERATOR, PARTICLE.)

For the same chemical element, the number of neutrons may vary without affecting the identity of the element. Versions of an element with different numbers of neutrons are called ISOTOPEs of that element. The sum of the number of protons and the number of neutrons, which is the number of nucleons in the nucleus, is called the ATOMIC MASS, or atomic mass number. Thus isotopes are characterized by both their atomic number and their atomic mass. The isotope of carbon denoted carbon-14 has six protons and eight neutrons, while the most abundant isotope of carbon, carbon-12, possesses six protons and six neutrons.

The nucleons of a nucleus are bound together through the STRONG INTERACTION. This is a short-range FORCE that does not extend outside the nucleus. The strong interaction force overcomes the protons' mutual electric repulsion and maintains the structure of the nucleus.

When one or more electrons are removed from or added to an atom, i.e., when the number of electrons does not equal the atomic number, the resulting entity is electrically charged and is called an ION. Due to their electric charges, ions are affected by electric and magnetic FIELDs and also experience relatively strong electric forces among themselves. (*See* MAGNETISM.)

See also BOHR THEORY; PAULI EXCLUSION PRINCIPLE.

atomic mass The total number of NUCLEONs in a NUCLEUS of an ISOTOPE, i.e., the number of its PROTONs plus the number of NEUTRONs, is the atomic mass (also called atomic mass number or mass number) of that isotope. Atomic mass is what distinguishes the various isotopes of the same ELEMENT. There are three isotopes of hydrogen, for example. The lightest and, by far, the most abundant has an atomic mass of 1. Its nucleus consists only of a single proton. The next massive isotope is deuterium, whose atomic mass is 2. The deuterium nucleus, called a DEUTERON, comprises one proton and one neutron. Deuterium is a stable isotope. The rarest isotope is tritium, with atomic mass 3, whose nucleus contains one proton and three neutrons. This isotope is radioactive.

See also ATOM; RADIOACTIVITY.

atomic number The number of PROTONs in each NUCLEUS of a chemical ELEMENT is the element's atomic number. Since a nucleus's protons are its electrically charged component (the NEUTRONs, which are electrically neutral, being its other constituent), the atomic number determines the charge of the nucleus and the chemical identity of the element. In other words, elements are characterized by their atomic number. For hydrogen the atomic number is 1, for helium 2, for lithium 3, . . ., for uranium 92, and so on. Uranium is the naturally occurring element with the highest atomic number. Elements with higher atomic numbers have been produced in the laboratory by causing the nuclei of lighter elements to collide and coalesce with the aid of particle accelerators. (*See* ACCELERATOR, PARTICLE.)

atomic weight The atomic weight of a chemical ELEMENT is a number whose value equals the MASS, expressed in grams, of one MOLE of the element in its naturally occurring form. A mole consists of AVOGADRO'S NUMBER of ATOMs, which is $6.02214199 \times 10^{23}$ atoms. The atomic weight of hydrogen is 1.0079, while that of carbon is 12.011, for example. The atomic weights of the elements can be found in the PERIODIC TABLE of the elements that appears in Appendix III, as well as in every introductory chemistry textbook and many introductory physics textbooks.

aurora This is the effect of colored lights appearing in the night sky, most commonly near the Earth's magnetic poles. Energetic, electrically charged particles emitted by the Sun in the SOLAR WIND are channeled by the Earth's magnetic FIELD to the polar regions. There they collide with the air MOLECULES and ATOMS in the upper layers of the atmosphere, exciting and ionizing the atoms and molecules, which then emit LIGHT, which forms the aurora.

See also CHARGE; ELECTRICITY; ENERGY; GEOMAGNETISM; IONIZATION; MAGNETISM.

Avogadro, Amedeo (1776–1856) Italian *Physicist* His law concerning gases made Amedeo Avogadro famous. AVOGADRO'S LAW, put forth in 1811, states that equal VOLUMEs of all GASes at the same TEMPERATURE and PRESSURE have the same number of constituent particles (i.e., ATOMs or MOLECULEs). The law was not immediately accepted, due to uncertainty about the nature of the constituent particles and to doubts about their very existence. The number of constituent particles in one MOLE of substance, $N_A = 6.02214199 \times 10^{23}$ per mole (mol^{-1}), is named AVOGADRO'S NUMBER in his honor. Avogadro originally studied law and started practicing it, but he became so attracted to physics and mathematics that he studied them privately and made them his career beginning in 1806, when he was appointed professor of physics at the Academy of Turin, Italy.

Amedeo Avogadro studied the properties of gases in the late 18th/early 19th centuries and determined that equal volumes of all gases at the same temperature and pressure have the same number of constituent particles (atoms or molecules). ***(Edgar Fahs Smith Collection, University of Pennsylvania Library)***

Avogadro's law Named for the Italian physicist AMEDEO AVOGADRO, Avogadro's law states that equal VOLUMEs of all GASes at the same TEMPERATURE and PRESSURE have the same number of constituent particles (i.e., ATOMs or MOLECULEs). In other words, the volume of a gas is proportional to the number of constituent particles in it, and the proportionality coefficient is the same for all gases. That can be expressed as:

$$V/n = \text{constant}$$

where V denotes the volume of a gas, n is the number of atoms or molecules, and the constant has the same value for all gases at the same temperature and pressure. Equivalently:

$$V_1/n_1 = V_2/n_2$$

where the subscripts refer to any two states of the gas that have the same temperature and pressure. Avogadro's law is strictly valid only for ideal gases, but real gases obey it to a good approximation as long as the DENSITY is not too high and the temperature is not too low. (*See* GAS, IDEAL.)

Avogadro's law is a special case of the ideal gas law:

$$pV = nRT$$

where p denotes the pressure in pascals (Pa), equivalent to newtons per square meter (N/m^2), V is in cubic meters (m^3), n denotes the amount of gas in MOLES (mol), T is the absolute temperature in kelvins (K), and R is the gas constant, with value 8.314472 joules per mole per kelvin (J/[mol·K]).

The validity of Avogadro's law was a step toward universal recognition of the atomic nature of matter.

Avogadro's number Named for the Italian physicist AMEDEO AVOGADRO, Avogadro's number is the number of ATOMS in 12 grams (i.e., in one MOLE) of the ISOTOPE carbon-12 and equals $6.02214199 \times 10^{23}$ per mole (mol^{-1}). It is, by definition, the number of atoms or molecules in one mole of *any* substance.

B

Balmer formula Named for the Swiss mathematician Johann Jakob Balmer, the Balmer formula gives the WAVELENGTHs of a series of visible spectral lines of hydrogen, called the Balmer series. These lines result from transitions of atomic ELECTRONs to the second ENERGY level (which is the first above the GROUND STATE) from higher-energy levels. The formula is:

$$1/\lambda = R(1/2^2 - 1/n^2)$$

where λ denotes the wavelength of the light in meters (m) emitted during transition from the nth energy level ($n > 2$) to the second, and R is the Rydberg constant, whose value is 1.09737316×10^7 inverse meters (m^{-1}). This formula is a special case of the general formula for the wavelength of the spectral line associated with the transition in hydrogen from the nth atomic energy level to the mth:

$$1/\lambda = R(1/m^2 - 1/n^2)$$

See also ATOM; BOHR THEORY; LIGHT; SPECTRUM.

Bardeen, John (1908–1991) American *Physicist* Born in Madison, Wisconsin, John Bardeen became a pioneer in SEMICONDUCTOR physics and in SUPERCONDUCTIVITY. He studied electrical engineering at the University of Wisconsin and received an M.S. degree in 1929. His graduate studies at Princeton University introduced him to SOLID-STATE PHYSICS, and he received a Ph.D. in physics in 1936. In 1945 Bardeen joined the Bell Telephone Laboratories, now

John Bardeen was a solid-state physicist and shared two Nobel Prizes in physics: in 1956 for the transistor and in 1972 for a theory of superconductivity. ***(AIP Emilio Segrè Visual Archives)***

BELL LABS, where, together with W. H. Brattain, he discovered the TRANSISTOR effect in 1947. He was appointed professor of electrical engineering and of physics at the University of Illinois, Urbana, in 1951, and remained there until retirement. While at the University of Illinois, Bardeen investigated both semiconductors and superconductivity. During 1956–57, Bardeen, L. N. Cooper, and J. R. Schrieffer developed a very influential theory of superconductivity. He continued working and publishing original papers into his 83rd year, the year of his death.

Bardeen shared the 1956 Nobel Prize in physics with W. H. Brattain and W. Shockley "for their researches on semiconductors and their discovery of the transistor effect," and the 1972 Nobel Prize in physics with L. N. Cooper and J. R. Schrieffer "for their jointly developed theory of superconductivity, usually called the BCS-theory."

baryon The baryons are a class of ELEMENTARY PARTICLES that comprises the NUCLEONS (i.e., the PROTON and the NEUTRON, the constituents of atomic NUCLEI) and the hyperons. The latter can be viewed as more massive versions of nucleons that do not form part of natural nuclei. All baryons consist of three QUARKS each, one quark of each color, where the nature of the constituent quarks determines the type of baryon. For every baryon there exists a corresponding antibaryon, composed of three corresponding antiquarks. The baryons form a subclass of the HADRONS, which are the elementary particles that are composed of some combination of quarks and antiquarks and are affected by the STRONG INTERACTION. Among the baryons are the PROTON (p), the NEUTRON (n), lambda-zero (Λ^0), the sigmas (Σ^+, Σ^-, Σ^0), and omega-minus (Ω^-).

See also ANTIMATTER.

base quantity A physical quantity that serves to define base UNITS is a base quantity. Base units are defined directly from measurement and are not derived solely from other units. All units in a unit system are derived from a limited number of base units and supplementary units. In the formation of SI units, the base quantities are length, MASS, TIME, electric CURRENT, TEMPERATURE, amount of (i.e., number of elementary entities constituting a) substance, and luminous INTENSITY. From these are defined the base units of kilogram, second, kelvin, ampere, MOLE, and candela, respectively. (*See* ELECTRICITY; INTERNATIONAL SYSTEM OF UNITS; LIGHT.)

Specifically:

- The *meter* (m) is defined as the length of the path traveled by light in VACUUM during a time interval of 1/299,792,458 of a second.
- The *kilogram* (kg) is defined as the mass of a platinum-iridium prototype kept at the INTERNATIONAL BUREAU OF WEIGHTS AND MEASURES (BIPM) in Sèvres, France.
- The *second* (s) is the duration of 9,192,631,770 PERIODs of microwave electromagnetic RADIATION corresponding to the transition between the two hyperfine levels in the GROUND STATE of the ATOM of the ISOTOPE cesium-133. (*See* ELECTROMAGNETISM.)
- The *kelvin* (K) is 1/273.16 of the temperature interval between absolute zero and the triple point of water (the temperature at which ice, liquid water, and water vapor coexist).
- The *ampere* (A) is the constant electric current that would produce a FORCE of exactly 2×10^{-7} newtons per meter length (N/m), if it flowed in each of two thin straight parallel conductors of infinite length, separated by 1 meter in VACUUM.
- The *mole* (mol) is the amount of substance that contains as many elementary entities (such as atoms, MOLECULES, IONs, ELECTRONs, etc., which must be specified) as there are atoms in exactly 0.012 kilogram of the ISOTOPE carbon-12.
- The *candela* (cd) is the luminous intensity in a given direction of a source that emits electromagnetic radiation of FREQUENCY 5.40×10^{14} hertz (Hz) at radiant intensity in that direction of 1/683 watt per steradian (W/sr).

battery Any device for storing ENERGY and allowing its release in the form of an electric CURRENT is called a battery. Batteries store energy almost exclusively in the form of chemical energy, although other forms are possible, such as rotational KINETIC ENERGY. The current is obtained from a pair of terminals on the battery, designated positive (+) and negative (–), where the positive terminal is at a higher electric potential than the negative. A battery may be composed of one or more individual units, called cells. Batteries and cells are characterized by their ELECTROMOTIVE FORCE (emf, commonly called VOLTAGE in this connection), designated in volts (V), and their CHARGE-carrying capability, indicated in terms of

Batteries of various voltages, sizes, and types. The 9-volt battery on the far left has been opened to show that it is composed of cells: six cells of 1.5 volts each, connected in series. As an indication of the scale of the figure, the height of the case of the tall battery in the center is 9.7 centimeters (about 3.8 inches). *(Photo by Frost-Rosen)*

current × time, such as ampere-hours (A·h) or miliampere-hours (mA·h). This indication of charge capability means that the battery is capable of supplying a steady current I for time interval t, where the product It equals the indicated rating. An automobile battery normally supplies an electric current at a voltage of 12 V and might be rated at 500 A·h. Such a battery could supply a current of 250 A, say, for two hours. Some batteries are intended for one-time use, while others are rechargeable. Attempts to recharge a nonrechargeable battery can be dangerous, possibly resulting in explosion. (*See* ELECTRICITY; POTENTIAL, ELECTRIC; ROTATION.)

Similar batteries or cells can be connected in series (the positive terminal of one to the negative terminal of another), giving a combined battery whose emf equals the sum of individual emfs (equivalently, the emf of one unit times the number of units). Four 1.5-volt batteries in series are equivalent to a six-volt battery. The charge-carrying capability of the combination equals that of an individual unit. Alternatively, they can be connected in parallel (all positive terminals connected together and all negative terminals connected together), whereby the combination has the emf of an individual unit and a charge capability that is the sum of those of all the units (or, that of one unit times the number of units). Four 1.5-volt batteries in parallel are equivalent to a 1.5-volt battery, but one that can supply a current that is four times the current that can be drawn from any one of the four. (Dissimilar batteries can be combined in series, but should *not* be connected in parallel.)

A battery is equivalent to an ideal (zero-RESISTANCE) emf source in series with a RESISTOR, whose resistance equals the internal resistance of the battery. If a battery possesses an emf of E and internal resistance r, the actual potential difference, V, correctly called voltage, between its terminals is given by:

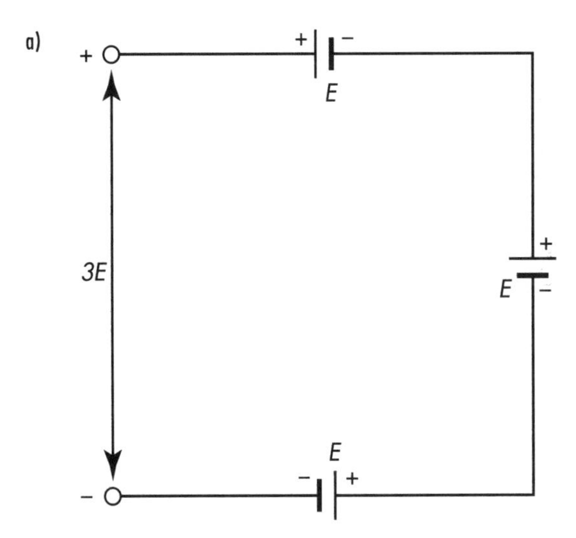

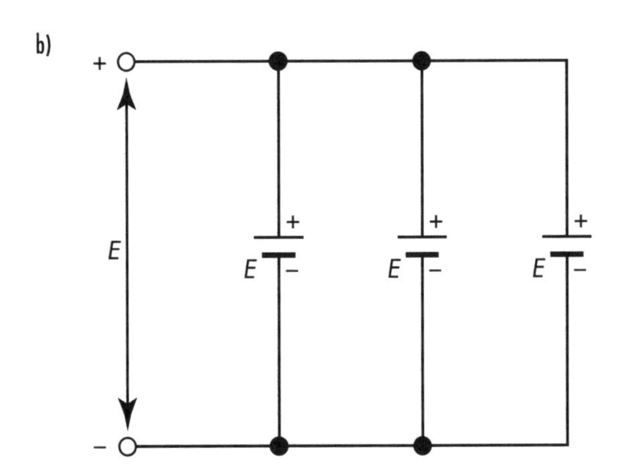

Three similar batteries in series and parallel combinations. (a) The batteries are connected in series. The emf of the combination equals three times the emf of a single battery. (b) A parallel connection. The emf of the combination is the same as that of each battery.

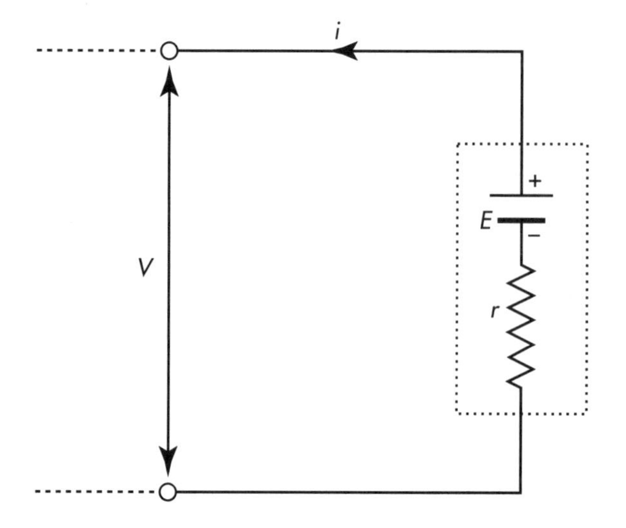

A battery is equivalent to an emf source in series with a resistor. The terminal voltage, *V*, is given by *V* = *E* − *ir*, where *E* and *r* denote, respectively, the emf and internal resistance of the battery, and *i* is the current being drawn from the battery by the circuit to which it is connected.

$$V = E - ir$$

when the circuit is drawing current i from the battery.

beats The effect called beats is the periodic alternating increase and decrease of amplitude of OSCILLATION of a system due to the combined effect of a number of oscillations of differing FREQUENCIES occurring in the system simultaneously. Usually, the term refers to two simultaneous frequencies. More specifically, in ACOUSTICS, beats are the periodic alternating increase and decrease of INTENSITY, termed *beating*, of the perceived single tone when two tones of sufficiently close frequencies are sounded simultaneously. In such a case, given two sounded frequencies, f_1 and f_2, the frequency of the beats is $|f_1 - f_2|$, the absolute value of the difference of the two frequencies, while the perceived beating tone has frequency $(f_1 + f_2)/2$, the average of the two. When the two sounded frequencies are sufficiently different, beats are not perceived; the listener simply hears the two individual tones simultaneously. In the intermediate region of frequency difference, neither a beating single tone nor two individual tones are perceived, but rather an auditory effect that is often described as "rough" or "grating."

Bell Labs Founded in 1925, Bell Labs forms the research and development arm of Lucent Technologies, Inc., specializing in communications technology and networking. It employs nearly 10,000 research and development employees in more than a dozen countries, including the United States. Bell Labs is granted an average of just over two patents every working day, totaling over 30,000 patents since its inception. Its research and development have been recognized by the 11 Nobel Prize awards in physics for work done at the laboratories—six Nobel Prizes in physics (1937, 1956, 1977, 1978, 1997, 1998), four Nobel Prize awards in physics (1964, 1981, 1996) and one in chemistry (1996)—given to scientists who once worked at Bell Labs. Scientists working at Bell Labs have also received numerous other awards in science and technology.

The accomplishments of Bell Labs in physics and in related fields are numerous. The first TRANSISTORS were invented there between 1947 and 1952, and the mathematical foundations of information theory and digital communications were built there in the late 1940s. Much of the world's technical foundation and infrastructure in data networking, mobile and optical communications, and computing and software technology was invented and developed by Bell Labs. Technological milestones that were introduced by the laboratories include stereo recording; sound motion pictures; the first long-distance TV transmission; the first fax machine; the Touch-Tone™ telephone; several generations of modems; communications satellites; LASERS; solar cells; cellular telephony; light-wave communication systems; the charge-coupled device (CCD) (which enables digital imaging); and software that operates, maintains, and manages sophisticated public and private communications networks. (*See* ELECTROMAGNETISM; LIGHT; WAVE.)

Research at Bell Labs is organized into six divisions: Physical Sciences, Computer Science and Software, Mathematical Sciences, Network Applications and Services, Optical Networking, and Wireless Networking. The Physical Sciences Division is active in the areas of THEORETICAL PHYSICS, biological physics, materials research, optical physics, SEMICONDUCTOR physics research, and nanotechnology research. In these fields, Bell Labs has made recent contributions to CONDENSED-MATTER PHYSICS, understanding of materials, processing technology, organic electronics,

Beauty: Only in the Eye of the Beholder?

"They performed a beautiful experiment on liquid helium." "His theory of high-temperature superconductivity is a real beauty." "She discovered a beautiful law for the low-pressure behavior of such systems." "Now, that's a beautiful idea!" Such expressions are common among physicists. It would seem there is a lot of beauty in physics.

But what does *beauty* have to do with physics? Why should aesthetics enter the picture? After all, physics is a *rational* study of nature. Physics is carried out with reason, logic, and mathematics. Like all branches of science, physics strives for objectivity, and what could be more subjective than beauty, which, as is claimed, lies in the eye of the beholder?

In spite of all that and irrational as it might be, physicists do indeed find beauty in physics. Physics does possess an aesthetic component. Other things being equal, a physicist will always prefer a beautiful theory to an ugly one. In fact, a physicist might very well prefer a beautiful theory, even when an uglier one fits the data better! Many physicists will admit that the pleasure they derive from their profession contains a large aesthetic component, and for some that component dominates.

What, then, is beauty in physics? And in particular, what makes a theory beautiful? What is it that arouses in a physicist the feeling of beauty? It is generally agreed that the principal ingredients of beauty for a physicist are simplicity, generality, and unification.

Let us start with *simplicity,* perhaps the hardest of the three properties to pin down precisely. At least for physics, simplicity might best be correlated with a small number of conceptual ingredients. As an illustration, consider the law, discovered by GALILEO GALILEI, that the distance, d, that a uniform sphere rolls from rest down a straight, inclined track during a time interval, t, is proportional to the square of the time interval:

$$d = bt^2$$

It follows from this that the instantaneous speed, v, of the rolling sphere is proportional to the elapsed time:

$$v = 2bt$$

This law is considered extremely simple. It is, perhaps, the simplest law that might be imagined for the given situation. Consider various alternatives. The proportionality coefficient, b, might have different values for different horizontal orientations of the plane. Then there would be a different version of the law for each orientation: a north value of b, a northeast value, an east value, etc. That is clearly more complicated than a single value of b for all directions. Or, the dependence of d on t might not be a power dependence. It might be, for instance, an exponential or logarithmic dependence. Here physicists and mathematicians alike agree that such a dependence would be more complicated than a power law.

If it is a power law, then it could have been a sum of terms with different powers, such as $bt^2 + ct^{1/2}$. Again, clearly this is more complicated than a single term. Well, if it is a single term, why not something like $bt^{2.067}$? Clearly, integer powers are mathematically simpler than others. So if it is an integer power, then how about negative integers? That would not work, since negative powers of time give an infinite value of d at $t = 0$ and decreasing d as time progresses, which clearly does not describe what happens. Among the positive-integer powers, the simplest is unity, with d proportional to t. But that describes constant-speed motion, while the rolling ball clearly starts with zero initial speed and accelerates. Thus, the simplest positive-integer power of time that accounts for this kind of motion is the second power, which is just what is observed.

For another example of simplicity, we turn to ALBERT EINSTEIN's general theory of relativity. This is one of a number of proposed theories of GRAVITATION, the universal force of attraction between all pairs of bodies in the universe, and has been well confirmed as the winner. Although all those theories might appear overwhelmingly complicated, among physicists who deal with such matters, Einstein's theory is generally perceived as also taking the prize with regard to simplicity. Thus, it would stand as the preferred theory, even if the experimental data were ambiguous. (*See* RELATIVITY, GENERAL THEORY OF.)

SYMMETRY contributes to simplicity. In a symmetric situation, there exists equivalence among certain aspects of the situation. In the rolling-sphere example, having the same value for b in all horizontal directions is symmetry, symmetry under all rotations about a vertical axis. Thus, the more symmetric the situation, the simpler it is. That is because more aspects are equivalent, which is simpler than the aspects being completely different. Compare the shapes of a sphere and a cow. The former is clearly the simpler by far. As for symmetry, the sphere possesses symmetry under all rotations about any axis through its center, as well as reflection symmetry through any plane passing through its center. All its orientations and reflection images are equivalent. The symmetry of a cow, on the other hand, consists merely of left-right reflection symmetry through the front-back vertical plane down the center. A cow's shape is equivalent solely to its single reflection image.

Now consider *generality* as a beauty-enhancing property. Generality is easier to describe than simplicity: the

(continues)

Beauty: Only in the Eye of the Beholder? *(continued)*

more general a category, the greater the number of natural phenomena it encompasses. An experiment whose result has broad implications is more general than one that reflects only on a particular case. A law is more general, as it covers a wider range of situations. The law in the above rolling-sphere example is very general indeed. It is valid for all spheres of all materials and all sizes, not merely for ball bearings, marbles, or bowling balls.

For an additional example, consider JOHANNES KEPLER's laws of planetary motion and compare them with ISAAC NEWTON's laws of motion and law of gravitation. The former, as they were originally stated, deal with the solar system. However, they are actually more general than that and are valid also for the moon systems of multimoon planets. Newton's laws, however, deal not only with the solar system and moon systems, but with *all* objects and systems of objects. Thus, the latter are far more general than the former. Indeed, NEWTON'S LAWS form a theory of—i.e., an explanation for—both KEPLER'S LAWS and Galileo's rolling-sphere law.

Unification goes beyond generality. A unifying law or theory must not only encompass a range of phenomena, but must also show that they are, in reality, only different aspects of the same phenomenon. Take Kepler's laws, for instance. Before Kepler, the motions of the Sun's various planets appeared to have nothing to do with each other. Kepler unified the motions by showing that they are all particular cases of the same phenomenon and obey the same laws. Or consider JAMES CLERK MAXWELL's equations that form a theory of ELECTROMAGNETISM. This theory unifies ELECTRICITY and MAGNETISM and shows that they are two aspects of the single phenomenon of electromagnetism. (*See* MAXWELL'S EQUATIONS.)

Those, then, are the three properties that mostly affect physicists' perception of beauty in physics: simplicity, generality, and unification. As mentioned previously, physicists always prefer and strive for beautiful laws and theories, even when less beautiful ones appear to better fit the data. Moreover, it turns out that correct laws and theories are inevitably beautiful! Whatever it is that causes physicists to perceive beauty as they do, it somehow attunes them to nature itself. In brief: *Nature prefers beauty.* Why nature does so is quite a mystery. Here is an example.

PAUL ADRIEN MAURICE DIRAC developed a very beautiful theory of the ELECTRON called the DIRAC EQUATION. The theory predicted the existence of another type of ELEMENTARY PARTICLE, one having the same MASS, SPIN, and magnitude of electric CHARGE as the electron, but with opposite sign of electric charge—positive rather than negative. At the time, the only other known types of elementary particle were the PROTON and the NEUTRON, and neither fitted the specifications of Dirac's predicted particle. Thus the theory was considered false. Nevertheless, Dirac did not abandon his theory, and the eventual discovery of the POSITRON, the electron's antiparticle, proved him right. The moral, according to Dirac, is that "it is more important to have beauty in one's equations than to have them fit experiment." (*See* ANTIMATTER.)

One who did not follow Dirac's way and lived to regret it was ERWIN SCHRÖDINGER. Schrödinger devised a beautiful theory to explain atomic phenomena, but when he applied it to the electron in the hydrogen ATOM, the simplest atomic system, he obtained results that were in disagreement with experiment. Then he noticed that a rough approximation to his equation gave results that agreed with experimental observations. So he published his approximate theory, now called the SCHRÖDINGER EQUATION, which is a much less beautiful theory than the original. Due to his delay, the original theory was published by others and credited to them. What had happened was this. The original, beautiful theory was not appropriate for the type of elementary particle called a FERMION, and the electron is a fermion. Rather, the theory was suitable for BOSONS, a type of elementary particle that had not yet been experimentally discovered. The approximate, uglier theory was insensitive to the difference between fermions and bosons and turned out to be fairly accurate when applied to the hydrogen atom.

As a final example: when Einstein proposed his general theory of relativity, there existed very little experimental evidence in its support. Since then, much more evidence has accumulated, and the theory is now well confirmed. One can imagine Einstein's repeating to himself at the time, "This theory is too beautiful to be wrong!" He would have been absolutely right, of course: the general theory of relativity is widely considered to be one of the most beautiful theories in physics.

and other new devices, some of which have opened up new fields of research. Examples include areas of nanotechnology, quantum cascade lasers, new nanopatterning techniques, and "soft" condensed-matter technology. This division is also engaged in other diverse areas of research, including the active study of quantum information processing, biomimetics research (on how nature creates structure and

Bell Labs's Arno A. Penzias (right) and Robert W. Wilson shared the Nobel Prize in physics in 1978. Penzias and Wilson were cited for their discovery of the faint cosmic microwave background radiation remaining from the "big bang" explosion that gave birth to the universe some 15 billion years ago. The antenna they used appears in the background. ***(Courtesy of Lucent Technologies Inc./Bell Labs)***

mechanics), and work on large-scale mapping of the distribution of dark matter and ENERGY in the UNIVERSE.

The URL of the Bell Labs website is http://www.bell-labs.com/. Bell Labs's mailing address is Lucent Technologies, Corporate Headquarters, 600 Mountain Avenue, Murray Hill, NJ 07974.

See also BIOPHYSICS; COSMOLOGY; OPTICS; QUANTUM MECHANICS; QUANTUM PHYSICS.

Bernoulli, Daniel (1700–1782) Swiss *Mathematician* Born into a mathematical family, Daniel Bernoulli is best known for showing that as the flow SPEED of a FLUID increases, the fluid's PRESSURE decreases. This is one consequence of BERNOULLI'S EQUATION, named after him. Bernoulli held various teaching positions around Europe during his life. Although he was considered a mathematician, Bernoulli possessed great physics insight, and his

Daniel Bernoulli, an 18th-century mathematician, showed that as the flow speed of a fluid increases, the fluid's pressure decreases. ***(The Burndy Library, Dibner Institute for the History of Science and Technology, Cambridge, Massachusetts)***

investigations included oscillating systems (such as vibrating strings), ocean tides, MAGNETISM, the KINETIC THEORY of GASes, and various astronomical and nautical topics. He won the annual prize of the French Academy 10 times for his wide-ranging work.

See also OSCILLATION; VIBRATION.

Bernoulli's equation Named for the Swiss mathematician-physicist DANIEL BERNOULLI, Bernoulli's equation is an equation describing the steady, irrotational flow of an incompressible, nonviscous FLUID. In greater detail, the meaning of the constraints is this: a *steady* flow is one in which the flow pattern does not change over time. Irrotational flow is flow in which the fluid particles do not rotate about their own axes (although the fluid as a whole might flow in a circular path). The term *incompressible* means that the DENSITY of the fluid does not change as it flows, no matter how the PRESSURE might vary. A nonviscous fluid is one in which there are no FORCEs due to VISCOSITY (i.e., no internal FRICTION). Bernoulli's equation is an idealization. Nevertheless, it is useful in many real-world situations, when the nonideal character of a fluid and its flow can be ignored to sufficiently good approximation. (*See* HYDRODYNAMICS; ROTATION.)

Bernoulli's equation is an expression of the WORK-ENERGY THEOREM, whereby the WORK done on a small volume of flowing fluid equals the change in its total mechanical ENERGY, i.e., in the sum of its KINETIC ENERGY (related to its flow SPEED) and its POTENTIAL ENERGY (proportional to its altitude). One form of the equation is that, as a small volume of fluid flows along, the quantity:

$$p + \rho v^2/2 + \rho g h$$

remains constant. Here p denotes the fluid pressure in pascals (Pa); ρ is the (constant) density of the fluid in kilograms per cubic meter (kg/m^3); v is the flow speed of the small volume in meters per second (m/s); g is the ACCELERATION due to gravity, whose nominal value is 9.8 meters per second per second (m/s^2); and h indicates the altitude, or height, of the small volume from some

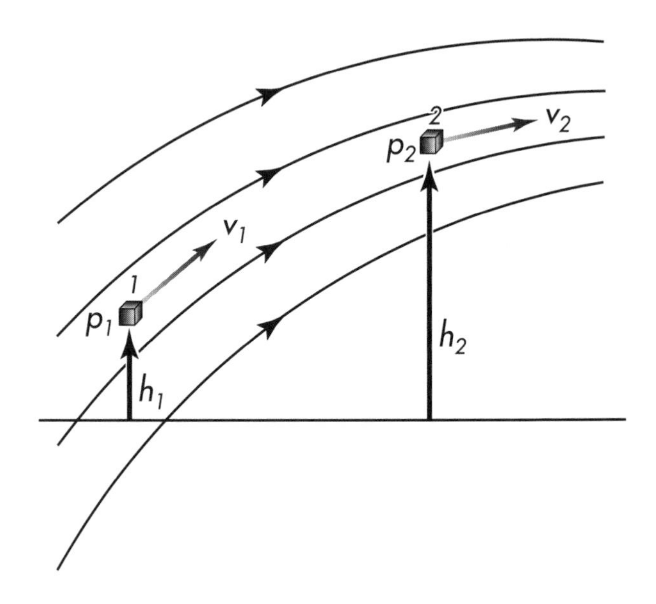

Bernoulli's equation: for a small volume of an incompressible, nonviscous fluid in steady, irrotational flow, the quantity $p + \rho v^2/2 + \rho g h$ remains constant along the flow. The variable p denotes the pressure of the fluid, ρ the fluid density, v the flow speed, g the acceleration due to gravity, and h the height from some reference level. The figure shows stream lines describing the flow of a fluid and a small volume of fluid that flows from location 1 to location 2. For this volume, $p_1 + \rho v_1^2/2 + \rho g h_1 = p_2 + \rho v_2^2/2 + \rho g h_2$.

reference level in meters (m). Bernoulli's equation can equivalently be cast in the form:

$$p_1 + \rho v_1^2/2 + \rho g h_1 = p_2 + \rho v_2^2/2 + \rho g h_2$$

where the subscripts 1 and 2 refer to the values of the quantities at any two points along the flow. (*See* GRAVITATION.)

One of the very useful applications of Bernoulli's equation is in horizontal flow, when $h_1 = h_2$. In this case the equation reduces to:

$$p_1 + \rho v_1^2/2 = p_2 + \rho v_2^2/2$$

which tells us that as the flow speed increases, the fluid pressure must decrease, and the converse. This relation is important in the calculation of LIFT forces on aircraft.

beta decay This is a form of RADIOACTIVITY, whereby an atomic NUCLEUS spontaneously emits an ELECTRON, also called a BETA PARTICLE, and an antineutrino of electron type, thus transmuting into a daughter nucleus with the same ATOMIC MASS as that of the parent nucleus and an ATOMIC NUMBER greater by one. (*See* ATOM.) The basic process of beta decay is the conversion of a NEUTRON within the nucleus to a PROTON, which remains inside the nucleus, and an electron and an antineutrino, which leave:

$$n \rightarrow p + e^- + \bar{\nu}_e$$

where the symbols denote, from left to right, neutron, proton, electron, and electron-type antineutrino. This process is governed by the WEAK INTERACTION. An example of beta decay is the decay of the radioactive ISOTOPE carbon-14, ${}^{14}_{6}C$ (atomic number 6, atomic mass 14)—used to determine the age of archaeological objects—to the stable isotope nitrogen-14, ${}^{14}_{7}N$ (atomic number 7, atomic mass 14):

$${}^{14}_{6}C \rightarrow {}^{14}_{7}N + e^- + \bar{\nu}_e$$

(*See* ANTIMATTER; NEUTRINO.)

The ENERGY difference between the parent and daughter nuclei emerges as the total energy of the emitted electron and antineutrino. That energy can be apportioned differently between the electron and the antineutrino in any individual decay process. So the electrons emitted by the radioactive decay of a sample of carbon-14, for example, will possess a range of energies not exceeding the ${}^{14}_{6}C$-${}^{14}_{7}N$ energy difference. In each individual decay, the antineutrino will carry the remainder of the ${}^{14}_{6}C$-${}^{14}_{7}N$ energy difference not taken by the electron. The penetrating power of the electrons emitted by beta decay is strongly dependent on their energy.

A related form of radioactive decay is inverse beta decay, when a proton in a nucleus spontaneously converts to a neutron (which remains in the nucleus) and a POSITRON and electron-type neutrino, which are emitted as follows:

$$p \rightarrow n + e^+ + \nu_e$$

where e^+ and ν_e denote the positron and the electron-type neutrino, respectively. An example of inverse beta decay is the decay of the radioactive isotope nitrogen-13, ${}^{13}_{7}N$ (atomic number 7, atomic mass 13), to the stable isotope carbon-13, ${}^{13}_{6}C$ (atomic number 6, atomic mass 13):

$${}^{13}_{7}N \rightarrow {}^{13}_{6}C + e^+ + \nu_e$$

Whereas free neutrons, not confined to a nucleus, undergo beta decay, free protons cannot undergo inverse beta decay. That is because the MASS of the neutron is greater than that of the proton, so no energy is available for the decay. Within a nucleus, on the other hand, there might be available energy. (*See* MASS ENERGY.)

beta particle The name given to the ELECTRON emitted through the BETA DECAY of a NUCLEUS is beta particle. In that decay, a NEUTRON in the nucleus spontaneously converts to a PROTON, which remains in the nucleus, and an electron and electron-type antineutrino, which leave the nucleus. The name is a relic from the early years of the study of RADIOACTIVITY.

See also ANTIMATTER; NEUTRINO.

big bang This is the name given to the primordial explosion from an extremely dense, hot state that is taken, according to a class of cosmological models, to be the origin of the UNIVERSE. Big-bang cosmological models were devised to explain the observations that the universe appears to be expanding and to be permeated by a COSMIC MICROWAVE BACKGROUND. According to those models, the universe has been expanding and cooling ever since the big bang. As for the origin of the big bang itself, there are various ideas being proposed. These include the big bang resulting from a quantum fluctuation in some fundamental substrate and the big bang being caused by the COLLISION, within a space of extra DIMENSIONS, of our SPACE-TIME with another.

Most commonly, the big bang is thought of not as an explosion *within* SPACE, as a bomb might explode, but rather as an explosion of *space itself*. However, there are also ideas that big bangs are occurring at various times and at various locations in an overarching space, producing universes, all of which together might be called a multiverse. Our universe would then be one of the universes making up the multiverse. No observer in a universe could observe any of the other universes.

See also COSMOLOGY; DENSITY; HEAT; QUANTUM PHYSICS; TEMPERATURE.

binding energy In general, the binding energy of a system is the ENERGY required to separate the system into its component parts. The term is most commonly used for NUCLEI, where the binding energy is the energy required to decompose a nucleus into the NUCLEONs, (i.e. the PROTONs and NEUTRONs) that it comprises. Conversely, the binding energy of a nucleus is the amount of energy released when its constituent nucleons come together to form the nucleus. The binding energy of a nucleus is the energy equivalent of its mass defect. The latter is the difference between the sum of the MASSes of the constituent nucleons and the mass of the nucleus. The binding energy per nucleon is a useful measure of stability for a nucleus: the greater the binding energy per nucleon, the greater the stability of the nucleus. Among all the isotopes, the binding energy per nucleon peaks to a maximum in the vicinity of atomic number 26 and mass number 56, around the isotope iron-56, $^{56}_{26}\text{Fe}$. Thus, the most nuclearly stable element is iron. Heavier nuclei may split, or undergo FISSION, to produce more stable nuclei with atomic numbers closer to that of iron, while releasing energy. On the other hand, lighter nuclei may coalesce, or undergo FUSION, to form heavier nuclei, closer in atomic number to iron, again with the release of energy. Iron, due to its maximal nuclear stability, is the most abundant of the heavier elements in the universe. (*See* MASS ENERGY.)

Here is an example of a binding energy calculation. The masses of the proton and the neutron are 1.67262×10^{-27} kilogram (kg) and 1.67493×10^{-27} kg, respectively. Deuterium is a stable isotope of hydrogen, whose nucleus, the deuteron, consists of one proton and one neutron. The deuteron's mass is 3.34358×10^{-27} kg. The sum of the masses of its constituent proton and neutron is 3.34755×10^{-27} kg, which is greater than its own mass. The difference, the mass defect of the deuteron, is:

$$(3.34755 \times 10^{-27}\text{ kg}) - (3.34358 \times 10^{-27}\text{ kg}) = 0.00397 \times 10^{-27}\text{ kg}$$

Now we use ALBERT EINSTEIN's mass-energy relation, $E = mc^2$, to convert the mass defect to its energy equivalent, which is the binding energy of the deuteron. Here m denotes the mass defect in kilograms (kg), E its energy equivalent in joules (J), and c the SPEED OF LIGHT, with approximate value 3.00×10^8 meters per second (m/s). The result is:

$$(0.00397 \times 10^{-27}\text{ kg})\,(3.00 \times 10^8\text{ m/s})^2 = 3.57 \times 10^{-13}\text{ J}$$

We then divide by two to obtain the binding energy per nucleon, 1.79×10^{-13} J.

biophysics This is the branch of physics that deals with the application of physics to biological phenomena. The gamut of phenomena runs from the whole organism, and even societies of organisms, down to the cellular and subcellular level. On the large scale, biophysics deals with the MECHANICS of bones and muscles, the OPTICS of eyes, and the THERMODYNAMICS of organism metabolism, for example. At the other end of the range, biophysics considers such as the HYDRODYNAMICS of the swimming of certain protozoa by flagellum action and the thermodynamics of cell metabolism.

Biot-Savart law Named for the French physicists Jean-Baptiste Biot and Félix Savart, the Biot-Savart law shows how all electric CURRENTs contribute to the magnetic FIELD at every point in space. Specifically, it relates the magnitude and direction of an infinitesimal current element to the magnitude and direction of its contribution to the magnetic field at any point. Then the total magnetic field at that point is obtained by summing up, or integrating, over all current elements. The equation is:

$$d\mathbf{B} = \frac{\mu}{4\pi}\frac{i\,d\mathbf{s} \times \mathbf{r}}{r^3}$$

where $d\mathbf{s}$ is a differential VECTOR element of CONDUCTOR length in meters (m) in the direction of the current i in amperes (A) flowing through it; $\mathbf{r}$ is the vector pointing from that element to the point where the magnetic field is being calculated, the field point, whose magnitude, r, is the distance between the two in meters (m); $d\mathbf{B}$ is the infinitesimal vector contribution to the magnetic field at the field point in teslas (T) due to the infinitesimal current element (the current flowing in the infinitesimal element of conductor length); and μ is the magnetic

PERMEABILITY of the medium in which this is taking place in tesla·meters per ampere (T·m/A). This equation gives both the magnitude and the direction of the contribution. (*See* ELECTRICITY; MAGNETISM.)

For the magnitude alone, the equation reduces to:

$$dB = \frac{\mu}{4\pi}\frac{i\,ds \sin\theta}{r^2}$$

where dB is the magnitude of the infinitesimal contribution to the magnetic field at the field point in teslas (T), ds is the magnitude of the differential element of conductor length in meters (m), and θ is the angle between the vectors $d\mathbf{s}$ and $\mathbf{r}$. The direction of the contribution is determined in this manner. Consider a straight line along the vector $d\mathbf{s}$ (i.e., the tangent to the conductor at the point of the current element). Consider a circle, perpendicular to and centered on this line, that passes through the field point. Consider the tangent to the circle at the field point and endow this tangent with a direction as follows. Imagine grasping the circle at the field point with your right hand, in such a way that your fingers pass through the plane of the circle in the same direction as does the current in the current element. Then your thumb is pointing in the direction of the tangent. (Alternatively, you can place your right fist at the current element with thumb extended in the direction of the current. Then your curled fingers define a rotation sense for the circle and thus a direction of the tangent to the circle at the field point.) The direction of the tangent is the direction of $d\mathbf{B}$, the infinitesimal contribution to the magnetic field at the field point.

When the field is being produced in VACUUM (or, to a good approximation, in air), μ is then the magnetic permeability of the vacuum, denoted μ_0, whose value is $4\pi \times 10^{-7}$ T·m/A. The corresponding vacuum forms of the above two equations are:

$$d\mathbf{B} = \frac{\mu_0}{4\pi}\frac{i\,d\mathbf{s} \times \mathbf{r}}{r^3}$$

$$dB = \frac{\mu_0}{4\pi}\frac{i\,ds \sin\theta}{r^2}$$

An example of use of the Biot-Savart law is to find the magnitude of the magnetic field, B, in teslas (T) at perpendicular distance r in meters (m) from a long straight wire carrying current i in amperes (A). Integration gives the result:

$$B = \frac{\mu_0}{4\pi}\frac{i}{r}$$

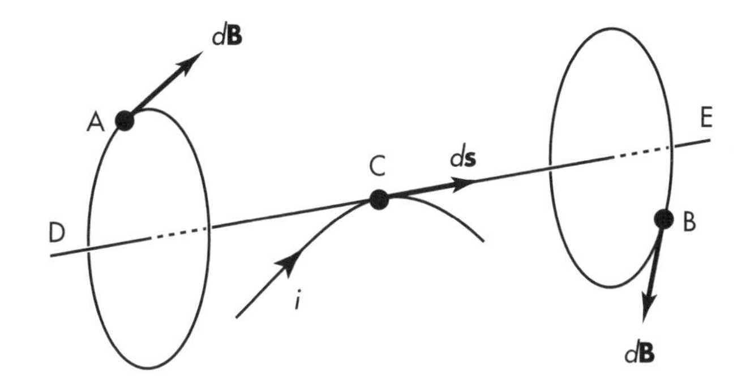

The direction of the contribution to the magnetic field, *d*B, at arbitrary field points A and B due to an infinitesimal element of electric current at point C, according to the Biot-Savart law. The symbol *d*s denotes the differential vector element of conductor length. It is tangent to the conductor at C and points in the sense of the current *i*. The straight line DE is a continuation of *d*s. The direction of *d*B is tangent to the circle that passes through the field point, is perpendicular to line DE, and whose center lies on DE. The sense of *d*B for the illustrated sense of the current is shown.

The Biot-Savart law can also be expressed in terms of the magnetic field, **B**, produced by an electric charge, q, moving with velocity **v**, giving:

$$\mathbf{B} = \frac{\mu_0}{4\pi}\frac{q\,d\mathbf{v} \times \mathbf{r}}{r^3}$$

where **r** is the separation vector from the charge to the field point.

See also AMPÈRE'S LAW.

BIPM *See* INTERNATIONAL BUREAU OF WEIGHTS AND MEASURES.

birefringence The possession of two INDEXes OF REFRACTION is termed birefringence. Such a situation occurs typically in anisotropic CRYSTALS (i.e., crystals that do not possess the same physical properties in all directions) when the SPEED of LIGHT propagation through the material depends on direction. In this case, a beam of light impinging on the surface of the crystal refracts into two beams, each obeying SNELL'S LAW with its own index of refraction, where the two beams are perpendicularly plane-polarized. Calcite, a crystal of calcium carbonate, is well known for this effect. Isotropic materials might be made birefringent through the application of a directional external influence, such as a FORCE or an electric or magnetic FIELD.

See also ELECTRICITY; KERR EFFECT; MAGNETISM; POLARIZATION; REFRACTION.

black body Also spelled *blackbody;* this is the name given to a body that absorbs all electromagnetic RADIATION impinging on it. Although a black body is an idealization, it can be approximated by a small hole in the wall of a cavity, when the cavity and its enclosure are in thermal EQUILIBRIUM at some TEMPERATURE. Then practically all electromagnetic radiation falling on the hole from the outside will be absorbed by the cavity. The concept of a black body is an important one for the theory of the INTERACTION of electromagnetic radiation with matter. (*See* ELECTROMAGNETISM; HEAT.)

A black body emits electromagnetic radiation of all WAVELENGTHS, with a particular wavelength for which the INTENSITY of radiation is maximal and the intensity tapering off for greater and lesser wavelengths. The wavelength of maximal intensity, λ_{max}, is inversely proportional to the black body's absolute temperature, T, according to Wien's displacement law:

$$\lambda_{max}T = 2.898 \times 10^{-3} \text{ m·K}$$

where the wavelength is in meters (m) and the absolute temperature in kelvins (K). This underlies the effect that real bodies, such as a chunk of iron or a star, at higher and higher temperatures appear to glow successively red, orange, yellow, white, blue, and violet. At extremely low or high temperatures, they might not be visible at all, since most of their radiated energy would then be in the infrared or ultraviolet range, respectively, and thus invisible to the human eye. (*See* LIGHT.)

The total POWER emitted by a black body at all wavelengths is proportional to the fourth power of the absolute temperature. That is expressed by the STEFAN-BOLTZMANN LAW:

$$P = \sigma T^4$$

where P denotes the electromagnetic power (energy per unit time) radiated per unit area in watts per square meter (W/m^2), and σ is the Stefan-Boltzmann constant, whose value is 5.67051×10^{-8} W/(m^2·K^4).

Nothing about black body radiation—neither the distribution of intensity over wavelengths (the black-body SPECTRUM), Wien's law, nor the Stefan-Boltzmann law—is understood in terms of CLASSICAL PHYSICS. Early in the 20th century, MAX KARL ERNST LUDWIG PLANCK, making the nonclassical assumption that energy is exchanged between the electromagnetic radiation and the matter composing the body only in discrete "bundles," derived the formula correctly describing the black-body spectrum:

$$\Delta E = \frac{8\pi ch}{\lambda^5} \frac{\Delta\lambda}{e^{\frac{hc}{kT\lambda}} - 1}$$

Here ΔE denotes the energy density (energy per unit volume) of the electromagnetic radiation in joules per cubic meter (J/m^3) in a small range of wavelengths $\Delta\lambda$ around wavelength λ for a black body at absolute temperature T. The symbols h, c, and k represent, respectively, the PLANCK CONSTANT, the SPEED OF LIGHT, and Boltzmann's constant, with values $h = 6.62606876 \times 10^{-34}$ joule·second (J·s), $c = 2.99792458 \times 10^8$ meters per second (m/s), and $k = 1.3806503 \times 10^{-23}$ joule per kelvin (J/K). ALBERT EINSTEIN helped to better understand the concept of discrete energy "bundles" underlying this formula, and that led to the development of QUANTUM MECHANICS. Both Wien's displacement law and the Stefan-Boltzmann law follow from Planck's formula for the black-body spectrum.

black hole This is an object whose gravity is so strong that nothing, not even light or other electromagnetic radiation, can escape from it. According to ALBERT EINSTEIN's general theory of relativity, bodies of sufficiently high DENSITY will exert gravitational FORCES on other bodies and on electromagnetic RADIATION, including LIGHT, in their vicinity, so that neither the other bodies nor the radiation will ever be able to leave that vicinity. A surface, called the event horizon, surrounds the black hole and determines the fate of other bodies and of radiation. A body or light ray originating within the event horizon will never pass through the event horizon and descends inexorably toward the center of the black hole. A body or light ray heading toward the event horizon from outside is drawn in, becomes trapped, and suffers the fate just described. However, a body or light ray that is outside the event horizon may escape if its direction is appropriate. As a rule of thumb, derived from the idealized situation of a nonrotating, spherically symmetrical black hole, the event horizon can be thought of as a spherical surface surrounding the black hole and centered on it, with radius of about 3 kilometers (km) times the MASS of the black hole expressed in terms of the mass of the Sun. Thus a five-solar-mass black hole possesses an event horizon with radius of about 15 km and is wholly contained within that radius. The idea of five solar masses compacted within a sphere of radius 15 km gives an impression of the tremendous density of a black hole (around 10^{19} kilograms per cubic meter [kg/m^3])

and of the gigantic gravitational forces holding it together. (*See* ELECTROMAGNETISM; GRAVITATION; RELATIVITY, GENERAL THEORY OF.)

Since they do not emit light, black holes cannot be seen directly. Nevertheless, they can be detected in two ways. One way is by their gravitational effect on nearby bodies, such as when a black hole is one member of a pair of stars rotating around each other, called a binary-star system. Then the motion of the other member of the pair gives information on the black hole. Another way to detect black holes is through the X RAYs emitted by matter falling into them. Black holes have been observed by both means.

Black holes are understood to form as the final stage of the evolution of stars whose masses are greater than about 25 times the mass of the Sun. Such stars, after catastrophically throwing off much of their material, end their lives as black holes of at least some three solar masses. In addition to black holes as end-of-life stellar remnants, black holes also lie at the centers of GALAXIES. They are of mammoth size, possess masses of millions of stars, and continually grow by sucking in more stars. (*See* ASTROPHYSICS.)

In spite of their not emitting anything from inside their event horizon, black holes can nevertheless "evaporate," as was discovered theoretically by the British physicist STEPHEN WILLIAM HAWKING. The process is understood to commence with the spontaneous production of a virtual particle-antiparticle pair just outside of the black hole's event horizon. Normally, a virtual particle-antiparticle pair would immediately mutually annihilate and disappear, as if nothing had ever happened. Such goings-on characterize the quantum VACUUM and are a consequence of the HEISENBERG UNCERTAINTY PRINCIPLE. This principle allows the "borrowing" of ENERGY for the creation of short-lived, virtual particle-antiparticle pairs and other particles, as long as the particles self-annihilate and the "loan" is repaid within a sufficiently short time interval. However, near the event horizon of a black hole, it might happen that one member of such a pair becomes sucked into the black hole before the pair can mutually annihilate. Then the other member becomes a real particle and leaves the black hole's vicinity, as if it had been emitted. As a result, the black hole loses energy, in the amount carried away by the escaping particle, and thus loses mass. In this manner, black holes eventually "evaporate." The relative rate of evaporation is inversely related to the mass of the black hole. Thus, the effect is negligible for large-mass black holes, such as those that develop from the death throes of stars.

See also ANTIMATTER; ELEMENTARY PARTICLE; MASS ENERGY; QUANTUM PHYSICS.

Bohr, Niels Henrik David (1885–1962) Danish *Physicist* One of the major contributors to the founding of QUANTUM PHYSICS, Niels Bohr is well known for the BOHR THEORY of the hydrogen ATOM. He received his doctorate in physics in 1911 from the University of Copenhagen, Denmark, and went on to investigate atomic and nuclear phenomena. In 1920, Bohr founded the Institute for Theoretical Physics in Copenhagen and served as its director for the rest of his life, except for a period during World War II. After the occupation of

Niels Bohr was one of the founding fathers of quantum physics and was awarded the 1922 Nobel Prize in physics. His model of the hydrogen atom was the first to incorporate quantum ideas. *(AIP Emilio Segrè Visual Archives, Margrethe Bohr Collection)*

Denmark by German forces, Bohr—whose mother was Jewish—escaped to the United States via Sweden and England in 1943. There he worked on the development of a nuclear-FISSION ("atomic") bomb, as part of the Manhattan Project, at what is now LOS ALAMOS NATIONAL LABORATORY. Bohr was awarded the Nobel Prize in physics in 1922 "for his services in the investigation of the structure of atoms and of the radiation emanating from them."

See also NUCLEUS.

Bohr theory In the early 20th century, Danish physicist NIELS HENRIK DAVID BOHR did pioneering work in the structure and behavior of ATOMS by devising a theory, known as the Bohr theory, or the Bohr model, for the simplest atom, the hydrogen atom. What Bohr was faced with was this. It was then known that atoms emit and absorb light and other electromagnetic RADIATION only at very definite WAVELENGTHs, known as spectral lines, which are characteristic of each kind of atom. Spectroscopists had amassed a tremendous amount of such data. As for the atoms themselves, their structure was known as a positively charged NUCLEUS, about 10^{-15} meters (m) in size, surrounded by a cloud of ELECTRONs, with overall size of some 10^{-10} m. It was not at all clear what the electrons were doing or how they might be organized. It was natural to assume the electrons were in orbit around the nucleus, much like the planets around the Sun in the SOLAR SYSTEM, since otherwise it was expected that the electric attraction between each electron and the nucleus would cause the electrons to fall into the nucleus. But in that case, the electrons in their orbital motion would be continually undergoing centripetal ACCELERATION, and it was known that they should then emit electromagnetic radiation, lose ENERGY, and fall into the nucleus. Thus, CLASSICAL PHYSICS could only describe atoms as unstable—while they are obviously generally stable—and, during their brief lifetimes, emitting electromagnetic radiation in continuous ranges of wavelengths, despite all the spectroscopic data on discrete wavelengths to the contrary. (*See* ELECTRICITY; ELECTROMAGNETISM; SPECTRUM.)

Bohr took a cue from MAX KARL ERNST LUDWIG PLANCK's theory of the BLACK BODY spectrum and introduced the quantum idea of discreteness in what had classically been considered to be continuous. For his description of the hydrogen atom, Bohr made three postulates: (1) The electron is in a circular orbit around the nucleus and the atom is normally in a stationary state, in which it does not radiate and lose energy as the electron undergoes centripetal acceleration. Thus, a stationary state has a definite energy associated with it. (2) The possible stationary states are only those for which the ANGULAR MOMENTUM of the electron's motion equals $nh/(2\pi)$, where n is an integer, called the principal QUANTUM NUMBER, and h is the PLANCK CONSTANT, which equals $6.62606876 \times 10^{-34}$ joule·second (J·s). Thus the angular momentum, instead of having any of a continuous range of values, is quantized, by being allowed to possess values that are only integer multiples of $h/(2\pi)$. (3) The atom can "jump" from one stationary state to another by emitting or absorbing one quantum of electromagnetic energy, or PHOTON (as it is called today), whose energy ΔE must equal the energy difference between the two states:

$$\Delta E = |E - E'|$$

where E and E' are the respective energies of the two stationary states, and the absolute value keeps the photon's energy positive. The FREQUENCY f of the emitted or absorbed radiation in hertz (Hz) relates to the energy ΔE of the photon in joules (J) by:

$$\Delta E = hf$$
$$f = \Delta E/h$$

The wavelength λ in meters (m) is consequently related by:

$$\frac{1}{\lambda} = \frac{f}{c} = \frac{\Delta E}{ch}$$
$$\Delta E = ch/\lambda$$

where c is the SPEED OF LIGHT and equals 2.99792458 $\times 10^8$ meters per second (m/s). (*See* LIGHT; QUANTIZATION; QUANTUM PHYSICS; WAVE.)

Except for these postulates, the classical rules for MECHANICS and ELECTRICITY were assumed to apply. It can then be shown that the energy of the atom in the nth stationary state, E_n, is:

$$E_n = -chR/n^2$$

where R is the Rydberg constant, whose value is R = 1.09737316×10^7 inverse meters (m^{-1}). The reference level of zero energy is taken here to be when the atom is ionized (i.e., when the electron is removed from it), $n = \infty$. For finite n, the atom then has negative energy.

In the GROUND STATE, the atom's lowest-energy state, when $n = 1$, the value of the energy is:

$$E_1 = -chR/1^2 = -2.18 \times 10^{-18} \text{ J}$$
$$= -13.6 \text{ electron volts (eV)}$$

Thus a photon of energy 2.18×10^{-18} J = 13.6 eV or more can ionize a hydrogen atom. (*See* IONIZATION.)

From the result for the energy of a state follows the general formula for the wavelength of the spectral line of emission or absorption associated with the transition in hydrogen from the *n*th atomic energy level to the *m*th:

$$1/\lambda = R\,|1/m^2 - 1/n^2|$$

This formula is very well confirmed by spectroscopic data. In the special case of $m = 2$ and $n > 2$, this gives a series of visible spectral lines, called the Balmer series. (*See* BALMER FORMULA.) All other transitions are either in the infrared or in the ultraviolet and are not visible.

The radius of the electron's orbit when the atom is in the *n*th stationary state, r_n, can also be obtained and is found to be:

$$r_n = (5.3 \times 10^{-11})n^2 \text{ m}$$

Thus the radius of the hydrogen in its ground state is:

$$r_1 = 5.3 \times 10^{-11} \text{ m}$$

in accord with experiment.

The Bohr model is applicable, with appropriate modifications, to other one-electron atoms or atoms that are effectively so. These include singly ionized helium as well as atoms in the first column of the PERIODIC TABLE, which have a single electron in the outer shell. In spite of its success, the Bohr model could not be extended to more complicated atoms without quite arbitrary assumptions. It became clear that the model's postulates were not all correct and that its success was to some extent accidental. It has been superseded by QUANTUM MECHANICS, which is a very successful theory for a tremendous range of phenomena and has not yet been found to be falsified by experiment in any way. Nevertheless, Bohr's theory served as an important step toward reaching an understanding of the nature of atoms, and to this day it remains useful as an aid to the visualization of atomic structure.

boiling The PHASE TRANSITION of a substance, whereby its LIQUID form, or liquid PHASE, changes into a GAS, or vapor, is boiling. Boiling is achieved by adding HEAT (i.e., thermal ENERGY) to the liquid, by reducing its PRESSURE, or by both together. Boiling occurs when the substance's VAPOR PRESSURE is equal to or greater than the pressure of the liquid (normally atmospheric pressure, but see below). The TEMPERATURE at which a liquid boils at standard atmospheric pressure is called its boiling point. For example, the boiling point of water (a liquid at room temperature) is 100 degrees Celsius (°C), that for lead (a solid at room temperature) 1620°C, and for propane (a gas at room temperature) −42°C. (*See* STATE OF MATTER.)

The boiling point rises as the pressure on the liquid is increased and falls with reduced pressure. An example of that is boiling water in a pressure cooker, which is a strong, sealed vessel and allows the pressure inside to rise above atmospheric pressure. In pressure cookers, water boils at temperatures considerably higher than 100°C. Alternatively, water can be caused to boil at room temperature by reducing the pressure on it, such as by means of a vacuum pump. The boiling point rises with the addition of solutes to the liquid. The boiling point of water can be raised above the normal 100°C by dissolving table salt in the water.

As a liquid is heated, its temperature rises until the boiling point is reached. Additional heating no longer raises the temperature, but rather the added thermal energy goes into the conversion of the liquid to a gas. At the molecular scale, according to the KINETIC THEORY of matter, a rise in temperature means an increase in the average speed of the molecules of the liquid, which are constantly moving about, colliding with each other and with the walls of the container, and possess a range of speeds. At the boiling point, the fastest of the molecules have just enough KINETIC ENERGY to overcome the cohesive FORCEs holding the liquid together, and they escape as a gas. Further heating of the liquid produces more high-speed molecules that escape to the vapor phase, leaving unchanged the average speed of the remaining molecules and consequently also the temperature of the liquid. The amount of heat required to fully convert a sample of liquid to a gas at the boiling point is the HEAT OF VAPORIZATION of the sample. (*See* COHESION; MOLECULE.)

Boiling is facilitated by the presence of impurities and sharp points, edges, and crevices, called nucleating sites, around which bubbles of the gas phase can form. In the absence of such boiling aids, and keeping the liquid very still, it is possible to heat it gently and slowly raise its temperature to above the boiling point—or lower its boiling point to below its temperature by reducing its pressure—without boiling taking place. In such a state, the liquid is said to be superheated. The superheated state of a liquid is quite unstable, and introducing a nucleating site or jarring the liquid can immediately initiate sudden, sometimes explosive, boiling.

Boltzmann, Ludwig (1844–1906) Austrian *Physicist* Famous as a father of the field of STATISTICAL MECHANICS, Ludwig Boltzmann studied how the macroscopic properties of MATTER are related to the properties of the ATOMs, ELECTRONs, MOLECULEs, etc. that form its constituents. He received his doctorate in 1866 from the University of Vienna, Austria, for a thesis on the KINETIC THEORY of GASes. Boltzmann held a number of professorial positions in Austria and elsewhere in Europe, ending up at the University of Vienna. His work in statistical mechanics involved extensive use of probability, and he proposed a probabilistic definition of ENTROPY that explained the second law of THERMODYNAMICS. Boltzmann's ideas were controversial at the time, since the very notion of atomic structure of matter was not yet universally accepted. He briefly visited the United States shortly before his death and died of depression-induced suicide without knowing that experimental results were already verifying his work.

See also BOLTZMANN DISTRIBUTION; MAXWELL-BOLTZMANN STATISTICS; STEFAN-BOLTZMANN LAW.

Ludwig Boltzmann was a 19th/20th-century pioneer in the field of statistical mechanics, which relates the macroscopic properties of matter to the properties of its constituent particles (atoms, electrons, molecules, etc.) and the forces among them. ***(AIP Emilio Segrè Visual Archives, Physics Today Collection)***

Boltzmann distribution Named for the Austrian physicist LUDWIG BOLTZMANN, the Boltzmann probability distribution states that for a system that is in thermal EQUILIBRIUM at absolute TEMPERATURE *T*, the probability, *p(E)*, of finding the system in a state that has ENERGY *E* is given by:

$$p(E) = \frac{1}{Z} e^{-\frac{E}{kT}}$$

Here *E* is in joules (J), *T* is in kelvins (K), and *k* is the Boltzmann constant with value $1.3806503 \times 10^{-23}$ joule per kelvin (J/K). The denominator, *Z*, called the partition function, is determined so as to make the sum of probabilities of all states of the system (i.e., the probability that the system is in *some* state) equal to one. So when we denote the energy of state *i* by *E(i)*:

$$Z = \sum_i e^{-\frac{E(i)}{kT}}$$

where the summation is over all states of the system. (*See* HEAT.)

The sum can instead be performed over the system's energy levels, E_j. Let $g(E_j)$ denote the degeneracy of energy level E_j (i.e., the number of states that have energy E_j). Then the partition function can be expressed equivalently as:

$$Z = \sum_j g(E_j) e^{-\frac{E_j}{kT}}$$

If the energy range is continuous, rather than discrete, the probability of the system having energy *E* is replaced with the probability that the system possesses energy in the small range from *E* to *E* + *dE*, and the summation in the partition function is replaced by suitable integration.

See also FREE ENERGY.

Born, Max (1882–1970) German *Physicist* One of the founders of QUANTUM PHYSICS, Max Born received his doctorate in physics from the University of Göttingen, Germany, in 1907. He continued his physics investigations, including work on THERMODYNAMICS, SOLID-STATE PHYSICS, and quantum physics, while holding professorial positions at a number of institutions in Germany, mostly at the University of Göttingen. Born was forced to flee Germany in 1933, after which he held positions at Cambridge University, England, and the University of Edinburgh, Scotland. He returned to Germany in 1953. Born was awarded half the Nobel Prize in physics in 1954 "for his fundamental research in quantum mechanics, especially for his statistical interpretation of the wavefunction."

Max Born was one of the founders of quantum physics and received the 1954 Nobel Prize in physics for his fundamental work. *(AIP Emilio Segrè Visual Archives, Physics Today Collection)*

Bose-Einstein statistics The statistical rules governing any collection of identical BOSONS (particles having an integer value of SPIN) are called Bose-Einstein statistics, named for the Indian physicist Satyendra Nath Bose (1894–1974) and German/Swiss/American physicist ALBERT EINSTEIN. These statistics are based on (1) the absolute indistinguishability of the identical particles and (2) any number of the particles being allowed to be in the same quantum state, including to possess the same QUANTUM NUMBERS. (*See* QUANTUM PHYSICS.)

One result of Bose-Einstein statistics is that in a system of identical bosons in thermal EQUILIBRIUM at absolute TEMPERATURE T, the probability for a particle to possess energy in the small range from E to $E + dE$ is given by $f(E)\,dE$, where $f(E)$, the probability distribution function, is:

$$f(E) = \frac{1}{Ae^{\frac{E}{kT}} - 1}$$

Here E is in joules (J), T is in kelvins (K), f is dimensionless, and k denotes the Boltzmann constant, whose value is $1.3806503 \times 10^{-23}$ joule per kelvin (J/K). The dimensionless coefficient A is determined by the type of system. For PHOTONs or PHONONs, for instance, the probability distribution takes the simple form:

$$f(E) = \frac{1}{e^{\frac{E}{kT}} - 1}$$

Note that this distribution function goes to infinity as the energy goes to zero and decreases to zero as the energy increases. (*See* DIMENSION; HEAT.)

Bose-Einstein statistics predicts that at sufficiently low TEMPERATUREs, very close to 0 K, a well-isolated collection of identical bosons can become so mutually correlated that they lose their individual identities and form what amounts to a single entity. That state is called a Bose-Einstein condensate and has been created in laboratories.

See also BOLTZMANN DISTRIBUTION; FERMI-DIRAC STATISTICS; MAXWELL-BOLTZMANN STATISTICS.

boson A boson is any particle—whether an ELEMENTARY PARTICLE or a composite particle, such as a NUCLEUS or an ATOM—that has an integer value of SPIN (i.e., 0, 1, 2, . . .). This type of particle is named for Indian physicist Satyendra Nath Bose (1894–1974). All the particles that mediate the various INTERACTIONs among the elementary particles are bosons. In particular, they are the GLUONS, which mediate the STRONG INTERACTION among QUARKS; the intermediate vector bosons W^+, W^-, and Z^0, which mediate the WEAK INTERACTION; and the PHOTON, which mediates the ELECTROMAGNETIC INTERACTION. These are all spin-1 bosons and are called vector bosons. The graviton, the mediator of the gravitational interaction, is a boson with spin 2. At the level of atomic NUCLEI, the NUCLEONS constituting the nuclei are held together through the mediation of PIONS, which are spin-0 bosons. (More fundamentally, this interaction derives from the strong interaction among quarks, mediated by gluons, since nucleons and pions consist of quarks.) (*See* GRAVITATION.)

Collections of identical bosons are governed by the rules of BOSE-EINSTEIN STATISTICS. At sufficiently low TEMPERATURES, very close to 0 K, a well-isolated collection of identical bosons can become so mutually correlated that they lose their individual identities and form what amounts to a single entity. That state is called a Bose-Einstein condensate and has been created in laboratories.

See also FERMION.

Boyle's law Named for the Irish physicist and chemist Robert Boyle (1627–91), Boyle's law states that for a fixed amount of any GAS at constant TEMPERATURE, the VOLUME and PRESSURE are inversely proportional to each other:

$$pV = \text{constant}$$

where p denotes the gas's pressure and V its volume. That can also be expressed as:

$$p_1V_1 = p_2V_2$$

where the subscripts refer to any two states of the gas that have the same amount and temperature. Boyle's law is strictly valid only for ideal gases, but real gases obey it to a good approximation as long as the DENSITY is not too high and the temperature not too low. (*See* GAS, IDEAL.)

Boyle's law is a special case of the ideal gas law:

$$pV = nRT$$

where p is in pascals (Pa), equivalent to newtons per square meter (N/m^2), V is in cubic meters (m^3), n denotes the amount of gas in MOLES (mol), T is the absolute temperature in kelvins (K), and R is the gas constant, with value 8.314472 joules per mole per kelvin [J/(mol·K)].

See also AVOGADRO'S LAW; CHARLES'S LAW; BAY-LUSSAC'S LAW.

Bragg's law Bragg's law, named for the British physicist William Lawrence Bragg (1890–1971), deals with the DIFFRACTION of X-RAYS by CRYSTALS. When a beam of X-rays strikes a crystal, the crystal's structural components—ATOMS, IONS, or MOLECULES—scatter the X-rays in all directions. The waves scattered from all the components undergo INTERFERENCE with each other. Only in certain directions will there be constructive interference, so that only in those directions will X-rays emerge from the crystal. The process can be viewed as if X-rays were being reflected from families of parallel crystal planes, with the angles of reflection equal to the corresponding angles of incidence. Bragg's law relates the angle between the X-ray beam and the planes of a parallel family, called the Bragg angle, to the WAVELENGTH of the X-rays and the spacing between adjacent "reflecting" crystal planes:

$$2d \sin \theta = n\lambda$$

where d is the distance between adjacent planes (in any unit of length), θ is the Bragg angle, and λ is the wavelength of the X-rays (in the same unit of length as d). The positive integer n (= 1, 2, 3, . . .) indicates the order of constructive interference. Bragg's law is the foundation of X-ray CRYSTALLOGRAPHY, which is the determination of the structure of crystals and of their components from the diffraction pattern of scattered X-rays that results when an X-ray beam of known wavelength is directed at the crystal. (*See* REFLECTION.)

bremsstrahlung A German word meaning "braking radiation," *bremsstrahlung* is the term for the X-RAYS that are produced when high-speed ELECTRONS strike material targets, normally made of METAL. The process producing the X-rays is the deceleration of the electrons by their electromagnetic INTERACTION with the NUCLEI of the target. X-rays result, since electrically charged particles that are accelerating or decelerating produce electromagnetic RADIATION. The X-rays are emitted in the direction of the electrons' motion. Bremsstrahlung is one source of X-rays in X-ray imaging devices, such as medical and dental X-ray machines.

See also ACCELERATION; CHARGE; ELECTRICITY; ELECTROMAGNETISM.

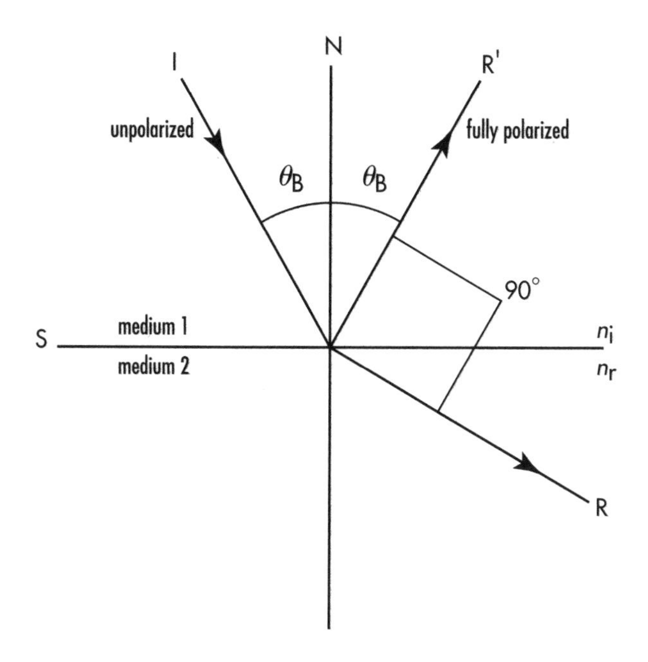

Brewster's law states that when unpolarized light I in a medium where the index of refraction is n_i (medium 1) falls on the interface with a medium of refractive index n_r (medium 2), the reflected ray R′ is fully plane polarized when it makes an angle of 90° with the refracted ray R. That occurs when the angle of incidence—and of reflection—equals the Brewster angle, given by $\theta_B = \tan^{-1} (n_r/n_i)$. Then the direction of the electric field in the reflected ray is parallel to the interface surface (and perpendicular to the plane of incidence, the plane of the figure). The interface surface is denoted S, while N denotes the normal (perpendicular to the interface surface).

Brewster's law Named for the English physicist David Brewster (1781–1868), Brewster's law states that for unpolarized LIGHT in air (or in VACUUM) falling on the surface of a material with INDEX OF REFRACTION n, the reflected ray is fully plane polarized when the angle of incidence (and of REFLECTION) equals the *Brewster angle* θ_B, where :

$$\theta_B = \tan^{-1} n$$

This formula is a special case of the more general relation for full plane polarization of the reflected ray of light, in a medium having index of refraction n_i, incident upon a medium with index of refraction n_r:

$$\theta_B = \tan^{-1} (n_r/n_i)$$

When the incident ray is at the Brewster angle, so that the reflected ray is fully polarized, the reflected ray and the refracted ray are perpendicular to each other. (*See* POLARIZATION.)

Even more generally, unpolarized light at *any* angle of incidence is at least partially plane polarized by reflection. The reflected ray is always poorer than the incident ray in the proportion of polarization component having its electric FIELD parallel to the plane of incidence. At the Brewster angle that component is completely absent, leaving only the polarization component whose electric field is perpendicular to the plane of incidence (i.e., parallel to the reflecting surface). (*See* ELECTRICITY; ELECTROMAGNETISM.)

Broglie, Louis-Victor Pierre Raymond de (1892–1987) French *Physicist* Famous for his proposal, later proved correct, that particles possess a WAVE nature, Louis de Broglie did much to further the understanding of WAVE-PARTICLE DUALITY. He studied

Louis de Broglie proposed that matter particles, such as electrons, possess a wave character, in analogy with the particle nature of light waves, manifested by photons. He was awarded the 1929 Nobel Prize in physics for his theoretical discovery. *(AIP Emilio Segrè Visual Archives)*

for a science degree, which he received in 1913, and was then drafted into military service for the duration of World War I, 1914–18. After the war, de Broglie specialized in THEORETICAL PHYSICS with a special interest in QUANTUM PHYSICS. In 1924, he received his doctorate from the University of Paris, France, for a thesis in which he proposed that the behavior of particles of MATTER can be better understood by attributing to them a WAVELENGTH that is inversely proportional to the magnitude of their LINEAR MOMENTUM. This idea was confirmed with the 1927 discovery of the DIFFRACTION of ELECTRONs by CRYSTALs. (The electron microscope is based on the wave character of electrons.) He continued his investigations in various aspects of quantum physics until his death. In 1929 de Broglie was awarded the Nobel Prize in physics "for his discovery of the wave nature of electrons."

Brookhaven National Laboratory (BNL) Established in 1947 with the initial mission of carrying out research on peaceful uses of the ATOM, Brookhaven National Laboratory is a U.S. government laboratory, under the Department of Energy, located on eastern Long Island, New York. Today BNL is a multipurpose research laboratory, where studies are performed in physics, chemistry, biology, medicine, applied science, and advanced technology. The laboratory's own staff of scientists, engineers, and support personnel numbers about 3,000, while more than 4,000 researchers visit the facilities each year.

BNL boasts four Nobel Prize awards in physics, all in the field of ELEMENTARY PARTICLEs: three of them (awarded in 1976, 1980, and 1988) were for experimental work performed there, and one (in 1957) was for a theory developed at BNL.

The laboratory's research in elementary-particle and NUCLEAR PHYSICS is being enhanced by its new Relativistic Heavy Ion Collider (RHIC) and accompanying detectors, presently the world's largest particle accelerator for nuclear-physics research. Using superconducting-magnet technology, the RHIC accelerates IONs of heavy atoms, such as gold, to extremely high ENERGIES and causes them to collide with each other head-on. Such interactions should reproduce the conditions that are thought to have existed during the first few microseconds in the life of the UNIVERSE, according to the big-bang model. This is expected to provide insight into the nature of QUARKs and GLUONs. (*See* ACCELERATOR, PARTICLE; BIG BANG; COSMOLOGY; MAGNET; SUPERCONDUCTIVITY.)

Other facilities and accomplishments of BNL include:

- *The National Synchrotron Light Source.* This facility provides electromagnetic RADIATION in the X-ray, ultraviolet, and infrared ranges for research. It has served for studying properties of materials, including MAGNETISM and superconductivity, for deciphering the molecular structure of proteins and viruses, and for constructing microscopic machines. (*See* ELECTROMAGNETISM; MOLECULE; X RAY.)
- *Medical physics and chemistry.* Scientists have probed brain activity and chemistry in their study of addiction, mental illness, and aging. L-dopa was developed to treat Parkinson's disease, and work is proceeding to find treatments for other diseases as well.
- *Technology transfer.* BNL works with industry on commercial applications of innovative technologies. Examples include improved X-ray imaging and manufacture of micromachines.
- *Advanced computing.* The laboratory designed and developed a supercomputer for processing large volumes of scientific data and for very lengthy theoretical calculations.
- *Microscopy.* The scanning transmission electron microscope can image single heavy atoms, while the transmission electron microscope studies the structure of materials. (*See* ELECTRON MICROSCOPY.)
- *Environment.* The laboratory has worked on the development of chemical asbestos abatement and of bacteria for the treatment of waste.
- *Nanotechnology.* The Center for Functional Nanomaterials will fabricate and study materials on the scale of nanometers ($\sim 10^{-9}$ meters). On that scale, properties of materials differ from those of bulk materials and could underlie new technologies.
- *Counterterrorism.* BNL has developed and continues to develop technologies that contribute to counterterrorism efforts.
- *Global climate.* Scientists study the effects of air pollutants on global temperature, plant growth, and human health.
- *Nuclear safety.* BNL's experts participate in international efforts to safeguard nuclear materials, limit the spread of nuclear weapons, and improve nuclear-reactor safety.

The website URL of Brookhaven National Laboratory is http://www.bnl.gov/. The laboratory's mailing address is P.O. Box 5000, Upton, NY 11973-5000.

A view of some of the superconducting magnets of the new Relativistic Heavy Ion Collider particle accelerator at Brookhaven National Laboratory. As ionized gold atoms travel along the collider's 3.8-kilometer-long tunnel at nearly the speed of light, 1,740 of these magnets guide and focus the particle beams. ***(Courtesy of Brookhaven National Laboratory)***

Brownian motion Named for the Scottish botanist Robert Brown (1773–1858), Brownian MOTION is the random "dancing" motion of small particles suspended in a FLUID. Examples are pollen grains in water and dust particles in air. The motion comes about through the continual impacts the particles receive from the MOLECULEs of the fluid. Brownian motion was explained by ALBERT EINSTEIN and serves as evidence for the KINETIC THEORY of MATTER.

bubble chamber The bubble chamber is a device for observing the MOTION, properties, and INTERACTIONs of ionizing particles. It is a container of LIQUID, often hydrogen, with a viewing window. Immediately before the particles of interest pass though the bubble chamber, the pressure of the liquid is reduced to make the liquid superheated. The particles passing through the liquid collide with its ATOMs or MOLECULEs and ionize them, losing KINETIC ENERGY as they go, and leave tracks of IONs along their paths. The ions serve as nucleating sites and GAS bubbles immediately form on them. In that manner, tracks of bubbles appear in the liquid, which can be photographed and later analyzed. The length and bubble density of each track serve to measure the properties of and to identify the particle that generated the track. Often, a magnetic FIELD is applied, which causes bending of the paths of electrically charged particles and, consequently, curving of their bubble tracks. The curvature of the tracks is used to deduce additional information about the particles. The direction of track bending shows the sign of the particle's charge.

See also BOILING; CHARGE; ELECTRICITY; IONIZATION; MAGNETISM.

buoyancy The effect that a FLUID exerts a lifting FORCE on an object totally or partially immersed in the fluid is called buoyancy. The magnitude of that force, called buoyant force, equals the magnitude of the WEIGHT of the fluid displaced by the object, according to ARCHIMEDES' PRINCIPLE. If the magnitude of the buoyant force acting on a totally submerged object is less than the magnitude of the object's weight, the object will sink. This occurs when the DENSITY of the object is greater than that of the fluid. When the object's density is less than the fluid's, the object will rise if totally submerged. If the fluid is a liquid, the object will then float partially submerged at the liquid's surface, in such a manner that the submerged part of the object displaces a quantity of liquid whose weight equals the body's weight. If the density of the object equals the density of the fluid, the object will hover at any height in the fluid, neither sinking nor rising.

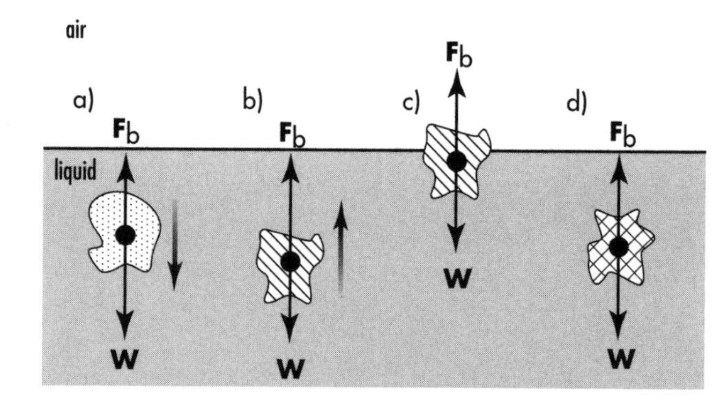

Buoyancy is the effect of a lifting force—the buoyant force—acting on an object that is wholly or partially submerged in a fluid. The magnitude of the buoyant force equals that of the weight of the fluid that the object displaces, by Archimedes' principle. The figure shows an object in a liquid. (a) The density of the object is greater than the density of the liquid, so the magnitude of the object's weight $|\mathbf{W}|$ is greater than the magnitude of the buoyant force $|\mathbf{F}_b|$, $|\mathbf{W}| > |\mathbf{F}_b|$, and the object sinks. (b) The density of the object is less than the density of the liquid. Then when the object is fully submerged, $|\mathbf{F}_b| > |\mathbf{W}|$ and the object rises. (c) It reaches equilibrium when it is floating on the surface and partially submerged to the extent that $|\mathbf{F}_b| = |\mathbf{W}|$. (d) If the object's density equals that of the liquid, the object will be in equilibrium when wholly submerged at any depth, since then $|\mathbf{F}_b| = |\mathbf{W}|$.

C

calorimetry The measurement of HEAT and of heat transfer is calorimetry. It is the method of measuring various thermal properties, such as HEAT CAPACITY, HEAT OF FUSION, HEAT OF VAPORIZATION, and heat of combustion. Calorimetry involves the use of a calorimeter, a well-insulated container in which heat transfers can take place and TEMPERATUREs can be measured. The foundation of calorimetry is the conservation of ENERGY, according to which any change in the internal (thermal) energy of an object inside a calorimeter is accompanied by an equal and opposite change in the internal energy of the other objects there. (*See* CONSERVATION LAW; INSULATOR.)

In a typical measurement of heat capacity, for example, a metal object whose heat capacity is being measured is inserted into a water-filled calorimeter. The object and the water must start out at different temperatures, and after insertion, they are allowed to reach a common EQUILIBRIUM temperature. From knowledge of the MASSes and the initial and final temperatures of the object, the water, and possibly also parts of the calorimeter, the heat capacity of the object and the specific heat capacity of the metal it is made of can be calculated. For measurement of the heat of combustion of some material, the material is actually burned inside a special type of calorimeter, called a bomb calorimeter. From the knowledge of masses and of initial and final temperatures, the amount of thermal energy produced by the combustion can be found.

capacitance The ability of objects to store electric CHARGE or to maintain a charge separation is referred to as capacitance. In the first case, when a body is charged with electric charge Q, it acquires electric potential V. It is found that V is proportional to Q, so their ratio:

$$C = Q/V$$

is constant. This ratio, C, is the capacitance of the body. Note that a (relatively) high capacitance means that the body can take much charge with little increase of its electric potential, while a low-capacitance body needs only a small charge to cause it reach a high potential. (*See* ELECTRICITY; POTENTIAL, ELECTRIC.)

In the case of charge separation, we are concerned with devices that possess two conducting parts, commonly called "plates," that are electrically insulated from each other. When charge Q is transferred from one plate to another, a potential difference, or VOLTAGE, V develops between the plates. As before, V is proportional to Q, and their ratio C, defined as above and called the capacitance of the device, is constant. In this case, rather than becoming charged, the device remains electrically neutral, since charge is merely moved from one of its plates to the other. The device maintains a charge separation (rather than a charge), with one plate bearing a positive charge and the other carrying an equal-magnitude negative charge. Similar to the first case, a (relatively) high capacitance means that the body can take a large charge separation with little increase of voltage, while a low-capacitance device needs only a small charge separation to cause it to develop a considerable voltage. Devices that are manufactured to have definite capacitances are called CAPACITORS. Capacitors are essential components of electronic circuits. (*See* CONDUCTION; INSULATOR.)

The SI UNIT of capacitance is the farad (F), equivalent to a coulomb/volt (C/V). For many practical purposes, the farad is a very large unit, so the microfarad ($\mu F = 10^{-6}$ F) and picofarad ($pF = 10^{-12}$ F) are often used.

capacitor Any electric device that is constructed for the purpose of possessing a definite CAPACITANCE is a capacitor. Every capacitor consists of two electric CONDUCTORS, called "plates," separated by an INSULATOR, the DIELECTRIC. In use, electric CHARGE is transferred, or separated, from one plate to the other, whereby the plates develop a VOLTAGE, or potential difference, between them, which is proportional to the transferred charge. The capacitance C of a capacitor is defined as the (constant) ratio of the transferred charge Q to the concomitant voltage V:

$$C = Q/V$$

and indicates the capacitor's ability to maintain a greater charge separation with lower voltage. Its SI UNIT is the farad (F), while voltage is in volts (V) and charge is in coulombs (C). When "charged," a capacitor is actually electrically neutral, with one plate positively charged and the other carrying an equal-magnitude negative charge. The dielectric maintains a constant, uniform, controlled distance between the plates and, as an insulator, keeps the positive and negative charges separated. The dielectric also contributes to the capacitor's capacitance through its DIELECTRIC CONSTANT. (*See* ELECTRICITY.)

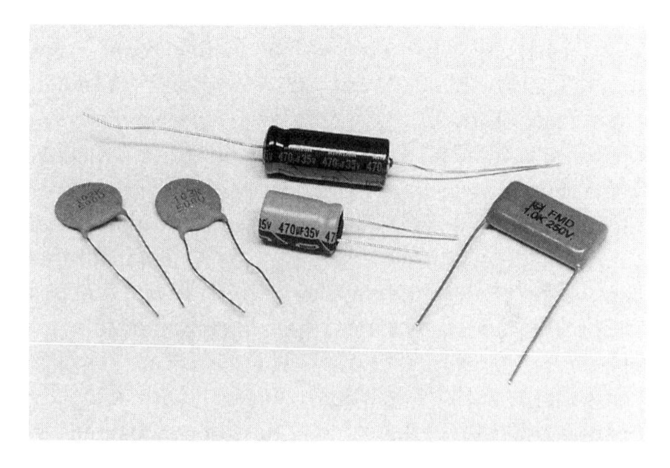

A collection of different kinds of capacitors. The capacitance and maximal allowed voltage are indicated on each. To set the scale of the figure, the length of the case of the capacitor at the top is 2.5 centimeters (1 inch). ***(Photo by Frost-Rosen)***

The "plates" of a capacitor are very often indeed plates, i.e., are thin compared with their other dimensions and are of uniform thinness. If they are parallel to each other and are large compared with the distance between them, as is very often the case, the capacitance of the capacitor is given by:

$$C = \kappa\varepsilon_0 A/d$$

where A is the AREA of each plate in square meters (m^2), d the distance between the plates in meters (m), κ the dielectric constant of the dielectric, and ε_0 the PERMITTIVITY of the vacuum, whose value is $8.85418782 \times 10^{-12}$ $C^2/(N{\cdot}m^2)$. Thus, the capacitance can be increased by increasing the plate area, decreasing the distance between the plates, and using a dielectric with higher dielectric constant.

Capacitors are normally specified by and labeled with two numbers: their capacitance and their maximal allowed voltage. The latter is the highest voltage at which the capacitor can be safely operated without danger of its breaking down and burning out. It is derived from the dielectric strength of the dielectric, which is the greatest magnitude of electric FIELD that the dielectric can tolerate without breaking down. Capacitors are essential for electronic circuits and are ubiquitous in them.

In an alternating current (AC) circuit, the IMPEDANCE of a capacitor is called capacitive reactance and has the value:

$$X_C = \frac{1}{2\pi fC}$$

where X_C is in ohms (Ω), f is the FREQUENCY in hertz (Hz), and C is the capacitance in farads (F). The instantaneous voltage across a capacitor lags behind the instantaneous current through it by a quarter cycle ($\pi/2$ radians, or 90°). (*See* REACTANCE.)

capillary flow Capillary flow is the spontaneous flow of liquids in very narrow tubes, called capillaries. It is caused by an imbalance between the attraction of the liquid's molecules at the liquid-tube-air boundary

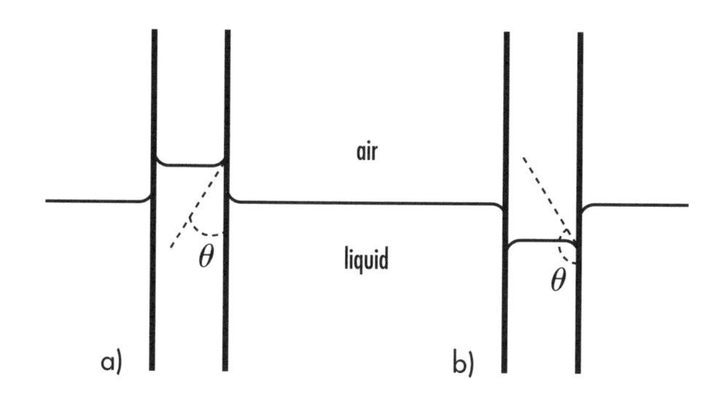

Two cases of capillary flow in a vertical tube as viewed in vertical cross-section. (a) The contact angle, θ, between the edge of the meniscus and the capillary wall is less than 90°; the meniscus is concave as viewed from above; and the liquid rises in the capillary. Water and glass behave in this way. (b) When the contact angle is greater than 90°, the meniscus is convex, and the liquid in the capillary descends. An example of this is mercury and glass. Capillary tubes are very narrow, and the scale of the figure is exaggerated for the sake of clarity.

to the tube (ADHESION) and their attraction to the liquid (SURFACE TENSION, or COHESION). If the former dominates over the latter, the liquid will tend to be drawn into the tube, such as water in a glass capillary. If the latter is greater, however, the liquid will tend to be ejected from the tube, as exemplified by liquid mercury in a glass capillary.

In the situation of a vertical capillary whose end is dipped into a liquid, capillary flow can cause the level of the liquid in the tube to be different from its level outside, either higher or lower, according to whether the adhesion is greater or lesser than the surface tension, respectively, as described above. The surface of the liquid in the tube, called the meniscus, will not be flat in those cases but will be concave or convex, respectively, as viewed from its top. For a round capillary tube of radius r in meters (m), the rise in height h in meters (m) for a liquid of density ρ in kilograms per cubic meter (kg/m^3) and surface tension γ in newtons per meter (N/m) is given by:

$$h = \frac{2\gamma \cos\theta}{\rho g r}$$

where θ is the angle between the surface of the meniscus at its edge and the surface of the capillary. This angle is measured below the meniscus. For a concave meniscus (such as forms on water in glass), $0 \leq \theta < 90°$, giving a positive cosine and a positive h, indicating a rise of the liquid in the capillary. For a convex meniscus (in the example of mercury in glass), $90° < \theta \leq 180°$, so the cosine and h are negative, with the liquid lower in the tube than outside. In the intermediate case of a flat meniscus, when $\theta = 90°$ and $h = 0$, adhesion and surface tension balance and there is no capillary flow. In the formula, g denotes the ACCELERATION due to gravity, whose nominal value on the surface of the Earth is 9.8 meters per second per second (m/s^2).

See also GRAVITATION.

Carnot cycle Named for the French engineer Sadi Nicolas Léonard Carnot (1796–1832), the Carnot cycle is the cycle undergone by an idealized HEAT engine called the Carnot engine. The latter is a reversible engine constructed of a gas-filled cylinder and piston that can either be in contact with a high-TEMPERATURE heat reservoir (such as a steam supply) at absolute temperature T_H, be in contact with a low-temperature heat reservoir (such as the cool water of a river) at absolute temperature T_L, or be thermally insulated from its surroundings. Reversibility means that every stage of the engine's cycle proceeds infinitely slowly and that the cycle can be run in reverse. FRICTION is assumed to be absent. (*See* REVERSIBLE PROCESS.)

These are the four stages of the Carnot cycle in order of their performance: (1) The engine is at temperature T_H. It is put in contact with the high-temperature reservoir and undergoes isothermal expansion, absorbing heat Q_H at temperature T_H and performing work on its surroundings. (2) The engine is then thermally isolated and undergoes adiabatic expansion, again doing work on the surroundings, as its temperature decreases from T_H to T_L. (3) With the engine at temperature T_L, it is put in contact with the low-temperature reservoir and undergoes isothermal compression, during which it discharges heat Q_L to the reservoir at temperature T_L and has work performed on it by the surroundings. (4) Finally, the engine is again thermally isolated and undergoes adiabatic compression, as work is done on it by the surroundings and its temperature increases from T_L to T_H. When it has thus reverted to its initial state, it has completed the cycle and is ready to repeat it. (*See* ADIABATIC PROCESS; ISOTHERMAL PROCESS.)

After completing a cycle, the Carnot engine has done net work W on its surroundings, has absorbed heat Q_H, and has discharged heat Q_L. According to the first law of thermodynamics (the WORK-ENERGY THEOREM):

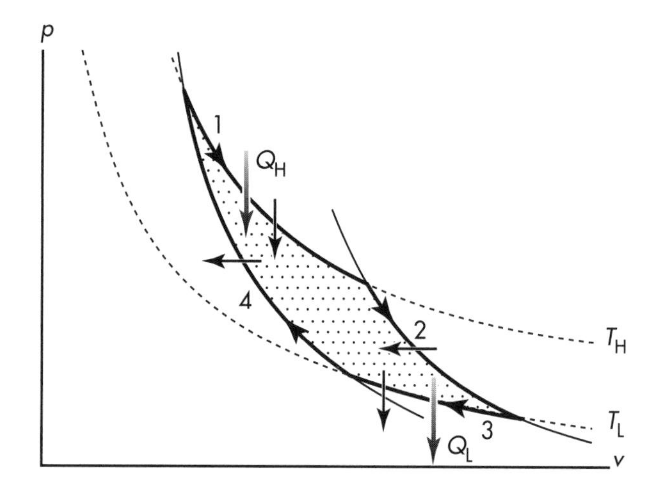

A Carnot cycle of an ideal heat engine operating between two temperatures, as represented in a pressure-volume *(p-V)* diagram. The isotherms are labeled T_H (higher temperature) and T_L (lower temperature). The other two curves are adiabats. The direction and amount of heat flow during the isothermal stages are indicated by arrows labeled Q_H (into the engine) and Q_L (into the surroundings). The direction of work performed during each stage of the cycle is shown by an unlabeled arrow (pointing inward for work done on the engine, pointing outward for work done on the surroundings). The four stages of the cycle are: (1) isothermal expansion at temperature T_H, (2) adiabatic expansion to temperature T_L, (3) isothermal compression at temperature T_L, and (4) adiabatic compression back to temperature T_H. The total work performed by the engine on the surroundings during a single cycle equals the enclosed area and has the value $Q_H - Q_L$.

$$W = Q_H - Q_L$$

This can be expressed as follows. Of the total heat Q_H that entered the engine, Q_L merely flowed through it and left it, while the remainder, $Q_H - Q_L$, was converted by the engine to work. (*See* CONSERVATION LAW; THERMODYNAMICS.)

The efficiency of any heat engine is defined as the ratio of the net work performed to the heat absorbed, giving:

$$e = \frac{W}{Q_H} = \frac{Q_H - Q_L}{Q_H} = 1 - \frac{Q_L}{Q_H}$$

For a Carnot cycle, the efficiency can be expressed in terms of the temperatures of the reservoirs and be shown to equal:

$$e_C = 1 - \frac{T_L}{T_H}$$

(*See* THERMAL EFFICIENCY.)

A number of conclusions can now be drawn. One is that the efficiency of a Carnot engine depends solely on the temperatures of the reservoirs and does not depend on the particulars of the engine, such as the nature of the gas in the engine. Another conclusion is that because T_L is less than T_H and both are positive, $e_C < 1$, so the efficiency of a Carnot engine is always less than 100 percent. By studying the expression for e_C, we see that the efficiency is increased by lowering the temperature of the low-temperature reservoir (into which the engine discharges heat) and raising that of the high-temperature reservoir (from which the engine absorbs heat). In the hypothetical case of lowering T_L to 0 kelvins (K) the efficiency would reach 100 percent, but absolute zero temperature is not attainable.

Although the Carnot engine is an idealization, it has practical importance in that it can be shown that no realizable heat engine operating between two fixed temperatures can have an efficiency higher than that of a Carnot engine operating between the same temperatures. In this manner e_C sets an upper limit on practically achievable engine efficiencies.

When a Carnot engine is run in reverse, it becomes a heat pump, also referred to as a refrigerator. Net work is then performed by the surroundings (say, by an electric motor) on the engine, which absorbs heat from the cool reservoir and discharges heat to the warm reservoir. The engine is often called a refrigerator when one is interested in its cooling effect. In this case the cool reservoir might be the interior of a kitchen refrigerator, and the warm reservoir might be the air in the kitchen or the interior of a car and the outside air. If it is heating one is interested in, then the warm reservoir might be the interior of the house, and the cool reservoir could be the outside air or the bottom of a nearby lake.

In the case of a heat pump, the heat discharged by the engine, Q_H, equals the sum of the heat absorbed, Q_L, and the work done on the engine, W:

$$Q_H = Q_L + W$$

This can be understood as having heat Q_L "pumped" from the cool to the warm reservoir, while the work W done to bring about the "pumping" is converted to heat and discharged to the warm reservoir as well. For a heat pump, a coefficient of performance can be defined as the ratio of the heat removed from the cool reservoir to the work invested, giving:

$$\eta = \frac{Q_L}{W} = \frac{Q_L}{Q_H - Q_L}$$

For a Carnot cycle the coefficient of performance can be expressed in terms of the temperatures of the reservoirs and be shown to equal:

$$\eta_C = \frac{T_L}{T_H - T_L}$$

As stated previously, no realizable heat pump operating between two fixed temperatures can have a coefficient of performance greater than that of a Carnot engine operating as a heat pump between the same temperatures.

center of mass The center of mass of a system is a point at which the MASS of the system can be considered to be located for certain purposes. For those purposes, the system behaves as would a point particle of the same mass located at the system's center of mass. If a FORCE acts on a free body along a line through the body's center of mass, the force will not affect the body's ROTATION. Instead, it will only cause ACCELERATION in accord with ISAAC NEWTON's second law of MOTION, as if the body were a point particle at the position of the center of mass. A general force acting on a body at rest causes it to move such that its center of mass moves in the direction of the force, and the body itself rotates around its center of mass. (*See* NEWTON'S LAWS.)

If a system is composed of point particles of masses m_i, situated at locations with coordinates (x_i, y_i, z_i), the coordinates of the center of mass (x_{cm}, y_{cm}, z_{cm}) are given by:

$$x_{cm} = \Sigma m_i x_i / M$$
$$y_{cm} = \Sigma m_i y_i / M$$
$$z_{cm} = \Sigma m_i z_i / M$$

where:

$$M = \Sigma m_i$$

is the total mass of the system. (For a system made of continuous material, the summations are replaced by appropriate integrations.)

For a body of uniform DENSITY, the center of mass is located at the center of SYMMETRY, if the body possesses a center of symmetry. The center of mass of a uniform body of each of these shapes, as examples, is at the geometric center, which is the center of symmetry: sphere, cube, straight rod.

When calculating the location of the center of mass of a system, it might be convenient first to conceptually break the system down into parts. Then find the center of mass of each part; replace each part with a point particle of the part's total mass, located at the part's center of mass; and calculate the center of mass of that collection of fictitious point particles.

With respect to an INERTIAL REFERENCE FRAME, the center of mass of an isolated system moves with constant VELOCITY (i.e., moves with constant SPEED in a straight line).

center-of-mass reference frame This REFERENCE FRAME for a given system is a reference frame in which the system's CENTER OF MASS is at rest. The advantage of such a reference frame is that the system's total LINEAR MOMENTUM is zero in that frame. As viewed in their center-of-mass reference frame, two colliding particles have oppositely directed linear momenta of equal magnitudes both before and after their COLLISION, even if—in the reference frame of the laboratory—one particle is initially stationary with the other striking it.

central force A FORCE that is always directed toward a fixed point is called a central force. An example is the gravitational force of attraction by the Sun acting on every body in the SOLAR SYSTEM. Those forces are all directed toward the center of the Sun. Similarly, the electric force of attraction by a NUCLEUS acting on every ELECTRON of an ATOM is a central force directed toward the nucleus. (*See* ELECTRICITY; GRAVITATION.)

Under the influence of a central force, a body moves with constant ANGULAR MOMENTUM. If the magnitude of a central force is inversely proportional to the square of the body's distance from the center (as is the case for gravitational and electric attraction), the body's path will have the form of either an ellipse (or a circle as a special case), a parabola, or a hyperbola, depending on the body's speed.

centrifugal force *See* CENTRIPETAL FORCE.

centripetal force Any FORCE that keeps a body in CIRCULAR MOTION at constant SPEED is termed a centripetal force. Such a force is directed toward the center of the circular path and has magnitude:

$$F_c = mv^2/r$$

where m is the mass of the body in kilograms (kg), v is its speed in meters per second (m/s), and r is the radius of the circular path in meters (m). The UNIT of force is the newton (N).

For a tie-in with the ACCELERATION associated with such motion, called centripetal acceleration, note that this acceleration is directed toward the center of the circular path and has constant magnitude:

$$a_c = v^2/r$$

where acceleration is in meters per second per second (m/s^2). According to ISAAC NEWTON's second law of MOTION, the force causing this acceleration, which is the centripetal force, is also directed toward the center, and its magnitude is:

$$F_c = ma_c = mv^2/r$$

(*See* NEWTON'S LAWS.)

Centripetal force is not an additional kind of force. It is simply the term applied to *any* force that causes constant-speed circular motion (i.e., brings about centripetal acceleration), be it a gravitational force, an electric force, a magnetic force, or a TENSION force in a string, for example. Centripetal force is a special case of CENTRAL FORCE. (*See* ELECTRICITY; GRAVITATION; MAGNETISM.)

The term *centrifugal force* is often associated with this kind of motion as a force supposedly pulling the body away from the center. The body does, indeed, tend to pull away from the center, but not because of any force. To clarify the situation, imagine you are swinging a weight around in a horizontal circle at constant speed. The weight is attached to a string, and you are holding the other end of the string in your hand. You are pulling on the string, and the string is pulling on the weight, keeping it in its circular path. The centripetal force on the weight is the string's force on it, due to the tension in the taut string. Your hand's force on the string produces the string's tension. It is not incorrect to say that your hand's force is a centripetal force too, since it is keeping in circular motion the body consisting of the weight and the string.

Let us now put Newton's third law of motion into play. While the string is pulling the weight inward, the weight is pulling the string outward. And while your hand is pulling the string inward, the string is pulling your hand outward, which you feel very well. These outward forces are called centrifugal forces. They are not applied forces, but rather reaction forces, which arise, by Newton's third law, in reaction to real, acceleration-causing forces. Note that they act on the string and on your hand, but not on the weight.

There is no force pulling the weight outward. Its tendency to move away from the center is not due to any force but to the effect of INERTIA, as expressed by Newton's first law of motion. With no force acting on the weight, Newton's first law states that it would move at constant speed in a straight line. In fact, if you let go of the string, the weight does not fly radially outward. Rather, it continues moving in a straight line (for simplicity of discussion, we ignore its falling), tangent to the circle at the moment of release. (This is the principle of operation of that ancient weapon called the sling.) To reiterate, what is called centrifugal force is a reaction to a centripetal force, and its source is fundamentally just the inertial tendency of a body to move in a straight line.

What is also called centrifugal force is often invoked as an "explanation" for acceleration that is observed in a noninertial, rotating reference frame. Imagine a ball initially at rest on the floor of a carousel. As the carousel turns, the ball tends to move radially outward. As viewed in the REFERENCE FRAME of the carousel, the ball is undergoing acceleration. If Newton's first law of motion is applied, some force must be causing the acceleration, and that force is termed "centrifugal force," an outward-pulling force. This is a fictional force, since there is no real force pulling the ball outward. Newton's first law is invalid in any noninertial reference frame, such as the carousel's, so no force is needed as a cause for what appears to be accelerated motion in that reference frame. (*See* REFERENCE FRAME, INERTIAL; ROTATION.)

See also CORIOLIS ACCELERATION.

Čerenkov radiation When an electrically charged particle moves through a medium at a SPEED greater than the SPEED OF LIGHT in the medium, it emits light, called Čerenkov radiation, named for the Russian physicist Pavel Alekseyevich Čerenkov (also spelled Cherenkov) (1904–90). That is the source of the bluish light that can be seen in the water of water-cooled nuclear REACTORS. This effect is used in modern ELEMENTARY PARTICLE detectors, in which the passage of energetic charged particles through a liquid is detected and studied via their Čerenkov radiation. The light is "seen" by photomultipliers surrounding the detector chamber, which convert the light to electric signals and send them to a computer for processing. (*See* CHARGE; ELECTRICITY; INDEX OF REFRACTION; NUCLEAR PHYSICS.)

An effect analogous to Čerenkov radiation is the generation of a "bow wave" by a boat traveling faster than the speed of surface WAVEs on water. Another analogous effect is the "sonic boom" produced by aircraft flying faster than the speed of SOUND in air.

See also MACH NUMBER; SHOCK WAVE.

CERN *See* EUROPEAN ORGANIZATION FOR NUCLEAR RESEARCH.

Chandrasekhar, Subrahmanyan (1910–1995) American *Astrophysicist* Known to all as Chandra, Subrahmanyan Chandrasekhar was one of the foremost astrophysicists of the 20th century. Born in what was then India and is now Pakistan, he received his doctorate in astrophysics from Cambridge University, England, in 1933. In 1937, Chandra joined the faculty of the University of Chicago, Illinois, where he remained for the rest of his life. His investigations included such topics as HYDRODYNAMICS, MAGNETOHYDRODYNAMICS, the structure and evolution of stars, the general theory of relativity, and BLACK HOLEs. In 1983, Chandra was awarded half the Nobel Prize in physics "for his theoretical studies of the physical processes of importance to the structure and evolution of the stars." (*See* ASTROPHYSICS; RELATIVITY, GENERAL THEORY OF.)

Subrahmanyan Chandrasekhar, known as Chandra, was a leading astrophysicist, who studied black holes and the structure and evolution of stars, among other matters. He shared the 1983 Nobel Prize in physics. ***(AIP Emilio Segrè Visual Archives, Physics Today Collection)***

chaos This term refers to a situation where the evolution of a physical system over time is sensitive to INITIAL CONDITIONS. This means that the results of its evolving from two similar initial states can be very different. Since there are always limitations on the precision of measurements—both practical limitations, which can be reduced in principle, and inherent limitations, which are unavoidable—the states of systems cannot be determined with total certainty. Thus, chaos makes long-term prediction difficult, if not impossible. A common example of that is the weather, for which predictions over times longer than about a week have a low-to-vanishing level of accuracy. The mathematical treatment of chaos is based on the description of systems by means of their PHASE SPACE. (*See* QUANTUM PHYSICS.)

charge Most generally, charge is a physical quantity that an ELEMENTARY PARTICLE may possess that determines whether and how it is affected by some INTERACTION. The values of the various charges of an elementary particle depend on the type of particle and are part of the identity of the particle. Charge is found to be quantized, whereby it can only possess certain values. (*See* QUANTIZATION.)

The most familiar charge is electric charge, which is involved with the particle's participation in the ELECTROMAGNETIC INTERACTION. A particle with zero value of electric charge is called neutral. Such a particle is not affected by the electric FIELD nor does it affect that field. Neither, when it is in motion, does it affect nor is it affected by the magnetic field. A charged particle, on the other hand, does affect and is affected by the electric field and, when in motion, affects and is affected by the magnetic field. The strength of the interaction is proportional to the amount of charge that the particle carries. The amounts that nature seems to allow are positive and negative integer multiples of, as well as a third and two-thirds of, a fundamental unit of charge. Thus the charges of the elementary particles are commonly specified simply as +1, −2, $+^1/_3$, $-^2/_3$, etc. For example, the electric charge of the ELECTRON is −1, that of the PROTON is +1, the up QUARK has electric charge $+^2/_3$, and all the NEUTRINOs are neutral (i.e., each has electric charge 0). (*See* ELECTRICITY; MAGNETISM.)

The SI unit of electric charge is the coulomb (C). This unit is very large compared with the fundamental unit, which is commonly denoted by *e*, and whose value is $1.60217738 \times 10^{-19}$ C.

The additional kinds of charge are not commonly referred to as "charge." Some are called "number," like BARYON number and LEPTON number. Others are simply strangeness, charm, color, etc. Many of those charges have positive and negative integer values. Baryon number has, in addition, the value of one-third for quarks. The color charge of quarks is not numerical and has the values red, green, and blue. For all the charges mentioned or alluded to so far, an antiparticle has the negative of the corresponding values of its corresponding particle, while the values of the color charge of the antiquarks are called antired, antigreen, and antiblue. A particle that is its own antiparticle, such as the neutral PION and the PHOTON, has zero values for all such charges. (*See* ANTIMATTER.)

An important property of those charges—indeed, the reason many of them were introduced at all—is that they obey CONSERVATION LAWs. During processes in which elementary particles are created and destroyed, the algebraic sum (i.e., taking the sign into account) of various of the charges might not change in value. Which charges are or are not conserved depends on the interaction involved. Electric charge, baryon number, and lepton number seem to be conserved by all elementary-particle interactions.

As a demonstration, consider the process of spontaneous decay of a free NEUTRON:

$$n \rightarrow p + e^- + \bar{\nu}_e$$

where n, p, e^-, $\bar{\nu}_e$ denote, respectively, neutron, proton, electron, and antineutrino of electron type. This process is governed by the weak interaction. The neutron is electrically neutral, so the total electric charge on the left-hand side is 0. The proton carries electric charge +1 and the electron −1. All neutrinos and antineutrinos are electrically neutral. So the total electric charge on the right-hand side is zero also, and electric charge is conserved in this process. The neutron and proton each have baryon number +1, while that of the electron and antineutrino is 0. Total baryon number +1 becomes total baryon number +1, and the process conserves total baryon number. As for lepton number, its value for the neutron and proton is 0, while the electron has value +1 and the antineutrino −1. So both the initial and the final total lepton numbers are 0, and we have conservation again.

A further important property of those charges is that they simply add up algebraically for conglomerations of elementary particles. Thus the electric charge of a helium NUCLEUS is +2, since it is composed of two protons, each of charge +1, and two neutrons, which are neutral. Going one step further, the electric charge of a helium ATOM is 0, as it comprises its nucleus, with charge +2, and two electrons, each carrying charge −1. The baryon number of a helium atom is +4, since protons and neutrons each have baryon number +1 and for electrons the number is 0. Similarly, the helium atom possesses lepton number +2 due to its two electrons, each with lepton number +1, with the protons and neutrons having lepton number 0.

A charge of a different kind is MASS. Every type of elementary particle is characterized by the value of its mass, which is a positive number. However, the masses allowed by nature for the elementary particles are not integer multiples of some fundamental mass unit and there is no negative mass. Neither does the mass of an assembly of elementary particles equal the sum of the masses of the constituents. (That has to do with BINDING ENERGY.) Mass is not conserved in the sense that the sum of masses of the particles entering a process always equals the sum of masses of the products. Nevertheless, ENERGY is conserved, so the mass equivalent of energy is conserved. A particle's mass measures, among other things, the degree to which the particle affects the gravitational field and is affected by it. (*See* GRAVITATION; MASS ENERGY.)

Charles's law Also called GAY-LUSSAC'S LAW, Charles's law, named for the French scientist Jacques Alexandre César Charles (1746–1823), states that at constant PRESSURE the VOLUME of a fixed amount of any GAS is directly proportional to its absolute TEMPERATURE:

$$V/T = \text{constant}$$

where V denotes the volume of the gas and T is its absolute temperature. That can also be expressed as:

$$V_1/T_1 = V_2/T_2$$

where the subscripts refer to any two states of the gas that have the same amount and pressure. Charles's law is strictly valid only for ideal gases, but real gases obey it to a good approximation as long as the DENSITY is not too high and the temperature not too low. (*See* GAS, IDEAL.)

Charles's law is a special case of the general law for ideal gases:

$$pV = nRT$$

where p denotes the pressure in pascals (Pa), equivalent to newtons per square meter (N/m^2), V is in cubic meters (m^3), n denotes the amount of gas in MOLEs (mol), T is in kelvins (K), and R is the gas constant, with value 8.314472 joules per mole per kelvin (J/mol·K).

See also AVOGADRO'S LAW; BOYLE'S LAW.

circular motion As the name straightforwardly implies, circular motion is MOTION of a particle along a circular path (i.e., a path of constant radius confined to a plane). Since the direction of such motion is continuously changing, whether or not the SPEED is changing, the particle is undergoing continuous ACCELERATION, called centripetal acceleration. This acceleration is directed toward the center of the circular path. Its magnitude, a_c, is related to the particle's speed, v, and the (constant) radius of the circular path, r, by:

$$a_c = v^2/r$$

Acceleration is in UNITs of meters per second per second (m/s^2), speed is in meters per second (m/s), and radius is in meters (m).

The centripetal acceleration is brought about by some CENTRIPETAL FORCE, a center-directed force, whose magnitude is:

$$F_c = ma_c = mv^2/r$$

where m denotes the MASS of the particle. The force is in newtons (N) and the mass in kilograms (kg).

If the particle's speed is changing as well, its acceleration has an additional component, a tangential component, perpendicular to the centripetal acceleration, and is given by:

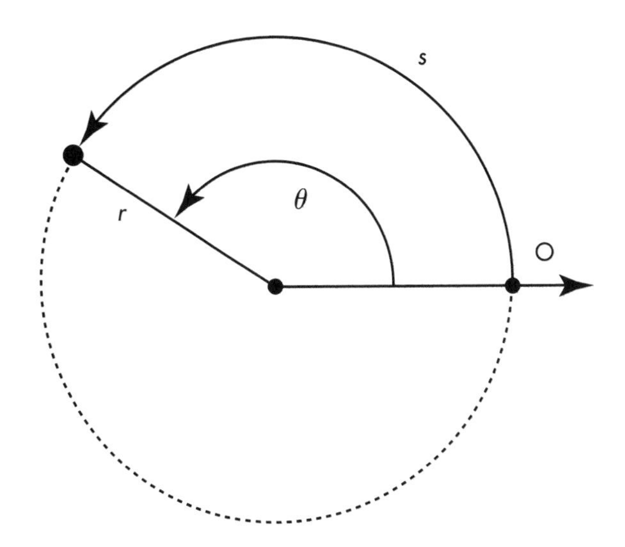

Circular motion of a particle. The radius of the circle is *r*. The particle's angular displacement *θ* (in radians) from the indicated reference direction relates to the length of circular arc *s*, from the reference point O to the particle's position, by *θ* = *s/r*.

$$a_t = dv/dt$$

Thus the magnitude of the total acceleration of a particle in circular motion is:

$$a = \sqrt{a_c^2 + a_t^2} = \sqrt{(v^2/r)^2 + (dv/dt)^2}$$

Circular motion can also be described by a number of angular quantities. Angular displacement, θ, is the position of the particle in terms of its angle around the circular path, as measured from some reference direction. It relates to the length of the circular arc from the corresponding reference point on the circle to the particle's position, s, by:

$$\theta = s/r$$

where θ is in radians (rad) and s is in meters (m). Angular speed (or velocity), ω, in radians per second (rad/s) is defined as:

$$\omega = d\theta/dt$$

which relates it to the speed by:

$$\omega = \frac{d\theta}{dt} = \frac{d}{dt}\left(\frac{s}{r}\right) = \frac{1}{r}\frac{ds}{dt} = \frac{v}{r}$$

Angular acceleration, α, defined as:

$$\alpha = d\omega/dt$$

relates to the tangential acceleration by:

$$\alpha = \frac{d\omega}{dt} = \frac{d}{dt}\left(\frac{v}{r}\right) = \frac{1}{r}\frac{dv}{dt} = \frac{a_t}{r}$$

The unit of angular acceleration is radian per second per second (rad/s^2).

In terms of these angular quantities, the magnitude of the centripetal acceleration is:

$$a_c = \omega^2 r$$

the magnitude of the centripetal force is:

$$F_c = m\omega^2 r$$

and the KINETIC ENERGY in joules (J) is:

$$E_k = mv^2/2 = m(\omega r)^2/2 = m\omega^2 r^2/2$$

(*See* DISPLACEMENT; VELOCITY.)

For circular motion at constant speed, the motion is periodic, with PERIOD (the time of one revolution, which is the circumference divided by the speed) in seconds (s):

$$T = \frac{2\pi r}{v} = \frac{2\pi r}{\omega r} = \frac{2\pi}{\omega}$$

and FREQUENCY in hertz (Hz):

$$f = \frac{1}{T} = \frac{1}{2\pi/\omega} = \frac{\omega}{2\pi}$$

classical physics The term *classical,* as it refers to physics, is used in two different ways. In common discourse, speaking historically, it stands in opposition to "modern" and means the physics of the 19th century. More specifically, this means physics that does not involve relativity or quantum considerations. So it is the physics of situations in which SPEEDs are not too high, MASSes and ENERGIES neither too great nor too little, and distances and durations neither too small nor too large. (*See* QUANTUM PHYSICS; RELATIVITY, GENERAL THEORY OF; RELATIVITY, SPECIAL THEORY OF.)

Among physicists, however, classical physics stands in contrast only to quantum physics, while relativistic effects are included in it. Quantum physics is characterized by the PLANCK CONSTANT, $h = 6.62606876 \times 10^{-34}$ joule·second (J·s), which is nature's elementary unit of ACTION. While nature is fundamentally quantum in character and quantum physics seems to give an excellent understanding of it, classical physics can be valid to high accuracy for situations involving actions that are large compared to h, what is referred to as the classical domain. This includes everyday phenomena, of course, and furthermore ranges over lengths, durations, and masses that are not too small, meaning generally larger, say, than the atomic and molecular scales. (*See* ATOM; MOLECULE.)

Some of the characteristics of classical physics, which, to a good approximation, reflect properties of nature in the classical domain, are these:

Definiteness. Every physical quantity that is relevant to a physical system possesses a definite value at every time. Even if we do not know the value of such a quantity, it nevertheless *has* a definite value.

Determinism. The state of a physical system at any instant uniquely determines its state at any time in the future and, accordingly, is uniquely determined by its state at any time in the past. It follows that a full understanding of the working of nature and a complete knowledge of a system's state at any time allow, in principle, prediction of the system's state at any time in the future and retrodiction of its state at any time in the past. Limitations on such predictability and retrodictability, even when the LAWS OF NATURE are known, are the result of incomplete knowledge of the system's state. In practice, such incompleteness is unavoidable, since all measurements possess only finite precision. (And even if infinite precision were hypothetically possible, there exist inherent quantum uncertainties in the values of physical variables. [*See* HEISENBERG UNCERTAINTY PRINCIPLE].)

Continuity. Processes occur in a continuous manner, so that physical quantities that vary in time do so continuously, with no sudden jumps. Also, the ranges of allowed values of physical quantities are continuous: if a physical quantity can possess a certain value, then it can also take values as close to that value as one might like.

Locality. What happens at any location is independent of what happens at any other location, unless some influence propagates from one location to the other, and the propagation of influences occurs only in a local manner. The latter means that influences do not instantaneously span finite distances. (Such a hypothetical effect is known as "action at a distance.") Rather, the situation at any location directly affects only the situations at immediately adjacent locations, which in turn directly affect the situations at *their* immediately adjacent locations, and so on. In this way, an influence propagates at finite speed from its cause to its effect.

Wave-particle distinction. A WAVE is a propagating disturbance, possibly characterized by such as FREQUENCY and WAVELENGTH. It possesses spatial extent, so is not a localized entity. A particle, on the other hand, is localized. It is characterized by MASS, VELOCITY, ENERGY, etc. Waves and particles are distinct from each other and bear no relation to each other.

Particle distinguishability. All particles are distinguishable from each other, at least in principle.

clock This is any device for measuring TIME. It does so by producing and counting (or allowing us to count) precisely equal time units. A PENDULUM swings with cycles of equal time durations. The most modern precision clocks, which are called atomic clocks, count time units based on the FREQUENCY of the electromagnetic RADIATION emitted by ATOMs, when the latter undergo transitions between certain states of different ENERGY. (*See* BOHR THEORY; ELECTROMAGNETISM.)

But how can we be sure the time units generated by a clock are indeed equal? How do we know that a pendulum's swings are of equal duration or that the OSCILLATIONs of an electromagnetic WAVE emitted by an atom have constant PERIOD? Clearly there is circular reasoning, if we use a clock to assure the equality of another clock's time units.

What it boils down to is that we are forced to make the *assumption* that a clock's time units are equal. We do not do that for just any device, but only for those for which we have good reason to think it true. In the example of the pendulum, at the end of each swing cycle the bob returns to momentary rest at its initial position, whereupon it starts a new cycle. As far as we can tell, it always starts each cycle from precisely the same state. Why, then, should it not continue through each cycle in precisely the same manner and take precisely the same time interval to complete each cycle?

If we assume that is indeed the case and the pendulum's time units are equal, we find we obtain a simple and reasonable description of nature based on this equality. Convinced that nature should be describable in simple terms, we then feel justified in our assumption. For more sophisticated clocks, such as atomic clocks, the reasoning is similar, but correspondingly more sophisticated. We end up *defining* the cycles of electromagnetic radiation as having equal duration. Thus the time interval of one second is defined as the duration of 9,192,631,770 periods of microwave electromagnetic radiation corresponding to the transition between two hyperfine levels in the GROUND STATE of the atom of the ISOTOPE cesium-133. (*See* BASE QUANTITY.)

coherence The property of OSCILLATIONs that they possess a definite and regular PHASE relationship among themselves is called *coherence.* Oscillations occurring at two different locations might or might not be coherent. Or, oscillations taking place at different times at the same location may be considered as to their coherence. For coherence, it is necessary, but not sufficient, that the oscillations under consideration have the same FREQUENCY. Coherence is most often achieved by the oscillations being produced by the same source. The extent of coherence is then limited by the coherence of the source, i.e., by the extent to which the source maintains a fixed phase relationship with itself over time. Thus a source of oscillations is characterized by its coherence time, the maximal time interval during which it maintains a definite phase relationship in its oscillations.

The coherence of WAVEs, too, depends on the coherence of their source or sources. For electromagnetic RADIATION, a LASER is a particularly coherent source (i.e., has a long coherence time) and thus produces highly coherent radiation. One can speak of the coherence length of a wave. That is the maximal distance along the direction of propagation over which the wave maintains a definite phase relationship. A wave's coherence length is directly related to the coherence time of the wave source. It is just the coherence time of the source multiplied by the propagation SPEED of the wave. Two or more sources can be made to be phase-locked, whereby coherence of the waves from the sources is assured, within the limitations of the sources' coherence times.

See also ELECTROMAGNETISM.

cohesion The FORCEs that hold MATTER together are designated cohesion. It is the mutual attraction of the ATOMs or MOLECULEs that constitute a material. Those forces are electromagnetic in nature. The forces that constrain the molecules of a LIQUID to remain within the VOLUME of the liquid, or maintain the shape of a SOLID, are cohesion forces.

See also ADHESION; ELECTROMAGNETISM; SURFACE TENSION.

collision The term *collision* has very much the same meaning as in everyday use, with the extension that colliding objects do not have to actually come into contact with each other, but instead can interact in other ways. Two ELECTRONs might collide by passing each other closely while interacting through the ELECTROMAGNETIC and WEAK INTERACTIONs. For another example, the maneuver in which a spacecraft swings closely around a planet and thereby gains SPEED is a collision between the spacecraft and the planet through the gravitational interaction. On the other hand, colliding automobiles, for instance, *do* affect each other through contact. In the aftermath of a collision, the colliding objects might or might not emerge unscathed, and additional objects might be produced. After the flyby, the spacecraft and the planet obviously still retain their identities. But the results of collisions of atomic NUCLEI, say, might involve free NUCLEONs and various lighter nuclei. Or, colliding nuclei might merge and fuse into heavier nuclei. And the properties of the fundamental INTERACTIONs are investigated by causing ELEMENTARY PARTICLES to collide and studying the particles that are produced by the collision. (*See* GRAVITATION.)

For collisions in isolated systems, certain CONSERVATION LAWs are always valid. In other words, for systems that are not influenced by their surroundings, certain physical quantities are always found to have the same the total values after a collision as they had before the collision. ENERGY, LINEAR MOMENTUM, ANGULAR MOMENTUM, and electric CHARGE are the most commonly known quantities that are always conserved. Other quantities might also be conserved, depending on the interactions involved. (*See* ELECTRICITY.)

Although energy is always conserved in collisions in isolated systems, the KINETIC ENERGY of the collision results might not equal that of the colliding objects. In a head-on car crash, for example, almost all the initial kinetic energy of the colliding cars is converted to energy in the form of HEAT, SOUND, and energy of metal deformation. Those collisions in which kinetic energy *is* conserved are called elastic collisions. An example is a planetary flyby, as described above. If the spacecraft gains speed, and thus kinetic energy, from such an encounter, it is at the expense of the kinetic energy of the planet. Collisions in which kinetic energy is not conserved are called inelastic collisions. When the colliding objects stick together and do not separate after collision, there is maximal loss of kinetic energy. That is termed a completely inelastic collision. In the extreme, for instance, when two bodies of equal mass and speed collide head on and stick together, the bodies stop dead, and all their initial kinetic energy is lost.

complementarity This refers to WAVE-PARTICLE DUALITY and is the term for the fact that WAVE phenomena and particle phenomena both inherently possess aspects of each other and that a complete description of a phenomenon involves both its wave and particle aspects. Thus the two aspects are complementary, in that both are needed for a complete description. That might seem paradoxical, since waves and particles are very different. Particles are localized in space and are characterized by ENERGY and MOMENTUM. Waves are spread out in space and can be characterized by WAVELENGTH and FREQUENCY, which are irrelevant for particles. Moreover, waves are not characterized by localized energy and momentum. Nevertheless, every phenomenon possess both aspects, and there are relations between the energy and momentum of an individual particle in the particle aspect, on the one hand, and the frequency and wavelength of the wave aspect, on the other. Either aspect can be revealed by using an appropriate apparatus, but demonstrating one aspect suppresses the other. NIELS HENRIK DAVID BOHR pioneered the notion of complementarity.

See also QUANTUM PHYSICS.

compound A substance composed of two or more chemical ELEMENTs that are bound together through interatomic FORCEs is termed a compound. The elements that compose a compound cannot be separated by physical means, but decomposition can be effected by chemical processes. The properties of a compound are characteristic of the compound and are generally very different from the elements of which it is constituted. Table salt, for example, whose chemical designation is sodium chloride, is a compound of the elements sodium and chlorine. Whereas at room temperature sodium is a soft metal and chlorine is a greenish gas, table salt is a white brittle solid. Ordinary sugar, whose chemical name is sucrose, is a compound of the elements carbon, hydrogen, and oxygen. Sugar, like salt, is a white brittle solid at room temperature. Carbon, depending on its form, can be black (as soot or graphite) or clear and hard (as diamond), and hydrogen and oxygen are gases.

See also ATOM; MOLECULE.

Compton effect In the scattering of X-RAYS or GAMMA RAYS from the more loosely bound ELECTRONS of matter, the Compton effect, named for the American physicist Arthur Holly Compton (1892–1962), is the increase of WAVELENGTH of the scattered RADIATION from that of the incident rays. This is described as a COLLISION between a PHOTON and an electron at rest (compared with the energy of the photon, the electron can be considered to be at rest), whereby the photon's initial ENERGY and LINEAR MOMENTUM are shared with the electron, which is ejected from the ATOM. Thus the incident photon loses energy, resulting in an increase of wavelength of the radiation. (*See* WAVE; WAVE-PARTICLE DUALITY.)

The incident radiation can be scattered at any angle from its initial direction. In terms of its scattering angle, θ, the wavelength decrease is:

$$\lambda_2 - \lambda_1 = \lambda_C(1 - \cos\theta)$$

where λ_2 and λ_1 are the wavelengths in meters (m) of the scattered and incident radiation, respectively. The quantity:

$$\lambda_C = \frac{h}{mc}$$

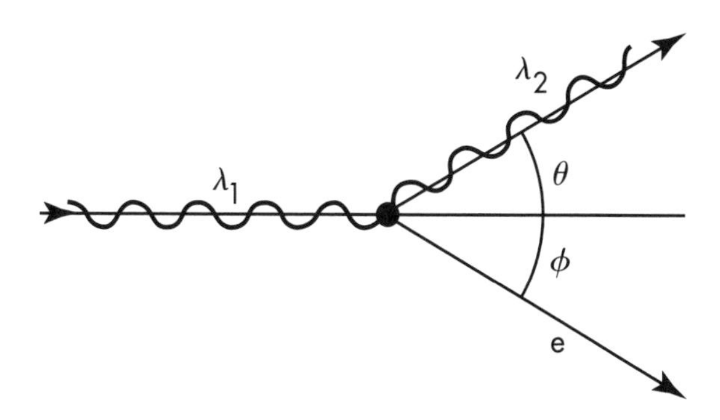

Compton effect. When X-rays or gamma rays scatter from an atomic electron (considered to be at rest), the wavelength of the scattered radiation is greater than that of the incident radiation. The incident and scattered radiation are labeled by their wavelengths λ_1 and λ_2, respectively. The initially resting electron is represented by the black circle. The directions of the scattered radiation and ejected electron *e* make angles θ and φ, respectively, with the direction of incident radiation. For any scattering angle θ, the increase in wavelength is given by $\lambda_2 - \lambda_1 = \lambda_C(1 - \cos\theta)$ and the electron's direction by $\cot\varphi = (1 + \lambda_C/\lambda_1)\tan(\theta/2)$. λ_C denotes the Compton wavelength of the electron, given by $\lambda_C = h/mc$, where *h,m,* and *c* denote, respectively, the Planck constant, the electron mass, and the speed of light.

is known as the Compton wavelength of the electron, where *h, m,* and *c* denote, respectively, the PLANCK CONSTANT, the MASS of the electron, and the SPEED OF LIGHT, whose values are $h = 6.62606876 \times 10^{-34}$ joule·second (J·s), $m = 9.1093897 \times 10^{-31}$ kilogram (kg), and $c = 2.99792458 \times 10^8$ meters per second (m/s). Thus $\lambda_C = 2.4263 \times 10^{-12}$ m. The angle, ϕ, at which the electron is ejected, measured from the direction of the incident radiation, is given by

$$\cot\phi = (1 + \lambda_C/\lambda_1)\tan(\theta/2)$$

condensation The process of PHASE TRANSITION from a GAS to a LIQUID or a SOLID is called condensation. Note that the term *vapor* is often used for a gas in contact with its liquid or solid PHASE.

Condensation occurs because the distribution of KINETIC ENERGY among the constituent particles (ATOMS or MOLECULES) of the gas leaves some of them with sufficiently low energy that they can stick together when they collide or remain in the condensed phase when they impinge upon it. The rate of condensation increases with lower gas TEMPERATURES. That happens because lower temperature means there is a greater fraction of low-energy particles in the gas. (*See* KINETIC THEORY.)

See also EVAPORATION; FREEZING.

condensed-matter physics Condensed MATTER is matter in the SOLID or LIQUID state. The physics of condensed matter investigates those states of matter. In the case of solids, subjects of investigation include HEAT conduction, CONDUCTION of ELECTRICITY, SUPERCONDUCTIVITY, crystalline structure, and SEMICONDUCTORS. SOLID-STATE PHYSICS is the subfield of condensed-matter physics that specializes in only the solid state of matter. For the liquid state, one might study BOILING, SURFACE TENSION, and LIQUID CRYSTALs, for example. (*See* CRYSTAL.)

conductance The inverse of RESISTANCE is termed *conductance* and indicates the ability of an electric component to conduct an electric CURRENT. Its SI unit is the siemens (S), equivalent to the inverse ohm (Ω^{-1}).

See also CONDUCTION; ELECTRICITY.

conduction The term *conduction* refers either to conduction of HEAT or conduction of electric CURRENT. In the first case, heat conduction, or thermal conduction, is the transfer of heat ENERGY by MATTER without concomitant transfer of the matter. The mechanism of heat conduction depends on the nature of the heat energy. In FLUIDS (i.e., in GASes and LIQUIDS), the constituent MOLECULES or ATOMS are free to move about within the confines of the material. Heat is the random motion of the molecules or atoms, whereby they move more or less freely between COLLISIONS with the container walls and with each other. Heat conduction takes place through the DIFFUSION of higher-speed molecules or atoms away from regions where they are concentrated (to be replaced by lower-speed ones, with no net transfer of matter). This type of conduction is relatively efficient compared with that typical of solids. (*See* ELECTRICITY.)

In solids, the constituent molecules or atoms are to a large extent confined to fixed positions, and heat is the random OSCILLATION of the constituents about their positions. Conduction of heat takes place by the more energetically oscillating molecules or atoms nudging their neighbors into greater activity. In some solids, typically in metals, there are what are known as free electrons, which are ELECTRONS that are not bound to any particular atom and behave as a gas confined to the volume of the material. They contribute strongly to heat conduction in the manner described for a fluid. Thus metals are typically relativity good CONDUCTORS of heat.

Additional ways in which heat energy is transferred are convection and RADIATION. In convection, the transfer is accomplished through the motion of the material itself, such as when heated air from a furnace is blown through ducts into the rooms of a house. Heat transfer by radiation proceeds through electromagnetic radiation. All matter constantly radiates electromagnetic radiation, at the expense of its heat energy, and absorbs at least some of any electromagnetic radiation impinging on it, thereby increasing its heat energy. In this way, heat is transferred among bodies via electromagnetic radiation. (*See* BLACK BODY; ELECTROMAGNETISM.)

Electrical conduction is the transfer of electric CHARGE through matter (i.e., the occurrence of an electric current). In solids that have free electrons—such solids are called electrical conductors—the motion of the free electrons accomplishes that. Since free electrons in solids contribute strongly also to heat conduction, as mentioned above, electric conductors—typically metals—are also good conductors of heat, compared with other solids. In SEMICONDUCTORS, electrons can be made free with the investment of a relatively small amount of energy that boosts them to the CONDUCTION BAND. That also produces positively charged HOLES. The electrons and the holes, together, carry electric charge through a semiconductor. In gases and in electrolytic solutions, charge may be carried by electrons and by positive and negative IONs.

See also RESISTANCE; SUPERCONDUCTIVITY.

conduction band When many ATOMs are in close proximity and are interacting with each other, such as in a SOLID, the ENERGY levels allowed to the ELECTRONs in the outer shells become very much more numerous than for individual atoms. The energy levels are so numerous that they can be considered to be continuous, rather than discrete, and are described in terms of energy "bands," ranges of energy. Electrons in the GROUND STATE are said to be in the VALENCE BAND. Electrons with sufficient energy to move freely among the relatively fixed atoms—such electrons are called free electrons—are said to be in the conduction band. If the two bands overlap or almost overlap, then the material is endowed with naturally occurring free electrons and has low RESISTIVITY. Such a material is called a CONDUCTOR. If there is a large energy gap between

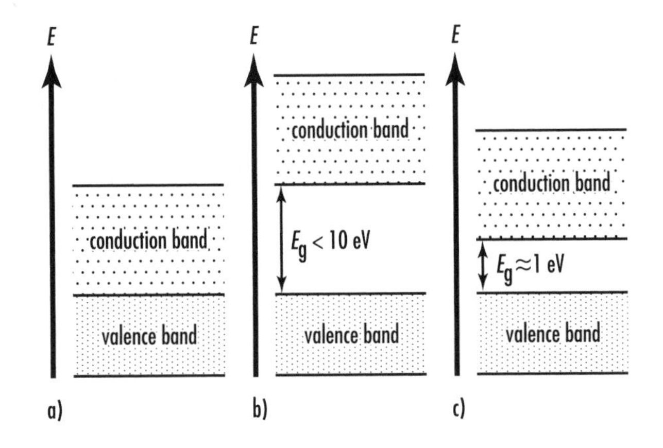

The conduction band of a solid represents the range of energies at which electrons can move about freely. The valence band is the range of energies of electrons in the ground state. The energy axis is labeled *E*. (a) When the bands overlap or almost overlap, the material is a conductor. (b) When the energy gap E_g between the bands is large, greater than about 10 eV, the material is an insulator. (c) Semiconductors have intermediate energy gaps, on the order of 1 eV.

the valence band and the conduction band, greater than, say, around 10 electron volts (eV), very few electrons will occupy the conduction band, and the material's resistivity will be high. It is called an INSULATOR. With intermediate energy gaps of about 1 eV, electrons can be boosted into the conduction band with relatively small investments of energy. Such materials are called SEMICONDUCTORS.

See also CONDUCTION; INTERACTION.

conductivity The inverse of RESISTIVITY is conductivity and indicates a material's intrinsic ability to conduct an electric CURRENT, independently of the material's shape. Its SI unit is siemens per meter (S/m), equivalent to 1/(ohm·meter) ($\Omega^{-1}\cdot m^{-1}$). (*See* CONDUCTION; ELECTRICITY.)

See also SUPERCONDUCTIVITY.

conductor A material of low RESISTIVITY, or equivalently, of high CONDUCTIVITY, is called a conductor. Thus, a conductor is a material that allows a relatively easy passage of electric CURRENT. (*See* ELECTRICITY.)

See also CONDUCTION; CONDUCTION BAND; SUPERCONDUCTIVITY.

conservation law When a physical quantity maintains a constant value in an isolated system as the system evolves in time, the physical quantity is said to be conserved. When that is the case for all systems, or even for a class of them, a conservation law is said to apply to the physical quantity under consideration. The best known of all conservation laws is the law of conservation of ENERGY, stating that for all isolated systems, the total energy of the system remains constant in value over time (although it might change its form). This law is indeed considered to be valid for all systems, at least as far as is presently known. Additional conservation laws that are understood to hold for all systems are those for LINEAR MOMENTUM, ANGULAR MOMENTUM, electric CHARGE, BARYON number, and LEPTON number. According to the first of these, for example, the vector sum of the linear momenta of all the components of any isolated system does not change its value (its magnitude and direction) as the system evolves in time. (*See* ELECTRICITY.)

Note that the meaning of the term *conservation* in physics is very different from its everyday meaning. Ordinarily "energy conservation" means saving energy by using it sparingly and avoiding waste. But in physics, the meaning is as explained above. There *is* a connection, however. Due to the law of conservation of energy, energy cannot be obtained from nothing. Its use is always at the cost of depleting resources. Thus it makes sense to use energy sparingly in order to allow energy resources to last for as long as possible.

There exists a relation between every conservation law and a SYMMETRY of the LAWS OF NATURE, also called an INVARIANCE PRINCIPLE. Conservation of energy, for instance, is related to the invariance of the laws of nature in time. The fact that the laws of nature do not change over time is related to the fact that energy is conserved. Similarly, conservation of linear momentum is related to the invariance of the laws of nature in space, that is, to the fact that the laws of nature are the same everywhere. Similar relationships hold for the other change conservation laws.

conservative force A FORCE that does zero total WORK on a particle that moves in a closed path is termed a conservative force. An equivalent formulation is that the work that a conservative force does on a particle that moves between two points is independent of the path that the body follows between those points. As a result, a POTENTIAL ENERGY can be associated with a conservative force. (Indeed, a third definition of a conservative force is a force with which a potential energy can be associated.) What that means is this. Imagine that a force does some positive work on a particle, thus increasing the particle's KINETIC ENERGY. By the law of conservation of ENERGY, the particle's gain in kinetic energy is at the expense of an energy decrease of some source. For a conservative force, and only for such a force, the source's energy can be regained in full by having the particle return to its initial state. Then the force will do negative work on the particle of the same magnitude as before, the particle will lose the same amount of kinetic energy it previously gained, and the source will be replenished. In the potential-energy picture, it is the potential energy that serves as the source. The process can then be viewed as an initial conversion of potential energy to kinetic, followed by a total reconversion of kinetic energy back to potential. (*See* CONSERVATION LAW.)

Two examples of a conservative force are the gravitational force and the electric force. For the former there is gravitational potential energy, and for the latter there is electric potential energy. From these, the gravitational and electric potentials are derived. As a demonstration of the gravitational case, consider an ideal PENDULUM displaced from EQUILIBRIUM and released from rest. As the pendulum descends, the gravitational force between it and the Earth does positive work on it, whereby it gains kinetic energy and achieves maximal kinetic energy as it passes through its equilibrium point. When the pendulum is ascending, however, the gravitational force does negative work, causing the pendulum to lose kinetic energy until it comes to momentary rest at its initial height. In terms of gravitational potential energy, the pendulum starts with zero kinetic energy and some value of potential energy. As it descends, its potential energy decreases and its kinetic energy increases, such that their sum remains constant (conservation of energy). When the pendulum passes through its equilibrium point, its potential energy is lowest and its kinetic energy highest. As the pendulum ascends, its kinetic energy converts back into potential energy, and at the end of its swing it again has zero kinetic energy and the same amount of potential energy it started with. (*See* ELECTRICITY; GRAVITATION; POTENTIAL, ELECTRIC.)

Any force that is not conservative is called a nonconservative force. Some such forces perform only negative work and convert other forms of energy into HEAT. Common examples of those are FRICTION and VISCOSITY forces. In the above pendulum example, if the pendulum is realistic rather than ideal, so that friction in the suspension bearing and air viscosity are in play, the pendulum will eventually come to rest at its equilibrium position. The energy given it by the work that was done in initially displacing it from its equilibrium position is eventually dissipated as heat.

constants, fundamental The fundamental constants of nature are the values of physical quantities that both seem to play a fundamental role in nature's working and appear not to change over TIME. Here are some of them.

SPEED OF LIGHT, which is the propagation speed of electromagnetic RADIATION in VACUUM:

$$c = 2.99792458 \times 10^{8} \text{ m/s}$$

PLANCK CONSTANT, which characterizes quantum phenomena:

$$h = 6.62606876 \times 10^{-34} \text{ J·s}$$

Magnitude of the electric CHARGE of the ELECTRON:

$$e = 1.602176462 \times 10^{-19} \text{ C}$$

Gravitational constant:

$$G = 6.67259 \times 10^{-11} \text{ N·m}^2/\text{kg}^2$$

GAS constant:

$$R = 8.314472 \text{ J/(mol·K)}$$

AVOGADRO'S NUMBER:

$$N_A = 6.02214199 \times 10^{23} \text{ mol}^{-1}$$

PERMEABILITY of the VACUUM:

$$\mu_0 = 4\pi \times 10^{-7} \text{ T·m/A} = 1.25663706143 \times 10^{-6} \text{ T·m/A}$$

PERMITTIVITY of the VACUUM:

$$\varepsilon_0 = 1/(\mu_0 c^2) = 8.85418781762 \times 10^{-12} \text{ C}^2/(\text{N·m}^2)$$

Boltzmann constant:

$$k = 1.3806503 \times 10^{-23} \text{ J/K}$$

MASS of the ELECTRON:

$$m_e = 9.10938188 \times 10^{-31} \text{ kg}$$

MASS of the PROTON:

$$m_p = 1.67262158 \times 10^{-27} \text{ kg}$$

MASS of the NEUTRON:

$$m_n = 1.67492716 \times 10^{-27} \text{ kg}$$

Rydberg constant:

$$R = 1.0973731568549 \times 10^{7} \text{ m}^{-1}$$

Stefan-Boltzmann constant:

$$\sigma = 5.67051 \times 10^{-8} \text{ W/(m}^2\text{·K}^4)$$

(*See* ELECTRICITY; ELECTROMAGNETISM; GRAVITATION; QUANTUM PHYSICS.)

It is an open question whether the fundamental constants are indeed constant over time and have the same values in different locations. Although that is generally assumed to be the case, and on the whole appears to be true, there are theoretical ideas to the effect that some of the constants might be changing concomitantly with the evolution of the UNIVERSE, and occasionally there are reports of evidence for such change. There exists a continual theoretical effort to understand why the fundamental constants are constant (if indeed they are) and why they possess the values they do.

Cooper pair At least some manifestations of SUPERCONDUCTIVITY can be understood through the notion of Cooper pairs, named for the American physicist Leon N. Cooper (1930–). A Cooper pair is a pair of electrically conducting ELECTRONS in a crystalline SOLID that, rather than behaving independently, are linked via a long-range INTERACTION that is mediated by the crystal LATTICE. In this manner, as explained by QUANTUM MECHANICS, the pairs can move through the material completely freely, unimpeded by the lattice, by any impurities in it, or by each other. That results in zero electrical RESISTANCE. (*See* CONDUCTION; CRYSTAL; ELECTRICITY.)

coordinate system A grid in space that serves as a REFERENCE FRAME for the specification of locations by means of sets of numbers, called coordinates, is a coordinate system. The most commonly known among coordinate systems is the Cartesian coordinate system of three-dimensional SPACE, which consists of three mutually perpendicular straight axes bearing equal length scales. Cartesian coordinate axes are usually denoted as x, y, and z, and the coordinates of a point denoted as (x, y, z). Other coordinate systems are often convenient. They include curvilinear systems and nonorthogonal systems. For three-dimensional space, for example, a spherical coordinate system or a cylindrical coordinate system is often used. The former is related to the latitude-longitude coordinate system used for the surface of the Earth. (*See* DIMENSION.)

The coordinates of a point in a spherical coordinate system are (r, θ, φ). They are related to the (x, y, z) coordinates by:

$$x = r \sin \varphi \cos \theta$$
$$y = r \sin \varphi \sin \theta$$
$$z = r \cos \varphi$$

and:

$$r = \sqrt{x^2 + y^2 + z^2}$$
$$\theta = \tan^{-1} \frac{y}{x}$$
$$\varphi = \cos^{-1} \frac{z}{\sqrt{x^2 + y^2 + z^2}}$$

In a cylindrical coordinate system, the coordinates of a point are (r, θ, z), which are related to (x, y, z) by:

$$x = r \cos \theta$$
$$y = r \sin \theta$$
$$z = z$$

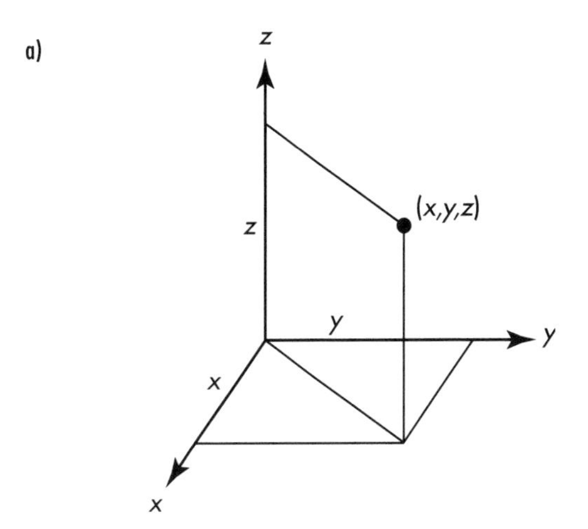

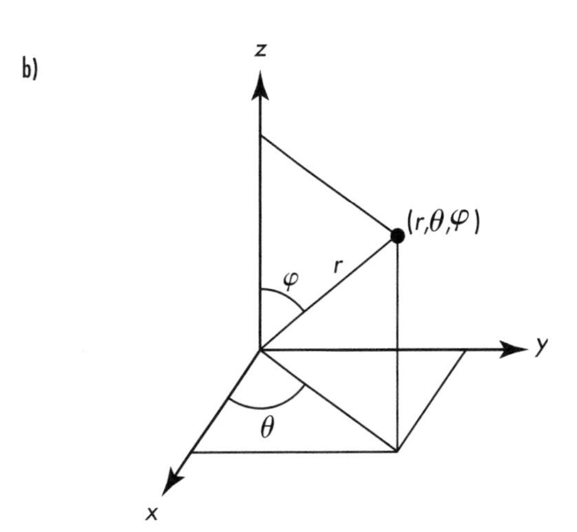

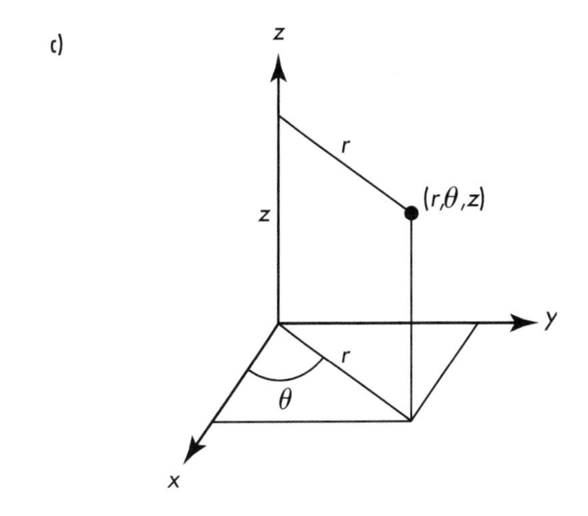

The coordinates of the same point in three-dimensional space are given in three coordinate systems. (a) Cartesian coordinate system, consisting of three mutually perpendicular axes: *(x, y, z)*. (b) Spherical coordinate system: *(r, θ, φ)*. (c) Cylindrical coordinate system: *(r, θ, z)*.

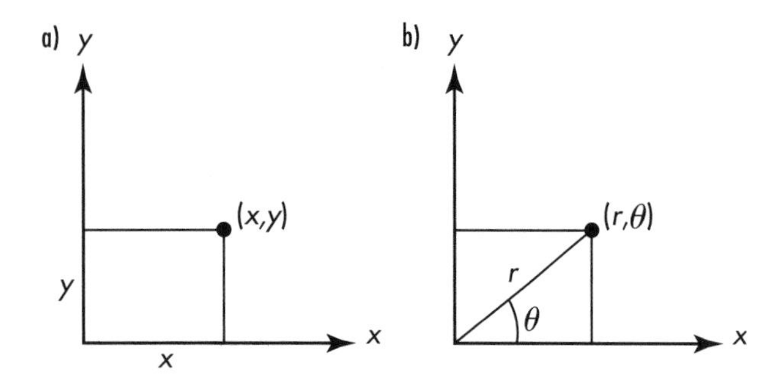

The coordinates of the same point in two-dimensional space are given in two coordinate systems. (a) Cartesian coordinate system, consisting of two mutually perpendicular axes: (*x*, *y*). (b) Polar coordinate system: (*r*, *θ*).

and

$$r = \sqrt{x^2 + y^2}$$
$$\theta = \tan^{-1}\frac{y}{x}$$
$$z = z$$

For a plane, which is a two-dimensional space, in addition to the common Cartesian coordinate system of x- and y-axes, a polar coordinate system is often used. The coordinates of a point in a Cartesian coordinate system are (x, y) and in a polar system (r, θ). They are related by:

$$x = r \cos \theta$$
$$y = r \sin \theta$$

and:

$$r = \sqrt{x^2 + y^2}$$
$$\theta = \tan^{-1}\frac{y}{x}$$

Coordinate systems are also used for abstract spaces. One such use is for four-dimensional SPACE-TIME. In space-time coordinate systems, one of the four axes indicates time, while the other three serve for space. When all axes are straight and mutually perpendicular, we have a Minkowskian coordinate system, especially useful for ALBERT EINSTEIN's special theory of relativity. Einstein's general theory of relativity makes use of curvilinear coordinate systems called Riemannian coordinate systems. Abstract spaces might have nothing to do with space or with time. States of a GAS might be represented by points in an abstract two-dimensional volume-pressure space, specified with reference to an orthogonal coordinate system consisting of a VOLUME axis and a PRESSURE axis, each marked with an appropriate scale.

See also RELATIVITY, GENERAL THEORY OF; RELATIVITY, SPECIAL THEORY OF.

Coriolis acceleration Moving bodies appear, to observers on Earth, to undergo an ACCELERATION that has no corresponding FORCE causing it, as would be required by ISAAC NEWTON's first law of MOTION. This acceleration is called Coriolis acceleration, named for the French physicist Gaspard Gustave Coriolis (1792–1843), and the effect is called the Coriolis effect. Due to its ROTATION about its axis, the Earth is not an inertial reference frame. This means that an observer on Earth, using a COORDINATE SYSTEM attached to the Earth, finds that Newton's laws of motion appear to be violated. In particular, if a body moves on the surface of the Earth, it appears to acquire, merely by virtue of its motion, an acceleration to its right, if it is in the Northern Hemisphere, or to its left, in the Southern Hemisphere. That affects, for example, moving air masses and weather predictions. Since we are conditioned by Newton's second law of motion to expect accelerations always to be caused by forces, we sometimes find references to "Coriolis forces," which are the fictitious forces that are associated with Coriolis accelerations in an Earth-based coordinate system. By generalization, the terms *Coriolis acceleration* and *Coriolis force* are used also for fictitious motion-dependent accelerations and forces in any rotating (and thus noninertial) reference frame.

See also NEWTON'S LAWS; REFERENCE FRAME, INERTIAL.

corona discharge A corona discharge is a glow, possibly including lightning streaks, in the air (or other GAS) surrounding an electrically charged CONDUCTOR. It is caused by the "breakdown" of the air in the conductor's vicinity, whereby the electric FIELD produced by the electric charge of the conductor is sufficiently strong to ionize air molecules and generate a PLASMA. When the IONs and ELECTRONs recombine, light is emitted. The freely moving ions and electrons offer conducting paths for electric CURRENTs to flow to and from the charged conductor. Such currents may appear as lightning flashes. Steady discharge currents can produce a buzzing, or humming, noise, while lightning is accompanied by characteristic snapping and crackling sounds. Moist air is especially conducive to corona discharge, so the effect is often seen and heard in the vicinity of high-voltage electric lines when the air is very humid. Corona discharge is especially pronounced at sharp points and edges of the charged conductor, since the electric field is especially strong there. (*See* CHARGE; CONDUCTION; ELECTRICITY; IONIZATION.)

cosmic microwave background The Earth is found to be immersed in electromagnetic RADIATION, called the cosmic microwave background, or cosmic background radiation, that has no apparent source and whose SPECTRUM is characteristic of that emitted by a BLACK BODY at the absolute TEMPERATURE of 2.73 kelvins (K). It is assumed that this radiation permeates all of SPACE (and is not merely a fluke of some special situation of the Earth). The cosmic microwave background is generally understood to be a remnant from an early stage in the evolution of the UNIVERSE, when the universe was much hotter and denser, following a cosmic explosion from an extremely dense, hot state called the BIG BANG. As the universe expanded, the radiation cooled to its present temperature. The cosmic microwave background is commonly taken as evidence for the big bang. By this view, in observing the cosmic background radiation, we are "seeing" the actual glow of the big bang. Over some 15 billion years since the big bang, the WAVELENGTHs of that glow have stretched, due to the expansion of the universe, from the extremely short wavelengths characteristic of the big bang's very high temperature to the present-day wavelengths that correspond to a very cold 2.73 K. (*See* COSMOLOGY; DENSITY; ELECTROMAGNETISM.)

The cosmic microwave background is observed to be extremely uniform in all directions. Tiny fluctuations have been detected in it, which are generally assumed to reflect small inhomogeneities in the cosmic distribution of matter soon after the big bang. Such inhomogeneities are thought to have been involved in the formation of the earliest GALAXIES.

cosmic ray The high-ENERGY particles and RADIATION that reach the Earth from space and from the Sun are called cosmic rays. Their energy range is about 10^8–10^{19} electron volts (eV). They consist mainly of PROTONs but also include ALPHA PARTICLES (helium NUCLEI), heavier nuclei, ELECTRONs, and GAMMA RAYS (energetic PHOTONs). Cosmic rays collide with the nuclei of atmospheric ATOMs, a process that produces secondary rays. The latter might decay or might, in turn, collide and generate further rays, and so on. Thus, a single cosmic ray can result in a shower of particles and rays hitting the surface of the Earth.

cosmic string This is a hypothetical, long one-dimensional singularity, or "defect," in SPACE-TIME that involves a high concentration of ENERGY and thus produces a strong gravitational FIELD. If cosmic strings exist in nature, they might play a role in the large-scale structure of the UNIVERSE by attracting GALAXIES to their vicinities, which would result in long lines of galaxies.

See also DIMENSION; GRAVITATION.

cosmology The study of the UNIVERSE, the cosmos, as a whole is called cosmology. In particular, cosmology deals with the large-scale structure and evolution of the universe. One of the most important theoretical tools of modern cosmology is the cosmological principle, which states that, at sufficiently large scales, the universe is homogeneous (i.e., possesses the same properties everywhere). Another way of stating this is that an observer at any location in the universe would look into the sky and see very much the same thing that we see. In other words, the Earth has no privileged position in the universe, and what we observe can be taken to be representative of the situation in general. The cosmological principle is an assumption, but a necessary one for us to make any progress in understanding the universe. Another important theoretical tool of modern cosmology is ALBERT EINSTEIN's general theory of relativity. (*See* RELATIVITY, GENERAL THEORY OF.)

The present picture of the large-scale structure of the universe is one of great complexity. Many GALAXIES are grouped into clusters, and clusters are grouped into superclusters. There seem to exist tremendous sheets of galaxies, many millions of light-years in size. Between the sheets, there appear to be gigantic voids, which are empty of galaxies. Galaxy clusters are found where the sheets intersect. As far as is presently known, galaxies and clusters affect each other only through gravitational attraction, which is thus considered responsible for molding the universe's structure. (*See* GRAVITATION.)

Cosmology also deals with the evolution of the universe, from its beginning—if, indeed, it had a beginning—to its eventual fate. (Cosmogony is the study of the origin of the universe in particular.) One of the most widely accepted determinations in cosmology is that the universe is expanding: all galaxies are, on the average, moving apart from each other. This would seem to offer two possibilities. Either the universe will expand forever, or it will eventually reverse its expansion and collapse into itself. Based on the gravitational attraction among all the galaxies acting as a retarding FORCE, it would be expected that the rate of expansion is decreasing. Then, depending on the amount of MASS in the universe, which determines the gravitational attraction, the cosmic expansion might eventually come to a halt and reverse itself, leading to what is often called the "big crunch." However, recent observations seem to indicate that not only is the rate of expansion not decreasing, but it is actually increasing (i.e., the cosmic expansion is accelerating). Much theoretical investigation is taking place into possible causes for that.

The general expansion of the universe strongly suggests that in the past all MATTER was closer together and the universe was denser. Evolution scenarios along these lines have the universe starting its existence in an explosion from an extremely dense, hot state, called the BIG BANG, and expanding and cooling ever since. It is generally, but not universally, accepted among cosmologists that the evolution of the universe is following some version of a big-bang model. (*See* COSMIC MICROWAVE BACKGROUND; DENSITY.)

A possible alternative to the big-bang idea is based on the perfect cosmological principle, stating that not only would an observer at any *location* in the universe see very much the same as what we see, but so would also an observer at any *time* during the universe's evolution. In other words, the universe should remain the same over time (at sufficiently large scales). Such a scenario is called a steady-state model. The constant density required thereby and the observed expansion of the universe are reconciled through the assumption that new matter is continuously being created. However, the big-bang model of universe evolution is most commonly accepted at present, and the perfect cosmological principle is in disrepute. (*See* STEADY-STATE PROCESS.)

Present times are exciting ones for those interested in cosmology. Advancing technologies are allowing observations at larger and ever larger distances, which, due to the finite SPEED OF LIGHT, are revealing the secrets of the universe from earlier and even earlier times in its past. New evidence is continually casting doubts on old notions, while active cosmologists are coming up with new ideas. Some of those ideas sound quite weird, such as dark matter, noninteracting matter, dark energy, COSMIC STRINGS, and extra DIMENSIONS. However, they might not yet be weird enough to fully comprehend what the universe is up to. The universe might be much weirder than we think. Might it be even weirder than we *can* think?

Coulomb, Charles Augustin de (1736–1806) French *Engineer, Physicist* Best know today for his work in ELECTRICITY, and in particular for COULOMB'S LAW, Charles Augustin de Coulomb has been honored by having the SI UNIT of electric CHARGE, the coulomb (C), named after him. Coulomb was trained as in engineer and graduated in 1761. Thereafter, he worked as an engineer, mostly as a military engineer, but he also devel-

Known today for Coulomb's law in electricity, Charles Augustin de Coulomb was an 18th-century physicist who investigated mechanics, electricity, and magnetism. *(AIP Emilio Segrè Visual Archives, E. Scott Barr Collection)*

oped and published his ideas in theoretical MECHANICS, making use of mathematical methods that were more sophisticated than most engineers of his time were used to. A work on theoretical mechanics, *Théorie des Machines Simples,* that Coulomb published in 1781 won him the grand prize of the Academy of Sciences, Paris, France, for that year and allowed him to stop working as an engineer (except for consulting jobs) and to devote his life to physics. His subsequent investigations included important work in electricity and MAGNETISM.

Coulomb's law Named for the French physicist CHARLES AUGUSTIN DE COULOMB, Coulomb's law gives the electric FORCES that two point electric CHARGES exert on each other. The forces on the two charges are oppositely directed along the line joining the charges and are equal in magnitude. The force on either charge is toward the other charge if the charges are of opposite sign: unlike charges attract each other. If the charges are of the same sign, the force on either charge is directed away from the other charge: like charges repel each other. The magnitude of the forces is proportional to the product of the magnitudes of the two charges and inversely proportional to the square of the distance between them. In a formula:

$$F = kq_1q_2/r^2$$

where F denotes the magnitude of the force in newtons (N) on either charge, q_1 and q_2 are the magnitudes of the two charges in coulombs (C), and r is the distance between them in meters (m). The value of the proportionality constant k depends on the medium in which the charges are immersed. In VACUUM $k = k_0 = 8.98755179 \times 10^9$ N·m²/C². Otherwise:

$$k = k_0/\kappa$$

where κ denotes the DIELECTRIC CONSTANT of the medium and measures the extent to which the medium, by means of its electric POLARIZATION, reduces the magnitude of the forces between the charges. (*See* ELECTRICITY.)

Note that F is the magnitude of the forces and is thus a positive number (or zero). A negative value for F is meaningless. That is why only the magnitudes of the charges—ignoring their signs—are used in the formula. The directions of the forces are not obtained from the formula for F but, rather, from the rule that like charges repel and unlike charges attract. Nevertheless, the formula for F is sometimes used with the signed charges. In that case, according to the repulsion-attraction rule, a positive or negative F indicates repulsion or attraction, respectively.

The constant k is often written in the form:

$$k = \frac{1}{4\pi\varepsilon} = \frac{1}{4\pi\kappa\varepsilon_0}$$

where $\varepsilon = \kappa\varepsilon_0$ is the PERMITTIVITY of the medium and ε_0 the permittivity of the vacuum, whose value is $8.85418782 \times 10^{-12}$ C²/(N·m²). Accordingly, the formula for the magnitude of the electric forces that two point charges exert on each other, according to Coulomb's law, is often written in the form:

$$F = \frac{1}{4\pi\kappa\varepsilon_0}\frac{q_1q_2}{r^2}$$

critical point Also called the critical state, the critical point of a substance is the condition under which the substance's LIQUID and GAS PHASES possess the same DENSITY and are, in fact, indistinguishable. The density, PRESSURE, and TEMPERATURE at the critical point are referred to as the critical density, critical pressure, and critical temperature. At temperatures higher than a substance's critical temperature, increasing the pressure on the substance makes it denser, but at no pressure is there a distinct gas-liquid PHASE TRANSITION. So the critical temperature is a measure of the strength of the cohesive FORCEs holding the liquid phase together. In contrast, at any temperature lower than the critical temperature, increasing the pressure of the gas phase makes the gas denser only until the VAPOR PRESSURE of the liquid phase for that temperature is reached. Attempting to compress the gas further—by reducing its volume—does not increase the pressure, but it does cause some of the gas to change to liquid. Additional volume reduction converts even more gas to liquid. Only when all of the gas phase has undergone transition to liquid will further compression actually raise the pressure.

As an example, the critical temperature of carbon dioxide is 304.3 kelvins (K), or 31.3 degrees Celsius (°C). Above that temperature, no amount of pressure can convert gaseous carbon dioxide to a liquid. Below 304.3 K, gaseous carbon dioxide can be liquified. The required pressure is below the critical pressure of 7.38 megapascals (MPa) and decreases with decreasing temperature.

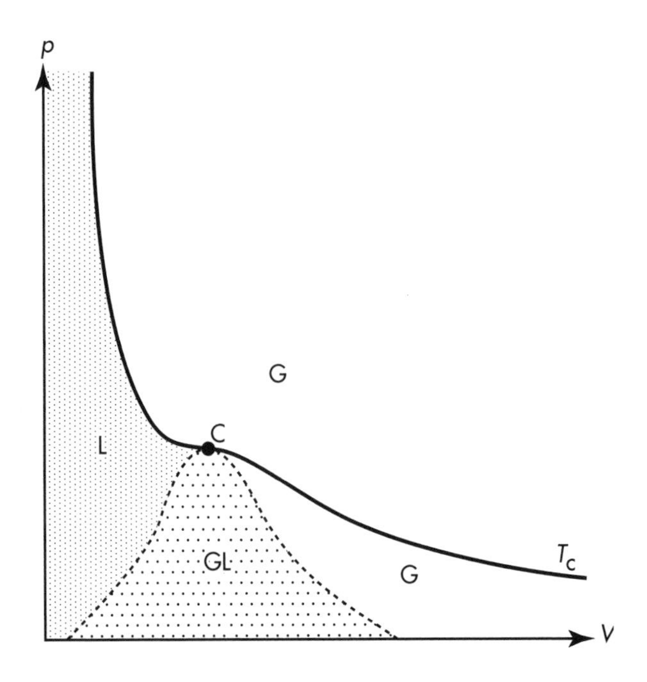

The critical point of a substance, at which the liquid and gas phases are indistinguishable, is labeled C in a typical pressure-volume (*p-V*) phase diagram. Above the critical temperature, which is the temperature at the critical point, no amount of pressure will bring about a gas-liquid phase transition. The critical-temperature isotherm is denoted T_c. Regions of the diagram are labeled according to whether under those conditions only a gas can exist (G), only a liquid can exist (L), or gas and liquid can exist in equilibrium (GL). (The diagram does not extend to conditions in which a solid phase might exist.)

cryogenics This is the branch of physics that deals with very low TEMPERATURES, including methods for cooling to those temperatures and the properties and behavior of MATTER under those conditions. Low-temperature physicists struggle with the fundamental constraint of absolute zero temperature, 0 kelvin (K). Not only are absolute zero and lower temperatures unobtainable in principle, according to the second law of thermodynamics, but as absolute zero temperature is approached, lowering the temperature further becomes increasingly more difficult. Nevertheless, temperatures as low as millikelvins have been achieved. (*See* THERMODYNAMICS.)

Temperature is a measure of the random molecular and atomic motion of matter, so the lower the temperature, the weaker is that motion. In the study of matter at very low temperatures, subtle effects can become manifest that otherwise would be destroyed by the random motion. One such effect, requiring extremely low temperatures, is the Bose-Einstein condensate. SUPERCONDUCTIVITY is another such effect, but it can appear at higher temperatures, depending on the material involved. (*See* ATOM; BOSE-EINSTEIN STATISTICS; HEAT; MOLECULE.)

crystal A crystal is a SOLID in which the constituting particles—ATOMS, MOLECULES, or IONS, as the case may be—possess a regularly repeating spatial arrangement called the crystal structure. The geometric points at which the particles are located in a crystal structure are referred to collectively as the crystal LATTICE, although this term is often used loosely to mean crystal structure. CRYSTALLOGRAPHY is the study of crystals.

A solid in the crystalline state often exhibits a regular external appearance. A common everyday example of that are grains of table salt, which are crystals of sodium chloride and are built of a regular structure of alternating sodium and chloride ions. In general, crystals can possess different physical properties in different directions, a condition called anisotropy. The speed of LIGHT propagation in a crystal might depend on direction, as might a crystal's electric RESISTANCE. (*See* ELECTRICITY.)

See also POLYCRYSTALLINE SOLID.

crystallography The field of physics that deals with CRYSTALS, in particular with their structure, properties, and formation, is called crystallography, which is a subfield of SOLID-STATE PHYSICS.

In a crystal the constituent ATOMS, MOLECULES, or IONS are located at the points of a crystal LATTICE, which is a regularly repeating spatial arrangement of geometric points. Thus, the structure of a crystal is determined to a large extent by its lattice. Crystal lattices are the subject of investigation of theoretical, or mathematical, crystallography. They are classified according to their SYMMETRY properties, which give 230 different types. Those can be grouped into seven classes or crystal systems. The latter have to do with the relations among three VECTORS that represent the repeat distances of the lattice in three independent directions. The full structure of a crystal is determined by a unit cell, which is a minimal set of constituents that can serve as the crystal's structural element through its repetition in three independent directions. X-ray crystallography is the experimental determination of crystal structure by analyzing the DIFFRACTION pattern of X-RAYS that are aimed at the crystal.

Physical properties of crystals include DENSITY, electric RESISTANCE, SPECIFIC HEAT, and the speed of LIGHT propagation. Some of a crystal's properties—electric resistance and speed of light propagation, for example—may depend on direction. Theoreticians attempt to calculate the physical properties of crystals from fundamental principles. The results are compared with the actual situations, as measured by experimentalists, to test the accuracy of the theoretical models. (*See* ELECTRICITY.)

Crystal formation, both in nature and in the laboratory, is an additional subject of interest to crystallographers. Due to the importance and extensive use of crystalline silicon in the SEMICONDUCTOR industry, the production of larger and more nearly perfect silicon crystals is of great commercial interest.

Curie, Pierre (1859–1906) French *Physicist* Known for his work in CRYSTALLOGRAPHY and MAGNETISM, Pierre Curie is famous as well for research on RADIOACTIVITY performed together with his wife, Marie (Sklodowska) Curie. In 1895, he obtained his doctorate in physics from the Sorbonne, France, and was there appointed professor of physics. Curie's crystallography work led to the discovery of piezoelectric effects, and he introduced SYMMETRY considerations into the theoretical treatment of crystals and magnetism. CURIE'S LAW and the Curie TEMPERATURE are named after him. Work with his wife led to the discovery of the radioactive ELEMENTs radium and polonium. In 1903, the Curies shared half the Nobel Prize in physics "in recognition of the extraordinary services they have rendered by their joint researches on the radiation phenomena discovered by Professor Henri Becquerel."

See also FERROMAGNETISM; PARAMAGNETISM; PIEZOELECTRICITY.

Shown here delivering a lecture, Pierre Curie made important contributions to crystallography and the understanding of magnetism and performed research on radioactivity together with his wife, Marie (Sklodowska) Curie. They shared half of the 1903 Nobel Prize in physics. ***(AIP Emilio Segrè Visual Archives)***

Curie's law This law, discovered by the French physicist PIERRE CURIE in 1895, has to do with paramagnetic materials. It relates the magnetic dipole moment that is induced in a unit VOLUME of such a material, which is called the material's magnetization, to the magnetic FIELD in which the experimental specimen of the material is placed and the absolute TEMPERATURE of the material. The relation is:

$$M = C(B/T)$$

where M denotes the magnitude of the magnetization in amperes per meter (A/m), B is the magnitude of the magnetic field in teslas (T), T is the absolute temperature in kelvins (K), and C is a constant whose value depends on the material. As for the physical significance of the law, M is directly related (in fact, proportional) to B, which is the result of the magnetic field's aligning the atomic MAGNETIC DIPOLEs that, when aligned, add up to make the material's bulk magnetic dipole moment. The inverse relation (inverse proportionality in this case) of M to T follows from the disrupting effect of thermal agitation on the alignment of the atomic magnetic dipoles. (*See* HEAT; MAGNETISM; MOMENT, MAGNETIC DIPOLE; PARAMAGNETISM.)

The validity of Curie's law is limited to some range of B/T starting at zero, since, although B/T might have any value, the range of possible values of M is limited by the material's magnetic saturation, when all its atomic magnetic dipoles are aligned.

current The term *current* refers to the flow of any physical quantity. As a physical quantity itself, the current is the amount of the flowing physical quantity that passes a fixed point per unit time. The current of a beam of ATOMs, for example, might be 2 million atoms per second.

The most common use of the term is for the flow of electric CHARGE, which is electric current, defined as the net amount of charge flowing past a point per unit time. If net charge Δq in coulombs (C) passes a point during time interval Δt in seconds (s), the average current during the time interval is:

$$i = \Delta q/\Delta t$$

The SI UNIT of electric current is the ampere (A), or coulomb per second (C/s). Taking the limit of infinitesimal time interval, we have the instantaneous current:

$$i = dq/dt$$

(*See* ELECTRICITY.)

Any net flow of charge forms an electric current. The most commonly known is the flow of ELECTRONs in a CONDUCTOR, such as a copper wire. Examples of other forms of electric current are: the flow of IONs in an electrolyte, an ion beam in VACUUM, the flow of electrons and ions in a CORONA DISCHARGE and in lightning, and the flow of electrons and HOLEs in a SEMICONDUCTOR. (*See* CONDUCTION.)

An electric current that constantly flows in the same direction, such as that produced by a BATTERY, is called a direct current (DC). An alternating current (AC) is one that continually changes its direction of flow, in particular when its instantaneous value oscillates as a sinusoidal function of time:

$$i = i_{max} \sin (2\pi f t + \beta)$$

Here i_{max} denotes the maximal value of the alternating current, f the FREQUENCY of oscillation in hertz (Hz), t the time in seconds (s), and β the phase shift in radians (rad). The effective current, i_{eff}, of an alternating current is the value of the direct current that, when it flows through a RESISTOR, would produce heat at the same rate as does the alternating current. Its value is:

$$i_{eff} = i_{max}/\sqrt{2}$$

(*See* OSCILLATION.)

damping The action of FRICTION or DRAG on an oscillating system, causing the amplitude of oscillation to decrease in time, is called damping. The damping FORCE is a dissipative force, which performs negative WORK on the oscillating system, thus diminishing its mechanical energy (the sum of its KINETIC and POTENTIAL ENERGIES) and converting it irreversibly to HEAT. In an electromagnetic or electromechanical oscillator, electric RESISTANCE in the circuit causes damping. Damping tends to reduce the FREQUENCY of oscillation from what it would be in the absence of damping. In practice, damping cannot be completely avoided. (*See* DISSIPATION; ELECTRICITY; ELECTROMAGNETISM; OSCILLATION; REVERSIBLE PROCESS.)

In those situations where oscillation can occur but is not desired, damping is introduced for the purpose of reducing it or preventing it altogether. An automobile should best not be set oscillating on its suspension springs by every bump in the road, which would make the ride very uncomfortable. Neither should the pointer of a GALVANOMETER oscillate for long about its equilibrium position. That would make it hard to read. The amount of damping that brings the system to rest in its equilibrium position in the shortest amount of time is called critical damping. The system is then said to be critically damped. It comes to rest in a time that is on the order of its oscillation PERIOD. Less than critical damping, when the system is underdamped, allows the system to oscillate as it comes to rest. Overdamping, on the other hand, prevents oscillation but causes the system to creep back to equilibrium, taking longer than necessary to do so.

decay Decay is the decrease of some physical quantity over time. One example is radioactive decay, in which the amount of radioactive material diminishes over time, as does the INTENSITY of the accompanying RADIATION. Another example is the decay of the amplitude of a damped HARMONIC OSCILLATOR. (*See* DAMPING; RADIOACTIVITY.)

It is very often the case, and is so for radioactivity and damped harmonic oscillations, that the rate of decay of a quantity is proportional to the quantity itself. That results in exponential decay, according to which the value of the quantity, N, is given as a function of time by the expression:

$$N = N_0 \, e^{-\gamma t}$$

where N_0 denotes the value of the decaying quantity at time zero, t denotes the time in seconds (s), and γ is a constant, called the decay constant, in inverse seconds (s^{-1}), that is characteristic of the process. For radioactivity, N might represent the amount of undecayed radioactive material, or it might represent the intensity of radioactive radiation. Or, N might represent the amplitude of a damped harmonic oscillator, for example.

Another way of expressing exponential decay is:

$$N = N_0 \, e^{-t/\tau}$$

where τ is the TIME CONSTANT in seconds (s) and is the reciprocal of the decay constant. The time constant is the time interval required for N to decrease to $1/e$ of its value (i.e., for its value to decrease by a factor of approximately 2.718). Note that after a time interval of at least five time constants, N is reduced to less than 1 percent of its initial value. For radioactive decay, it is common to use the half-life instead of the time constant. The half-life is the time interval in which half the radioactive material decays, or in which the intensity of radiation decreases to half its value. The half-life, $\tau_{1/2}$, equals $\tau \ln 2 \approx 0.693\tau$. So exponential decay can be expressed in terms of the half-life as:

$$N \approx N_0 e^{-0.693t/\tau_{1/2}}$$

Note that after at least six half-lives, less than 1 percent of the initial amount of radioactive material is still undecayed.

Decay is also used to describe any process in which a system spontaneously changes from a higher-ENERGY state to a lower-energy one. Radioactivity is an example of decay in this sense also, since unstable NUCLEI are spontaneously transforming from higher- to lower-energy states and emitting radiation in the process.

decay process Also called a nuclear decay process, a decay process is the spontaneous transformation of an unstable NUCLEUS (the parent nucleus) into another nucleus (the daughter nucleus) or other nuclei or into a lower-ENERGY state of the same nucleus. Usually a decay process is accompanied by the emission of one or more particles. (*See* RADIOACTIVITY.)

Among the possible decay processes, four are common enough to have been given special names. They are:

ALPHA DECAY. The nucleus emits a helium nucleus, called an ALPHA PARTICLE in this context, and transforms into a nucleus with ATOMIC NUMBER lower by two and ATOMIC MASS lower by four than those of the parent nucleus.

BETA DECAY. The nucleus emits an ELECTRON, also called a BETA PARTICLE, and an electron-type antineutrino, whereby the daughter nucleus has the same atomic mass as does the parent and an atomic number greater by one. (*See* ANTIMATTER; NEUTRINO.)

Inverse BETA DECAY. The nucleus emits a POSITRON, the antiparticle of the electron, and an electron-type neutrino. The daughter nucleus has the same atomic mass as does the parent and an atomic number smaller by one.

GAMMA DECAY. The decaying nucleus is itself the result of some nuclear process and is produced in an excited state, i.e., in a state that possesses higher energy than does the GROUND STATE of that nucleus. The nucleus emits a photon, referred to as a GAMMA RAY, thereby releasing energy and transforming to a lower-energy state, possibly to its ground state. There is no change of atomic number or atomic mass. (*See* NUCLEAR PHYSICS.)

degree of freedom Each possible independent mode of MOTION of a system is called a degree of freedom. A point particle, for example, has three degrees of freedom, which correspond to motions in each of any three independent directions in three-dimensional SPACE, say, the mutually perpendicular x, y, and z directions. A system of N point particles then has three degrees of freedom for each particle, giving a total of $3N$ degrees of freedom for the whole system. (*See* DIMENSION.)

A rigid body of finite size possesses—in addition to the three possible independent translational DISPLACEMENT motions of its CENTER OF MASS that a point particle has also—the possibility of ROTATION about its center of mass. The possible rotational motion about an axis in each of any three independent directions—again, say, the x, y, and z directions—through the body's center of mass adds another three degrees of freedom, for a total of six. Thus a system of N bodies has $6N$ degrees of freedom. The system consisting of the Sun together with the nine planets of the SOLAR SYSTEM possesses $6 \times 10 = 60$ degrees of freedom, six for each of the 10 bodies.

The notion of degree of freedom is used somewhat differently when deriving properties of matter from its atomic makeup, such as when calculating the SPECIFIC HEAT of a GAS according to the KINETIC THEORY. Then a degree of freedom is an independent mode for an atom or molecule to possess a significant amount of ENERGY. Thus, the atoms of a monatomic gas, such as helium or neon, although they can both move in space and rotate about their centers of mass, are each considered to possess only the three translational degrees of freedom. That is because the amounts of rotational kinetic energy they normally acquire are not significant, due to their relatively small MOMENTS OF INERTIA. (*See* ATOM; MOLECULE; STATISTICAL MECHANICS.)

On the other hand, each molecule of a diatomic gas, such as nitrogen (N_2) or carbon monoxide (CO), does have rotational degrees of freedom, but only two. They are possible rotations about any two axes through the center of mass and perpendicular to the molecule's own axis (which is the line connecting the centers of the two atoms), since the molecule's moment of inertia with respect to such axes thereby allows it to possess significant rotational kinetic energy. Rotations about the molecule's axis, however, do not normally endow the molecule with a significant amount of rotational kinetic energy, since its moment of inertia with respect to its axis is too small. There exists an additional degree of freedom due to the molecule's possible vibrational motion along its axis. (Imagine the two atoms connected by a spring and oscillating lengthwise.) Thus a diatomic molecule possesses three translational degrees of freedom, along with two rotational and one vibrational degrees of freedom, for a total of six degrees of freedom. (*See* OSCILLATION; VIBRATION.)

A nonrigid macroscopic body, which can bend, twist, expand, contract, etc., is, for the purpose of dealing with its MECHANICS, considered to possess an infinite number of degrees of freedom. A stretched string, such as on a guitar or a cello, is a simple example. A whole building can serve as another, more complex example. A FIELD, too, such as the electromagnetic field or the gravitational field, is treated as a system with an infinite number of degrees of freedom. Such systems can, among other behaviors, oscillate in various ways and carry WAVES. (*See* ELECTROMAGNETISM; GRAVITATION.)

density In its most common use, the term *density* refers to the ratio of MASS to VOLUME, or volume per unit mass. It is an intrinsic property of a homogeneous material, independent of volume or shape. A material's density is found by dividing the mass of any specimen of the material by the specimen's volume. If a material is not homogeneous, its local density at any point is the mass of an infinitesimal sample of the material at that point divided by the sample's volume. The average density of a body is the body's mass divided by its volume. The SI UNIT of density is kilogram per cubic meter (kg/m^3).

The density of liquid water at 4 degrees Celsius (°C), for example, is 1.000×10^3 kg/m^3. For comparison, the density of Styrofoam, 0.03×10^3 kg/m^3, is only about 1/30th of water's, while that of platinum, 21.5×10^3 kg/m^3, is more than 20 times the density of water. Many gases at everyday TEMPERATURES and at atmospheric PRESSURE have densities of only around 0.1–2 kg/m^3. At the other extreme, the density of nuclear matter that forms an atomic NUCLEUS is about 10^{17} kg/m^3, and that of a BLACK HOLE of one solar mass is around 10^{19} kg/m^3. In the middle range, the average density of the Earth is 5.25×10^3 kg/m^3. (*See* ATOM.)

More generally, a density is the ratio of any spatially distributed physical quantity to a volume, area, or length. If electric CHARGE is distributed over a surface, for example, one speaks of the surface density of charge, the charge per unit area, at a point on the surface. Or the energy content of electromagnetic RADIATION is expressed as energy density, energy per unit volume, at any point in space. Mass, too, can be distributed over a surface or along a line. Paper of 20 pound (lb) "weight," for instance, has a surface mass density, or mass per unit area, of 0.075 kg/m^2. A violin string might have a linear density, mass per unit length, of 0.0003 kg/m. (*See* ELECTRICITY; ELECTROMAGNETISM.)

deuteron This is a particle consisting of one PROTON and one NEUTRON. It forms the NUCLEUS of deuterium, a naturally occurring, stable ISOTOPE of hydrogen, whose ATOMIC NUMBER is 1 and ATOMIC MASS is 2.

diamagnetism The INDUCTION, by an external magnetic FIELD, of atomic MAGNETIC DIPOLES in materials that do not possess them intrinsically is diamagnetism. The material's resulting magnetic dipole moment is oppositely directed to the magnetic field. Such a material is called diamagnetic. Thus a diamagnetic material tends to partially cancel the magnetic field acting on it, so its PERMEABILITY is less than the permeability of the VACUUM, whose value is $4\pi \times 10^{-7}$ T·m/A. The effect is weak, however, so the permeability of a diamagnetic material is not much less than the permeability of the vacuum. A sample of diamagnetic material is (weakly) repelled from a pole of a MAGNET and thus behaves in a manner opposite to that of ferromagnetic and paramagnetic materials, which are attracted to magnetic poles. Actually, all materials are subject to diamagnetism, but the effect is masked by the much stronger effects of FERROMAGNETISM and PARAMAGNETISM. (*See* ATOM; MAGNETISM; MOMENT, MAGNETIC DIPOLE.)

dielectric The term *dielectric* is a synonym for electric INSULATOR, which is a nonconductor of ELECTRICITY. As an historic relic, *dielectric* is the preferred term when referring to the insulating material that separates the plates of a CAPACITOR. A dielectric is characterized by its DIELECTRIC CONSTANT and also by its dielectric strength, which is the greatest magnitude of external electric FIELD it can tolerate without breaking down. (*See* CONDUCTOR; PERMITTIVITY.)

dielectric constant The ratio of the PERMITTIVITY of an electric INSULATOR, or DIELECTRIC, to the permittivity of the vacuum is called the dielectric constant of the material. It is the factor by which the dielectric reduces the magnitude of the electric FORCE between charges immersed in the material. Accordingly, it is the factor by which the material reduces the magnitude of an external electric FIELD within itself. As a result, when a dielectric is used to fill the gap in a CAPACITOR, the CAPACITANCE of the capacitor is increased by the factor of the dielectric constant from the value it would have without the dielectric (i.e., in VACUUM). (*See* COULOMB'S LAW; ELECTRICITY.)

This effect of a dielectric is achieved by means of its electric POLARIZATION, whereby its atomic or molecular ELECTRIC DIPOLEs are aligned by an external electric field. This produces an internal electric field that is oppositely directed to the external field and partially cancels it inside the dielectric. A material that does not normally possess atomic or molecular dipoles might still be polarized by an external electric field. That happens through the field's creating dipoles by causing a separation of positive and negative electric CHARGEs in each of the material's ATOMs or MOLECULEs.

Air under ordinary conditions, for example, has a dielectric constant very close to 1, the dielectric constant of the vacuum. While materials such as rubber, paper, and glass possess dielectric constants in the 2–10 range, water has an especially high dielectric constant of around 80.

diffraction Inseparable from INTERFERENCE and actually caused by it, diffraction is the bending of WAVEs around obstacles. It is a property of waves of all kinds. As an instant, portable demonstration of diffraction of LIGHT, bring two adjacent fingers very close to each other, perhaps touching, so that there are tiny gaps between them, and look through such a gap at the sky or any other light source. If the gap is small enough, you will see a pattern of dark and light stripes or spots, depending on the shape of the gap. That is the diffraction pattern produced by the light bending around the edges of the gap. The bending of waves around an obstacle is the result of constructive interference of nonobstructed waves in directions that would be blocked by straight-line propagation.

Due to diffraction, the shadow of an opaque object is not sharp, but has fringes at its edge, if it is examined in sufficient detail. In particular, the shadow of a circular object, for example, possesses a system of concentric darker and lighter rings extending both outward and inward. At the very center of the shadow is found a bright spot. Again due to diffraction, the image of an opening in an opaque obstacle, when light passes through it and strikes a screen, can have a very complex structure. In this way diffraction limits the RESOLVING POWER, the ability to form sharp images, of any optical system.

As an example, consider the case of light of WAVELENGTH λ passing through a narrow slit of width d and

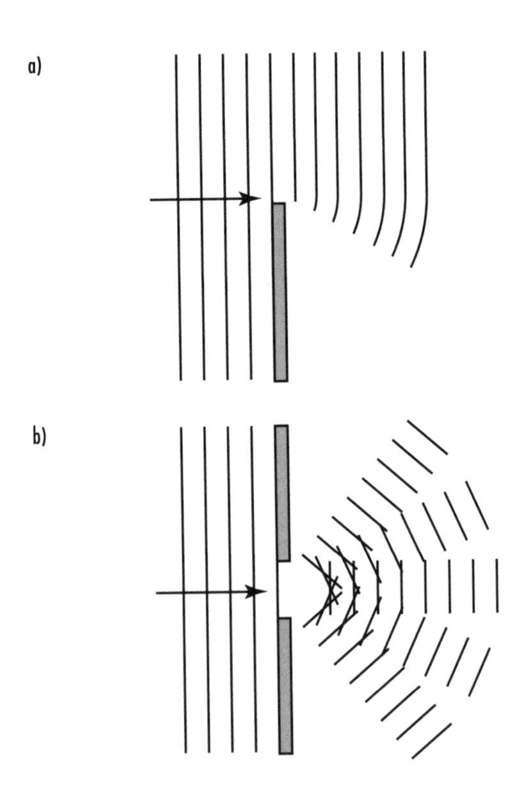

Diffraction is the bending of waves around obstacles. (a) Straight wavefronts moving in the indicated direction bend around a long obstacle. An example is ocean waves impinging on a jetty. (b) Straight wavefronts incident on a slit bend around the edges and fan out in certain directions.

striking a screen at distance L from the slit. If the light propagated in straight lines, the image on the screen would be a faithful image of the slit. However, due to diffraction, the image is spread out in the direction of the slit's width and has dark regions in it. At the center of the diffraction pattern is a strip of width $(2/L)$ (λ/d), which is flanked on both sides by strips of half that width separated by darkness.

Note in the example, and it is true in general, that the strength of the diffraction effect is proportional to the ratio of the wavelength of the waves to the size of the obstacle or of the opening (λ/d in the example). For a given geometry, diffraction becomes more pronounced for longer wavelengths, while for the same wavelength the effect is stronger for smaller obstacles and openings. As an easy demonstration of wavelength dependence, note that you can stand to the side outside an open doorway to a room and hear what is going on in the room while not being able to see inside. The SOUND waves emanating from the room, with wavelengths more or less in the centimeter-meter range, are diffracting around the edge of the doorway. The light waves from the room, on the other hand, whose wavelengths are about a million times shorter, around 5×10^{-7} meter, do not diffract effectively. You can use the finger demonstration to confirm the dependence of diffraction on opening size by varying the size of the gap and observing what happens to the diffraction pattern.

diffusion In one of its senses, diffusion is the spreading-out process that results from the random motion of particles. After the release of perfume, as an example, the odor can soon be detected some distance away. That is due to the diffusion of the perfume MOLECULES through the air molecules by their random wandering as they are knocked around in all directions. Similarly, a drop of milk in a glass of water is soon spread out and diluted. The molecules of the milk travel randomly among the water molecules, continually colliding with them and getting redirected at each collision. In a SOLID, too, diffusion can occur, even though the elementary constituents of a solid—its ATOMS, molecules, or IONS—are relatively fixed in position. Given sufficient time, it is found that foreign particles—say, radioactive tracer atoms—introduced into a solid at a location in it diffuse throughout the material. Since heat is the random motion of particles, its CONDUCTION through matter is by diffusion. (*See* RADIOACTIVITY.)

The rate of net material transfer by diffusion between two points in a diffusive medium is proportional to the difference between the DENSITIES of the material at the two points. In reality, matter is diffusing from each point to the other. But there is more diffusion from a higher concentration to a lower than vice versa, so the net flow is from higher to lower. In this manner, heat flows by conduction from a region of higher TEMPERATURE to one of lower temperature, with the rate of flow proportional to the temperature difference.

An isolated system in which diffusion is taking place will eventually reach a uniform state, with no further net flow occurring. This is a state of EQUILIBRIUM, however, and not a static situation. Diffusion is actually continuing between any pair of locations as before, but the rates of diffusion in both directions are equal, so there is no net diffusion.

Another sense of diffusion is the scattering of LIGHT as it travels through a translucent medium or is reflected from a rough surface. Frosted light bulbs, for instance, produce diffuse light compared with that from clear light bulbs, in which the glowing filament can be seen. In this sense, light is diffused when its direction of propagation is spread out into a range of directions. In situations where sharp shadows are to be avoided, diffuse light is sometimes preferred because it softens the edges of shadows.

dimension The possibility of assigning a measure to something is termed a dimension. For instance, we can assign a measure to an object's position along the x-axis, its x-coordinate, so we have one dimension there. The same holds true for the y- and z-axes, which gives us two more dimensions. An object's x-, y-, and z-coordinates together fully specify the object's location in SPACE. Since three numbers, and only three, are required to specify a location in space, we say that space is three-dimensional. Similarly, only a single number is required to designate when an event occurs. This is another dimension, the dimension of TIME. We can now combine space and time as SPACE-TIME. Four space-time coordinates completely specify where and when an event takes place. So space-time is four-dimensional. (*See* COORDINATE SYSTEM.)

The concept of dimension extends beyond space and time. The possibility of measuring TEMPERATURE, for example, gives us the dimension of temperature. We can speak also of a dimension of MASS, for instance.

Similarly, we have a dimension of FORCE, which can actually be considered to consist of three dimensions, since, as a VECTOR, a force needs three numbers—its three components with respect to some coordinate system—for its full specification.

Here is a further example. As a vector, a particle's VELOCITY has three dimensions. A particle's behavior, according to NEWTON'S LAWS, is fully determined by a specification of both the particle's position and its velocity at any time. Now, consider an abstract space with six axes: the usual x-, y-, z-axes, together with one axis for each component of the particle's velocity, the v_x-, v_y-, v_z-axes. A point in this space represents a state of the particle that completely determines its behavior, while a curve in this space represents the particle's behavior over time: its position and velocity over time. This abstract space is an example of a six-dimensional space. We go even beyond that and consider a system of N interacting particles. By generalizing the previous reasoning, we can describe the behavior of this system by means of an abstract space with six axes for every particle. Here we have a $6N$-dimensional space. For five particles, for instance, it would be a space of 30 dimensions.

Dimensional Analysis

Dimensional analysis is the reduction of dimensions of physical quantities to some basic set and using the results to check the validity of, or even to predict, relations among physical quantities. Here are some examples of dimensional reduction. Velocity is spatial displacement per unit time, so in this approach its dimension can be expressed as [length][time]$^{-1}$, where [. . .] denotes the dimension of whatever is inside the brackets. ACCELERATION is change of velocity per unit time, giving a dimension of ([length][time]$^{-1}$)/[time] = [length][time]$^{-2}$. By Newton's second law of motion, force equals mass times acceleration, so force has the dimension [mass] × ([length][time]$^{-2}$) = [length][time]$^{-2}$[mass]. And so on. (*See* NEWTON'S LAWS.)

As a very simple application of dimensional analysis, we can ask the dimension of the gravitational constant G in Newton's law of gravitation:

$$F = Gm_1m_2/r^2$$

where F denotes the magnitude of the force of attraction that two point bodies exert on each other, m_1 and m_2 are the bodies' masses, and r is the distance between the bodies. Dimensional analysis is based on the equality of the dimensions of the quantities on either side of an equation. Thus:

$$[F] = [G][m_1m_2/r^2]$$

For the left-hand side, as we saw earlier:

$$[F] = [\text{length}][\text{time}]^{-2}[\text{mass}]$$

The dimension of the right-hand side is:

$$[G][m_1m_2/r^2] = [G][\text{length}]^{-2}[\text{mass}]^2$$

We equate the dimensions of the two sides:

$$[\text{length}][\text{time}]^{-2}[\text{mass}] = [G][\text{length}]^{-2}[\text{mass}]^2$$

and solve for $[G]$. The result is:

$$[G] = ([\text{length}][\text{time}]^{-2}[\text{mass}])/([\text{length}]^{-2}[\text{mass}]^2)$$
$$= [\text{length}]^3[\text{time}]^{-2}[\text{mass}]^{-1}$$

(*See* GRAVITATION.)

A dimensionless quantity is one whose dimensional reduction yields unity. Such a quantity takes the same value in every system of UNITs. The ratio of the MASS of the PROTON to that of the ELECTRON equals about 1,836. This is a dimensionless quantity and has the same value whether the masses are both expressed in kilograms, in grams, in ounces, etc.

dipole *See* ELECTRIC DIPOLE; MAGNETIC DIPOLE.

Dirac, Paul Adrien Maurice (1902–1984) British *Physicist* Best known among physicists for the equation that bears his name, Paul Dirac was led by the DIRAC EQUATION to the prediction of the existence of antiparticles. His first papers in THEORETICAL PHYSICS were completed in 1925, only a year after commencing his studies as a research student at the University of Cambridge, England. Dirac received his doctorate in physics in 1926. QUANTUM PHYSICS and the connection between QUANTUM MECHANICS and the special theory of relativity formed the focus of Dirac's investigations for the rest of his life. He developed the Dirac equation, which governs the quantum mechanics of ELECTRONS and other ELEMENTARY PARTICLES of SPIN 1/2 that possess MASS. Dirac's attempt to understand the implications of his equation led to the prediction of antiparticles, such as the POSITRON. Most of Dirac's academic life was spent at the University of Cambridge. He shared the 1933 Nobel Prize in physics with ERWIN

Paul Dirac, shown lecturing on the hydrogen molecule, is famous for the Dirac equation, which implied the existence of antiparticles before they were discovered experimentally. He shared the 1933 Nobel Prize in physics. ***(AIP Emilio Segrè Visual Archives)***

SCHRÖDINGER "for the discovery of new productive forms of atomic theory."

See also ANTIMATTER; RELATIVITY, SPECIAL THEORY OF.

Dirac equation In the early 20th century, British physicist PAUL ADRIEN MAURICE DIRAC, in his attempt to wed the principles of QUANTUM MECHANICS with those of the special theory of relativity, devised the equation that bears his name. This equation describes an electrically charged point particle that has SPIN 1/2, such as the ELECTRON. The Dirac equation predicted the existence of the electron's antiparticle, the POSITRON.

See also ANTIMATTER; CHARGE; ELECTRICITY; RELATIVITY, SPECIAL THEORY OF.

disorder As the term implies, *disorder* is the lack of ORDER. That means the lack of structure, organization, hierarchy, differences, discrimination, and correlation. Conversely, disorder has to do with disorganization, anarchy, equivalence, equality, and independence. Disorder is closely related to the notion of randomness. It is often associated with CHAOS, at least in nontechnical usage, but the latter has a different technical meaning. Disorder is a relative concept. Just as there is no absolute order, there is no absolute disorder. Thus, situations can be compared as to their degrees of disorder. Consider liquid water and ice, two different states of the same substance, whose chemical symbol is H_2O. Whereas the water MOLECULEs are moving about randomly in the LIQUID, in the SOLID, which is a CRYSTAL, they are constrained to an orderly array of positions. There is considerably more organization and structure in the solid than in the liquid. So liquid water has a higher degree of disorder than does ice.

Imagine a sample of GAS in a container, as another example. In one state, the gas is uniform, with the same TEMPERATURE in all parts of the container. In another state, the faster-moving molecules happen to be in one half of the container and the slower ones in the other, so that the two halves are at different temperatures. (The latter situation is not one of EQUILIBRIUM and will quickly evolve into a uniform state.) The nonuniform state possesses differences and discrimination that the uniform state does not have. So the uniform state is the more disordered.

Disorder is directly related to ENTROPY and to SYMMETRY.

See also STATISTICAL MECHANICS; THERMODYNAMICS.

dispersion The dependence of the SPEED of propagation of a WAVE on the FREQUENCY (or, equivalently, on the WAVELENGTH) of the wave is called dispersion. Thus, in a dispersive medium, waves of different frequencies travel at different speeds. A mathematical expression of dispersion is called a dispersion relation. In the case of LIGHT waves, that is manifested in the dependence of the INDEX OF REFRACTION on the frequency (or wavelength). And this, in turn, means that effects that depend on the index of refraction are color-dependent. In particular, the angle of REFRACTION of a beam of light as it passes from one medium to another depends on the indexes of refraction of the media, according to SNELL'S LAW. Since the speed of propagation of light in glass, for example, is indeed frequency dependent, a glass LENS focuses different colors of light differently. A glass PRISM can be used to separate out the various colors (or, equivalently, frequencies or wavelengths) that compose a beam of light by bending each color through a different angle, thus creating a SPECTRUM.

See also ABERRATION.

displacement Displacement is the separation between two points in SPACE expressed as a VECTOR. Specifically, the displacement of point B from point A is the vector whose base is at A and whose head is at B. In other words, the displacement vector between B and A is a vector whose magnitude is the distance between A and B and whose direction is the direction of B from A.

The position vector, associated with a point in space, is the vector whose magnitude equals the point's distance from the origin of the COORDINATE SYSTEM and whose direction is the direction of the point as viewed from the origin. In terms of the position vectors of points A and B, which we denote by $\mathbf{r}_A$ and $\mathbf{r}_B$, respectively, the displacement of B from A, which we denote by $\mathbf{r}_{BA}$, is the vector difference:

$$\mathbf{r}_{BA} = \mathbf{r}_B - \mathbf{r}_A$$

Alternatively, the position vector of point B is obtained by vectorially adding the displacement vector to the position vector of A:

$$\mathbf{r}_B = \mathbf{r}_A + \mathbf{r}_{BA}$$

Note that, whereas position vectors depend on the location of the coordinate origin, displacement vectors are origin independent.

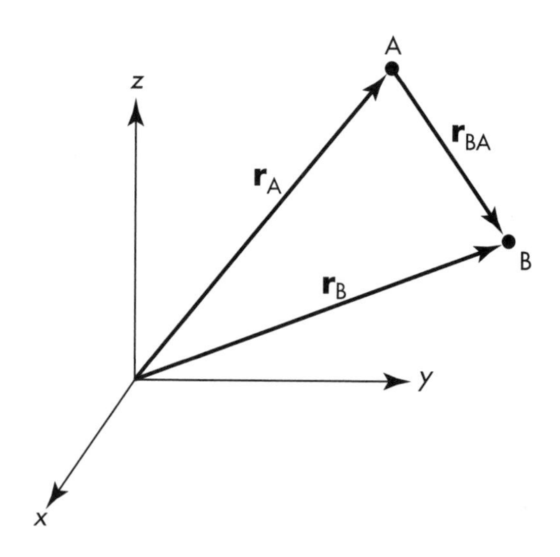

The displacement of point B from point A is specified by the displacement vector $\mathbf{r}_{BA}$, pointing from A to B. This vector is the vector difference between $\mathbf{r}_B$ and $\mathbf{r}_A$, the respective position vectors of points B and A: $\mathbf{r}_{BA} = \mathbf{r}_B - \mathbf{r}_A$. Whereas position vectors depend on the location of the coordinate system origin, displacement vectors are origin-independent.

In similar vein, an angular displacement can be defined as the difference between two angles designating orientation about the same fixed axis:

$$\theta_{BA} = \theta_B - \theta_A$$

and:

$$\theta_B = \theta_A + \theta_{BA}$$

The concept of displacement plays an essential role in the definition of VELOCITY, SPEED, and ANGULAR SPEED. The term is also generalized to refer to TIME, where it means the temporal separation of two instants.

See also KINEMATICS; ROTATION.

dissipation The conversion of nonthermal forms of ENERGY into HEAT is termed dissipation. Dissipation is not a REVERSIBLE PROCESS. That is because, although nonthermal forms of energy can, at least theoretically, be converted completely into heat, the total conversion of heat to nonthermal energy forms is impossible in principle, according to the second law of THERMODYNAMICS.

In mechanical systems dissipation typically occurs through FORCEs of FRICTION and VISCOSITY, referred to as dissipative forces. The initial POTENTIAL ENERGY of a PENDULUM that is displaced and released from rest is gradually dissipated through bearing friction and air viscosity, and the pendulum eventually comes to rest in its EQUILIBRIUM position. Dissipative forces always perform negative work, thus reducing the amount of mechanical energy, which is the sum of potential and KINETIC ENERGIES. In electric systems, electric energy is dissipated by the passage of electric CURRENTs through RESISTANCEs, which generates heat. (*See* ELECTRICITY; MECHANICS.)

Dissipation occurring during the propagation of energy through a medium results in ABSORPTION of the propagating energy and thus in attenuation of the INTENSITY of propagation.

See also CONSERVATIVE FORCE.

domain, magnetic *See* MAGNETIC DOMAIN.

Doppler effect The Doppler effect is the dependence of the observed FREQUENCY of a periodic WAVE on the MOTION of the wave source and of the observer toward or away from each other. The Doppler effect is named for the Austrian physicist Christian Johann Doppler (1803–53). Motions that do not alter the distance between the source and the observer do not create a Doppler effect. Any change in frequency due to the Doppler effect is called a Doppler shift.

The effect can be heard whenever a honking car, whistling train locomotive, or siren-sounding emergency vehicle passes by. The pitch we hear is higher as the source approaches, then quickly shifts to lower as the source recedes. One use of the Doppler effect is to detect motion and measure its SPEED. For that, the observer emits radio or LIGHT waves, which are reflected back by the object under observation. Any difference between the frequency of the returning wave and that of the emitted wave indicates that the object is in motion. If the former is higher or lower than the latter, the object is moving, respectively, toward or away from the observer. The speed of the object's approach or recession can be calculated from the magnitude of frequency difference. (*See* REFLECTION.)

Let f_s denote the frequency of the wave source in hertz (Hz). For illustrative purposes let us imagine that the source is emitting a sequence of pulses at this frequency (i.e., f_s pulses per unit time). (What it is really emitting at this frequency are cycles of OSCILLATION.) As each pulse is emitted, it leaves the source at the wave propagation speed determined by the medium. The distance between adjacent pulses propagating through the medium represents the WAVELENGTH of the wave.

Moving Source

If the observer is at rest with respect to the medium of wave propagation and the source is moving away from the observer, the source is emitting pulses at increasingly greater distances from the observer. That increases the distance between adjacent pulses propagating toward the observer and thus "stretches" the wavelength of the wave reaching the observer. There exists a fundamental relation for all periodic waves:

$$f\lambda = v$$

where f, λ, and v denote, respectively, the frequency, the wavelength in meters (m), and the propagation speed in meters per second (m/s) of the wave. According to this relation, for a fixed propagation speed, the frequency and wavelength are inversely proportional to each other. Thus, an increase in wavelength causes a lowering of the frequency, which results in the observer's detecting a frequency that is lower than that of the source, f_s. On the other hand, if the source is moving toward the observer, the wavelength is correspondingly "squeezed," resulting in an observed frequency higher than f_s.

Note that if the source approaches the observer at the same speed as the wave propagates in the medium, the wavelength is squeezed down to zero, since the source is then emitting pulses on top of each other. All the pulses reach the observer at the same time. This phenomenon is known as a SHOCK WAVE.

Moving Observer and Combined Effect

Now let the source remain at rest with respect to the medium, while the observer moves toward it. While at rest, the observer receives pulses at the frequency of the source, f_s. Moving toward the source, the observer "intercepts" the next pulse before it would otherwise have reached him, which decreases the time interval between receipt of adjacent pulses compared to when he is at rest. That causes the observer to receive more pulses per unit time than the source is emitting, which means an observed frequency higher than f_s. If the observer is moving away from the source, the pulses have to "chase" him. He then receives pulses at greater time intervals than if he were at rest, meaning fewer pulses per unit time than the source is emitting. That translates into an observed frequency lower than f_s.

Note that if an observer is moving away from the source at the speed of wave propagation or greater, a wave emitted while this motion is occurring will never overtake her. She will then observe no wave at all.

When both the source and the observer are moving with respect to the medium, the Doppler effects are combined. That might enhance the effect or diminish it. Assume that the source and observer are moving, if they are moving at all, along the same straight line. Assume also that the source is moving away from the observer at speed v_s in meters per second (m/s) with respect to the medium and that the observer is moving toward the source at speed v_o in meters per second (m/s) with respect to the medium. Let f_s denote the frequency of the source, as above, and f_o the observed frequency. Then the observed frequency relates to the emitted frequency as follows:

$$f_o = f_s \frac{1 + v_o / v}{1 + v_s / v}$$

The speeds v_s and v_o are taken as positive when the motions are in the directions specified. If either motion is in the opposite direction, the corresponding speed is taken as negative. This formula is valid only if the observer is actually receiving waves from the source. Then all the effects mentioned above follow from the formula. When the source and observer are moving in such a way that the distance between them remains constant, i.e., when $v_o = v_s$, the effects cancel each other and there is no net Doppler effect; the observer observes the frequency of the source, f_s.

If v_o and v_s are both small compared with the speed of wave propagation, v, the formula for the Doppler effect takes the form:

$$f_o \approx f_s (1 + u/v)$$

Here u denotes the relative speed of the source and observer in meters per second (m/s) and takes positive values when they are approaching each other and negative values when they are moving apart.

Red and Blue Shifts

In the case of light or any other electromagnetic RADIATION, there is no material medium of wave propagation. That requires a modification of the above considerations. The qualitative conclusions, however, are similar. When the source is receding from the observer, the observed frequency is lower than that of the source. This is called a RED SHIFT, since lowering the frequency (or equivalently, increasing the wavelength) of visible light shifts it toward the red end of the SPECTRUM. The light, or other electromagnetic radiation, is then said to be red-shifted. On the other hand, a blue shift results when the source is approaching the observer. The formula that relates the observed frequency to the source frequency for electromagnetic waves is:

$$f_o = f_s \sqrt{\frac{1 + u/c}{1 - u/c}}$$

As before, u denotes the relative speed of the source and observer, with the same sign convention, and c is the SPEED OF LIGHT, whose value is 2.99792458×10^8 meters per second (m/s). For u small compared to c, this formula reduces to a form similar to the small-speed formula presented earlier:

$$f_o \approx f_s(1 + u/c)$$

(*See* ELECTROMAGNETISM.)

According to the above considerations, motion of the source perpendicular to the line connecting the observer and the source should not produce a Doppler effect, since the distance between the two is not changing. Nevertheless, due to the relativistic effect of TIME DILATION, the source's motion in any direction whatsoever causes a decrease in its frequency, as observed by the observer. This effect is termed the relativistic Doppler effect, or relativistic red shift. It becomes significant only at speeds that are a sufficiently large fraction of the speed of light. If the source is moving at speed u in a direction perpendicular to the line connecting it to the observer, then the observed frequency and source frequency are related by:

$$f_o = f_s\sqrt{1 - u^2 / c^2}$$

The low-speed approximation for this formula is:

$$f_o \approx f_s\left(1 - \frac{u^2}{2c^2}\right)$$

(*See* RELATIVITY, SPECIAL THEORY OF.)

drag The retarding effect of FORCES of FRICTION and VISCOSITY acting on a body moving through a FLUID is called drag. Such forces are called drag forces. Those forces are dissipative, performing negative work on the body, and thus reducing its mechanical energy (the sum of its KINETIC and POTENTIAL ENERGIES) by converting it irreversibly to HEAT. (*See* DISSIPATION; REVERSIBLE PROCESS.)

As an example, for a blunt body moving sufficiently rapidly through air, the magnitude of the drag force is given by:

$$\text{Drag force} = D\rho Av^2/2$$

where the drag force is in newtons (N), ρ denotes the DENSITY of the air in kilograms per cubic meter (kg/m^3), A is the effective cross-section area of the body in square meters (m^2), and v is the body's speed in meters per second (m/s). The symbol D denotes the drag coefficient, which is dimensionless. (*See* DIMENSION.)

dynamics In MECHANICS, dynamics is the study of the causes of MOTION and of its change, i.e., the study of the effect of FORCES and TORQUES on motion. More generally, dynamics is the study of the evolution of physical systems and its causes.

Classical mechanical dynamics is based on NEWTON'S LAWS of motion, especially on his second law of motion, which relates an applied force, as a cause, to a change in MOMENTUM or to an ACCELERATION, as an effect. For classical ELECTROMAGNETISM, it is MAXWELL'S EQUATIONS that describe the dynamics. And the dynamics of classical systems involving both mechanics and electromagnetism is comprehended through Newton's laws, Maxwell's equations, and the LORENTZ FORCE. (*See* CLASSICAL PHYSICS; NEWTON, ISAAC.)

The dynamics of physical systems is described more precisely by QUANTUM MECHANICS and QUANTUM ELECTRODYNAMICS. The dynamics of the universe is the subject of COSMOLOGY.

See also KINEMATICS; STATICS.

E

Eddington, Arthur Stanley (1882–1944) British *Astronomer, Astrophysicist* Among the first to grasp the significance of ALBERT EINSTEIN's general theory of relativity, Arthur Eddington is famous for his experimental confirmation of a prediction of the theory. He received an M.A. degree from the University of Cambridge, England, in 1905. Eddington's research was mainly in the areas of stellar structure and evolution and the general theory of relativity. In 1919, he led an expedition to western Africa for the purpose of photographing stars in the Sun's vicinity during a total solar eclipse. Comparison of the stars' apparent positions in his photographs with their positions measured in ordinary circumstances confirmed Einstein's prediction about the bending of light rays (from the stars, in this case) as they pass close to a massive object (the Sun). After working at the Royal Observatory in Greenwich, England, he joined the University of Cambridge as professor of astronomy and remained there for the rest of his life.

See also RELATIVITY, GENERAL THEORY OF.

Arthur Eddington, an astronomer and astrophysicist, is famous for experimentally confirming in 1919 the prediction of the general theory of relativity that light (from stars, in his measurements) is deflected as it passes near a massive body (the Sun). ***(AIP Emilio Segrè Visual Archives)***

Einstein, Albert (1879–1955) German/Swiss/American *Physicist* The best-known of all physicists, Albert Einstein is most famous for his theories of relativity. He received his doctorate in THEORETICAL PHYSICS from the University of Zurich, Switzerland, in 1905. While holding various temporary jobs, including that of a technical expert in the Swiss patent office, Einstein continued his fruitful investigations in theoretical physics. From 1909, he held a number of professorial positions at various European institutions, and in 1932, he joined the Institute for Advanced Study at Princeton, New Jersey, as a researcher and remained there until his death. Besides developing his theories of relativity, Einstein did theoretical research in QUANTUM PHYSICS (including the PHOTOELECTRIC EFFECT) and STATISTICAL MECHANICS (including BROWNIAN MOTION). His work during his Princeton years was devoted to unsuccessful attempts to unify the laws of physics. Einstein made use of his celebrity to speak out on social issues, for international peace, and in support of the State of Israel. He was even offered the presidency of Israel (which he declined). In 1921, Einstein was awarded the Nobel Prize in physics, not for relativity, but "for his services to theoretical physics, and especially for his discovery of the law of the photoelectric effect." It is generally agreed that, of all physicists, Einstein has had the greatest influence on modern physics.

See also LAWS OF NATURE; RELATIVITY, GENERAL THEORY OF; RELATIVITY, SPECIAL THEORY OF.

Albert Einstein, the best-known of all physicists, is famous for his theories of relativity. Among physicists, he is known also for his theoretical work in quantum physics and in statistical mechanics. He was awarded the 1921 Nobel Prize in physics. ***(AIP Emilio Segrè Visual Archives)***

elasticity The term *elasticity* describes the tendency of materials to resist deformation and, after being deformed, to regain their original size and shape after the cause of deformation is removed. Here we are referring to SOLID materials, although LIQUIDs and GASes possess elastic properties as well, with regard to VOLUME change. The cause of deformation is called STRESS, while the response of the material to stress, the deformation itself, is STRAIN. Many materials do indeed return to their initial configuration, at least if the stress is not too large. These are termed elastic materials. Examples are steel, wood, some plastic, bone, and rubber. Materials that do not return to their original dimensions, such as chewing gum, caulk, putty, and ice cream, are designated as plastic materials.

The behavior of elastic materials under increasing stress is typically as follows. For a range of stress from zero up to some limit, called the proportionality limit, the strain is proportional to the stress. The extension of a coil spring, for example, is proportional to the stretching FORCE, up to some maximal force. The proportionality of strain to stress is the content of HOOKE'S LAW, so materials with a significantly high proportionality limit can be said to obey Hooke's law (with the understanding that above the proportionality limit, Hooke's law becomes invalid for them). Although many elastic materials do obey Hooke's law, some, such as rubber, do not, or do so only approximately.

Materials that obey Hooke's law can be characterized by a number of quantities called moduli of elasticity, all of which are defined as the ratio of the stress to the resulting strain:

$$\text{Modulus} = \frac{\text{stress}}{\text{strain}}$$

In all cases, the stress is force per unit area and its UNIT is the pascal (Pa), equivalent to newtons per square meter (N/m^2). The strain is always a dimensionless ratio of a change in a quantity to the value of the quantity. So the unit of the modulus of elasticity is the same as that of stress, the pascal. (*See* DIMENSION.)

For linear deformation caused by TENSION or compression, Young's modulus is the ratio of the tensile stress, the longitudinal force per unit cross-section area, to the tensile strain, the relative change in length. If a longitudinal force of magnitude F causes a length change ΔL in a length L, then Young's modulus, Y, is defined as:

$$Y = \frac{F/A}{\Delta L/L}$$

where A denotes the cross-section area. Here F is in newtons (N), A in square meters (m^2), and ΔL and L are in the same units of length. F and ΔL are taken to have the same algebraic sign.

For volume change under pressure, the bulk modulus, B, is defined as the ratio of the PRESSURE, which is the bulk stress, to the relative change of volume, the bulk strain:

$$B = -\frac{p}{\Delta V/V}$$

Here p denotes the pressure in pascals (Pa), V the volume, and ΔV the volume change, where ΔV and V are in the same units of volume. A minus sign is introduced to make B positive, since a positive pressure causes a decrease of volume (i.e., a negative value for ΔV).

An additional modulus is the shear modulus, G, for SHEAR deformation. It is defined as the ratio of the shear stress, the force per unit parallel area, to the shear strain, the angle of deformation:

$$G = \frac{F/A}{\gamma}$$

Here F denotes the shear force in newtons (N), A the area parallel to the force over which the force is acting, in square meters (m^2), and γ the deformation angle in radians (rad).

For a range of stress above the proportionality limit, although the strain is not proportional to the stress, the material still behaves elastically. If the stress is removed, the material will still regain its original configuration. That holds for stresses up to the elastic limit. Beyond the elastic limit, the ATOMS or MOLECULES that constitute the material rearrange themselves, and the deformation becomes permanent. In this case, the material behaves in a plastic manner. As an example, think of a paper clip. In normal use it stays within its elastic limit and successfully holds sheets of paper together. If used to clip too many sheets, it becomes bent open and no longer does its job.

Above the elastic limit is the yield point, the stress at which the material starts to flow, its strain increasing over time with no further increase of stress. That leads to the breaking point, the stress at which the material ruptures.

A POTENTIAL ENERGY can be defined for elastic deformation, since the deforming force is then a CONSERVATIVE FORCE. That is especially simple as long as Hooke's law is obeyed. For a spring, for instance, the applied force, of magnitude F, in newtons (N) and the spring's resulting displacement, x, in meters (m) are related by:

$$F = kx$$

for any force below the proportionality limit, where k is the spring constant in newtons per meter (N/m). The corresponding potential energy, E_p, in joules (J) of a spring that has been stretched or compressed through displacement x is:

$$E_p = kx^2/2$$

electric dipole A pair of equal-magnitude and opposite electric CHARGEs at some distance from each other, or a charge configuration that is equivalent to that, is called an electric dipole. A water MOLECULE is an electric dipole, since the oxygen ATOM's affinity for electrons brings about a concentration of negative charge in its vicinity, leaving a region of net positive charge where the hydrogen atoms are. Because of the geometric structure of the water molecule (specifically, that the three atoms do not lie on a straight line), charge separation creates a natural electric dipole. In situations where there normally is no electric dipole, one might be created by an electric field, which tends to pull positive charges one way and negative charges the other way, producing charge separation. An electric dipole is characterized by its electric dipole moment. Since an electric dipole is electrically neutral, it suffers no net FORCE in a uniform electric FIELD. Rather, a uniform electric field exerts a TORQUE on an electric dipole, whose magnitude depends on the dipole's orientation in the field. In a nonuniform field, however, a net force can act on a dipole, in addition to the torque.

See also ELECTRICITY; MOMENT, ELECTRIC DIPOLE.

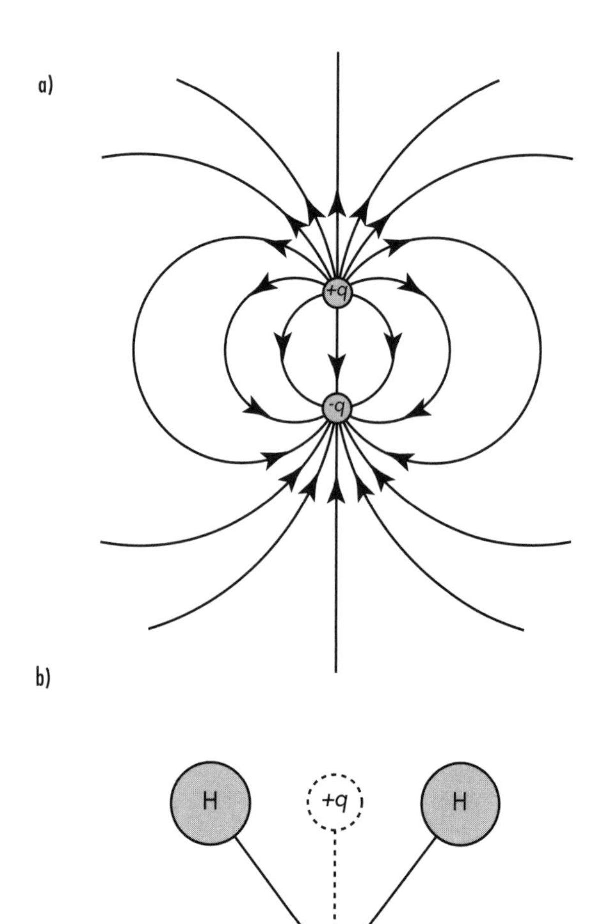

An electric dipole is a pair of separated equal-magnitude opposite electric charges or a situation equivalent to that. (a) An idealized electric dipole comprising charges +*q* and −*q*. A number of electric field lines are shown for it. (b) A water molecule, H_2O, which is a natural electric dipole. There is a concentration of negative charge at the oxygen atom and a center of positive charge between the hydrogen atoms. The equivalent electric dipole is shown in dashed lines.

electric dipole moment *See* MOMENT, ELECTRIC DIPOLE.

electricity The phenomenon of electricity is based on the existence in nature of electric CHARGEs, which exert FORCES on each other in accord with COULOMB'S LAW. At the level of ELEMENTARY PARTICLEs, from which all MATTER is formed, it is found, for example, that an ELECTRON and a PROTON attract each other, electrons repel each other, as do protons, while neutrons are immune to the effect. Coulomb's law is found to be valid for those forces when electrons are assigned one fundamental unit of negative electric charge, protons one fundamental unit of positive electric charge, and neutrons are declared neutral (i.e., free of electric charge).

Force and Field

Electric forces are described by means of the electric FIELD, which is a VECTOR quantity that possesses a value at every location in SPACE and can vary over time. The electric field is considered to mediate the electric force, in the sense that any charge, as a source charge, contributes to the electric field, while any charge, as a test charge, is affected by the field in a manner that causes a force to act on the charge. The value of the electric field at any location is defined as the magnitude and direction of the force on a positive unit test charge at that location. Its SI UNIT is newton per coulomb (N/C) or volt per meter (V/m).

The force, **F**, in newtons (N) that acts on an electric charge, *q*, in coulombs (C) as a test charge, at a location where the electric field has the value **E**, is:

$$\mathbf{F} = q\mathbf{E}$$

The contribution that a charge *q*, as a source charge, makes to the electric field at any point, called the field point, is as follows. The magnitude of the contribution, *E*, is given by:

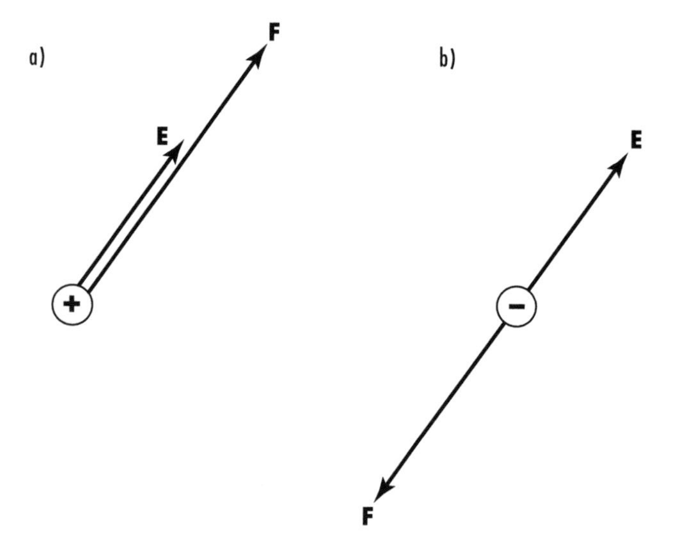

The direction of the force **F** acting on a point test charge in an electric field **E**. (a) When the test charge is positive, the force is in the direction of the field. (b) The force is directed opposite to the field in the case of a negative test charge.

$$E = \frac{1}{4\pi\kappa\varepsilon_0}\frac{|q|}{r^2}$$

where $|q|$ denotes the magnitude of the source charge (i.e., its absolute value, ignoring its sign), and r is the distance from the location of the charge to the field point. The constant ε_0 denotes the PERMITTIVITY of the vacuum, whose value is $8.85418781762 \times 10^{-12}$ $C^2/(N \cdot m^2)$, κ is the DIELECTRIC CONSTANT of the medium in which all this is occurring, and $\varepsilon = \kappa\varepsilon_0$ is the permittivity of the medium. The direction of the contribution is away from the source charge if the latter is positive, and it is toward the source charge if it is a negative charge. The total electric field is the vector sum of the contributions from all the source charges. Note that in calculating the electric field and the force on a test charge, the test charge cannot at the same time serve as a source charge too. In other words, a charge cannot affect itself. The relation between the electric field and the source charges that produce it is also expressed by GAUSS'S LAW.

An electric field line is a directed line in space whose direction at every point on it is the direction of the electric field at that point. Electric field lines make a very useful device for describing the spatial configurations of electric fields. The field lines for the electric field produced by a lone positive point charge, for example, are straight lines emanating from the charge and directed away from it. Only a single electric field line can pass through any point in space.

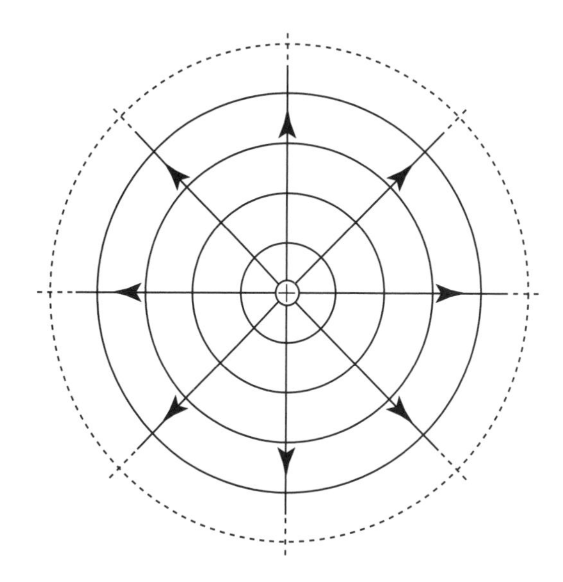

Electric field lines and equipotential surfaces for a positive point source charge, viewed in cross section. The outward-pointing radii indicate a number of field lines, and the concentric circles represent several equipotential surfaces, which are in reality concentric spherical surfaces. For a negative point source charge, the picture is similar, but with the field lines pointing inward.

Energy and Potential

The electric force is a CONSERVATIVE FORCE, so a POTENTIAL ENERGY can be defined for it: the potential energy of a charge at any location is the WORK required to move the charge from infinity to that location. From this, the electric potential, commonly referred to as the potential, at every location is defined as the potential energy of a positive unit test charge at that location. The electric potential is thus a SCALAR field. The potential difference, or VOLTAGE, between two locations is the algebraic difference of the values of the potential at the locations. The SI unit of potential and of potential difference is the volt (V). (*See* POTENTIAL, ELECTRIC.)

The potential energy, E_p, in joules (J) of a test charge, q, at a location where the potential has the value V is:

$$E_p = qV$$

The contribution, V, that a source charge, q, makes to the potential at any field point is:

$$V = \frac{1}{4\pi\kappa\varepsilon_0}\frac{q}{r}$$

where the symbols have the same meanings as given in the discussion of "Force and Field." Note that the signed charge appears in this formula. The total potential is the algebraic sum of the contributions from all source charges. As in the case of the electric field and its effect, here too, a test charge cannot also serve as a source charge at the same time.

A useful device for describing the spatial configuration of the electric potential is the EQUIPOTENTIAL SURFACE. This is a surface in space such that, at all points on it, the electric potential has the same value. The equipotential surfaces of a point source charge, as an example, are spherical shells centered on the charge. For any point in space, the electric field line passing through the point is perpendicular to the equipotential surface containing the point. A quantitative relation between the electric field and equipotential surfaces is that, for any pair of close equipotential surfaces: (1) the direction of the electric field at any point between them is perpendicular to them and pointing from the higher-potential surface to the lower, and (2) the magnitude of the electric field, E, at such a point is given by:

$$E = |\Delta V/\Delta s|$$

where ΔV and Δs are, respectively, the potential difference between the equipotential surfaces and the (small) perpendicular distance between them in meters (m).

Energy is stored in the electric field itself. The energy DENSITY (i.e., the energy per unit volume) in joules per cubic meter (J/m^3), at a location where the magnitude of the electric field is E, is given by:

$$\text{Energy density of electric field} = \varepsilon_0 E^2/2$$

Electric Dipole

A pair of equal-magnitude, opposite electric charges at some distance from each other forms an ELECTRIC DIPOLE, characterized by its electric dipole moment, **p**. The electric dipole moment is a vector pointing in the direction from the negative charge to the positive one, whose magnitude, p, in units of coulomb·meters (C·m), is given by:

$$p = |q|d$$

where $|q|$ denotes the magnitude of either of the charges in coulombs (C), and d is the distance between the charges in meters (m). An electric field causes a TORQUE to act on an electric dipole:

$$\boldsymbol{\tau} = \mathbf{p} \times \mathbf{E}$$

where $\boldsymbol{\tau}$ denotes the torque in units of newton·meters (N·m). A potential energy is associated with an electric dipole in an electric field. Its value is:

$$E_p = -\mathbf{p} \cdot \mathbf{E}$$

It is lowest when the dipole moment is aligned with the electric field (parallel), and it is highest when the dipole moment is pointing in the opposite direction (antiparallel). The potential-energy difference between the two orientations is $2pE$. (*See* MOMENT, ELECTRIC DIPOLE.)

The subfield of electricity that deals with charges at rest is called electrostatics. Charges may be in motion, however, and moving charges form electric CURRENTS. The additional forces, over and above the electric forces, that affect moving charges, or currents, due to the charges' motion are the subject of MAGNETISM.

Electric Current

Electric devices, such as BATTERIES, electric generators, RESISTORS, INDUCTORS, CAPACITORS, and TRANSISTORS, connected together by CONDUCTORS through which currents might flow, make up electric circuits. The analysis of electric circuits is performed with the help of such as OHM'S LAW and KIRCHHOFF'S RULES. Direct current (DC) is the situation when the currents flowing in a circuit maintain a constant direction or, if currents are not flowing, the voltages involved maintain a constant sign. When the voltages and currents in a circuit undergo regular reversals, the situation is one of alternating current (AC). In more detail, alternating current involves a sinusoidal variation in time of the voltages and currents of a circuit at some FREQUENCY, f, in hertz (Hz), whereby any voltage, V, or current, i, at time t in seconds (s) is given by:

$$V = V_0 \sin(2\pi ft + \alpha)$$
$$i = i_0 \sin(2\pi ft + \beta)$$

Here V_0 and i_0 denote the maximal values of the time-varying quantities V and i, respectively, which thus vary from V_0 and i_0 to $-V_0$ and $-i_0$ and back again f times per second. The parameters α and β, in radians (rad), denote possible phase shifts, whereby the varying voltage or current reaches its maximum at a different point in the cycle, or at a different PHASE, than does some reference voltage or current in the circuit. (*See* GENERATOR, ELECTRIC.)

Electricity and magnetism are two aspects of ELECTROMAGNETISM, which subsumes the two and, moreover, has them affecting each other.

electromagnet A device that acts as a MAGNET due to an electric CURRENT flowing through it is an electromagnet. Typically it is constructed of a coil of conducting wire wound around a core of ferromagnetic material, such that a magnetic FIELD results when an electric current flows through the coil. In the case of a long helical coil, called a solenoid, the magnitude of the magnetic field inside the coil is approximately:

$$B = \mu ni$$

and at its ends is about:

$$B = \mu ni/2$$

where B denotes the magnitude of the magnetic field in teslas (T), i the current in amperes (A), n the number of turns per unit length of coil in inverse meters (m^{-1}), and μ the PERMEABILITY of the core material in tesla·meters per ampere (T·m/A). In order to achieve especially high magnetic fields, the coil can be made of a superconductor, allowing the flow of very high currents.

See also CONDUCTION; CONDUCTOR; ELECTRICITY; FERROMAGNETISM; MAGNETISM; SUPERCONDUCTIVITY.

electromagnetic interaction The effect that matter has upon matter due to ELECTROMAGNETISM is termed the electromagnetic interaction. It is described by the theory of QUANTUM ELECTRODYNAMICS (QED), which is the quantum theory of electromagnetism and how it affects and is affected by MATTER. According to QED, the electromagnetic interaction is mediated by massless particles called PHOTONs, which are exchanged by the interacting ELEMENTARY PARTICLEs. Although the electromagnetic interaction weakens with distance, its range is infinite, which relates to the masslessness of the photon. (*See* MASS; QUANTUM PHYSICS.)

The electromagnetic interaction is one of the four presently recognized fundamental interactions, which also include GRAVITATION, the STRONG INTERACTION, and the WEAK INTERACTION. In the 19th century, JAMES CLERK MAXWELL unified ELECTRICITY and MAGNETISM to electromagnetism by showing that they are but two aspects of the latter. Similarly, in the 20th century, the electromagnetic and weak interactions were recognized as being derived from the ELECTROWEAK INTERACTION. Whereas electricity and magnetism are intimately involved with each other, the electromagnetic and weak interactions are quite distinct. For example, the three intermediating particles of the weak interaction possess mass, and two of them are electrically charged, while the photon is massless and electrically neutral. Related to this mass difference, the range of the weak interaction is finite (while that of the electromagnetic interaction is infinite). Moreover, the strengths of the two interactions are quite disparate. On the other hand, both interactions are of the same type, in that they are both described by GAUGE THEORIES. The unification of the two interactions to the electroweak interaction is only in the sense that, at sufficiently high ENERGIES (i.e., at high TEMPERATUREs), the two interactions should merge into one. In BIG BANG–type cosmological models, it is assumed that at early times in the life of the UNIVERSE, the temperature was so high that the electromagnetic and weak interactions that we know today were indeed merged into the electroweak interaction. As the universe cooled over time, the two interactions separated and evolved into what we find today. (*See* COSMOLOGY.)

electromagnetism Electromagnetism is the natural phenomenon that encompasses ELECTRICITY, MAGNETISM, and the effects they have on each other. While many everyday manifestations of electricity and magnetism give no indication that the two are related, they can affect each other strongly. For example, an electric CURRENT flowing through a coiled wire becomes a MAGNET, called an ELECTROMAGNET, which might unlock a car door or spin a computer's hard drive. On the other hand, in electric power plant generators and in automobile alternators, wire coils rotate near electromagnets and thereby produce electricity. This shows that electricity and magnetism are not independent phenomena but are intimately linked as constituents of the broader phenomenon of electromagnetism. (*See* GENERATOR, ELECTRIC.)

ALBERT EINSTEIN's special theory of relativity tells us that an observer will view a different mix of electric and magnetic effects, depending on her state of motion. If Amy is studying a purely electric effect, for example, then Bert, moving past Amy with sufficient speed, will see the same effect as a combination of electricity and magnetism. This, too, points to the deep connection between electricity and magnetism. (*See* RELATIVITY, SPECIAL, THEORY OF.)

Electricity is described in terms of electric CHARGEs and the electric FIELD. Magnetism involves no magnetic charges (at least as is presently known) and is described in terms of the magnetic field. Each field is a VECTOR quantity that extends over all space and at every location possesses some value (possibly zero), which might change over time. The electric and magnetic fields together constitute the electromagnetic field and are components of it.

The electric field is produced by electric charges, independently of whether the charges are at rest or in motion. The electric field in turn causes FORCEs to act on electric charges, again independently of whether the affected charges are at rest or in motion. The magnetic field is produced only by moving electric charges and causes forces only on electric charges in motion. So electric charges affect (i.e., cause forces on) each other through the mediation of the electromagnetic field.

The example of Amy and Bert can now be restated in terms of fields. Amy measures the forces on her test electric charge, both while keeping it at rest and while moving it in various ways. She finds a force acting on her test charge and discovers that it does not depend on the charge's motion. So she concludes that in her vicinity the electric field has some nonzero value, while the magnetic field is zero. Bert speeds by Amy and, as he passes her and while she is conducting her measurements, conducts similar experiments with his test charge. His finding is that in Amy's vicinity *both* fields possess nonzero values. The values of the electric and magnetic components of the electromagnetic field at the same time and place can have different values for observers in different states of motion.

Unification and Waves

A full and unified understanding of classical electromagnetism (i.e., electromagnetism without quantum effects) was achieved in the 19th century by JAMES CLERK MAXWELL and HENDRIK ANTOON LORENTZ. Based on the discoveries of earlier physicists (such as ANDRÉ-MARIE AMPÈRE, CHARLES-AUGUSTIN DE COULOMB, MICHAEL FARADAY, and Hans Christian Oersted), Maxwell formulated a set of equations that show (1) how the electric field is produced by electric charges and by the magnetic field's varying in time and (2) how the magnetic field is produced by moving electric charges (electric currents) and by the electric field's changing over time. Lorentz's contribution was a formula giving the force on an electric charge in any state of motion due to the electromagnetic field (i.e., due to the simultaneous effects of the electric and magnetic fields). This is called the LORENTZ FORCE. (*See* MAXWELL'S EQUATIONS; QUANTUM PHYSICS.)

It is especially noteworthy that Maxwell's equations predict the existence and describe the properties of WAVES in the electromagnetic field, called electromagnetic waves or electromagnetic RADIATION. These waves span a very wide range, or SPECTRUM, of radiations, which includes radio and TV waves, microwaves, infrared radiation, visible LIGHT, ultraviolet radiation, X RAYS, and GAMMA RAYS. What differentiates the various radiations is their FREQUENCY, or, correspondingly, their WAVELENGTH. Here is a brief table of the electromagnetic spectrum, showing approximate frequency and wavelength ranges.

Electromagnetic Spectrum

Common Name	Frequency Range	Wavelength Range
Radio and TV	Below 1 GHz	Above 30 cm
Microwaves	1 to 300 GHz	30 cm to 1 mm
Infrared	300 GHz to 430 THz	1 mm to 700 nm
Visible light	430 to 750 THz	700 to 400 nm
Ultraviolet	750 THz to 2.4×10^{16} Hz	400 to 13 nm
X rays	2.4×10^{16} to 5×10^{19} Hz	13 nm to 6×10^{-12} m
Gamma rays	Above 5×10^{19} Hz	Below 6×10^{-12} m

Electromagnetic waves are transverse waves. As such, they can be polarized. In VACUUM, all electromagnetic waves propagate at the same speed, called the SPEED OF LIGHT and conventionally denoted by c, with the value 2.99792458×10^8 meters per second (m/s). According to Maxwell's equations, the speed of light is related to the PERMITTIVITY of the vacuum, ε_0, and the PERMEABILITY of the vacuum, μ_0, by:

$$c = \frac{1}{\sqrt{\varepsilon_0 \mu_0}}$$

That is indeed the case, with $\varepsilon_0 = 8.85418781762 \times 10^{-12}$ C^2/(N·m^2) and $\mu_0 = 4\pi \times 10^{-7}$ T·m/A. The speed of light is found to have the same value no matter what the state of motion of the observer. This seemingly paradoxical effect led Einstein to the special theory of relativity and serves as one of the theory's cornerstones. (*See* POLARIZATION.)

QUANTUM ELECTRODYNAMICS (QED) is the name of the complete theory of electromagnetism, with the inclusion of quantum effects, which was developed in the 20th century. QED describes electromagnetic radiation in terms of flows of particles, called PHOTONs, and explains electric charges' affecting each other electromagnetically by exchanges of photons between them.

See also ELECTROMAGNETIC INTERACTION.

electromotive force (emf) The electromotive force is a potential difference, or VOLTAGE, in an electric circuit, that can supply ENERGY to the circuit and maintain electric CURRENTs in it. The source of emf is often chemical, as in a BATTERY, or electromagnetic, when the emf is induced by a changing magnetic FIELD, as in a generator. There are also other sources of emf, such as TEMPERATURE difference in a THERMOCOUPLE or PIEZOELECTRICITY.

The term *electromotive force* for a quantity that is not a FORCE is a relic of history. It remains from a time when the nature of electricity was less well understood than it is at present and, perhaps, the term *force* was used more loosely than it is today.

See also ELECTRICITY; ELECTROMAGNETISM; FARADAY'S LAW; GENERATOR, ELECTRIC; INDUCTION; MAGNETISM; POTENTIAL, ELECTRIC.

electron The electron is an ELEMENTARY PARTICLE belonging to the LEPTON family, whose other members are the muon, the tau, and the three types of NEUTRINO that correspond to the electron, the muon, and the tau. The electric CHARGE of the electron is negative and has the magnitude of one elementary unit (i.e., -1.60218×10^{-19} coulomb [C]). Its MASS is 9.10938×10^{-31} kilogram (kg) = 0.5110 MeV/c^2. The electron possesses a half unit of SPIN and is a FERMION. As with the other leptons, no structure has been detected for the electron, which is thus considered to be a point particle. The electron's corresponding antiparticle is the POSITRON. (*See* ANTIMATTER; ELECTRICITY.)

Electrons are a component of ordinary MATTER, since they, together with PROTONs and NEUTRONs, make up the ATOMs that constitute all ordinary matter. Whereas an atom's protons and neutrons are tightly confined to the relatively tiny NUCLEUS, the electrons form a "cloud" in the region surrounding the nucleus that defines the atom's volume. The flow of electrons in CONDUCTORs forms electric CURRENTs in those materials.

See also BOHR THEORY; CONDUCTION; CONDUCTION BAND.

electron microscopy Electron microscopy is a method of achieving higher RESOLVING POWER than is available in LIGHT microscopy. The resolving power of an image-forming system is $D/(1.22\lambda)$, where D denotes the effective diameter, or aperture, of the system, and λ is the WAVELENGTH of the RADIATION used to form the image, both specified in the same UNIT of length. Thus, the larger the aperture and the shorter the wavelength, the better the resolving power. The resolving power for light microscopes is limited by the practical limitations on the size of the system and by the shortest wavelength of visible light, about 8×10^{-7} meter. Electromagnetic radiation of shorter wavelengths is not conveniently manipulated and focused. In order to increase the resolving power, the WAVE nature of ELECTRONs is exploited, whereby electrons behave as waves of wavelength:

$$\lambda = h/p$$

where λ denotes the wavelength in meters (m), p is the magnitude of the electron's MOMENTUM in kilogram·meters/second (kg·m/s), and h is PLANCK'S CONSTANT, whose value is $6.62606876 \times 10^{-34}$ joule·second (J·s). Since electrons are electrically charged particles, they can easily be accelerated, manipulated, and focused through the use of electric and magnetic FIELDs. Raising the potential difference through which an electron beam is accelerated increases the electrons' KINETIC ENERGY, which implies greater momentum and, consequently, a shorter wavelength, according to the above relation. In this way, very short wavelengths, and hence high resolving powers, can be achieved. One limitation on the resolving power is that electrons with high kinetic energy tend to damage the object being imaged.

See also ACCELERATION; CHARGE; ELECTRICITY; ELECTROMAGNETISM; MAGNETISM; MATTER WAVE; OPTICS; POTENTIAL, ELECTRIC; WAVE-PARTICLE DUALITY.

electroweak interaction The electroweak interaction is the proposed unification of the ELECTROMAGNETIC INTERACTION and the WEAK INTERACTION, two of what are presently recognized as the four fundamental interactions among the ELEMENTARY PARTICLEs. (The other two are the STRONG INTERACTION and GRAVITATION.) The notion of unification of interactions is exemplified by the 19th-century work of JAMES CLERK MAXWELL, who unified ELECTRICITY and MAGNETISM to ELECTROMAGNETISM by showing that they are but two aspects of the latter. Similarly, in the 20th century, the electromagnetic and weak interactions were recognized as being derived from the ELECTROWEAK INTERACTION. Whereas electricity and magnetism are intimately involved with each other, the electromagnetic and weak interactions are quite distinct. For example, the three intermediating particles of the weak interaction possess MASS, and two of them are electrically charged, while the PHOTON, the mediator of the electromagnetic interaction, is massless and electrically neutral. Related to the difference in mass, the range of the weak interaction is finite, while that of the electromagnetic interaction is infinite. The strengths of the two interactions are quite disparate. On the other hand, both interactions are of the same type, in that they are both described by GAUGE THEORIES. (*See* CHARGE.)

The unification of the two interactions to the electroweak interaction is only in the sense that at sufficiently high ENERGIES (i.e., at high TEMPERATURES) the two interactions should merge into one. In BIG-BANG-type cosmological models, it is assumed that at early times in the life of the UNIVERSE, the temperature was so high that the electromagnetic and weak interactions that we know today were indeed merged into the electroweak interaction. As the universe cooled over time, the two interactions separated and evolved into what we find today. (*See* COSMOLOGY.)

element A substance that cannot be decomposed into simpler substances through chemical processes is called an element, or chemical element. The smallest constituent of an element is an ATOM. Every element is characterized by its ATOMIC NUMBER, which is the number of PROTONs in the nucleus of each of its atoms. The atomic number also gives the number of ELECTRONS in each neutral atom and determines the chemical properties and identity of the element. For hydrogen the atomic number is 1, for helium 2, for lithium 3, . . ., for uranium 92, and so on. Uranium has the highest atomic number among the naturally occurring elements. Elements with atomic numbers above 92 have been produced in the laboratory by causing the nuclei of lighter elements to collide and coalesce with the help of particle accelerators. The elements are arranged in the PERIODIC TABLE of the elements in order of ascending atomic number and reveal certain periodicities in their chemical and physical properties. (*See* ACCELERATOR, PARTICLE.)

The ATOMIC MASS of an atom is the sum of the numbers of protons and NEUTRONs in its nucleus. A substance with both a definite atomic number and a definite atomic mass is an ISOTOPE, which is thus a version of an element. Isotopes of the same element possess the same electron structure and the same chemical properties, although their physical properties may differ as a result of their different MASSes. Most naturally occurring elements are mixtures of isotopes. Natural carbon, for instance, with atomic number 6, contains a mixture of three isotopes, having atomic masses 12, 13, and 14, containing in each of their nuclei, in addition to six protons, six, seven, and eight neutrons, respectively. The first, carbon-12, is by far the most abundant isotope. Both carbon-12 and carbon-13 are stable isotopes, while carbon-14 is radioactive and is used for dating archeological specimens. (*See* RADIOACTIVITY.)

elementary particle Also called fundamental particles, the elementary particles include, strictly speaking, only the ultimate constituents of MATTER and the particles that mediate the fundamental INTERACTIONS among them. Somewhat more loosely, the term also includes the NUCLEONs, which constitute the NUCLEI of ATOMs, the particles that mediate the FORCEs holding the nucleus together, and other particles, similar to all those. For the looser meaning, the term *subatomic particles* is perhaps more appropriate.

According to present understanding of the elementary particles, they are divided into the two broad categories of matter particles and force particles. The former are all FERMIONs, so that the SPIN of each species is an odd multiple of $^1/_2$ (1/2, 3/2, . . .), while the latter are all BOSONs, whose spin is an integer (0, 1, . . .).

Restricting ourselves to the particles that do not appear to be composed of constituents (i.e., to what seem—at least for the present—to be truly fundamental particles), the matter particles are subdivided into QUARKs and LEPTONs, which are further subdivided into generations. The first generation consists of the up quark (denoted u), the down quark (d), the ELECTRON (e^-), and the electron-type neutrino (ν_e). The second generation is made up of the charm quark (c), the strange quark (s), the muon (μ), and the muon-type neutrino (ν_μ). The third generation of matter particles consists of the top (or truth) quark (t), the bottom (or beauty) quark (b), the tau (τ), and the tau-type neutrino (ν_τ). Lumping generations together, we have six flavors of quark (up, down, charm, strange, top, bottom) and six leptons (electron, electron-type neutrino, muon, muon-type neutrino, tau, tau-type neutrino). All of these possess a half unit of spin, and there is an antiparticle for each. (The antiparticles are denoted by adding a bar on top of the symbol of the corresponding particle, such as $\bar{u}$, $\bar{\mu}$, $\bar{\nu}_e$, etc. An exception to this rule is that the electron's antiparticle, the POSITRON, is often denoted e^+.) (*See* ANTIMATTER; NEUTRINO.)

The quarks all possess MASS and are electrically charged. Their electric CHARGES are fractions of the fundamental charge unit (which is the magnitude of the electron's charge): the up, charm, and top quarks carry charge $+^2/_3$, while the charge of the down, strange, and bottom quarks is $-1/3$. In addition, all quarks carry color charge, which has three values, denoted red, blue, and green. (The use of color names is a matter of physicists' whimsy and has nothing to do with real colors.) Quarks and antiquarks in various combinations make up the subatomic particles called HADRONs (see below). It appears that—except possibly under special, extreme conditions—quarks cannot exist independently outside of hadrons. (*See* ELECTRICITY.)

The leptons all have mass too, where the masses of the neutrinos are considerably smaller than those of the electron, muon, and tau. The latter carry one negative unit of electric charge each, while the neutrinos are electrically neutral. The electron is a stable particle, while the muon and the tau both DECAY. The neutrinos do not decay, but they do undergo the process of mixing, whereby, as they travel, they continuously convert into varying quantum mixtures of all three neutrino types. (*See* QUANTUM PHYSICS.)

The fundamental force particles are the particles that serve to intermediate the four fundamental interactions. They perform their mission by being exchanged between the interacting particles. The PHOTON (γ) is the intermediary of the ELECTROMAGNETIC INTERACTION. It is massless, electrically neutral, and has spin 1. It couples to (i.e., is emitted and absorbed by) particles possessing electric charge and/or magnetic dipole moment. The carriers of the STRONG INTERACTION, or color force, are the eight GLUONs (g). Like the photon, they are massless, electrically neutral, and possess spin 1. They carry various combinations of color charge. They couple only to particles carrying color charge (i.e., only to quarks and to themselves). (*See* MOMENT, MAGNETIC DIPOLE.)

The WEAK INTERACTION is mediated by the three intermediate vector bosons, which are named and denoted W^+, W^-, and Z^0. They all have mass and possess spin 1. The particles W^+ and W^- carry electric charge +1 and –1, respectively, while Z^0 is electrically neutral. They couple to all matter particles. Although it has not (yet) been experimentally detected, the graviton is the putative intermediary of GRAVITATION. It should be massless and possess spin 2.

We turn now to the hadrons, which are subatomic particles that consist of quarks and antiquarks in various combinations (as well as of the gluons that bind the quarks and antiquarks together). The hadrons are a family of particles that comprises the BARYONs, which are matter particles, and the MESONs, which are force particles. The baryons all consist of three quarks each. Among the baryons, the most common are the PROTON (p) and the NEUTRON (n), which are the building blocks of atomic nuclei. The proton consists of two up quarks and one down quark (uud), while the neutron is made of one up quark and two downs (udd). Both particles have mass and possess a half unit of spin. The proton carries electric charge +1, and the neutron is electrically neutral. The baryons include additional particles, such as those named and denoted Λ^0 (called lambda-zero), Ω^- (omega-minus), and various versions of Δ, Σ, and Ξ. The Λ^0, for example, consists of an up, a down, and a strange quark (uds), while the Ω^- is made of three strange quarks (sss).

The mesons are made of quark-antiquark pairs. Among the mesons, the pions (π^+, π^-, π^0) are most noted for their role in intermediating the strong interaction among protons and neutrons that holds nuclei together. The π^+ consists of an up quark and an antidown quark ($u\bar{d}$), the π^- is made of an antiup and a down quark ($\bar{u}d$), while the π^0 comprises a quantum mixture of up-antiup ($u\bar{u}$) and down-antidown ($d\bar{d}$). The pions π^+, π^-, π^0 possess mass and have spin 0. Their electric charge is +1, –1, and 0, respectively. There are additional kinds of meson as well, some of which are named and denoted η, η', J/ψ, Y, and various forms of the kaon (K). The η' (eta-prime), for instance, is composed of a strange-antistrange pair ($s\bar{s}$), and the J/ψ (jay-psi) comprises a charm-anticharm pair ($c\bar{c}$).

Among the hadrons, only the proton seems to be stable (although its long-term stability is put into question by certain elementary-particle theories). The neutron is basically unstable and decays via BETA DECAY. However, in nuclei the neutron can be stable, which fact allows the existence of stable nuclei.

To every species of hadron there corresponds an antiparticle. Its quark composition matches that of the corresponding particle, with every quark replaced by its antiquark and vice versa. The antiproton ($\bar{p}$), for example, consists of two antiups and an antidown ($\bar{u}\bar{u}\bar{d}$), the π^0 is its own antiparticle, and the π^+ and π^- are each other's antiparticles.

The elementary particles are characterized by additional kinds of charge to the ones mentioned above. Present understanding of the elementary particles and their interactions is based on GAUGE THEORIES, which are QUANTUM FIELD THEORIES that possess gauge SYMMETRY.

emf *See* ELECTROMOTIVE FORCE.

endothermic process A process that absorbs HEAT is termed an endothermic process. In EVAPORATION, for instance, the thermal ENERGY of the liquid is reduced—and the liquid cools—as molecules move from the LIQUID to the GAS PHASE. The liquid then absorbs heat from the environment.

See also EXOTHERMIC PROCESS.

energy The energy of a system is a measure of, and numerically equal to, the WORK that the system can perform. It is a SCALAR quantity, and its SI UNIT is the joule (J), which is the same as that of work (and is equivalent to the newton·meter [N·m]). Another unit of energy that is often used is the electron volt (eV). It is the amount of energy gained by an ELECTRON as it accelerates through a potential difference of one volt and equals 1.602176×10^{-19} J. This unit and its multiples are useful and commonly used for the microscopic domain of ATOMs, MOLECULEs, NUCLEI, and ELEMENTARY PARTICLEs. An additional, common unit of energy is the kilowatt-hour (kW·h), used mostly in connection with electric energy. It equals 3.6×10^6 J. (*See* ACCELERATION; ELECTRICITY; POTENTIAL, ELECTRIC.)

Energy has two basic forms, KINETIC ENERGY and POTENTIAL ENERGY. For a system with structure, i.e., a system composed of microscopic components such as atoms, molecules, and IONs, it is useful to distinguish between two manifestations of each form, internal energy and external energy. Energy is created as a result of work performed on a system. In addition, energy can be converted between its kinetic and potential forms (through the performance of work) as well as among the various forms of potential energy.

Kinetic Energy

Kinetic energy is energy due to MOTION. The kinetic energy of a point particle possessing MASS m in meters (m) and SPEED v in meters per second (m/s) is:

$$\text{Kinetic energy} = mv^2/2$$

Note that kinetic energy is relative, in that observers in different states of motion assign different values to the kinetic energy of the same particle, since they observe it as having different speeds.

Consider now a system with structure, such as a bowling ball. Its external kinetic energy is the sum of its translational kinetic energy and its rotational kinetic energy. The former is the system's kinetic energy due to the motion of its CENTER OF MASS, whereby the system is equivalent to a point particle of the same mass located at its center of mass. So for a system with total mass m, whose center of mass is moving with speed v, the translational kinetic energy is:

$$\text{Translational kinetic energy} = mv^2/2$$

If the system is rotating about its center of mass, it also possesses rotational kinetic energy, given by:

$$\text{Rotational kinetic energy} = I\omega^2/2$$

where ω denotes the magnitude of the system's angular velocity in radians per second (rad/s), and I is the system's MOMENT OF INERTIA in kilogram·meter^2 (kg·m^2) with respect to the axis of rotation. (*See* ROTATION.)

A system with structure can possess also internal kinetic energy, which appears as thermal energy, or HEAT, of the system. It is the kinetic energy of the random motion of the system's microscopic constituents. Of all forms of energy, heat is an exception to the definition of energy as a measure of the work a system can perform, since, according to the second law of THERMODYNAMICS, heat is not fully convertible to work.

Potential Energy

Potential energy is energy of any form that is not kinetic. Equivalently, it is energy of a system that is due to any property of the system except motion. It is called "potential" because it possesses the potential to produce motion (through the system's performing work), thus converting to kinetic energy. A form of potential energy can be associated with any CONSERVATIVE FORCE. There are many and diverse forms of potential energy. Here are some of them.

Elastic potential energy is the potential energy of a body due to its deformation. A stretched or compressed spring, for example, possesses this form of potential energy, whose value is given by:

$$\text{Elastic potential energy} = kx^2/2$$

where k is the spring constant in newtons per meter (N/m) and x the amount of stretching or compression in meters (m). Elastic potential energy is an internal energy. (*See* ELASTICITY.)

Gravitational potential energy is the potential energy associated with the gravitational force. Two particles that are attracting each other gravitationally possess gravitational potential energy, given by:

$$\text{Gravitational potential energy} = -Gm_1m_2/r$$

where G denotes the gravitational constant, whose value is 6.67259×10^{-11} N·m²/kg², m_1 and m_2 are the MASSES of the particles in kilograms (kg), and r is the distance between them in meters (m). (Here the reference separation, the separation for which the gravitational potential energy is defined as zero, is taken to be infinity. Hence, the gravitational potential energy is negative for finite separations.) In the case of a body near the surface of the Earth, the gravitational potential energy is:

$$\text{Gravitational potential energy} = mgh$$

where m is the mass of the body, g the acceleration due to gravity, whose value is approximately 9.8 meters per second per second (m/s²), and h denotes the vertical position in meters (m) of the body with respect to any reference level (at which the gravitational potential energy is taken to be zero). (*See* GRAVITATION.)

For a system of bodies, the gravitational potential energy associated with their mutual attraction is an internal energy. However, the gravitational potential energy of a body in the gravitational FIELD can be viewed as an external energy of the body.

Electric potential energy is the potential energy associated with the electric force. For two electrically charged particles, for instance, that are attracting or repelling each other electrically, the electric potential energy is given by:

$$\text{Electric potential energy} = \frac{1}{4\pi\kappa\varepsilon_0}\frac{q_1q_2}{r}$$

Here q_1 and q_2 are the (signed) charges of the particles in coulombs (C); r is the distance between them in meters (m); ε_0 is the PERMITTIVITY of the vacuum, whose value is $8.85418782 \times 10^{-12}$ C²/(N·m²); and κ is the DIELECTRIC CONSTANT of the medium in which the particles are located ($\kappa = 1$ for the VACUUM). (As for the gravitational potential energy of two particles, the reference separation is here taken to be infinity.) The electric potential energy of a system of charged particles is an internal energy. (*See* CHARGE; ELECTRICITY.)

The electric potential energy of a charged particle in the electric field is given by:

$$\text{Electric potential energy} = qV$$

where q denotes the charge in coulombs (C), and V is the electric potential in volts (V). For a charged body in the electric field, this is an external energy. (*See* POTENTIAL, ELECTRIC.)

Chemical potential energy is the energy that is stored in the chemical bonds of the materials that constitute a system and can be released and converted to other forms of energy through chemical reactions. The batteries we use in flashlights, electronic devices, and automobiles, for example, contain chemical potential energy, which, as we use the batteries, is converted into the electric potential energy for which we buy the batteries. The recharging of rechargeable batteries is the reconversion of electric potential energy to chemical. We exploit the chemical potential energy of gasoline by converting it into heat in internal combustion engines. Chemical potential energy is an internal energy.

Electromagnetic energy is the energy carried by electromagnetic WAVES. In this form, the Earth receives energy from the Sun. It is also the energy that is converted to heat in microwave ovens. Electromagnetic energy can be viewed as the energy of PHOTONS. This form of energy is not readily categorizable as internal or external. (*See* ELECTROMAGNETISM.)

See also RADIATION.

Other Forms of Energy

MASS ENERGY is the energy equivalent of MASS, according to ALBERT EINSTEIN's special theory of relativity. The value of the energy equivalent of mass m in kilograms (kg) is given by the formula:

$$\text{Mass energy} = mc^2$$

where c denotes the SPEED OF LIGHT, whose value is 2.99792458×10^8 meters per second (m/s). One way of obtaining this energy is by means of nuclear reactions, when the sum of masses of the reaction products is less than the sum of masses of the initial reaction participants. The "disappearing" mass is converted to kinetic energy of the products according to Einstein's formula. RADIOACTIVITY is one example of that. Another manifestation of mass energy is through particle-antiparticle annihilation. When an ELECTRON collides with a POSITRON, its antiparticle, they totally annihilate each other, and their mass is converted to electromagnetic energy in the form of a pair of PHOTONS. Mass energy is an internal energy. (*See* ANTIMATTER; NUCLEAR PHYSICS; RELATIVITY, SPECIAL THEORY OF.)

The term *mechanical energy* is used to refer to the sum of all forms of *external* kinetic and potential energy. That excludes heat, chemical potential energy, electromagnetic energy, and mass energy.

Energy obeys a CONSERVATION LAW, whereby in any isolated system, the total amount of energy, of all forms, remains constant over time. In such a case, the system's energy can change form over time. It might change from kinetic to potential and back again, for example, or from chemical to electric to heat. But the total amount does not change. (*See* WORK-ENERGY THEOREM.)

In certain situations in the quantum domain, energy is quantized: it cannot take any value, but is constrained to a set of values. An example of that is a quantum HARMONIC OSCILLATOR, whose allowed energy values are given by:

$$E_n = (n + 1/2)hf \text{ with } n = 1, 2, \ldots$$

Here E_n denotes the value of the energy in joules (J) of the harmonic oscillator's nth energy level, f is the classical FREQUENCY of the oscillator in hertz (Hz), and h represents the PLANCK CONSTANT, whose value is $6.62606876 \times 10^{-34}$ joule·second (J·s). Another example of a system whose energy is quantized is the hydrogen atom. Its energy levels, in electron volts, are given by:

$$E_n = -13.6/n^2 \text{ eV with } n = 1, 2, \ldots$$

(*See* BOHR THEORY; QUANTIZATION; QUANTUM PHYSICS.)

A further example of energy quantization is the fact that electromagnetic radiation cannot exchange energy with MATTER in any arbitrary amount, but only in integer multiples of the quantity hf, where f denotes the frequency of the electromagnetic wave. That is related to the quantum picture of electromagnetic radiation as consisting of photons—each of which possesses energy hf—and interacting with matter through the emission and absorption of photons.

entropy The term *entropy* describes a quantity that is a property of a macroscopic system and is a measure of the randomness, or DISORDER, of the system's microscopic constituents. Entropy helps to indicate the unavailability of internal ENERGY in the system for performing WORK. Entropy is a SCALAR quantity. Its SI UNIT is joule per kelvin (J/K). (*See* FREE ENERGY; STATISTICAL MECHANICS; THERMODYNAMICS.)

One of the various formulations of the second law of thermodynamics is: the entropy of an isolated system cannot decrease over time. Thus an isolated system evolves either with constant entropy or with increasing entropy. In the former case, the system's degree of disorder does not change, and neither does the amount of work it can perform from its internal energy content. Thus, the process is reversible. When entropy increases, however, the system's internal disorder increases, its ability to convert its internal energy to work decreases, and the process is irreversible. Note that as an isolated system evolves, it is possible for parts of the system to undergo a decrease in their entropy, but that is always at the expense of an entropy increase in other parts. (*See* REVERSIBLE PROCESS.)

The entropy of a nonisolated system can be increased or decreased by the flow of heat into it or from it, respectively. The increase in entropy—ΔS, in joules per kelvin (J/K) due to the amount of heat ΔQ in joules (J) flowing reversibly into the system at absolute TEMPERATURE T in kelvins (K)—is given by:

$$\Delta S = \Delta Q/T$$

For heat flowing out of the system, both ΔQ and ΔS are negative.

Entropy also has a probabilistic significance as a measure of the likelihood, or probability, of finding a macroscopic state of a system. This is related to the idea that the more disordered a system's microscopic components, the larger the number of microscopic states corresponding to the macroscopic state, and thus the more probable it is that the system is found in the macroscopic state. The precise relation between the entropy, S, of a macroscopic state and the number of microscopic states corresponding to it, W, is given by:

$$S = k \ln W,$$

where k is the Boltzmann constant, whose value is $1.3806503 \times 10^{-23}$ joule per kelvin (J/K).

A process that takes place without change of entropy is called an isentropic process. The term *isentrope* describes a graphic curve representing the relation between two variables of a system during such a process, such as between PRESSURE and VOLUME.

EPR The abbreviation *EPR* stands for Einstein-Podolsky-Rosen: the German/Swiss/American physicist ALBERT EINSTEIN, the American physicist Boris Podolsky (1896–1966), and the American/Israeli physicist Nathan Rosen (1909–95). In the first half of the 20th century, they proposed an experiment to test what seemed to be paradoxical aspects of quantum theory. Thus EPR often appears in expressions such as "EPR experiment" and "EPR paradox." (*See* QUANTUM PHYSICS.)

There are two components to the "paradox." First, according to quantum theory, a system does not possess a measurable physical characteristic until it is actually measured. If position, for instance, is measured, then a result is obtained, and the system can be said to have such and such a position at the time of measurement. But prior to the measurement, the state of the system might have been such that it could not be thought of as even possessing a position. A measurement, according to quantum physics, rather than merely revealing the existing, but unknown, value of a physical quantity characterizing the system, as in CLASSICAL PHYSICS, actually endows the system with the property being measured by putting it in a state characterized by the property. This idea was revolutionary at the time, and many physicists are still uncomfortable with it. The second component of the EPR "paradox" is that, as long as a system is isolated, all parts of it participate in its state, no matter how spread out the system becomes.

What EPR proposed is to have two particles created by one source, so that they are part of the same state, and fly apart at high speed. Then, when the particles are far apart, perform a measurement on one, and immediately afterward perform a measurement on the other. According to quantum theory, the first measurement should endow the whole system, *including the other particle,* with the property being measured. The second measurement, if designed properly, should detect whether or not the system is indeed in the state brought about by the first measurement. (Actually, the whole procedure must be repeated many times, and the correlation between the first and second measurement results must be examined.) The "paradox" is this. For the second particle measured to "know" that it is in a state that it was not in a very short time ago, it seems that some "influence" must have reached it from the first measurement. However, if the measurements are made sufficiently closely in time, the "influence" must travel faster than the SPEED OF LIGHT, and according to Einstein's special theory of relativity, no signal or information can do that. (*See* RELATIVITY, SPECIAL THEORY OF.)

In the late 20th century, the EPR experiment was first performed, and since then it has been performed very often. The result is clear: the predictions of quantum theory are confirmed. The two particles, produced by the same source and part of the same quantum state, are forever linked together in "quantum entanglement" no matter how far apart they are. The change of state brought about by the first measurement does indeed immediately affect the result of the second measurement through quantum entanglement. Contradiction with the special theory of relativity is avoided, however, because quantum entanglement does not allow the faster-than-light transmission of a signal or of information.

equation of state Any mathematical relation among the various physical quantities that specify a system's state is called an equation of state of the system. Such a relation shows that the quantities are not all independent and allows one of the quantities to be expressed in terms of the others.

For a common example of an equation of state, consider n moles of an ideal gas at PRESSURE p, absolute TEMPERATURE T, and filling VOLUME V. These four quantities are related by the ideal gas law:

$$pV = nRT$$

where R denotes the gas constant, whose value is 8.314472 joules per mole per kelvin (J/[mol·K]). Here p is in pascals (Pa), equivalent to newtons per square meter (N/m^2), and V is in cubic meters (m^3). The UNIT of n is the mole (mol), and T is in kelvins (K). This relation shows that pressure, volume, temperature, and amount of gas are not independent quantities for this system, and it allows any one of the four quantities to be expressed in terms of the other three. Given n, T, and V, for example, the pressure p must then be:

$$p = nRT/V$$

See also GAS, IDEAL.

equations of motion The equations of motion are, in general, a set of differential equations for the variables that specify the state of the system. These equations are used by taking a set of values for the variables and their first, and possibly higher, time derivatives at any given time—the set of values is called INITIAL CONDITIONS—and solving the equations for a solution that fulfills the initial conditions. The solution, in the form of the variables as functions of time, then describes the evolution of the system for the case when the system is in the state specified by the initial conditions at the given time. The solution describes both the system's evolution that leads to the specified state at the given time and the system's subsequent evolution from that state. This is the general meaning of equations of motion. Equations of motion are expressions of LAWS OF NATURE.

More particularly, if the system under consideration is a mechanical system consisting of a number of particles, the variables specifying the state of the system are the positions of the particles, say the x-, y-, z-coordinates of all the particles. Then, given the forces acting on and among the particles, ISAAC NEWTON's second law of motion provides the equations of motion in the form of second-order ordinary differential equations for the particle coordinates. Initial conditions for such a system consist of the values of all the particle coordinates and VELOCITIES (i.e., the coordinates' first derivatives) at a given time. The solution of the equations of motion has the form of the particle coordinates as functions of time that fulfill the initial conditions. (*See* NEWTON'S LAWS.)

We now express that more mathematically. Let the system consist of N particles, and let the mass of the ith particle be denoted by m_i; its coordinates by x_i, y_i, z_i; and the components of its velocity by v_{ix}, v_{iy}, v_{iz}, where:

$$v_{ix} = dx_i/dt$$
$$v_{iy} = dy_i/dt$$
$$v_{iz} = dz_i/dt$$

Denote the components of the force on the ith particle by F_{ix}, F_{iy}, F_{iz}. This force might depend on the position and velocity of the ith particle itself, on the positions and velocities of all the other particles, and might also otherwise vary with time. Let us indicate those dependences symbolically by:

$$F_{ix} = F_{ix}(x, y, z, v_x, v_y, v_z, t)$$
$$F_{iy} = F_{iy}(x, y, z, v_x, v_y, v_z, t)$$
$$F_{iz} = F_{iz}(x, y, z, v_x, v_y, v_z, t)$$
$$\text{for } i = 1, 2, \ldots, N$$

where t denotes the time. Newton's second law of motion then gives the equations of motion:

$$d^2x_i/dt^2 = m_iF_{ix}(x, y, z, v_x, v_y, v_z, t)$$
$$d^2y_i/dt^2 = m_iF_{iy}(x, y, z, v_x, v_y, v_z, t)$$
$$d^2z_i/dt^2 = m_iF_{iz}(x, y, z, v_x, v_y, v_z, t)$$
$$\text{for } i = 1, 2, \ldots, N$$

For this system, initial conditions are the specification of the coordinates and velocity components of all the particles at some time. Let us denote that time t_0; the values of the ith particle's coordinates at that time x_i^0, y_i^0, z_i^0; and the values of the components of its velocity at the same time v_{ix}^0, v_{iy}^0, v_{iz}^0. A solution of the equations of motion, describing the motion of the system for the case that at time t_0 the system was in the state specified by the initial conditions, might then take the form:

$$x_i = x_i(t)$$
$$y_i = y_i(t)$$
$$z_i = z_i(t)$$

where:

$$x_i(t_0) = x_i^0$$
$$y_i(t_0) = y_i^0$$
$$z_i(t_0) = z_i^0$$

and:

$$v_{ix}(t_0) = (dx_i/dt)(t_0) = v_{ix}^0$$
$$v_{iy}(t_0) = (dy_i/dt)(t_0) = v_{iy}^0$$
$$v_{iz}(t_0) = (dz_i/dt)(t_0) = v_{iz}^0$$
$$\text{for } i = 1, 2, \ldots, N$$

Let us now reiterate the above in compact, VECTOR notation. Let $\mathbf{r}_i$ denote the position vector of the ith particle. Its components are just the particle's coordinates x_i, y_i, z_i. Denote the velocity vector of the ith particle by $\mathbf{v}_i$. Its components are v_{ix}, v_{iy}, v_{iz}. Then:

$$\mathbf{v}_i = d\mathbf{r}_i/dt$$

Denote the force on the ith particle by $\mathbf{F}_i$. Its components are F_{ix}, F_{iy}, F_{iz}. The aforementioned dependences are then indicated symbolically by:

$$\mathbf{F}_i = \mathbf{F}_i(\mathbf{r}, \mathbf{v}, t) \text{ for } i = 1, 2, \ldots, N$$

The equations of motion are:

$$d^2\mathbf{r}_i/dt^2 = m_i\mathbf{F}_i(\mathbf{r}, \mathbf{v}, t) \text{ for } i = 1, 2, \ldots, N$$

For initial conditions $\mathbf{r}_i^0$ and $\mathbf{v}_i^0$ at time t_0, a solution of the equations of motion might take the form:

$$\mathbf{r}_i = \mathbf{r}_i(t)$$

where:

$$\mathbf{r}_i(t_0) = \mathbf{r}_i^0$$

and:

$$\mathbf{v}_i(t_0) = (d\mathbf{r}_i/dt)(t_0) = \mathbf{v}_i^0 \text{ for } i = 1, 2, \ldots, N$$

As a very simple example, consider the case of a single particle constrained to move in one dimension under the influence of a constant force, causing constant acceleration a. Denote the coordinate of the particle by x and its velocity by v. The equation of motion for this example is:

$$d^2x/dt^2 = a$$

Let the initial conditions be:

$$x(0) = x^0$$
$$v(0) = (dx/dt)(0) = v^0$$

i.e., the particle's position and velocity at time $t = 0$ have the values x^0 and v^0, respectively. The solution of the equation of motion that fulfills the initial conditions is:

$$x = at^2/2 + v^0t + x^0$$

equilibrium A state of equilibrium of a system is a state in which the system does not spontaneously evolve but, rather, maintains itself as it is. A PENDULUM at rest in its lowest position is in equilibrium, for example. If the pendulum is displaced from its lowest position by, say, an angle of 5°, it is not in a state of equilibrium because, if not prevented from doing so, it will immediately start moving (toward its lowest position).

It is useful to distinguish among three kinds of equilibrium: stable, labile, and unstable. The distinction is based on the behavior of the system when it is displaced slightly from its equilibrium state. If the system tends to return to its equilibrium state, the latter is a state of *stable* equilibrium. The lowest position of a pendulum is an obvious example of stable equilibrium. If, on the other hand, the system tends to evolve away from its equilibrium state when it is displaced slightly from it, the state is said to be one of *unstable* equilibrium. A rigid-rod pendulum in its highest position, with the bob balanced directly over the pivot, is in such a state. In everyday language, it is a state of precarious equilibrium. It can balance there, at least in principle. But the slightest nudge, a tiny breeze, or even a molecular motion, can cause the bob to spontaneously move away from that position.

Labile equilibrium is intermediate between stable and unstable equilibrium. When a system is displaced slightly from such a state, it does not tend to change at all with respect to the state. A ball resting on a level floor is an example of labile equilibrium. Displace it slightly to the side, and it is just as happy to remain in its new position as it was in its previous one. Labile equilibrium can be viewed as a continuous set of equilibrium states.

For isolated systems to which thermodynamic considerations are applicable, states of stable equilibrium are states of maximal ENTROPY and vice versa. That is due to the second law of THERMODYNAMICS, according to which isolated systems tend to increase their entropy. The uniform distribution of an isolated GAS in its container is an example of a state of stable equilibrium and of maximal entropy. Any nonuniform state, such as a higher DENSITY on one side of the container than on the other, possesses less entropy than the uniform state and spontaneously evolves to it in a very short time. (For systems that are in interaction with their environment, stable equilibrium corresponds to a minimum of the system's FREE ENERGY.)

Such systems might also be in unstable equilibrium. For instance, the condition of a superheated LIQUID is a state of unstable equilibrium. That is when its TEMPERATURE and PRESSURE are such that it normally boils, but the onset of BOILING is prevented by the absence of nucleating sites. It is indeed in equilibrium because, with no interference, it remains in its superheated state and does not tend to change. But a small disturbance, such as jarring the vessel or introducing a rough grain of sand, can immediately bring about sudden, sometimes violent, boiling.

The significance of equilibrium for such systems is that their macroscopic state does not change. Nevertheless, at the microscopic level changes *are* taking place. As an example, consider a mixture of liquid water and ice at the FREEZING point in a closed system. The mixture is in equilibrium, as neither the LIQUID nor the SOLID PHASE increases at the expense of the other. Yet the individual water MOLECULEs are hardly in equilibrium. Molecules are continuously leaving the solid for the liquid and vice versa. Such equilibrium is called dynamic equilibrium.

equipartition of energy The principle of equipartition of energy is used in calculating the SPECIFIC HEAT of matter from its atomic properties. The principle is that when HEAT is introduced into matter, it distributes itself among the various DEGREES OF FREEDOM of the constituent ATOMs, MOLECULEs, and IONs such that each degree of freedom of each atom, etc., possesses the same amount of ENERGY, which is $kT/2$. Here k denotes the Boltzmann constant, whose value is $1.3806503 \times 10^{-23}$ joule per kelvin (J/K) or 8.6173295×10^{-5} electron volt per kelvin (eV/K), and T is the absolute TEMPERATURE in kelvins (K). In this context, a degree of freedom is an independent mode for an atom, molecule, or ion to possess a significant amount of energy.

See also STATISTICAL MECHANICS.

equipotential surface A useful device for describing the spatial configuration of the electric potential is the equipotential surface. This is a surface in space such that at all points on it, the electric potential has the same value. The equipotential surfaces of a point source CHARGE, for example, are spherical shells centered on the charge. Or, the equipotential surfaces of a source charge distributed uniformly along an infinite straight line are cylindrical surfaces having the line as their axis. (*See* ELECTRICITY; POTENTIAL, ELECTRIC.)

For any point in space, the electric FIELD line passing through the point is perpendicular to the equipotential surface containing the point. A quantitative relation between the electric field and equipotential surfaces is that for any pair of close equipotential surfaces: (1) the direction of the electric field at any point between them is perpendicular to them and pointing from the higher-potential surface to the lower, and (2) the magnitude of the electric field at such a point, E, in newtons per coulomb (N/C), or equivalently in volts per meter (V/m), is given by:

$$E = |\Delta V/\Delta s|$$

where ΔV and Δs are, respectively, the potential difference between the equipotential surfaces in volts (V) and the (small) perpendicular distance between them in meters (m).

Moving a test charge along an equipotential surface requires no work.

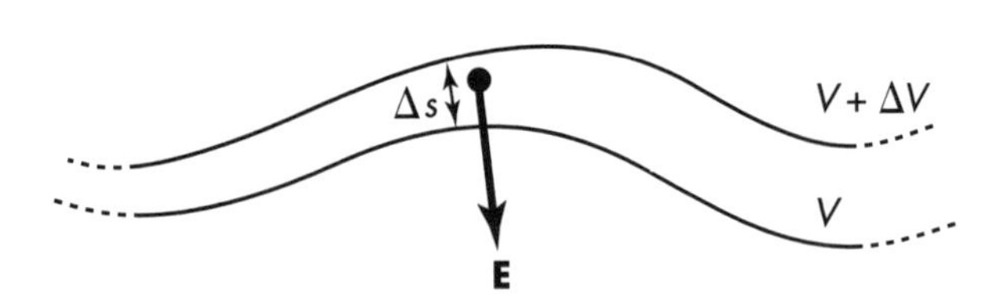

Parts of two close equipotential surfaces are shown in cross section. Their potentials are V (lower) and $V + \Delta V$ (higher). E is the electric field at a point between the surfaces. It is perpendicular to the surfaces, and it points from the higher-potential surface to the lower-potential one. Its magnitude, E, is given by $E = |\Delta V/\Delta s|$, where Δs is the (small) perpendicular distance between the surfaces.

equivalence principle The equivalence principle states that MASS, as it affects INERTIA—such as in ISAAC NEWTON's second law of MOTION—is equal to mass as it is affected by GRAVITATION—following Newton's law of gravitation, for instance. Experiments by Loránd Eötvös and, later, by others for the purpose of detecting a violation of this principle have come up with negative results. (*See* NEWTON'S LAWS.)

Another formulation of the equivalence principle is as follows. It is impossible to determine, by means of experiments carried out solely within a laboratory (i.e., no peeking outside the lab, for example), whether the room is at rest on the surface of a planet (so the masses in the lab are being affected by gravitation) or is undergoing constant acceleration (whereby the masses are exhibiting their inertial character) in a region where there is no net gravitational force. In both cases, if a body is attached to a spring anchored to the wall, the spring is stretched. An additional formulation is that it is similarly impossible to distinguish between the lab's being in a state of FREE FALL under the influence of gravitation and the lab's being at rest in a region where there is no net gravitational force acting. In both cases, a body floats freely around the laboratory. The occupants of a space station in orbit around the Earth experience a free-fall situation.

The equivalence principle was important for ALBERT EINSTEIN's development of his general theory of relativity and is central to the theory. (*See* RELATIVITY, GENERAL THEORY OF.)

escape speed The escape speed of an astronomical body, such as a planet, is the smallest SPEED that an object, perhaps a rocket, needs to leave the surface of the body and overcome the body's gravitational hold on it, i.e., to leave the body's surface and not fall back. (*See* GRAVITATION.)

Escape speed is most readily understood in terms of ENERGY. The gravitational potential energy in joules (J) of an object of MASS m on the surface of a spherical astronomical body of mass M, where both masses are in kilograms (kg), is:

$$E_p = -GMm/R$$

Here, G denotes the gravitational constant, whose value is 6.67259×10^{-11} N·m²/kg², and R is the radius of the body in meters (m). For this formula, the reference separation, i.e., the separation between the body and the object for which the gravitational potential energy is defined as zero, is taken to be infinity. For the object to just barely leave the surface and not fall back, it must be able (at least in principle) to reach an infinite height from the surface of the body with no energy to spare (i.e., to lose all its KINETIC ENERGY on the way). Thus, the object should be able to achieve infinite separation, meaning zero potential energy, and be there at rest, which is zero kinetic energy. Its final mechanical energy then is zero. According to the law of conservation of energy, its initial mechanical energy, on the body's surface, must also be zero. (*See* CONSERVATION LAW; POTENTIAL ENERGY.)

If the object is moving on the surface at escape speed v_e in meters per second (m/s), its kinetic energy is:

$$E_k = mv_e^2/2$$

Its mechanical energy on the surface is the sum of its kinetic and potential energies there, and we put that equal to zero:

$$mv_e^2/2 + (-GMm/R) = 0$$

Now we solve for v_e to find the escape speed:

$$v_e = \sqrt{\frac{2GM}{R}}$$

Note that m cancels out, so all objects on the surface of the same astronomical body have the same escape speed.

The escape speed from the surface of the Earth is about 11 kilometers per second.

European Organization for Nuclear Research (CERN) Founded in 1954, CERN (Organisation Européenne pour la Recherche Nucléaire—originally Conseil Européen pour la Recherche Nucléaire, hence the acronym) is the world's largest research laboratory for elementary-particle physics. It formed one of the first joint ventures of European countries, with the purpose of restoring European physics to its former prestige after

View of the hall in which the superconducting magnets for the Large Hadron Collider particle accelerator are assembled at the European Organization for Nuclear Research (CERN). These magnets control the direction and shape of the particle beam. The accelerator is scheduled for completion in 2007. ***(Courtesy of the European Organization for Nuclear Research [CERN])***

the ravages of World War II. Its membership now numbers 20 nations. CERN is located near Geneva, Switzerland, and its grounds and facilities straddle the border of France and Switzerland. The organization operates by supplying particle accelerators and detectors for the use of research groups, which come from universities and institutions of its member countries as well as of those many other countries, including the United States. CERN boasts three Nobel Prize awards (in 1984 and 1992) for work performed there. Also, the World Wide Web was invented at CERN. (*See* ACCELERATOR, PARTICLE; ELEMENTARY PARTICLE.)

The laboratory's more recent accomplishments include:

- Research on the ELECTROWEAK INTERACTION through the use of the Large Electron-Positron Collider. This was a particle accelerator that brought high-energy ELECTRONS and POSITRONS into head-on collision. (*See* ENERGY.)
- Discovery of the W and Z BOSONS in PROTON-antiproton collisions. (*See* ANTIMATTER.)
- The creation of antimatter ATOMS.
- Evidence for the existence of a new state of matter, the quark-gluon PLASMA, which is some 20 times denser than nuclear matter. This state is assumed to have existed during the first few microseconds of the evolution of the UNIVERSE, according to the big-bang model, before the universe cooled sufficiently for the formation of particles of ordinary MATTER. (*See* BIG BANG; COSMOLOGY; GLUON; NUCLEUS; QUARK.)
- Dismantling of the Large Electron-Positron Collider in 2000 to make way for the Large Hadron Collider. Upon its completion (due in 2007), this accelerator will probe more deeply into matter than ever before by studying the results of extra-high-energy head-on collisions of PROTONS and, separately, of lead IONS.

CERN's website URL is http://www.cern.ch/. Its mailing addresses are CERN, CH-1211 Genève 23, Switzerland; and Organisation Européenne pour la Recherche Nucléaire, F-01631 CERN Cedex, France.

evaporation The process of PHASE TRANSITION from a LIQUID or a SOLID to a GAS is called evaporation. The solid-to-gas transition is also called sublimation. (Note that the term *vapor* is often used for a gas in contact with its liquid or solid PHASE.)

Evaporation occurs because the distribution of ENERGY among the constituent particles (ATOMS, MOLECULES, or IONS) of the liquid or solid endows some of them with sufficient energy to overcome the forces constraining them to their phase. Of those, some leave the liquid or solid phase and join the gas phase. Since evaporation reduces the fraction of highest-energy particles in the liquid or solid, the average energy per particle decreases, which means that the TEMPERATURE decreases. Thus, evaporation brings about cooling of the liquid or solid phase. (*See* KINETIC THEORY.)

The rate of evaporation is greater at higher temperatures of the liquid or solid. That happens because higher temperature means that there is a larger fraction of particles with sufficient energy to leave the liquid or solid phase.

See also BOILING; CONDENSATION.

exothermic process Any process that produces HEAT is called an exothermic process. An example is ordinary combustion, which is a chemical process of oxidation, whereby heat is generated and flows into the environment.

See also ENDOTHERMIC PROCESS.

experimental physics Physics is broadly divided into two endeavors: experimental physics and THEORETICAL PHYSICS. The experimental physicists, or experimentalists, are in direct contact with the natural phenomena. They plan the experiments. Then they design, often manufacture, and construct and assemble the experimental apparatus for performing the experiments. They debug the apparatus and calibrate it. When all is operating smoothly and reliably, they then perform the experiments. That can require very many runs of each experiment. The experimentalists collect data as the experiment progresses, process the data to obtain meaningful results, and publish the results so others can (hopefully) confirm them or (unfortunately) prove them wrong. Confirmed results then become grist for the mill of the theoretical physicists, or theoreticians, who try to put the results in broader contexts, relate them to other experimental results, and gain an understanding of them.

In the field of ELEMENTARY-PARTICLE physics, there is a specialized intermediate endeavor of phenomenology. Phenomenologists are positioned somewhere between the experimentalists and the theoreticians. They study the experimental results and look for order and regularities. Then the theoreticians take over from there.

In different fields at different times the physics might be experiment driven or theory driven. In the former case, the experimentalists of the field take the initiative by examining interesting (to them), unexplained phenomena. The theoreticians are, so to speak, running along behind them, trying to make sense of the experimental observations. In theory-driven physics, on the other hand, the theoreticians are in the lead, devising theories of nature that not only explain what is already known, but predict new phenomena. The experimentalists' job is then to check those theories against the real world, to disprove the false ones and to confirm the more successful ones.

In most fields of physics, a PHYSICIST is either an experimentalist or a theoretician, although there are some exceptions, when someone is involved in both endeavors. To obtain a doctoral degree, a physicist usually does either an experimental or a theoretical project as a basis for the dissertation. Thereafter, most physicists maintain their status as either an experimentalist or a theoretician for the rest of their professional lives.

Faraday, Michael (1791–1867) British *Chemist, Physicist* A self-educated scientist, Michael Faraday is famous for his work in almost all the fields he investigated: electrochemistry, ELECTRICITY, MAGNETISM, and ELECTROMAGNETISM. The SI UNIT of CAPACITANCE, the farad (F), is named for him. The whole of Faraday's professional career was spent in research at the Royal Institution in London, England. His meticulous experiments resulted in, among other accomplishments, laws of electrolysis and improved understanding of the production of magnetism by electricity and the generation of electricity by means of magnetism. In addition, he was active in the popularization of science through his writing and lectures, for which he achieved great repute. Faraday's discoveries contributed to the foundation of modern technologies in electrochemistry, electric power generation, and communication.

Faraday's law Named for the British physicist and chemist MICHAEL FARADAY, Faraday's law of induction states that the ELECTROMOTIVE FORCE (emf) induced in a conducting coil is proportional to the rate of change in time of the magnetic FLUX through the coil. Expressed in a formula, Faraday's law is:

$$\mathcal{E} = -N\, d\Phi_m/dt$$

where $\mathcal{E}$ denotes the induced emf in volts (V), N is the number of turns in the coil, Φ_m is the magnetic flux through the coil in UNITs of webers (Wb) (equivalent to tesla·meter^2 (T·m^2)), and t denotes time in seconds (s). (*See* CONDUCTION; INDUCTION; MAGNETISM.)

Michael Faraday was a self-taught 19th-century scientist, who carried out important investigations in electricity, magnetism, and electrochemistry. ***(AIP Emilio Segrè Visual Archives, W. Cady Collection)***

If the magnetic flux undergoes a net change of $\Delta\Phi_m$ in the time interval Δt, then the average emf induced in a coil during the time interval is given by:

$$\mathcal{E}_{av} = -N\ \Delta\Phi_m/\Delta t$$

Magnetic FLUX can change in a number of ways. One way is by a change of the magnetic FIELD. Another is through a change of the AREA of the coil. And yet another is as a result of change in the orientation of the coil in the magnetic field or of the magnetic field with respect to the coil.

The minus sign in the formulas is of symbolic significance as a reminder to refer to LENZ'S LAW, which determines the polarity of the induced emf. The best way to use the formulas is to take the absolute value of the right-hand side, thus obtaining the emf as a positive number, and then determine the polarity of the emf in the circuit by means of Lenz's law.

feedback For a system that accepts input and converts it to output, some of the output can be "fed back" to the input, thus creating a feedback loop where the output affects the input. There are two important types of feedback, positive feedback and negative feedback. In the former, the output strengthens the input. That is an unstable situation, since a stronger output causes a stronger input, from which the system creates an even stronger output, and so on. Nevertheless, positive feedback is useful in the design of oscillators, which are devices (electrical, mechanical, or other) that produce OSCILLATIONS at a determined FREQUENCY. An example of unwanted positive feedback is when a microphone picks up too much of the speaker output in a sound amplification system. The resulting howls are the result of amplification gone wild for certain frequencies because of the amplifier circuit's positive feedback for those frequencies.

In negative feedback, the output suppresses the input. This approach is useful for stability and control. A well-designed audio amplifier, for example, uses negative feedback to equalize its response over a range of frequencies. In speed-control devices, such as the cruise control in an automobile, deviation from the desired speed causes a negative signal to return to the speed control, thus returning the speed to its desired value. As an additional example, consider quality electronic oscillators, which are electronic devices that generate oscillating VOLTAGEs and are designed with negative feedback for maintaining stable oscillation frequencies over time. When the operating frequency drifts from its desired value, the deviation is returned negatively to the frequency control to bring about a cancellation of the deviation. Autopilots in aircraft utilize negative feedback similarly.

Fermi, Enrico (1901–1954) Italian/American *Physicist* Famous for his work in NUCLEAR PHYSICS and ELEMENTARY PARTICLES, Enrico Fermi was among the last of the physicists who excelled in both THEORETICAL PHYSICS and EXPERIMENTAL PHYSICS. Following receipt of a doctorate in physics from the University of Pisa, Italy, in 1922, Fermi discovered what is now

Enrico Fermi, famous as both an experimental and a theoretical physicist, did important work in nuclear physics and in elementary-particle physics. He was awarded the 1938 Nobel Prize in physics. *(AIP Emilio Segrè Visual Archives, Fermi Film Collection)*

known as FERMI-DIRAC STATISTICS. (The type of particle called a FERMION is named after him.) Then he studied BETA DECAY and became an expert in the interaction of NEUTRONs with NUCLEI, which causes nuclear FISSION and FUSION. During this time, Fermi held positions at the University of Florence, Italy, and then at the University of Rome. In 1938 he was awarded the Nobel Prize in physics "for his demonstrations of the existence of new radioactive elements produced by neutron irradiation, and for his related discovery of nuclear reactions brought about by slow neutrons." Immediately upon receiving the prize, Fermi moved to the United States to escape the Fascist regime in Italy and its danger to his Jewish wife.

During 1939–42 he served as professor of physics at Columbia University in New York City, New York. The years 1942–46 were devoted to leading a team of physicists as part of the Manhattan Project, which was the U.S. government's World War II effort to develop a nuclear-fission ("atomic") bomb and nuclear ENERGY. The team achieved the first controlled nuclear chain reaction in 1942. In 1946, Fermi joined the Institute for Nuclear Studies of the University of Chicago, Illinois, where he remained until his death. There is a legend that while Fermi was traveling on a long car trip, his recently bought used car broke down. He managed to reach a nearby gas station, where he repaired the car so skillfully that the owner of the gas station offered Fermi a job as an auto mechanic.

Fermi-Dirac statistics The statistical rules governing any collection of identical FERMIONs (particles having a half-integer value of SPIN: 1/2, 3/2, . . .) are called Fermi-Dirac statistics, named for the Italian/American physicist ENRICO FERMI and the English physicist PAUL ADRIEN MAURICE DIRAC. This statistics is based on (1) the absolute indistinguishability of the identical particles and (2) no more than a single particle being allowed in the same quantum state, including possessing the same QUANTUM NUMBERS. (*See* PAULI EXCLUSION PRINCIPLE; QUANTUM PHYSICS.)

One result of Fermi-Dirac statistics is that in a system of identical fermions in thermal EQUILIBRIUM at absolute TEMPERATURE T, the probability for a particle to possess energy in the small range from E to $E + dE$ is given by $f(E)\,dE$, where $f(E)$, the probability distribution function, is:

$$f(E) = \frac{1}{e^{\frac{E-E_F}{kT}} + 1}$$

Here E and E_F are in joules (J), T is in kelvins (K), f is dimensionless, and k denotes the Boltzmann constant, whose value is $1.3806503 \times 10^{-23}$ joule per kelvin (J/K). The quantity E_F is the Fermi energy. It is the energy for which the probability distribution function takes the value 1/2:

$$f(E_F) = 1/2$$

Note that this distribution function has its maximum value (less than 1) at zero energy and decreases to zero as the energy increases. For temperatures close to absolute zero, the function is quite flat, with the value 1 for energies below the Fermi energy, dropping abruptly to zero at the Fermi energy, and remaining at this value for higher energies. That reflects the uniform filling of energy states from zero energy to the Fermi energy, while the higher-energy states are unoccupied. The states are filled uniformly from zero energy up because only a single particle is allowed in any state. The Fermi energy can then be understood as the highest particle energy in the system, when the system is close to absolute zero temperature. (*See* DIMENSION; HEAT.)

Fermi-Dirac statistics is crucial for the understanding of, for instance, solid CONDUCTORS, SEMICONDUCTORS, and INSULATORS. In SOLIDs there are a very large number of ENERGY levels available to the ELECTRONS (which are fermions) of the material's ATOMs. The energy levels possess what is known as a band structure, whereby, depending on the material, they group into ranges ("bands") with gaps of forbidden energy in between. Electrons in sufficiently high-energy states, states of the CONDUCTION BAND, can move freely throughout the solid and conduct ELECTRICITY. Without the constraint of Fermi-Dirac statistics, the material's electrons would tend to populate the lowest-energy state available to them, and much energy would be required to bring electrons into the conduction band. All materials would then be insulators. As it is, however, the electrons fill the available states one electron per state (where each of the two spin directions corresponds to a different state), which allows the highest-energy electrons to be in the conduction band (in conductors) or in states from which relatively small amounts of energy can raise them to the conduction band (in semiconductors). In insulators, the gap between the conduction band and the lower-energy band is wide, and the highest-energy electrons are below the gap. Thus a large amount of energy is needed to raise electrons to the conduction band.

See also BOLTZMANN DISTRIBUTION; BOSE-EINSTEIN STATISTICS; MAXWELL-BOLTZMANN STATISTICS; SOLID-STATE PHYSICS.

Fermi National Accelerator Laboratory (Fermilab) Founded in 1967 at Batavia, Illinois, as the National Accelerator Laboratory, today's Fermi National Accelerator Laboratory, named for ENRICO FERMI, is a U.S. government laboratory under the Department of Energy. Fermilab employs approximately 2,200 people, including students, and has almost 2,300 researchers from outside institutions performing experiments.

The Cockcroft-Walton preaccelerator provides the first stage of acceleration at Fermi National Accelerator Laboratory (Fermilab). Inside this device, hydrogen gas is ionized to create negative ions, each consisting of a proton with two electrons. The ions are accelerated by an electric field to an energy of 750,000 electron volts (750 keV), which is about 30 times the energy of the electrons in the electron beam inside a television picture tube. ***(Fermilab photo. Courtesy of Fermi National Accelerator Laboratory)***

The laboratory is devoted to advancing the understanding of the fundamental nature of MATTER and ENERGY by providing guidance and resources for researchers to conduct basic research at the frontiers of elementary-particle physics and related disciplines. To that end, Fermilab builds and operates accelerators, detectors, and other facilities. Fermilab is the largest elementary-particle physics laboratory in the United States and is second in the world only to the EUROPEAN ORGANIZATION FOR NUCLEAR RESEARCH (CERN). The laboratory's organization also includes groups devoted to THEORETICAL PHYSICS and theoretical ASTROPHYSICS. (*See* ACCELERATOR, PARTICLE; ELEMENTARY PARTICLE.)

Fermilab's Tevatron is the world's highest-energy particle accelerator and collider. In this facility, using superconducting-MAGNET technology, counter-rotating beams of PROTONs and antiprotons produce collisions that allow scientists to examine the most basic building blocks of MATTER and the FORCEs acting on them. Important discoveries at the laboratory include the top and bottom QUARKs and the tau-type NEUTRINO. (*See* ANTIMATTER; SUPERCONDUCTIVITY.)

The website URL of the Fermi National Accelerator Laboratory is http://www.fnal.gov/. The laboratory's mailing address is Fermilab, P.O. Box 500, Batavia, IL 60510-0500.

fermion Named for the Italian/American physicist ENRICO FERMI, a fermion is any particle—whether an ELEMENTARY PARTICLE or a composite particle, such as a NUCLEUS or an ATOM—that has a half-integer value of SPIN (i.e., 1/2, 3/2, . . .). All the elementary particles that form MATTER are fermions. The most common are the PROTONs and NEUTRONs, which are spin-$^1/_2$ particles and are the building blocks of atomic nuclei. Together with nuclei, ELECTRONs, which are spin-$^1/_2$ particles too, form atoms. Protons and neutrons consist of QUARKs, which are spin-$^1/_2$ particles as well.

Collections of identical fermions are governed by the PAULI EXCLUSION PRINCIPLE and by FERMI-DIRAC STATISTICS.

See also BOSON.

ferromagnetism The term *ferromagnetism* refers to the strong response of certain materials to an applied magnetic FIELD, by which the material becomes polarized as a MAGNET and may even remain a magnet when the external magnetic field is removed. In other words, ferromagnetic materials become strongly magnetized in the presence of a magnetic field and might even become permanent magnets. They are strongly attracted to the magnet producing the applied field. (*See* MAGNETISM; POLARIZATION.)

The mechanism of ferromagnetism is as follows. Each ATOM of a ferromagnetic material has a relatively large number of unpaired ELECTRONs, on the average, whose MAGNETIC DIPOLES (related to the electron's SPIN) endow it with a considerable magnetic dipole moment. There is strong coupling among the magnetic dipole moments of nearby atoms, which causes the atoms to spontaneously form local clusters, in each of which all the magnetic dipoles are aligned. Such clusters are called MAGNETIC DOMAINs and are observable under a microscope. Thus a domain serves as an elementary magnet. With no application of an external magnetic field, the domains are oriented randomly, so their contributions to the magnetic field cancel, and the bulk material is not magnetized. (*See* MOMENT, MAGNETIC DIPOLE.)

In the presence of an applied magnetic field, two processes can occur. The one that occurs in weaker applied fields is the growth of domains whose magnetic dipole moments are aligned with the applied field at the expense of the other domains. The magnetic fields of the aligned domains reinforce each other, and the material thus becomes a magnet. When the applied field is turned off, however, the domains randomize again, and the material loses its magnetism. This process occurs when paper clips or iron nails, for instance, are brought near magnets. They become magnets themselves, but only when the inducing magnet is nearby.

A process that requires stronger applied fields is the irreversible rotation of the alignment of domains. In this process, domains that are not aligned with the applied field have their magnetic dipole moments rotated toward alignment. The stronger the applied field, the better the resulting alignment. Again, the fields of the aligned magnetic dipole moments reinforce each other. In this process, the new alignments are retained when the applied field is removed. The material becomes a permanent magnet. This is the process by which permanent magnets, such as compass needles, refrigerator magnets, and loudspeaker magnets, are produced. It is also the process of magnetic recording on audio and video tapes and on magnetic data storage devices, such as computer diskettes and hard drives.

The induced magnetization is such that a ferromagnetic material strongly enhances the applied magnetic field. Accordingly, the magnetic PERMEABILITY of a ferromagnetic material is much larger, even as much as thousands of times larger, than the permeability of the VACUUM, whose value is $4\pi \times 10^{-7}$ tesla·meter per ampere (T·m/A). The value of the permeability of a ferromagnetic material depends on the magnitude of the applied field. (*See* INDUCTION.)

Since HEAT is the random motion of the atoms that constitute a material, heat acts in opposition to the coupling that aligns atoms to form magnetic domains. So heat is an enemy of magnetism. That is why we are warned to keep video cassettes and computer diskettes away from heat sources. Depending on the material, there is a TEMPERATURE, called the Curie temperature, above which ferromagnetism disappears. (*See* CURIE, PIERRE.)

See also DIAMAGNETISM; HYSTERESIS; PARAMAGNETISM; PERMEABILITY.

Feynman, Richard Phillips (1918–1988) American *Physicist* Recognized for his brilliance and unique personality, Richard Feynman is famous for his development of QUANTUM ELECTRODYNAMICS and his theoretical work in ELEMENTARY PARTICLES. The FEYNMAN DIAGRAM was his invention. Feynman received his Ph.D. in physics from Princeton University, New Jersey, in 1942. During 1941–45, he took part in the Manhattan Project, the U.S. government's World War II effort to develop a nuclear-FISSION ("atomic") bomb and nuclear ENERGY, moving to what is now LOS ALAMOS NATIONAL LABORATORY in 1943. After the war, Feynman served as professor of physics at Cornell University, in Ithaca, New York, during 1945–50 and then at the California Institute of Technology, in Pasadena, for the rest of his life. In 1965, he shared with Sin-Itiro Tomonaga and Julian S. Schwinger the Nobel Prize in physics "for their fundamental work in quantum electrodynamics, with profound consequences for the physics of elementary particles." Feynman played a major role, as a member of an investigating committee, in determining the cause of the deadly 1986 explosion on the space shuttle *Challenger.* (*See* NUCLEAR PHYSICS; NUCLEUS.)

The inventor of the Feynman diagram, Richard Feynman was a colorful character who is famous for developing quantum electrodynamics and for his work in elementary-particle physics. He is shown in his office, in front of quantum field theory equations. Feynman shared the 1965 Nobel Prize in physics for his work. ***(AIP Emilio Segrè Visual Archives, Physics Today Collection)***

Feynman diagram The invention of the American physicist RICHARD PHILLIPS FEYNMAN, Feynmann diagrams are pictorial representations of virtual processes that take part in physical processes, according to a certain approach to QUANTUM ELECTRODYNAMICS (QED). QED is the quantum theory of ELECTROMAGNETISM and has proved to be very successful. One approach to solving the equations of QED is to view a physical process as comprising an infinite set of idealized processes, all occurring simultaneously and each contributing in its way to the total physical process. The constituent processes are termed virtual processes, in that they cannot be isolated and observed as individual physical processes. The more complex the virtual process, the less its contribution to the physical process. Feynman diagrams represent virtual processes. This approach, called perturbative QED, was pioneered by Feynman and others in the middle of the 20th century. (*See* QUANTUM PHYSICS.)

As an example, consider the electric repulsion of two ELECTRONs from each other. It can be viewed as consisting of a simultaneous infinite sequence of increasingly complex virtual processes. The simplest virtual process is the exchange by the electrons of a single PHOTON, whereby one electron emits a photon that is absorbed by the other. This process contributes the most to electron-electron interaction. A more complex virtual process is the exchange of two photons. Another virtual process is the exchange of a single

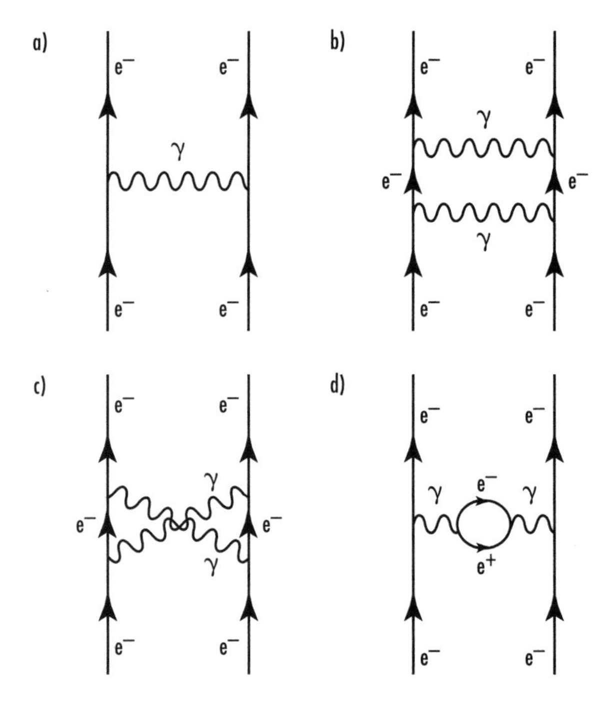

Feynman diagrams are graphic representations of virtual processes occurring among elementary particles, where such processes contribute to physical processes. The Feynman diagrams of the virtual processes that contribute most to the mutual electric repulsion of two electrons are shown. (a) One-photon exchange. (b) and (c) Two ways of two-photon exchange. (d) One-photon exchange, where the exchanged photon converts to an electron-positron pair that recombines to a photon. e^-, e^+, and γ denote an electron, a positron, and a photon, respectively. Time runs upward in the diagrams.

photon, but with the emitted photon converting to an electron-POSITRON pair that then recombines to produce a photon again, which is absorbed by the second electron. Further contributions come from virtual processes that are even more complex but that contribute correspondingly less to the physical process. The particles that materialize and dematerialize in virtual processes, that never achieve physical reality, so to speak, are called virtual particles. (*See* ELECTRICITY.)

The perturbative approach is applicable as well to the deceptively simple process of a free (i.e., not interacting) particle, such as an electron, at rest or moving at constant VELOCITY. The virtual processes in this case all contribute, in inverse relation to their complexity, to the properties of the particle, such as its MASS and electric CHARGE. The greatest contribution comes from a virtual noninteracting, or "bare," electron. Then the virtual electron can emit a virtual photon and reabsorb it. That is the next simplest virtual process available to it. Further, an electron can emit two virtual photons and reabsorb them. Also, an electron can emit a single virtual photon and reabsorb it, but in between the photon converts to a virtual electron-positron pair, which recombines back to a photon. Additional contributions come from virtual processes that are even more complex.

Feynman diagrams represent pictorially the mathematical expressions of the virtual processes that go into the complete description of physical processes in QED. They help physicists obtain an intuitive grasp of the virtual processes involved, as well as keep track of the virtual processes in order to avoid neglecting any mathematical contribution to the calculation. It has been through perturbative techniques and the use of Feynmann diagrams that physicists have obtained very accurate results from QED. Consequently, quantum electrodynamics is considered the best-confirmed theory in physics.

In addition to their service to QED, Feynman diagrams form useful tools in other QUANTUM FIELD THEORIES as well, including those of the WEAK INTERACTION and the STRONG INTERACTION.

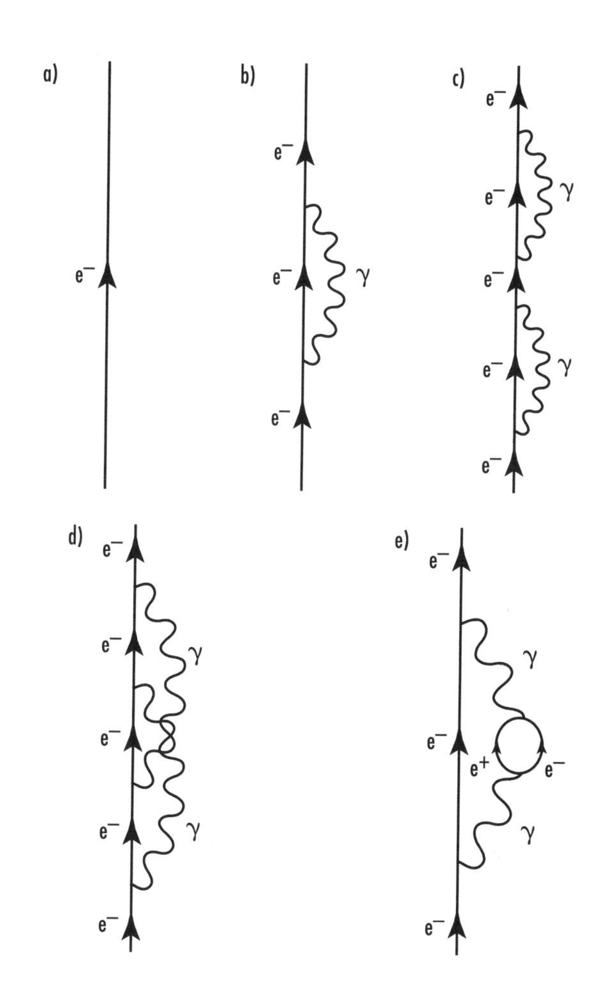

The Feynman diagrams of the virtual processes that make the greatest contributions to the properties of a free electron. (a) No interaction, or "bare" electron. (b) Emission and reabsorption of a single photon. (c) and (d) Two ways of emission and reabsorption of two photons. (e) Emission and reabsorption of a single photon, where the photon converts to an electron-positron pair that recombines to a photon. e^-, e^+, and γ denote an electron, a positron, and a photon, respectively. Time runs upward in the diagrams.

fiber optics Fiber optics involves the use of total internal reflection to transmit information and images through thin transparent fibers by means of LIGHT. The effect of total internal reflection from the walls of the fiber prevents the light that is propagating along the fiber from leaking out to any considerable extent. Due

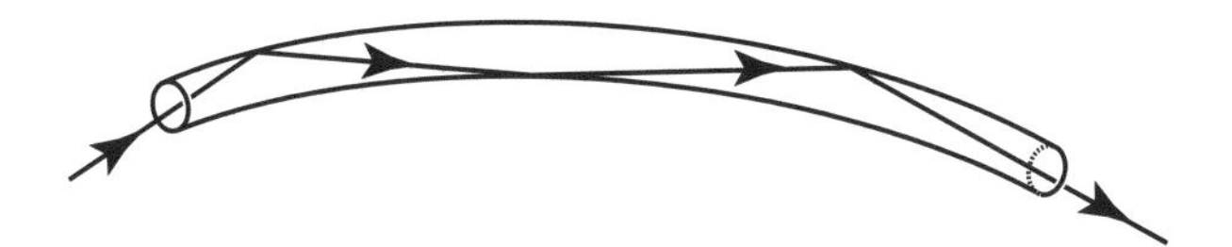

A ray of light is shown propagating through a segment of an optical fiber. The ray is confined to the interior of the fiber by total internal reflection at the fiber's wall.

to the small diameter of a fiber, most of the light that enters it at an end strikes the wall at a sufficiently large angle of incidence (i.e., greater than the critical angle) that it is totally internally reflected. Such total internal reflection is repeated again and again as the light propagates along the fiber. (*See* OPTICS; REFLECTION; REFRACTION.)

Data and conversations can be transmitted through a single fiber. For that, the world is encompassed by fiber optics networks, which make long- and short-distance communication fast and dependable. For the simultaneous transmission of entire images, a bundle of fibers is used. Such bundles are found, for example, in medical devices such as endoscopes, which allow the viewing of the internal condition and working of our bodies.

field In CLASSICAL PHYSICS, a field is a condition of SPACE by which MATTER interacts with matter. Matter both serves as a source of a field and is affected by it. Fields supply an answer to the question of action-at-a-distance: how does matter affect other matter when they are not in contact? How do two separated electrically charged bodies attract or repel each other? How do the Earth and Moon attract each other gravitationally? (*See* CHARGE; ELECTRICITY; GRAVITATION; INTERACTION.)

In the electric case, for example, the electric charge of either particle is considered to be the source of the electric field, which pervades all space. The field affects the other particle by causing a FORCE to act on it. The result is such that COULOMB'S LAW is valid. In general, all charges contribute to the electric field as source charges. When the effect of the field on a test charge is considered, the test charge itself does not serve also as a source charge (i.e., it does not act on itself). The situation is similar for moving charges interacting magnetically, which affect each other via the magnetic field. More generally, the electromagnetic field, of which the electric and magnetic fields are components, serves to mediate the ELECTROMAGNETIC INTERACTION. The electromagnetic field contains ENERGY and transmits energy and MOMENTUM by means of WAVES. (*See* ELECTROMAGNETISM; MAGNETISM.)

In like manner, the gravitational field mediates the gravitational force. All MASS contributes to the gravitational field. The effect of the field on a test mass is a force acting on it, the gravitational force. Again, a test mass is not considered to serve as a source for the field that is affecting it. Waves in the gravitational field transmit energy and momentum.

In QUANTUM PHYSICS, fields not only mediate the interactions among matter, but describe matter itself. According to quantum theory, every type of ELEMENTARY PARTICLE is described by a field, and every type of field possesses a particle manifestation as a certain mode of field behavior. The electromagnetic field, for instance, which is an interaction field, describes PHOTONs as the particle manifestation of the field. The GLUON field is the field of the STRONG INTERACTION and describes GLUONs as mediators of the strong interaction among QUARKs. Quarks are described by the quark field and ELECTRONs by the electron field, both of which are matter fields. And so on for all the interactions, with their corresponding mediating particles, and all types of matter particles.

See also FIELD THEORY; QUANTUM THEORY.

field emission The method of producing free ELECTRONs by causing them to be forced out of matter under the influence of a strong electric FIELD is termed field emission. The electric field in the vicinity of an electrically charged CONDUCTOR is especially strong near regions of high curvature of the surface of the conductor. To maximize the effect, a conducting wire is sharpened to a very fine point. The conductor is placed in a VACUUM chamber and given a negative potential (i.e., is charged negatively). For sufficient potential, the electric field at the tip of the point exerts forces on the electrons at the conductor's surface that overcome the forces confining them to the material and "tears" them out of the conductor. Quantum TUNNELING also plays a role in this process, allowing the emission of electrons that otherwise would not be released by the electric field. (*See* CHARGE; ELECTRICITY; POTENTIAL, ELECTRIC; QUANTUM PHYSICS.)

The field emission microscope is based on that process. The tip of the conductor's point is then given a regular, round shape. Since electrons in the conductor are most likely to be in the neighborhood of ATOMs, the electrons pulled from the conductor by field emission emanate mostly from the atoms that are near the surface. These electrons leave the surface perpendicularly to the surface and travel in straight lines to the chamber's walls, which are designed to fluoresce when electrons strike them, or otherwise record the electrons' arrival. Thus the microscope creates an image of the atomic structure of the material from which the conductor is made. (*See* FLUORESCENCE.)

Additional methods of causing electron emission from metals include the PHOTOELECTRIC EFFECT and THERMIONIC EMISSION.

field theory The mathematical study of the behavior of FIELDS is called field theory. In CLASSICAL PHYSICS, the fields studied include the classical INTERACTION fields, which are the electromagnetic field and the gravitational field. Among the fields that QUANTUM FIELD THEORY deals with are the four interaction fields (those of ELECTROMAGNETISM, GRAVITATION, the STRONG INTERACTION, and the WEAK INTERACTION) and the matter fields (such as the ELECTRON field, the QUARK field, etc.).

See also GAUGE THEORY; QUANTUM ELECTRODYNAMICS (QED).

fine structure constant In QUANTUM ELECTRODYNAMICS, which is the quantum theory of ELECTROMAGNETISM and its INTERACTION with MATTER, there appears a certain dimensionless expression, called the fine structure constant. This expression contains a number of physical constants and is taken as a measure of the strength of the ELECTROMAGNETIC INTERACTION in comparison with that of other interactions. The explicit form—in SI UNITS—and value of the fine structure constant, denoted by α, is:

$$\alpha = \frac{e^2}{2\varepsilon_0 hc} = 7.29737 \times 10^{-3} \approx 1/137$$

where e denotes the elementary electric CHARGE, whose value is 1.60218×10^{-19} coulomb (C); h is the PLANCK CONSTANT, with value 6.626076×10^{-34} joule·second (J·s); c denotes the SPEED OF LIGHT, whose value is 2.99792458×10^8 meters per second (m/s); and ε_0 is the PERMITTIVITY of the VACUUM, with value 8.85419×10^{-12} $C^2/(N{\cdot}m^2)$.

See also ELECTRICITY; ELECTROMAGNETIC INTERACTION; DIMENSION; QUANTUM PHYSICS.

fission The disintegration of atomic NUCLEI into two (usually, but possibly more) large parts is termed fission, or nuclear fission. It might be a spontaneous process, which is a rare form of RADIOACTIVITY. Or it might be induced as a nuclear reaction by bombarding the nuclei. (*See* ATOM; NUCLEAR PHYSICS.)

To bring about induced fission, nuclei are commonly bombarded with NEUTRONs. A neutron is absorbed into a nucleus, which then becomes a different ISOTOPE of the same ELEMENT and is often unstable, immediately undergoing fission. The products of the fission often include a number of free neutrons per disintegrating nucleus. If those neutrons are allowed to efficiently bombard additional nuclei, a self-sustaining chain reaction can occur, accompanied by a release of ENERGY in the form of KINETIC ENERGY of the reaction products. When the reaction is controlled, the released energy can be harnessed and used. A nuclear reactor is a device that does just that. An uncontrolled chain reaction results in an explosion. That is the mechanism of fission bombs, often called atom bombs, atomic bombs, or A-bombs. (*See* REACTOR.)

Here is an example of such a nuclear process, used in reactors and in bombs. The isotope uranium-235, denoted ${}^{235}_{92}U$ (ATOMIC NUMBER 92, ATOMIC MASS 235), absorbs a neutron, denoted ${}^{1}_{0}n$ (atomic number 0, atomic mass 1), and becomes uranium-236 in a highly unstable state, denoted ${}^{236}_{92}U^*$ (atomic number 92, atomic mass 236, with the asterisk indicating an unstable state). This immediately disintegrates into two nuclear fragments, commonly into barium-141 and krypton-92, denoted ${}^{141}_{56}Ba$ (atomic number 56, atomic mass 141) and ${}^{92}_{36}Kr$ (atomic number 36, atomic mass 92), respectively, with the release of three neutrons. The two-stage reaction is represented by:

$${}^{1}_{0}n + {}^{235}_{92}U \rightarrow {}^{236}_{92}U^* \rightarrow {}^{141}_{56}Ba + {}^{92}_{36}Kr + 3\,{}^{1}_{0}n$$

See also FUSION, NUCLEAR.

fluid Fluid is MATTER in any state that is not SOLID (i.e., either LIQUID or GAS). In other words, a fluid is matter that can flow. What differentiates a fluid from a solid is that, whereas in the latter the constituent particles (ATOMs, MOLECULEs, IONs) are constrained to fixed positions relative to each other, in a fluid the particles are sufficiently free that flow can occur. In a liquid, the particles move rather freely about, maintaining only a constant average distance from each other. A sample of liquid maintains a fixed volume, while flowing to take the shape of its container. The particles of a gas are even less constrained, and a gas flows so as to completely fill the volume of its container. (*See* STATE OF MATTER.)

fluorescence The emission of electromagnetic RADIATION by a material as a result of the material's excitation and simultaneously with the excitation is termed fluorescence. The mechanism of fluorescence is the absorption of ENERGY by the particles constituting the material (ATOMs, MOLECULEs, IONs) and the release of the energy as electromagnetic radiation through the particles' immediate transition to lower-energy states. Fluorescence is a special case of LUMINESCENCE, which is the general phenomenon of emission of electromagnetic radiation by excited materials, whether immediately or not. If the emission persists after the cessation of excitation, the effect is termed PHOSPHORESCENCE. (*See* ELECTROMAGNETISM.)

The excitation can be of various forms. Commonly, ultraviolet radiation excites the material, while visible LIGHT is emitted. Such is the case for the material that coats the inside of fluorescent tubes, where the ultraviolet radiation comes from the electric discharge through the GAS that fills the tube. When, as in this case, the excitation is by electromagnetic radiation, the WAVELENGTH of the emitted radiation is longer, and its FREQUENCY accordingly lower, than that of the absorbed radiation. Another possible source of excitation is energetic ELECTRONs. That is what occurs in cathode ray tubes (CRTs), such as television picture tubes and computer monitor tubes. There, accelerated electrons strike the coating on the inside of the picture surface, which consequently emits visible light. (*See* CONDUCTION; ELECTRICITY.)

flux The flux of a VECTOR field through a directed surface is the integral of the FIELD over that surface. Consider a vector field **F** and a surface. (The surface might be bounded by a curve, in which case it possesses an edge. Alternatively, the surface might be closed, such as the surface of a sphere, and have no edge. Then it will possess an inside and an outside.) Define a direction for the surface by thinking of one side of the surface as "from" and the other side as "to" and, at each point of the surface, imagining an arrow pointing from the "from" side to the "to" side. At each point of the surface, consider an infinitesimal element of surface area of magnitude dA and define a corresponding vector $d\mathbf{A}$ as the vector whose magnitude equals dA and that is perpendicular to the surface element in the direction of the "from-to" arrow. At each point of the surface, form the scalar product of the field vector **F**

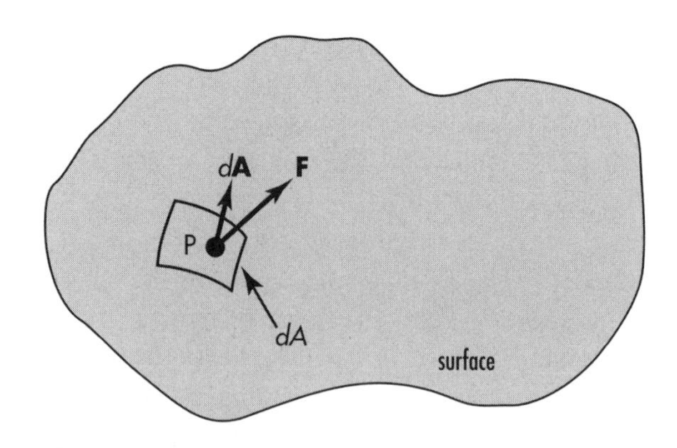

The figure shows a surface through which the flux of a vector field is to be calculated. An infinitesimal surface area element of magnitude *dA* is shown at an arbitrary point P on the surface. *d*A denotes the vector perpendicular to the surface at P with magnitude *dA*. Its chosen sense, from one side of the surface to the other, is the same for all points on the surface. F denotes the value of the vector field at P. The flux of the vector field through the surface is the integral of the scalar product F·*d*A over the whole surface.

and the surface-element vector $d\mathbf{A}$, $\mathbf{F}\cdot d\mathbf{A}$ (which equals $F\,dA\cos\theta$, where F is the magnitude of **F** and θ is the angle between **F** and $d\mathbf{A}$). Now, integrate this expression over the whole surface to obtain the flux of the field through the surface in the defined direction:

$$\Phi = \int_{\text{surface}} \mathbf{F} \cdot d\mathbf{A}$$

To help obtain a sense of what flux is about, imagine that the vector field **F** describes some flow (which is the origin of the word *flux*) by giving the VELOCITY of a LIQUID at every point. (Of course, in general, fields do not represent flows, but this model might be of help.) To make matters simple, let the flow be uniform and the surface bounded and flat. In other words, let **F** have the same value at every point, with constant magnitude (the flow SPEED in this model), F, and constant angle, θ, between the direction of flow and the perpendicular to the surface. Denote the area of the surface by A. Then the flux becomes simply:

$$\Phi = FA\cos\theta$$

This is the volume of liquid that flows through the surface per unit time. Note that the flux is directly proportional to both F and A: double the amount of liquid flowing through an opening by doubling the flow speed or doubling the area of the opening. Note the effect of direction: for given flow speed and area, the amount of liquid passing through the opening is maximal for flow that is perpendicular to the surface, $\theta = 0$, decreases as the flow is slanted away from the perpendicular, and vanishes for flow parallel to the surface, $\theta = 90°$. For angles greater than 90° and up to 108° the flux is increasingly negative, indicating flow in the negative sense, against the "from-to" arrow. For $\theta = 180°$ the flux is maximally negative. (*See* HYDRODYNAMICS.)

Flux is an important concept in ELECTRICITY and MAGNETISM. In particular, GAUSS'S LAW in electricity involves the electric flux, the flux of the electric field, **E**, through a closed surface. This flux is:

$$\Phi_e = \oint_{surface} \mathbf{E} \cdot d\mathbf{A}$$

The SI UNIT of electric flux is newton·meter² per coulomb (N·m²/C), where **E** is in newtons per coulomb (N/C) and $d\mathbf{A}$ is in square meters (m²).

In magnetism, FARADAY'S LAW has to do with the rate of change of the magnetic flux, the flux of the magnetic field, **B**, through a bounded surface. Magnetic flux is:

$$\Phi_m = \int_{surface} \mathbf{B} \cdot d\mathbf{A}$$

The SI unit of magnetic flux is the weber (Wb), equivalent to tesla·meter² (T·m²), and the magnetic field is in teslas (T).

Other fluxes are used in physics. One is radiant flux, which is the POWER transported by LIGHT or other electromagnetic RADIATION. Its SI unit is that of power, the watt (W). Another is luminous flux, the power transported through a surface by light, as corrected and evaluated for the light's ability to cause visual sensation in humans. The SI unit of luminous flux is the lumen (lm), equivalent to candela·steradian (cd·sr). The difference between luminous flux and radiant flux is that, while the latter involves all radiation energy, the former weighs the energy according to the strength of its visual effect. For instance, humans are insensitive to infrared and ultraviolet radiation. For visible light, we are most sensitive to a wavelength of about 560 nm, which produces the sensation of yellow-green. (Note that the radiant intensity of the Sun is maximal for this wavelength.) (*See* ELECTROMAGNETISM.)

Each of those fluxes has an associated surface DENSITY. The radiant flux density—also called irradiance, or radiant power density—on a surface is the radiant flux through or into the surface per unit area of surface. Its SI unit is watt per square meter (W/m²). Similarly, luminous flux density—also called illuminance or illumination—is the luminous flux through or into the surface per unit area of surface. Its SI unit is lumen per square meter (lm/m²).

See also INTENSITY, where luminous intensity and radiant intensity are discussed.

force Force is often defined as that which causes a body's ACCELERATION or causes change of a body's MOMENTUM. This is based on ISAAC NEWTON's second law of MOTION, which is expressed by:

$$\mathbf{F} = m\mathbf{a}$$

or:

$$\mathbf{F} = d\mathbf{p}/dt$$

Here **F** denotes force, which is a VECTOR quantity whose SI UNIT is the newton (N), m is the MASS in kilograms (kg) of the body on which the force acts, **a** the acceleration of the body in meters per second per second (m/s²), **p** the momentum of the body in kilogram·meters per second (kg·m/s), and t the time in seconds (s). The equations express the effect of a force acting on a body. The second equation is the more general, while the first is valid only for speeds that are sufficiently small compared with the SPEED OF LIGHT and thus applicable to everyday situations. (*See* NEWTON'S LAWS; RELATIVITY, SPECIAL THEORY OF.)

It would appear, then, that a force acting on a particle is detected and identified by the particle's acceleration, or change of momentum. Newton's second law of motion is valid only in inertial reference frames, which are those reference frames in which Newton's first law of motion is valid. Inertial reference frames are then those reference frames in which a particle that has no net force acting on it moves at constant SPEED in a straight line (or remains at rest, as the special case of zero speed) and vice versa. So the validity of the common definition of force, based on Newton's second law, hinges on inertial reference frames, whose definition involves force, according to Newton's first law. As a result, the common definition of force is flawed in that it refers to what it is defining. (*See* REFERENCE FRAME, INERTIAL.)

It seems that Newton's first law requires a primitive, underived understanding of a force as an obvious, readily identifiable physical effect. Apparently, Newton's thinking was that if a particle is not moving at constant speed in a straight line (or is spontaneously abandoning its state of rest)—but our examination of it reveals no physical influence affecting it—then we are not observing the particle in an inertial reference frame. In a decelerating automobile, for example, passengers appear to be accelerating toward the windshield or, if they are restrained by seat belts and cannot thus accelerate, they seem to be under the effect of a forward force. But there is no physical influence to be found that can cause this. In fact, a decelerating car is not an inertial reference frame. Once we have thus identified inertial reference frames, and only then, we can apply the second law. But there is no point in then defining force by the second law, since we have already made use of the concept in identifying inertial reference frames.

There is, indeed, arbitrariness in the definition of a force. That should not be of concern in elementary applications, but it needs to be taken into account in advanced physics. ALBERT EINSTEIN's general theory of relativity, for instance, abolishes the gravitational force. Instead, Newton's first law is modified to the effect that in the absence of a net nongravitational force, a body moves along a geodesic, which is a curve in SPACE-TIME that is the closest to a straight line. For situations in which Newton's law of gravitation is valid, the resulting motion of a body, according to Einstein, is that which would have been found nonrelativistically by using Newton's second law to calculate the effect of the gravitational force on the body's motion. What is a dynamic effect (involving force) in nonrelativistic physics is an inertial effect (involving the geometry of space-time) in the general theory of relativity. (*See* DYNAMICS; GRAVITATION; INERTIA; RELATIVITY, GENERAL THEORY OF.)

See also CONSERVATIVE FORCE; NONCONSERVATIVE FORCE; REACTION FORCE.

fractal This is a form that possesses the property of self-similarity: it has the same appearance when viewed at different scales. Approximate fractals turn up in various physical processes. The shape of coastlines and the vertical appearance of mountain ranges, as examples, are both approximately fractals. Fractal patterns can be generated mathematically and can be viewed on a computer screen or drawn. (*See* SYMMETRY.)

free energy The amount of HEAT (i.e., thermal ENERGY) contained in a system that can be converted into WORK is called the free energy of the system. According to the second law of THERMODYNAMICS, not all the heat contained in a system, also referred to as the system's internal energy, is available for conversion into work. The amount that is available depends on the type of conversion process being considered. So various "free energies" are defined for different processes. Any spontaneous changes that occur in a system under the conditions of each process involve a decrease of the corresponding free energy. So the appropriate free energy of a system interacting with its environment is minimal at EQUILIBRIUM. (Note that it is for isolated systems that equilibrium corresponds to maximal ENTROPY.) Free energy is a scalar quantity, and its SI UNIT is the joule (J). Two commonly used free energies are the following.

For a system that is interacting with its environment and undergoing a process in which its VOLUME and TEMPERATURE remain constant, the appropriate free energy is the Helmholtz free energy, named for the German scientist HERMANN LUDWIG FERDINAND VON HELMHOLTZ:

$$F = U - TS$$

Here U denotes the internal energy of the system in joules (J), T its absolute temperature in kelvins (K), and S its entropy in joules per kelvin (J/K). The Helmholtz free energy in a system can be expressed in microscopic terms through the system's partition function, Z, as:

$$F = -kT \ln Z$$

Here k denotes the Boltzmann constant, whose value is $1.3806503 \times 10^{-23}$ joule per kelvin (J/K), and Z is dimensionless. (*See* BOLTZMANN DISTRIBUTION; DIMENSION.)

In the case of a nonisolated system that undergoes a process in which its PRESSURE and temperature remain constant, the corresponding free energy is the Gibbs free energy, named after the American physicist JOSIAH WILLARD GIBBS, and is defined as:

$$G = F + pV = U + pV - TS$$

where p denotes the pressure in pascals (Pa) and V the volume in cubic meters (m^3).

free fall The state of a body in which the only FORCE acting on it is the gravitational force is free fall. A person in free fall has the perception of weightlessness, although not being truly weightless. Such a person is not able to detect by any measurement performed in her immediate vicinity whether she is truly weightless or in free fall. Such is the case for astronauts in an orbiting space laboratory, as an example. At the same point in space, all objects in free fall undergo the same ACCELERATION, called free-fall acceleration or gravitational acceleration. (Near the surface of an astronomical body, such as the Earth, that acceleration is termed acceleration due to gravity. Its value near the surface of the Earth is approximately 9.8 meters per second per second (m/s^2).) Those effects were among ALBERT EINSTEIN's considerations in developing the general theory of relativity. (*See* GRAVITATION; RELATIVITY, GENERAL THEORY OF; WEIGHT.)

freezing The process of freezing is the PHASE TRANSITION from a FLUID (i.e., LIQUID or GAS) to a SOLID. (Note that the term *vapor* is often used for a gas in contact with its solid [or liquid] PHASE.) Freezing occurs because the distribution of KINETIC ENERGY among the constituent particles (ATOMs or MOLECULES) of the fluid leaves some of them with sufficiently low energy that they can stick firmly together when they collide or remain stuck to the solid phase when they impinge upon it. The rate of freezing increases with lower TEMPERATUREs. That happens because lower temperature means there is a greater fraction of low-energy particles in the fluid or liquid. The process of freezing is in competition with the processes of MELTING and EVAPORATION. At temperatures higher than the freezing point, the latter win out, and the material does not solidify. When the temperature is lowered to the freezing point, the solid phase starts to form. As HEAT is further removed, the temperature no longer falls, but rather more of the fluid phase converts to a solid. When the phase transition is complete for the sample, further heat removal lowers the temperature. (*See* KINETIC THEORY.)

When the particles arrange themselves in an orderly manner upon freezing, the process is also one of crystallization. (*See* CRYSTAL.)

See also HEAT OF FUSION; STATE OF MATTER.

frequency For a periodic process, such as OSCILLATION, in which the same basic operation, the cycle, is repeated over and over, the number of cycles per unit time is the frequency of the process. The SI UNIT of frequency is the hertz (Hz), which is the number of cycles per second. The frequency, f, of a periodic process is the inverse of its PERIOD, T, which is the time duration of a single cycle:

$$f = 1/T$$

where the period is expressed in seconds(s). The angular frequency, ω, is also used and is defined as:

$$\omega = 2\pi f$$

Its SI unit is radian per second (rad/s).

frequency, natural A natural frequency of a system is any of the FREQUENCIES of free OSCILLATION of the system when it is displaced from a state of stable equilibrium. The tendency of a system to oscillate when displaced, or disturbed, from a state of stable EQUILIBRIUM is inherent to the nature of stable equilibrium. When a system is displaced from its state of stable equilibrium, it returns to that state. Due to its INERTIA, the system overshoots the equilibrium state, returns again, overshoots again, and so on. (Think of a swinging PENDULUM.) The exact character of such oscillation depends on the way the system is initially displaced from equilibrium. For each system, there is a certain number (which might be infinite) of ways of displacement that cause it to oscillate at a single frequency (rather than in a more complex way involving more than one frequency). Such oscillations are called the modes of oscillation, or normal modes, of the system, and the frequencies of the various modes are the natural frequencies of the system. The number of oscillation modes equals the number of DEGREES of FREEDOM of the system, in general. Every free oscillation of a system is actually some combination of its oscillation modes. Here are examples of modes of oscillation for mechanical systems (although the concept is much more general).

A simple pendulum, for example, has but a single mode of oscillation and, correspondingly, a single natural frequency given by (for small displacements from equilibrium):

$$f = \frac{1}{2\pi}\sqrt{\frac{g}{L}}$$

where f denotes the frequency in hertz (Hz), g the acceleration due to gravity, whose value is approximately 9.8

meters per second per second (m/s^2), and L the length of the pendulum in meters (m). (*See* ACCELERATION; GRAVITATION.)

A body oscillating due to a spring attached to it also possesses a single mode of oscillation. The natural frequency of this system is:

$$f = \frac{1}{2\pi}\sqrt{\frac{k}{m}}$$

where k is the spring constant of the spring in newtons per meter (N/m) and m denotes the body's MASS in kilograms (kg). (*See* ELASTICITY.)

A system consisting of two identical pendulums connected by a spring has two modes of oscillation and two natural frequencies. One mode has both pendulums swinging identically, so that the distance between them remains constant. Then the spring has no effect, and each pendulum swings as if it is not coupled to the other. The frequency of this mode, f_1, is the same as that of a single pendulum:

$$f_1 = \frac{1}{2\pi}\sqrt{\frac{g}{L}}$$

In the second oscillation mode, the pendulums oscillate in mirror-image fashion, alternately approaching and receding from each other, causing the spring to be compressed and stretched, respectively. The corresponding frequency, f_2, is higher than for the other mode:

$$f_2 = \sqrt{f_1^2 + \frac{1}{4\pi^2}\frac{2k}{m}}$$

A stretched uniform string is an example of a system possessing an infinite number of oscillation modes and, accordingly, an infinite number of natural frequencies, which are given by (for small transverse displacements from equilibrium):

$$f_n = \frac{n}{2L}\sqrt{\frac{F}{\mu}} \quad \text{for } n = 1, 2, \ldots$$

Here f_n denotes the nth frequency of the infinite sequence of natural frequencies, F is the TENSION of the string in newtons (N), L the length of the string in meters (m), and μ is the string's linear DENSITY in kilograms per meter (kg/m). Note that in this case (but not as a general rule), the natural frequencies are all integer multiples of the lowest one:

$$f_n = nf_1$$

When a stretched string is plucked or otherwise caused to oscillate, it will freely oscillate in some simultaneous combination of its oscillation modes, in general, depending on the manner of excitation.

Natural frequencies are also called resonant frequencies, since a system responds especially strongly when it is acted on by a continuing excitation at any one of its natural frequencies. The air in a shower stall, as an example, can undergo PRESSURE oscillations with an infinite number of natural frequencies. When a shower taker's singing hits any of those frequencies, a noticeable amplification is heard. (*See* RESONANCE; SOUND.)

The oscillation of continuous systems is a standing-wave phenomenon. (*See* WAVE.)

friction The term *friction* is a broad one and covers mechanisms and FORCES that resist and retard existing MOTION and that oppose imminent motion. They include forces of ADHESION and COHESION, as well as TURBULENCE, VISCOSITY, and other effects. When resisting actual motion, friction forces are dissipative. In spite of the great practical importance of friction, a full understanding of it and its reduction has not yet been achieved. (*See* DISSIPATION.)

Two commonly considered types of friction are solid-solid friction and solid-fluid friction:

Solid-solid Friction

This is friction between two bodies in contact. Its origin seems to be adhesion and impeding surface irregularities. Imagine two bodies, such as a book and a table top, in contact along flat faces, and imagine an applied force, parallel to the contact surface and tending to cause one body to move relative to the other. For a sufficiently weak applied force, there is no motion; the friction force matches the applied force in magnitude and opposes it in direction. This is *static friction.* But static friction has its limit. For some magnitude of the applied force, motion commences. The maximal value of the magnitude of the force of static friction is found, to a good approximation, to be proportional to the magnitude of the *normal force,* the force that is perpendicular to the contact surface and pushes the bodies together. For the book-on-table example, the normal force is the WEIGHT of the book. This maximal magnitude is also found to be practically independent of the AREA of contact. The proportionality constant is the *coefficient of static friction.* Thus

$$F_{\mathrm{f\ max}} = \mu_s F_N$$

where $F_{\mathrm{f\ max}}$ denotes the maximal magnitude of the static friction force, F_N is the magnitude of the normal force, and μ_s is the dimensionless coefficient of static friction for the two materials involved. This coefficient depends, as well, on the condition of the contact surfaces, such as whether they are smooth, rough, dry, wet, etc. (*See* DIMENSION.)

Once motion has been achieved, a friction force then acts to retard the motion. This friction is called *kinetic friction.* The direction of the force is parallel to the contact surface and opposite to the direction of the VELOCITY. Its magnitude is found, to a good approximation, to be independent of the SPEED and contact area and to be proportional to the normal force. Thus:

$$F_f = \mu_k F_N$$

where F_f is the actual magnitude of the friction force and μ_k is the dimensionless *coefficient of kinetic friction* for the two materials in contact and depends on the condition of the surfaces. Normally, for the same pair of surfaces in contact, $\mu_k < \mu_s$.

When a round body rolls on a surface, there is *rolling friction.* The origin of rolling friction is as follows. The weight of the rolling body causes distortions in the shape of the surface and the body by compression at the area of contact. When the body rolls and the area of contact advances on the surface and around the body, the distortions are continuously both created anew ahead and relieved behind. In the absence of any internal friction within the materials of the body and the surface, the decompression would aid the motion to the same extent that the compression hinders it. But the presence of internal friction, which cannot be avoided, makes the retarding force greater than the assisting one, and the net result is what is called rolling friction. As is the case for kinetic friction, the magnitude of the rolling friction force is proportional to the magnitude of the normal force:

$$F_f = \mu_r F_N$$

where μ_r denotes the dimensionless coefficient of rolling friction for the two materials involved, which is normally much smaller than the other two coefficients.

Solid-fluid Friction

This friction must be overcome, for example, in order to force blood through an artery or an aircraft through the air. It is caused by such factors as adhesion, viscosity, and turbulence. In the case of a liquid flowing through a uniform circular pipe, for instance, the effect can be seen through POISEUILLE'S LAW, which relates the VOLUME of liquid flowing through the pipe per unit time, called the volume flux, to the dimensions of the pipe, the viscosity of the liquid, and the difference of PRESSURES at the ends of the pipe. The relation is approximately valid in many useful situations:

$$\text{Volume flux} = \frac{\pi}{8} \frac{R^4}{L} \frac{\Delta p}{\eta}$$

In this relation, the volume flux is in cubic meters per second (m^3/s); R and L denote the pipe's radius and length, respectively, in meters (m); Δp is the difference of the pressures at the ends of the pipe in pascals (Pa), equivalent to newtons per square meter (N/m^2); and η represents the liquid's viscosity in newton·seconds per square meter ($N{\cdot}s/m^2$). What the relation is telling us is that the fluid in the pipe does not simply accelerate under the net force of the pressure difference, according to ISAAC NEWTON's second law of motion. Rather, there is an opposing friction force that balances the applied force and causes the fluid to flow at constant speed. (*See* FLUX; NEWTON'S LAWS.)

Another manifestation of solid-fluid friction is the drag force that acts on a solid body moving through a fluid, such as a dolphin through water or an aircraft through air. For a blunt body moving sufficiently rapidly through air, as an example, the magnitude of the drag force is given by:

$$\text{Drag force} = D\rho A v^2/2$$

where the drag force is in newtons (N), ρ denotes the DENSITY of the air in kilograms per cubic meter (kg/m^3), A is the effective cross-section area of the body in square meters (m^2), and v is the body's speed in meters per second (m/s). The symbol D denotes the drag coefficient, which is dimensionless. (*See* DRAG.)

fundamental constants *See* CONSTANTS, FUNDAMENTAL.

Fundamentality: What Lies Beneath?

Physics is the most fundamental of the sciences and forms the foundation of science's hierarchical structure. When explanations are needed to explain one branch of science by another, they inevitably lead to physics. While physics possesses explanatory power for, say, chemistry and biology, no other branch of science can offer explanations for physics. Interatomic forces, for instance, play an important role in chemistry, but chemistry needs physics for their explanation. In biology, the transmission of signals along the axons of neurons is an electric/chemical process, requiring both physics and chemistry for its understanding. But chemistry requires nothing of biology, and physics requires nothing of either of them. In that sense, physics occupies the foundation level of the hierarchy of science, with chemistry lying above it and biology above chemistry. Chemistry is more fundamental than biology, and physics more fundamental than both. (*See* ATOM; ELECTRICITY.)

Within physics itself, there exist various degrees of fundamentality. The properties of bulk MATTER are explainable by the behavior of matter at the microscopic level. From the properties of atoms, MOLECULES, IONS, ELECTRONS, and their INTERACTIONS, we can derive and understand the properties of METALS, LIQUIDS, etc. Thus, the former are more fundamental aspects of nature than are the latter. Stated in terms of physics fields, STATISTICAL MECHANICS, which is the study of large collections of particles, shows how the properties of macroscopic-scale matter, which are the subject of THERMODYNAMICS, are determined by the interactions of the particles. Statistical mechanics offers explanations for thermodynamics and is thus the more fundamental field.

In physics, as in science generally, a theory is an explanation. ISAAC NEWTON's laws of MOTION and law of GRAVITATION, for example, form a theory of mechanical phenomena. The latter include, among many others, JOHANNES KEPLER's laws of planetary motion and GALILEO GALILEI's laws of FREE FALL and rolling down straight, inclined tracks. JAMES CLERK MAXWELL's equations make up a theory for electromagnetic phenomena. QUANTUM MECHANICS, and in particular QUANTUM ELECTRODYNAMICS, form a theory of the structure of atoms. And a theory of the STRONG INTERACTION of ELEMENTARY PARTICLES is supplied by quantum chromodynamics. (*See* ELECTROMAGNETISM; KEPLER'S LAWS; MAXWELL'S EQUATIONS; MECHANICS; NEWTON'S LAWS.)

A theory uses a more fundamental aspect of nature to explain a less fundamental one. This, indeed, is an essential ingredient of a theory or of a proposed theory: what is explaining must be more fundamental than what is being explained. Newton's laws of motion and gravitation are valid for *all* mechanical systems and are consequently more fundamental than Kepler's laws, which are valid only for systems of planets and systems of moons. A hypothetical theory that took Kepler's laws as an explanation of Newton's would present a glaring absurdity, just as would an attempt to explain intermolecular forces by means of, say, the strength of materials, or as would a hypothetical theory of quantum electrodynamics in terms of atomic structure. The direction of explanation is always from the more to the less fundamental.

But why is that so? Perhaps we are going around in circles: A is more fundamental than B because it can serve as an explanation for B. Then it is not surprising that we use the more fundamental to explain the less fundamental. Since A explains B, we label A as the more fundamental. It simply becomes a matter of definition. Yet, there seems to be more to the matter than mere definition. When a physicist searches for a theory for some phenomenon, she does not just cast about randomly, find an explanation, and then label it more fundamental than the phenomenon simply because it successfully explains what needs to be explained. There are some aspects of nature that are clearly more fundamental than others, and it is among the former that a physicist looks for an explanation for one of the latter.

How does a physicist make the distinction? It is a matter of experience with explanatory power. When looking for a theory of B, a physicist has some preconception of those aspects of nature that are known to offer explanations for phenomena such as B. *Those* are the more fundamental ones. An explanation for B will surely come from among them. When B is finally understood to be explained by A, it is not this fact in itself that defines A as more fundamental than B. It is the fact that A explains not only B, but explains, or has the potential to explain, even more than B. Newton's laws, which explain Kepler's, can and do explain a lot more than just Kepler's laws. They underlie all the everyday mechanics—buildings, cars, sports—that we take for granted, for example. The explanatory power of interatomic forces is much broader than explaining merely the strength of materials. It includes, for instance, the behavior of GASes and the properties of molecules. Quantum electrodynamics explains much more than only atomic structure. It also extends to certain properties of elementary particles.

Physics possesses a hierarchical structure, just as science does. There are aspects and phenomena of nature that are considered to be more fundamental (residing at deeper levels) and those that are thought to be less fundamental (lying at higher levels). To find a theory for some phenomenon, a physicist always searches at levels that are deeper than the level of that which is to be explained. The structure of science has a foundation level, and it is occupied by physics. What about the structure of physics? Is there a deepest level in physics? Does nature possess aspects and phenomena of maximal fundamentality? They would serve as sources of explanation, but would not themselves require explanation or would have no explanation, at least no scientific explanation.

Throughout the evolution of physics, deeper levels have continually been discovered. At the start of the 20th century, bulk matter was found to be composed of atoms. That discovery revealed a level of nature that lies deeper than the level of bulk matter. Then the atom was found to possess a structure comprising a nucleus and electrons, and an even deeper level was discovered. Before long, the nucleus, too, was proved to have structure, being composed of protons and neutrons. An even deeper level was revealed. The proton and neutron (and other, related kinds of particles) are understood to be composed of quarks. This reveals a level yet deeper than that of the proton and nucleon. Have we finally reached the end of the process? There is so far no reason to think that we have. Does the process even have an end? That is a *very* big question.

In a parallel evolution, also taking place during the 20th century, physics passed from what is termed the classical picture to the current, quantum one. In the classical view, nature behaves deterministically and is composed of matter and FIELDS, which are distinct from each other. Nature also behaves in a local manner, in that what happens at a location affects only the immediate vicinity of the location, which then affects *its* immediate vicinity, and so on: no direct action at a distance. Effects propagate through space at finite speed. In the quantum picture, on the other hand, the behavior of nature is best described in a probabilistic way. What is determined, in general, is a range of possible outcomes and their corresponding probabilities, in contrast to the unique, definite outcome that is characteristic of the classical picture. Matter and fields are not distinct entities, but rather aspects of each other. Every field possesses a particle manifestation, while every particle reveals fieldlike behavior. Also, the quantum picture of nature involves nonlocality: the situations at different locations can be correlated in a way that cannot be understood as an effect propagating from one to the other. (*See* CLASSICAL PHYSICS; QUANTUM PHYSICS.)

The quantum picture offers a more fundamental understanding of nature than does the classical one and thus reveals a deeper level. This hierarchy parallels the structure-of-matter hierarchy described above. The quantum picture is required for an understanding of the levels of structure of matter. For this purpose, physics makes use of quantum mechanics and QUANTUM FIELD THEORY, including quantum electrodynamics and quantum chromodynamics, referred to earlier. Does nature possess a level even deeper than the quantum level? It appears very likely that it does. The reason for that involves SPACE-TIME, as we will see.

During the 20th century, physics evolved in a third way, as well. Until early in that century, SPACE and TIME were viewed as distinct from each other and observer independent. The latter means that distances and time intervals were assumed to have the same values for all observers, whatever their state of motion. ALBERT EINSTEIN's special theory of relativity changed all this by showing that for observers in different states of motion, space and time can become somewhat mixed, and distances and time intervals can have different values. The result was the understanding that space and time are better considered together as space-time, which is more fundamental than space and time separately. Space-time lies at a deeper level in the hierarchy of physics than that of space and time. (*See* RELATIVITY, SPECIAL THEORY OF.)

Up to that point, space, time, and space-time were serving as a "stage," so to speak, as a "backdrop" for events and processes. Space-time was passive. It "contained" matter, fields, interactions, etc., but it played no role in what was going on. Following on the heels of his special theory, Einstein proposed his general theory of relativity. This is a theory of gravitation that assigns space-time an active role. Rather than being a FORCE, gravitation was understood to be a matter of geometry, an effect of the curvature of space-time: matter determines the curvature of space-time, and space-time, in turn and through its curvature, directs the motion of matter. The result appears as if matter is attracting matter. Its active character endows space-time with a more fundamental role in nature than was previously assumed by passive space-time. Its level in the hierarchy is even deeper than that of passive space-time. (*See* RELATIVITY, GENERAL THEORY OF.)

Has physics reached the foundation of this hierarchy? The general consensus is that it has not, that there is further depth to be revealed. The point is that an active space-time should be subject to the same quantum effects as is the rest of nature. There does not yet exist a theory that successfully introduces space-time into the quantum picture. Such a theory would be a quantum theory of gravitation, also called a theory of QUANTUM GRAVITY. Physicists are awaiting the development of a theory that will reveal a hierarchy level that is deeper yet than both that of active space-time and the quantum level. At this deeper level, the character of nature is difficult to imagine. Some expectations of it are expressed in negatives, such as nontemporal, nonspatial, noncausal. In its extraordinary degree of fundamentality, this level of nature

(continues)

Fundamentality: What Lies Beneath?
(continued)

underlies space-time and the quantum picture. It should thus show how space-time and the quantum picture are related to each other, and it should offer explanations for both. An explanation for space-time? . . . for the quantum picture? Now, *that* is something to look forward to! But as deep as such a level is, does it form a foundation level, or does nature contain further depths? This is another *very* big question.

An additional hierarchy in physics extends toward the very large, out to the UNIVERSE and perhaps beyond. There was a time when the Earth was considered to be the center of everything. Then came the Copernican revolution, named for Nicolaus Copernicus (1473–1543), which gave the Earth the status of one of a number of planets revolving around the Sun, and the Sun was recognized as a star like the innumerable other stars. The stars, Sun, and SOLAR SYSTEM can be viewed as more fundamental than the Earth, with the Earth just an incidental planet among the solar planets and the Sun an incidental star among stars. The stars, Sun, and solar system form a hierarchy level that is deeper than the level of the Earth alone. Early in the 20th century, it became clear that the visible stars are not all that shines in the universe. They make up only a single GALAXY, the Milky Way galaxy, of which the Sun is a member, and this galaxy is but one of innumerable galaxies floating in space. The galaxies of the universe can be viewed as more fundamental than our galaxy, which is but an incidental member of them. The totality of galaxies lie at a deeper level than does the Milky Way galaxy.

During the 20th century, the galaxies were found to be concentrated in clusters, and the clusters were observed to form superclusters. A spatial arrangement of immense scale became apparent, with the galaxies distributed over sheets that border empty regions of space. We include all this structure in the hierarchy level of the galaxies. That seems to place the universe, comprising the totality of galaxies and everything else there is and might be—including radiation, dust, and putative dark energy—at the next deeper level of this hierarchy, more fundamental than the galaxies. (*See* COSMOLOGY.)

This level might reasonably be thought to form the foundation of its hierarchy. After all, is the universe not all there is? Recent attempts to understand the universe have led to proposals of even greater degrees of fundamentality, corresponding to even deeper levels of hierarchy. One possibility is that the universe, our universe, is but one of an infinite number of universes, forming a multiverse. Another proposal postulates additional DIMENSIONS to the four of space-time, with another universe, or other universes, existing in the extra dimensions (or, perhaps, in one extra dimension) and interacting with ours. A hierarchy level involving extra dimensions might connect with the structure-of-matter hierarchy at a deep level, since extra dimensions are being proposed in that connection as well. In addition, an extra-dimension level might tie in with the active-space-time level of the space-time hierarchy.

We see how fundamentality brings about a layering of physics into hierarchy levels, where the ordering takes place in a number of ways, corresponding to different hierarchies. We discussed the structure-of-matter hierarchy, the quantum hierarchy, the space-time hierarchy, and—the last presented—the cosmological hierarchy. In none of them is a definitive foundation level identified. That leaves plenty of work for physicists: in every one of the hierarchies, the current deepest level still needs further clarification. Possible "lateral" connections among hierarchies need to be investigated. And—the greatest challenge of all—there stands the eternal goal of physics, to uncover what lies beneath it all.

fusion, nuclear The joining, or fusing, of lighter nuclei into heavier ones is termed nuclear fusion. When the sum of MASSes of the products is less than the sum of masses of the reactants, ENERGY is produced in the form of KINETIC ENERGY of the products. (*See* MASS ENERGY; NUCLEAR PHYSICS.)

Nuclear fusion is the source of energy of the Sun and other stars and of fusion bombs (often called hydrogen bombs or H-bombs). Research is currently in progress with the goal of achieving controlled nuclear fusion for the purpose of energy production.

In the Sun and other stars, the major energy source is the conversion of hydrogen to helium. One such process, the proton-proton process, can be viewed as taking place in three steps. First, two hydrogen nuclei (PROTONs), denoted ${}^{1}_{1}H$ (ATOMIC NUMBER 1 and ATOMIC MASS 1), fuse to form a DEUTERON (the nucleus of deuterium, a stable, naturally occurring ISOTOPE of hydrogen, hydrogen-2), denoted ${}^{2}_{1}H$ (atomic number 1, atomic mass 2), a POSITRON, denoted e^+, and an ELECTRON-type NEUTRINO, denoted ν_e:

$${}^{1}_{1}H + {}^{1}_{1}H \rightarrow {}^{2}_{1}H + e^{+} + \nu_e$$

Second, the deuteron then fuses with another hydrogen nucleus, producing a nucleus of the isotope helium-3, denoted ${}^3_2H$ (atomic number 2, atomic mass 3), and a PHOTON (GAMMA RAY), denoted γ:

$${}^2_1H + {}^1_1H \rightarrow {}^3_2H + \gamma$$

Last, two helium-3 nuclei react to give a helium-4 nucleus (ALPHA PARTICLE, nucleus of the common isotope of helium), denoted ${}^4_2H$ (atomic number 2, atomic mass 4) and two hydrogen nuclei:

$${}^3_2H + {}^3_2H \rightarrow {}^4_2H + {}^1_1H + {}^1_1H$$

The overall reaction, obtained by adding up the third reaction, twice the first, and twice the second, is then:

$$4\,{}^1_1H \rightarrow {}^4_2He + 2e^+ + 2v_e + 2\gamma$$

This process generates energy mostly as photons and the kinetic energy of positrons and neutrinos.

Earth-based fusion, in bombs and hopefully in controlled sustained reactions, involves the conversion of isotopes of hydrogen into isotopes of helium. Some possible reactions are:

$${}^2_1H + {}^2_1H \rightarrow {}^3_1H + {}^1_1H$$
$${}^2_1H + {}^2_1H \rightarrow {}^3_2H + n$$
$${}^3_1H + {}^2_1H \rightarrow {}^4_2H + n$$
$${}^2_1H + {}^2_1H + {}^1_1H \rightarrow {}^4_2H + n + e^+ + v_e$$

The symbol ${}^3_1H$ denotes a triton, which is a nucleus of tritium, the isotope hydrogen-3, and n denotes a NEUTRON. In these processes, energy is released as the kinetic energy of the products.

G

Gabor, Dennis (1900–1979) Hungarian/British *Electrical engineer, Physicist* Dennis Gabor is known as the father of HOLOGRAPHY. Although very interested in physics, Gabor chose to study electrical engineering and received his doctorate in 1924 from the Technische Hochschule Berlin, Germany. Nevertheless, his work was almost wholly in applied physics, and he made numerous inventions of industrial significance. After receiving his degree, Gabor worked in German industry until Hitler came to power in 1933, when he returned to Hungary. He then moved to England in 1934 and did industrial work there, in electric lighting among other fields, until 1948. It was during this period that Gabor developed holography, which is the recording and storage of three-dimensional information in a two-dimensional medium and its subsequent reconstruction from the medium. In 1949, Gabor joined the Imperial College of Science & Technology in London, England, where in time he was appointed to the position of professor of applied electron physics, and where he remained until his retirement in 1967. Throughout that time, he carried out experimental and theoretical work in various areas of physics. In 1971, Gabor was awarded the Nobel Prize in physics "for his invention and development of the holographic method."

See also EXPERIMENTAL PHYSICS; THEORETICAL PHYSICS.

galaxy A galaxy is an assembly of stars containing millions or even billions of members along with dust and GASes. The galaxy in which the SOLAR SYSTEM resides is called the Milky Way galaxy. A galaxy is presumed to be held together by gravitational attraction. Observations of the motions of stars within galaxies indicate that the visible matter is insufficient to account for the gravitational effects. So it is commonly assumed

NGC 3310 is a spiral galaxy that blazes with extremely active star formation. It is at a distance of 59 million light-years from Earth. *(NASA and The Hubble Heritage Team [STScI/AURA])*

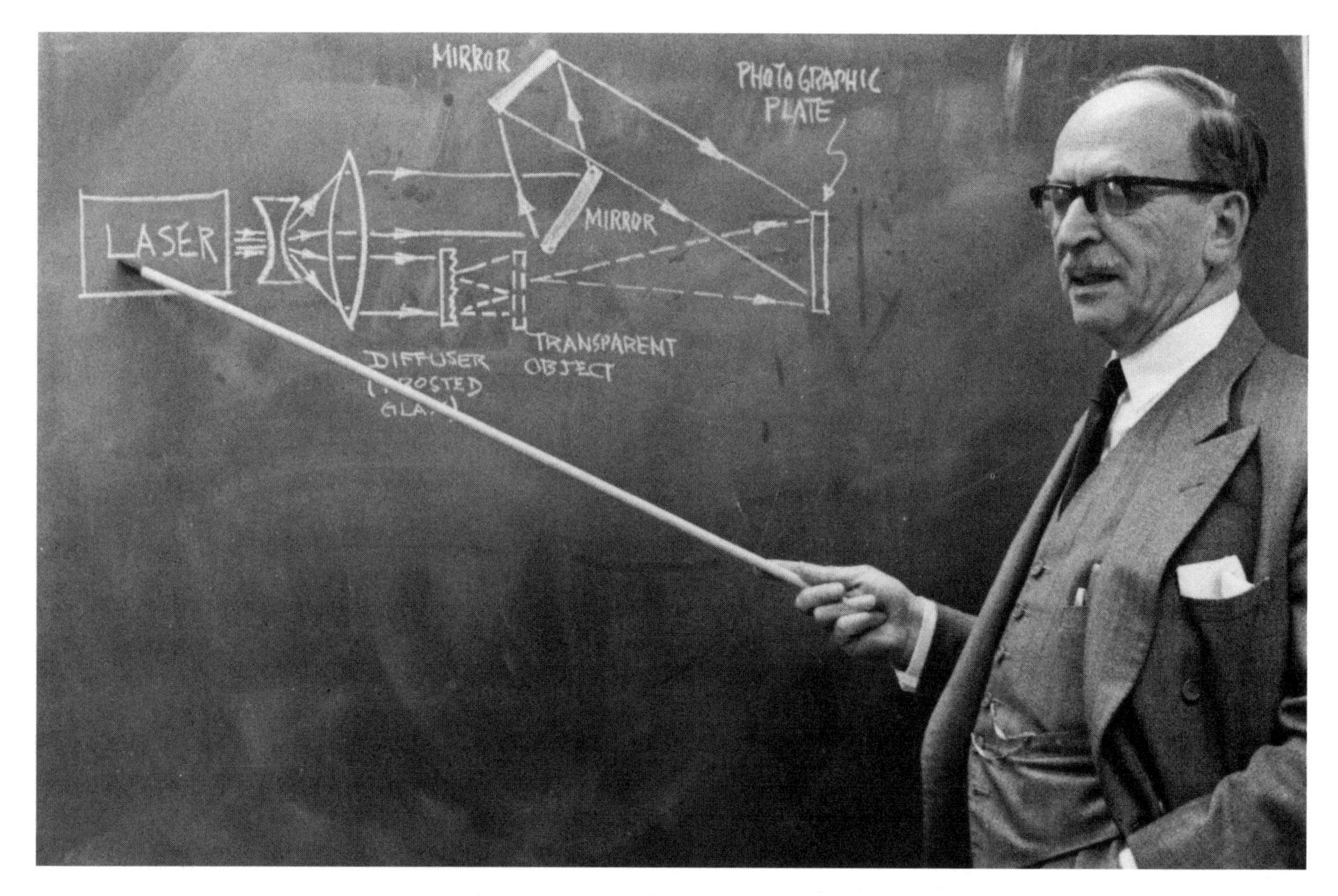

Dennis Gabor is the father of holography, for which he was awarded the 1971 Nobel Prize in physics. He is shown explaining the production of a hologram. ***(AIP Emilio Segrè Visual Archives, Physics Today Collection)***

that galaxies also possess a dark-matter component, whose identity is presently a mystery. Observations lead astronomers to believe that a gigantic BLACK HOLE lies at the center of most, if not all, galaxies. Galaxies are commonly found in gravitationally bound clusters, which in turn are often members of superclusters. (*See* GRAVITATION.)

The expansion of the UNIVERSE is observed as a moving apart of galaxies, while individual galaxies maintain their size. In this sense, the galaxy serves as a fundamental unit for COSMOLOGY.

See also HUBBLE EFFECT.

Galilei, Galileo (1564–1642) Italian *Astronomer, Physicist* Known simply as Galileo, this scientist obtained experimental results and developed ideas that were among those that ISAAC NEWTON built upon to obtain his laws of MOTION and GRAVITATION. Galileo was largely self-taught in mathematics and physics. He held various academic positions around what is now Italy and concurrently made astronomical observations, performed experiments on the motion of bodies, developed physics theories, wrote books, and publicly advocated the Copernican view of the SOLAR SYSTEM (that the planets and Earth revolve around the Sun). The latter brought Galileo into conflict with the Catholic Church, and in 1633 he was convicted of heresy and sentenced to house arrest for the rest of his life.

See also NEWTON'S LAWS.

Galilei transformation The mathematical relations that hold between the respective coordinates and TIME variables used by any two observers for whom NEWTON'S LAWS are valid are the Galilei transformations, named for the Italian astronomer and physicist GALILEO GALILEI. In other words, the coordinates and time variables referring to any pair of nonrelativistic inertial reference frames are related through a Galilei transformation. (*See* COORDINATE SYSTEM; REFERENCE FRAME, INERTIAL; RELATIVITY, SPECIAL THEORY OF.)

As an example of such a transformation, let coordinates $(x, y, z,)$ and time variable t refer to REFERENCE FRAME F, while (x', y', z') and t' refer to reference frame F′. Now, consider this case. Let frame F′ be moving at constant speed v with respect to frame F in F's positive x direction, and let their coordinate axes instantaneously coincide when the clocks at the respective origins both read 0. This is actually a general case, since whenever one reference frame is in constant-speed rectilinear motion with respect to another, their coordinate axes can always be displaced and rotated, and their clocks can always be reset so as to make this the case. Then, if an event is observed to occur at position (x, y, z) and at time t with respect to reference frame F, and if the same event is observed at position (x', y', z') and time t' with respect to reference frame F′, the Galilei transformation relating the variables is:

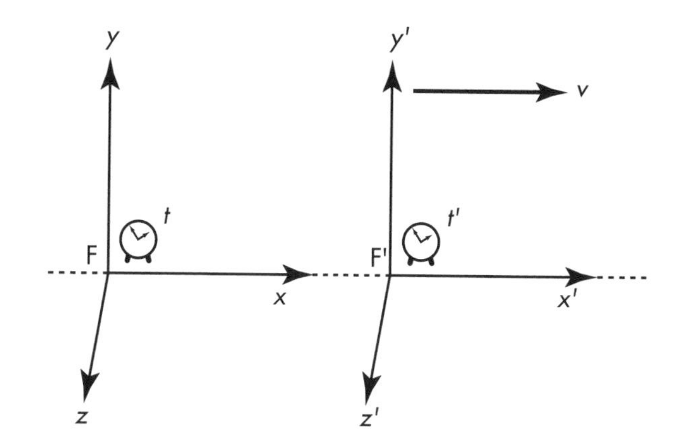

Galilei transformations relate the coordinates and time variables of two nonrelativistic inertial reference frames. The figure shows two such reference frames, F and F′, with coordinates (*x, y, z*) and (*x′, y′, z′*) and time variables *t* and *t′*, respectively. F′ is moving at constant speed *v*, which is much smaller than the speed of light, in the positive *x* direction with respect to F, and their coordinate axes coincide at some time, which is taken to be *t* = *t′* = 0.

Galileo was a 16th/17th-century scientist whose ideas and experimental results laid a foundation for later understanding of mechanics and gravitation. He was infamously persecuted by the religious authorities for publicly claiming (correctly) that the Earth and planets revolve around the Sun (rather than the planets and the Sun revolving around the Earth). *(AIP Emilio Segrè Visual Archives, Physics Today Collection)*

$$x' = x - vt$$
$$y' = y$$
$$z' = z$$
$$t' = t$$

or inversely:

$$x = x' + vt'$$
$$y = y'$$
$$z = z'$$
$$t = t'$$

Galilei transformations are obtained from corresponding POINCARÉ TRANSFORMATIONS or LORENTZ TRANSFORMATIONS in the limit of v being very small compared with the SPEED OF LIGHT.

galvanometer Named for the Italian physicist and physiologist Luigi Galvani (1737–98), a galvanometer is any measuring instrument for electric CURRENT whose operation is based on the force acting on an electric current in a magnetic FIELD. The type of galvanometer that is universally used in modern analog meters is the moving-coil galvanometer. In it, a small coil mounted in the magnetic field of a permanent MAGNET rotates when an electric current passes through it. This rotation is opposed by a rotary spring, so that an EQUILIBRIUM position is attained that is related to the amount of current. A pointer or mirror is attached to the coil. The value of the current through the galvanometer is shown by the position, on a calibrated scale, of the pointer or of a light beam reflected from the mirror. (*See* ELECTRICITY; MAGNETISM.)

When set up to specifically measure currents, a galvanometer is referred to as an ammeter. For highest sensitivity (i.e., for the measurement of the smallest currents), the current to be measured flows in full through the coil. To measure larger currents, a small RESISTOR, called a shunt resistor, is connected in parallel to the coil, so that only a fraction of the current passes through the coil. When a galvanometer is intended to measure VOLTAGES (i.e., potential differences), a large resistor is connected in series with the

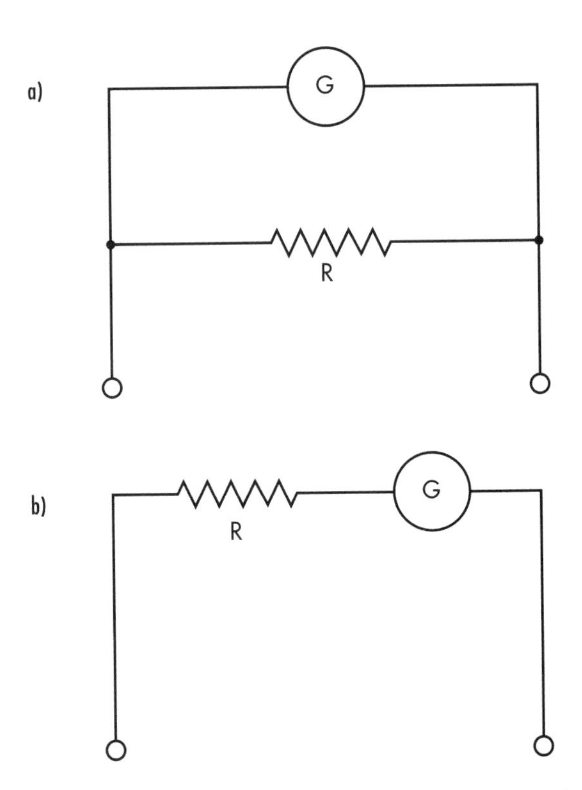

Converting a galvanometer to an ammeter and a voltmeter. (a) To be used as an ammeter, galvanometer G is connected in parallel with a shunt resistor, R. (b) For use as a voltmeter, galvanometer G is connected in series with a large resistor, R.

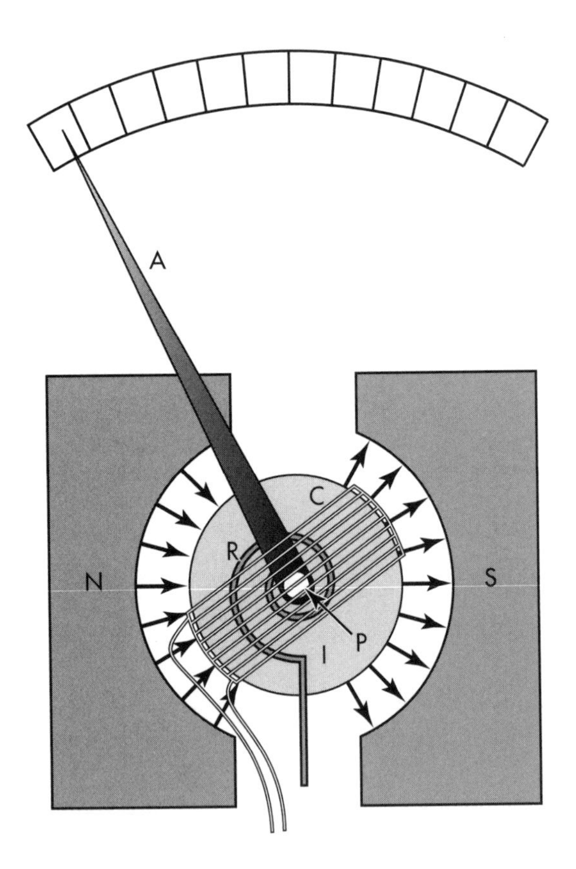

A schematic view showing the basic structure of a moving-coil galvanometer. N and S denote the poles of the permanent magnet. I is the cylindrical iron core, viewed end-on, that helps create a uniform, radial magnetic field, indicated by arrows. C denotes the coil, of rectangular shape, that rotates around the core about two pivots, one of which is seen and denoted P. R is the rotary spring that opposes the current-induced rotation of the coil. Pointer A is attached to the coil and rotates with it, indicating on the scale the magnitude of electric current flowing through the coil.

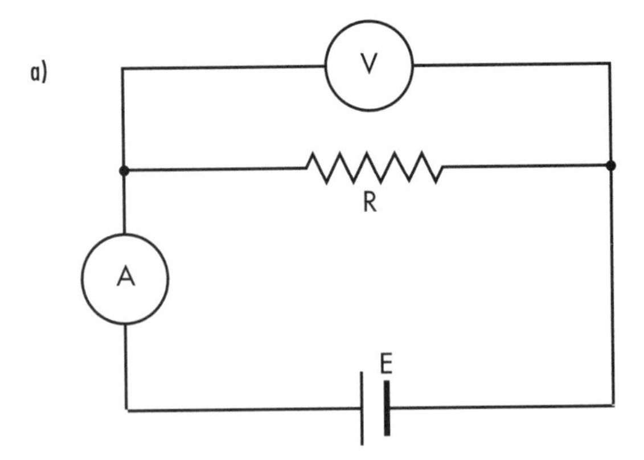

A correct way of connecting an ammeter, A, and voltmeter, V, to measure, respectively, the electric current through a resistor, R, and the voltage across the resistor. E denotes the emf source (such as a battery) of the circuit.

coil. Such a device is called a voltmeter. (*See* POTENTIAL, ELECTRIC.)

An ammeter is used by breaking the circuit where the current is to be measured and inserting the ammeter at the break, so that the current to be measured flows through the ammeter. Ideally, an ammeter should possess very low electric RESISTANCE compared with that of the circuit, so as not to affect the current that it is measuring. To measure a voltage between two points of a circuit, a voltmeter is simply connected directly between the two points, without breaking the circuit. To avoid affecting the voltage being measured, a voltmeter should have a large resistance compared with that of the circuit.

gamma decay Gamma decay is a type of RADIOACTIVITY in which a NUCLEUS does not alter its identity, but rather makes a transition from a higher to a lower ENERGY state while emitting a PHOTON, called also a GAMMA RAY. Gamma decay is governed by the ELECTROMAGNETIC INTERACTION. Among the various types of radiation from radioactivity, the photons produced by gamma decay are the most strongly penetrating. Gamma decay can accompany ALPHA DECAY and BETA DECAY when the latter leave the daughter nucleus in an excited state. Alpha decay of the radioactive isotope bismuth-212, $^{212}_{83}\text{Bi}$, to the isotope thallium-208, $^{208}_{81}\text{Tl}$, for example, can leave the latter in an excited state, from which it decays to its GROUND STATE by the emission of a photon.

The name of this type of decay is a relic from the early years of the study of radioactivity.

gamma ray The PHOTON emitted in the GAMMA DECAY of a NUCLEUS is called a gamma ray. By generalization, any high-energy photon—with ENERGY higher than that of X RAYS—can be referred to as a gamma ray. The name is a relic from the early years of the study of RADIOACTIVITY.

Gamow, George (1904–1968) Russian/American *Physicist* Widely known for his books popularizing physics, George Gamow's reputation as a physicist is based on his theoretical work in NUCLEAR PHYSICS and COSMOLOGY. He received his doctorate in physics in 1928 from the University of Leningrad. During 1928–34 Gamow held various fellowships in Europe and a professorship at the University of Leningrad, Russia. In 1934, he moved to the United States and served as professor of physics at The George Washington University in Washington, D.C., during 1934–56 and then at the University of Colorado at Boulder from 1956 until his death. Gamow's accomplishments in the field of nuclear physics include a theory of ALPHA DECAY and an understanding of the nuclear processes that supply ENERGY in stars. His cosmology studies led to the BIG-BANG theory for the origin and evolution of the UNIVERSE. Gamow's popular books include the *Mr. Tompkins* series, in which he explains various subjects of modern physics—such as relativity, curved SPACE-TIME, and the ATOM—through the dreamed adventures of the fictional Mr. Tompkins.

See also FUSION, NUCLEAR, RELATIVITY, GENERAL THEORY OF; RELATIVITY, SPECIAL THEORY OF; THEORETICAL PHYSICS.

The 20th-century theoretical physicist George Gamow is famous for his work in cosmology—leading to the big-bang theory—and in nuclear physics. He is widely known for his books popularizing physics. ***(AIP Emilio Segrè Visual Archives)***

gas A gas is one of the ordinary STATES OF MATTER. It is characterized by having no fixed shape and no fixed VOLUME: a gas expands to fill its container. Compared with LIQUIDs and SOLIDs, gases are relatively easy to compress. The constituents of a gas—its ATOMS or MOLECULES—move about very freely, colliding occasionally with each other and with the walls of its container. The air in which we are immersed on the surface of the Earth, for example, is a gas. When a gas is in contact with the liquid or solid PHASE of the same substance, it is often called a vapor. When water boils, the gas phase is often called water vapor. (*See* BOILING.)

See also GAS, IDEAL; VAN DER WAALS EQUATION.

gas, ideal An ideal gas is an idealization for real GASes. It is a hypothetical gas that fulfills the following conditions: its constituents (ATOMS or MOLECULES) have zero VOLUME and do not interact with each other or with the container walls except upon contact, whereupon the interaction is an instantaneous elastic COLLISION. So an ideal gas approximates those real-world situations in which the total volume of particles is much less than the volume of the gas, the durations of particle-particle and particle-wall INTERACTIONs are much shorter than the average free-flight times between collisions, and the collisions are sufficiently close to being elastic. Those simplifying assumptions lead to a very simple EQUATION OF STATE for an ideal gas, called the ideal gas law:

$$pV = nRT$$

where p denotes the PRESSURE in pascals (Pa), equivalent to newtons per square meter (N/m^2); V is the VOLUME in cubic meters (m^3); T is the absolute TEMPERATURE in kelvins (K); n is the amount of gas in MOLEs (mol); and R is the gas constant, whose value is 8.314472 joules per mole per kelvin (J/[mol·K]).

Real gases approach the conditions and behavior of an ideal gas when they are at low DENSITIES and high temperatures. At low densities, the total volume of the atoms or molecules is negligible compared with the volume of the gas, while at high temperatures, any noncontact interaction among the constituents and between them and the walls becomes insignificant. Accordingly, the equations of state of real gases approximate the ideal gas law under these conditions. On the other hand, at high densities and low temperatures, real gases can liquefy and solidify, whereas an ideal gas would remain in the gaseous state under all conditions.

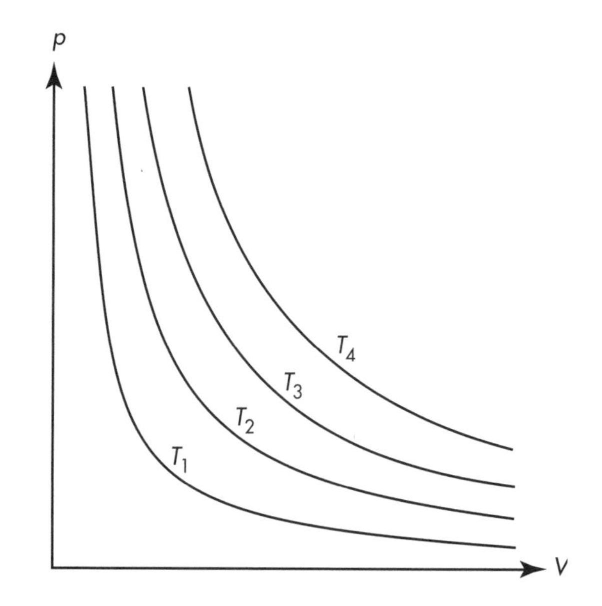

Ideal-gas isotherms in a pressure-volume (p-V) diagram for various temperatures $T_1 < T_2 < T_3 < T_4$. The equation of state for an ideal gas is $pV = nRT$, where n denotes the number of moles of the gas and R is the gas constant. The relation between p and V for constant temperature T—an isotherm—is represented by a hyperbola $pV =$ const.

See also KINETIC THEORY; VAN DER WAALS EQUATION.

gauge theory A QUANTUM FIELD THEORY of the ELEMENTARY PARTICLEs and their INTERACTIONs that is based on gauge symmetry is called a gauge theory. Gauge symmetry means that the mathematical description of the interactions remains unchanged when the FIELDs that represent the elementary particles are rather arbitrarily transformed into each other quite independently at every location and time. Expressed in other words, let different observers choose different standards of reference for the identity of the elementary particles, standards that can vary from point to point and from instant to instant even for the same observer. If, nevertheless, all such observers agree on the mathematical description of the interactions among the elementary particles, that description is said to possess gauge symmetry and to form a gauge theory. (*See* SYMMETRY.)

Gauge symmetry requires the existence of gauge force fields, which describe particles—called force particles—that transmit the interactions between matter particles. Since there are indeed such particles in nature, gauge theories might appear to be good candidates for correct theories of the elementary particles and their interactions. What is actually the case is that the requirement that elementary particle theories be gauge theories is so stringent that there remains almost no room for arbitrariness in the mathematical descriptions of the interactions. The descriptions thus obtained are consistent with nature. So nature seems to be correctly described at its fundamental level by gauge theories.

Consider the following example. QUARKs are known to carry a three-valued CHARGE called color charge. A gauge theory of quarks is one in which the color designation of quarks can rather arbitrarily vary from point to point and from instant to instant but whose description of quark processes is independent of such variations. Such a theory, it turns out, correctly predicts the existence of eight gluons and correctly describes the interactions between quarks and gluons and, through them, the quark-quark interactions. So the theory of the STRONG INTERACTION is almost completely determined by the requirement of gauge symmetry, i.e., by the requirement that it be described by a gauge theory.

Gauss's law Named for the German mathematician, astronomer, and physicist Karl Friedrich Gauss (1777–1855), Gauss's law states a relation between the electric FIELD and the electric CHARGEs that are its sources. More specifically, consider an arbitrary closed surface in SPACE called, for the present purpose, a Gaussian surface. Gauss's law relates the values of the electric field at all points of this surface to the net electric charge that is enclosed within the surface. (*See* ELECTRICITY.)

A closed surface possesses an inside and an outside. Divide the surface into infinitesimal surface elements. Let $d\mathbf{A}$ denote a vector that represents such an element at any point on the surface: the direction of $d\mathbf{A}$ at a point is perpendicular to the surface at the point and directed outward from the surface, while the magnitude of $d\mathbf{A}$, denoted by dA, is the infinitesimal AREA of the surface element. Let $\mathbf{E}$ denote the electric field at any point, with E denoting its magnitude. Form the scalar product between $d\mathbf{A}$ at any point on the surface and the electric field at the same point:

$$\mathbf{E}\cdot d\mathbf{A} = E\, dA \cos\theta$$

where θ is the angle between $d\mathbf{A}$ and $\mathbf{E}$. $\mathbf{E}\cdot d\mathbf{A}$ equals the product of the area of the surface element, dA, and the component of the electric field that is perpendicular to the element, $E \cos\theta$. Now sum (i.e., integrate) the values of $\mathbf{E}\cdot d\mathbf{A}$ over the whole surface. The result is the outward electric FLUX through the closed surface under consideration:

$$\Phi_e = \oint_{\text{surface}} \mathbf{E}\cdot d\mathbf{A}$$

Gauss's law states that the outward electric flux through an arbitrary closed surface equals the algebraic sum (taking signs into account) of electric charges enclosed within the surface, divided by the PERMITTIVITY of the medium in which it is all embedded. In symbols:

$$\oint_{\text{surface}} \mathbf{E}\cdot d\mathbf{A} = (1/\varepsilon)\Sigma q$$

where Σq denotes the algebraic sum of electric charges enclosed by the surface in coulombs (C), and ε is the permittivity of the medium. The latter equals $\kappa\varepsilon_0$, where κ is the DIELECTRIC CONSTANT of the medium and ε_0 is the permittivity of the vacuum, whose value is $8.85418782 \times 10^{-12}$ $C^2/(N\cdot m^2)$. Electric field is in UNITs of newtons per coulomb (N/C), equivalent to volts per meter (V/m), and the unit of area is the square meter (m^2). Note that while all the charges in the UNIVERSE

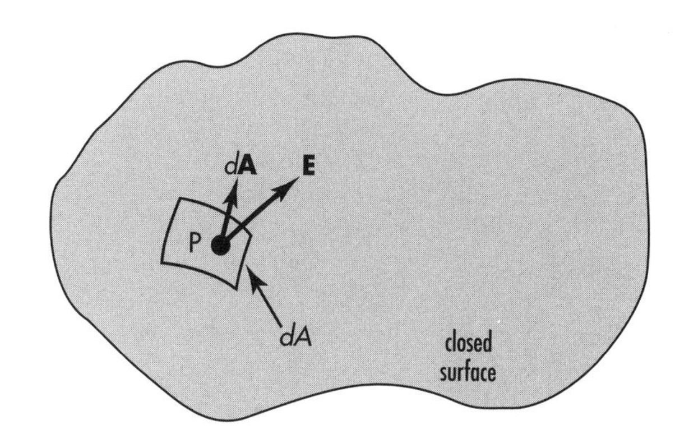

The figure shows a closed surface, called a Gaussian surface, to which Gauss's law is to be applied. An infinitesimal surface area element of magnitude *dA* is shown at an arbitrary point P on the surface. *d*A denotes the outward-pointing vector perpendicular to the surface at P with magnitude *dA*. E denotes the value of the electric field at P. The outward electric flux through the surface is the integral of the scalar product E·*d*A over the whole surface. According to Gauss's law, this electric flux equals the algebraic sum of all the electric charges enclosed within the surface, divided by the permittivity of the medium.

affect the electric field and determine its value at every point, only the charges enclosed by the surface enter Gauss's law. The effects of all the charges outside the Gaussian surface cancel out.

Although Gauss's law holds for arbitrary surfaces, in many applications of the law, certain Gaussian surfaces are more useful than others. As an example, we use Gauss's law to find the electric field caused by a single point charge. Consider the case of a single, positive, point electric charge, q. We ask the value (magnitude and direction) of the electric field, **E**, produced by the charge at some point. First, apply the following SYMMETRY argument. The lone source charge, as the cause of the electric field, is symmetric under all rotations about all axes through itself. By the symmetry principle, the electric field, which is an effect of the charge, must possess the same symmetry. As a result (we do not go into the details here), the magnitude of the electric field, E, can depend only on the distance of the field point from the charge, r, and the direction of the electric field is either everywhere radially away from the charge or everywhere radially toward it. In order to be able to solve for E, it must be possible to extract it from the integral in Gauss's law. That forces us to take for the Gaussian surface a spherical surface centered on the charge. Then, at all points on the surface, E has the same value, and **E** has the same direction with respect to $d\mathbf{A}$: either **E** points outward, $\theta = 0$, at all points, or **E** points inward, $\theta = 180°$.

Let us assume that **E** points outward, $\theta = 0$, at all points on the spherical Gaussian surface. The left-hand side of Gauss's law then becomes:

$$\begin{aligned}\Phi_e &= \oint_{\text{surface}} \mathbf{E} \cdot d\mathbf{A} \\ &= \oint_{\text{surface}} E\, dA \cos 0 \\ &= E \oint_{\text{surface}} dA\end{aligned}$$

The integral in the last line is simply the total area of the spherical surface, $4\pi r^2$, where r is the radius of the sphere. So the left-hand side of Gauss's law in this case equals:

$$\Phi_e = 4\pi r^2 E$$

The right-hand side for a single charge is:

$$(1/\varepsilon)\Sigma q = (1/\varepsilon)q$$

Gauss's law then gives:

$$4\pi r^2 E = (1/\varepsilon)q$$

Solve for E to obtain:

$$E = \frac{1}{4\pi\varepsilon}\frac{q}{r^2}$$

which is consistent, since both q and E are positive. So for positive q, the electric field at any point is directed radially away from the source charge, and its magnitude is given by the last equation, where r is the distance of the field point from the charge.

In the case of negative q, assuming that **E** points away from the source charge results in a sign inconsistency, which is resolved by assuming, instead, that the **E** points toward the charge. The final result is that for a point charge, q, of any sign, the magnitude of the electric field produced by the charge at distance r from it is given by:

$$E = \frac{1}{4\pi\varepsilon}\frac{|q|}{r^2}$$

where $|q|$ denotes the absolute value of q. The direction of the electric field is radially away from the source charge if the charge is positive, and it is radially toward the charge if the charge is negative. (Note that the magnitude and direction of the electric field can also be deduced from the definition of the electric field together with COULOMB'S LAW.)

See also MAXWELL'S EQUATIONS.

Gay-Lussac's law *See* CHARLES'S LAW, which is also know as Gay-Lussac's law. To further confuse matters, another GAS law is sometimes given the name "Gay-Lussac's law." This law, named for the French chemist and physicist Joseph Louis Gay-Lussac (1778–1850), states that for a fixed quantity of gas at constant VOLUME, the PRESSURE and absolute TEMPERATURE are directly proportional to each other:

$$p/T = \text{constant}$$

where p denotes the pressure of the gas and T its absolute temperature. Equivalently:

$$p_1/T_1 = p_2/T_2$$

where the subscripts refer to any two states of the gas that have the same amount and volume. Gay-Lussac's law is strictly valid only for ideal gases, but real gases obey it to a good approximation as long as the DENSITY is not too high and the temperature not too low. (*See* GAS, IDEAL.)

Gay-Lussac's law is a special case of the general law for ideal gases:

$$pV = nRT$$

where p is in pascals (Pa), equivalent to newtons per square meter (N/m^2); V is the volume in cubic meters (m^3); n denotes the amount of gas in MOLEs (mol); T is in kelvins (K); and R is the gas constant, with value 8.314472 joules per mole per kelvin [J/(mol·K)].

See also AVOGADRO'S LAW; BOYLE'S LAW.

Geiger counter Also called a Geiger-Müller counter and named for the German physicist Johannes (Hans) Wilhelm Geiger (1882–1945) and the German/American physicist Erwin Wilhelm Müller (or Mueller) (1911–77), a Geiger counter is a device for detecting and counting individual particles of ionizing RADIATION, in particular ALPHA PARTICLES (helium nuclei), BETA PARTICLES (ELECTRONs), and GAMMA RAYS (PHOTONs), such as are produced by RADIOACTIVITY. It is constructed of a cylindrical METAL tube with an electrode along its axis and a window, usually of mica, at one end and is filled with a low-PRESSURE gas. In operation, a VOLTAGE is applied between the central electrode and the tube, such that the former is the anode (positive) and the latter the cathode (negative). When an ionizing particle enters the tube through the window, it produces positive IONs and free electrons in the GAS. The ions accelerate toward the cathode. The electrons, in their acceleration toward the anode, attain sufficient KINETIC ENERGY to enable them to ionize additional gas molecules. The process repeats and generates an avalanche of electrons and ions, which creates a pulse of electric CURRENT in the circuit. The pulse can be used to produce an audible "click" and operate a counter. (*See* ELECTRICITY; IONIZATION.)

Gell-Mann, Murray (1929–) American *Physicist* Best known for his proposal of the "eightfold way,"

Murray Gell-Mann has worked mainly in theoretical elementary-particle physics and is known especially for his particle classification scheme, called the eightfold way, for which he was awarded the 1961 Nobel Prize in physics. He is shown explaining a point in quantum field theory. ***(AIP Emilio Segrè Visual Archives, Physics Today Collection)***

Murray Gell-Mann has worked mostly in the field of theoretical ELEMENTARY-PARTICLE physics. He received his Ph.D. in physics in 1951 from the Massachusetts Institute of Technology in Cambridge and has been serving on the faculty of the California Institute of Technology in Pasadena since 1955. Since 1984, Gell-Mann has also been involved with the Santa Fe Institute in Santa Fe, New Mexico, where he participates in the institute's study of complex systems. His work in elementary particles brought order to a confusing "zoo" of some 100 kinds of particles when he—and, independently, the Israeli physicist Yuval Ne'eman—proposed in 1961 a classification scheme that he called the eightfold way, based on SYMMETRY properties of the particles and of the STRONG INTERACTION. Its success led to the QUARK model. In 1969, Gell-Mann was awarded the Nobel Prize in physics "for his contributions and discoveries concerning the classification of elementary particles and their interactions."

See also THEORETICAL PHYSICS.

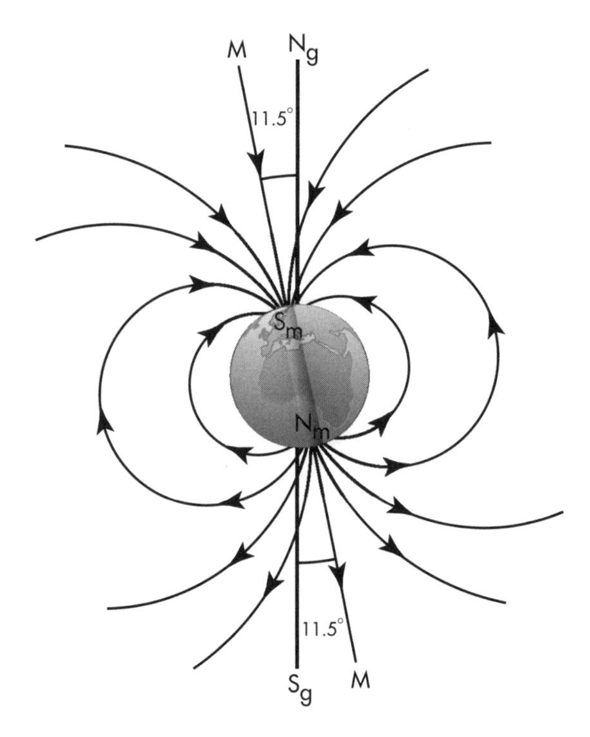

The Earth behaves as if a bar magnet were embedded at its center, where the magnet's axis M-M is presently tilted about 11.5° from the Earth's rotation axis N_g-S_g. Magnetic field lines outside the Earth are shown. The two locations where the field lines are perpendicular to the Earth's surface are the Earth's geomagnetic poles. The north (south) geomagnetic pole, located near the north (south) geographic pole N_g (S_g), is actually a south (north) magnetic pole, as indicated on the imaginary bar magnet by S_m (N_m).

generator, electric Any device for converting mechanical ENERGY to electric energy is called an electric generator. In the operation of a generator, a coil, the rotor, rotates in a magnetic FIELD, whereby a varying ELECTROMOTIVE FORCE (emf) of alternating polarity (i.e., an AC VOLTAGE) is induced in the coil. By suitable design, a generator can produce an emf of constant polarity (i.e., a DC voltage). Whenever an electric CURRENT is drawn from a generator, a TORQUE is needed to maintain the coil's rotation. Thus WORK performed on the coil is converted into electric energy. The source of the magnetic field can be one or more permanent MAGNETs or a fixed coil, the stator, in which the current that produces the field is taken from the rotor. In principle, a generator is an electric motor operated in reverse. Depending on the design and size of an electric generator, it might be referred to variously as an alternator, dynamo, or generator.

See also ELECTRICITY; INDUCTION; MAGNETISM.

geomagnetism The natural MAGNETISM of the Earth is termed geomagnetism. The magnetic FIELD of the Earth, detectable by a common magnetic compass, is similar to that which would be produced by a bar MAGNET located at the Earth's center and aligned about 11.5° from the Earth's ROTATION axis. The two locations on the surface of the Earth where the magnetic field lines are vertical are called the geomagnetic poles. Each is in the general neighborhood of one of Earth's geographic poles, the north geomagnetic pole in Greenland and the south geomagnetic pole in Antarctica. Since the north pole of a compass is, by definition, attracted to the north geomagnetic pole, the latter is actually a south magnetic pole in the terminology of magnetism, and magnetic field lines enter the Earth in the northern hemisphere. Similarly, the south geomagnetic pole is a north magnetic pole, and field lines emerge from the Earth in the southern hemisphere. Both the strength and direction of the Earth's magnetic field are continuously changing over time, as are the locations of the geomagnetic poles. Every million years or so, the polarity of the field reverses itself.

The origin of geomagnetism—and especially the causes of the continuous change in the Earth's magnetic field, the wandering of the geomagnetic poles, and the polarity flips—are little understood at present. It is clear that the high TEMPERATURES inside the Earth preclude anything like a permanent magnet existing there. So it is generally assumed that the Earth's magnetic field is caused by interior electric CURRENT loops, much as a magnetic field is produced by a current in a solenoid. (*See* ELECTRICITY.)

geophysics Geophysics is the branch of physics that studies the Earth, from its interior, through its surface, and out to its atmosphere. Geophysics can include such as atmospheric physics (and specifically meteorology, the study of weather), oceanography (the study of oceans), and seismology (the study of earthquakes and the propagation of SHOCK WAVES through the Earth). Seismologists are often involved in the search for underground resources, such as oil and ores.

Gibbs, Josiah Willard (1839–1903) American *Physicist* With a reputation in physics for his work on THERMODYNAMICS and STATISTICAL MECHANICS, Josiah Willard Gibbs has been honored by having the Gibbs FREE ENERGY named for him. In 1863, Gibbs received his Ph.D. in engineering from Yale University in New Haven, Connecticut. During 1863–71, he tutored at Yale and studied and traveled in Europe. Gibbs was appointed professor of mathematical physics at Yale in 1871, and he remained there until his death. He was a theoretical physicist whose major field of investigation was the foundations of thermodynamics and statistical mechanics and the connection between the two fields. Gibbs also did work on VECTOR analysis and on the electromagnetic theory of LIGHT. His first publications, in thermodynamics, appeared when he was 34 years old, which is unusually late for a physicist.

See also ELECTROMAGNETISM; THEORETICAL PHYSICS.

Josiah Gibbs was a 19th-century theoretical physicist who specialized in thermodynamics and statistical mechanics and investigated their foundations and the connection between the two fields. ***(AIP Emilio Segrè Visual Archives)***

glass Glass is a SOLID in which the constituent particles (ATOMS, MOLECULES) are not regularly arranged (i.e., a noncrystalline solid). As they are heated, glasses tend to soften over a noticeable range of TEMPERATURES before they become hot enough to actually flow. Thus glasses do not have well-defined MELTING points. They can be viewed as supercooled LIQUIDS. (*See* CRYSTAL; FREEZING; STATE OF MATTER.)

gluon Gluons are the FORCE particles associated with the STRONG INTERACTION, or color force, among QUARKS, which binds quarks together to form HADRONS, such as PROTONS, NEUTRONS, and PIONS. Gluons are BOSONS (as are all force particles), are massless and electrically neutral, and possess SPIN 1. There are eight types of gluon, distinguished by the value of their color CHARGE, as well as an antigluon corresponding to every gluon type. They couple only to particles carrying color charge (i.e., only to quarks and to themselves). (*See* ANTIMATTER; ELECTRICITY; ELEMENTARY PARTICLE; MASS.)

grand unified theory (GUT) Grand unified theory is the name of a theory of the ELEMENTARY PARTICLES and their INTERACTIONS that will unify the ELECTROWEAK INTERACTION and the STRONG INTERACTION. If and when a GUT is developed, it will form a further step in the unification of the elementary interactions.

Present understanding has ELECTRICITY and MAGNETISM unified as ELECTROMAGNETISM, or the ELECTROMAGNETIC INTERACTION, and the latter unified with the WEAK INTERACTION as the electroweak interaction. The awaited next stage in the unification process would be the formation of a GUT that encompasses the electroweak and the strong interactions. Among the hints that the latter two interactions are indeed two aspects of a more comprehensive interaction are the facts that both are described by GAUGE THEORIES and that the LEPTONs and the QUARKs are each classified into analogous generations.

Whereas the unification of electricity and magnetism is manifested under everyday conditions, the unification of electromagnetism and the weak interaction is fully expressed only at sufficiently high ENERGIES (or, equivalently, high TEMPERATUREs). The further unification of the electroweak and strong interactions to a GUT should attain full significance only at even higher energies and temperatures.

gravitation Gravitation is commonly understood as the FORCE of attraction between objects by virtue of their MASS. More generally, gravitation covers the forces operating between masses and/or ENERGIES that are independent of other intrinsic properties of the subjects of the INTERACTION. Gravitation appears to be the dominant force in cosmological and astronomical phenomena. It holds stars' planets in their orbits and planets' moons in theirs. The SOLAR SYSTEM and the Earth-moon system serve as examples. Gravitation binds stars into GALAXIES, galaxies into clusters, and the latter into superclusters. And gravitation appears to operate at the scale of the whole UNIVERSE, where it is thought to govern the rate of expansion of the universe and even to cause what seems to be an accelerating rate of expansion. Under certain conditions, gravitation can cause repulsion rather than the much more common attraction. (*See* COSMOLOGY.)

The term *gravity* is commonly, but not exclusively, used for the gravitational effect of an astronomical body, such as the Earth.

For a large range of conditions, gravitation is adequately described by ISAAC NEWTON's law of gravitation. It states that two point particles of masses m_1 and m_2 mutually attract each other with a force of magnitude:

$$F = Gm_1m_2/r^2$$

where F is in newtons (N), m_1 and m_2 are in kilograms (kg), r denotes the distance between the particles in meters (m), and G is the gravitational constant, whose value is 6.67259×10^{-11} N·m²/kg². Gravitation is a CONSERVATIVE FORCE, so a POTENTIAL ENERGY can be defined for it. The gravitational potential energy of two point particles as above is given by:

$$E_p = -Gm_1m_2/r$$

where E_p is in joules (J). *See* NEWTON'S LAWS.

As is the case for ELECTRICITY and MAGNETISM, so too can gravitation be described in terms of a FIELD, the gravitational field. The sources of this field are all masses and energy, and the field affects all masses and energy by causing forces to act on them.

Newton's law of gravitation is not adequate for all situations. The orbit of the planet Mercury, for example, is not consistent with Newton's law. Neither is the amount of deviation of a LIGHT ray as it passes near a massive object, such as the Sun, correctly predicted by Newton's law. A more widely applicable theory of gravitation is ALBERT EINSTEIN's general theory of relativity. It is formulated in terms of the gravitational field. The general theory of relativity has so far successfully passed all experimental tests. (*See* RELATIVITY, GENERAL THEORY OF.)

Gravitation is one of the four fundamental interactions among matter, together with the ELECTROMAGNETIC INTERACTION, the WEAK INTERACTION, and the STRONG INTERACTION. While the electromagnetic interaction and the weak interaction have been unified as the ELECTROWEAK INTERACTION, which, in turn, is expected to be unified with the strong interaction in a GRAND UNIFIED THEORY (GUT), the gravitational interaction is problematic in that regard. The theories of the other three interactions can easily be quantized (i.e., made consistent with the quantum character of nature). Gravitation is not so amenable. This seems to be intimately related to the fact that gravitation, as described by the general theory of relativity, involves SPACE-TIME in an essential way, while the other three interactions "merely" appear to play their roles on the space-time stage, so to speak, without affecting or being affected by space-time. So a quantization of gravitation is tantamount to a quantization of space-time itself. (*See* QUANTIZATION; QUANTUM PHYSICS.)

In the quantum picture of gravitation, the gravitational force is mediated by the exchange of gravitons.

According to present understanding, the graviton should be a massless, SPIN-2 particle, but it has not yet been discovered experimentally. (*See* ELEMENTARY PARTICLE.)

See also EQUIVALENCE PRINCIPLE; ESCAPE SPEED; FREE FALL.

gravitational lens The effect whereby LIGHT rays from a distant astronomical source are bent by an intervening massive astronomical object, such as a GALAXY or a galaxy cluster, on their way to an observer on Earth, causing the source to appear distorted—as several sources or as a (possibly incomplete) ring around the intervening object, called an Einstein ring—is gravitational lensing, with the intervening object serving as a gravitational lens. The bending of a light ray passing near a massive object is described by ALBERT EINSTEIN's general theory of relativity. Since a gravitational lens tends to bring to an observer light rays from the source that would otherwise not reach him, it tends to brighten the source's image and can even reveal sources that would otherwise be hidden by the intervening object. On the other hand, the presence of dark, massive astronomical objects can be detected through gravitational lensing, called microlensing in this case, by the temporary apparent brightening of a distant light source as a dark object passes through the line of sight to the source. (*See* LENS; RELATIVITY, GENERAL THEORY OF.)

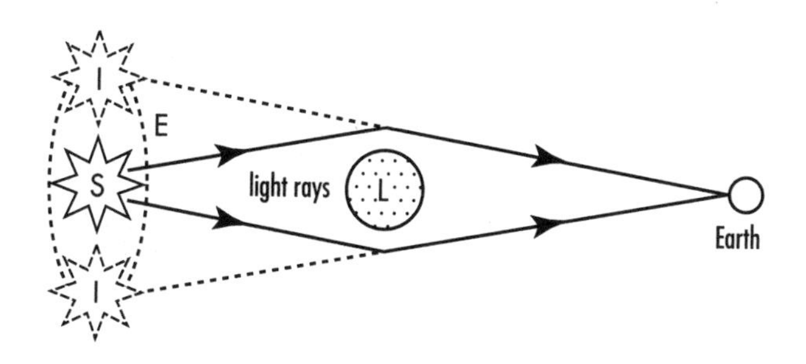

A gravitational lens is the effect in which a massive astronomical object, such as a galaxy, star, or dark object, bends light rays emanating from a distant luminous object, which could be a star or a galaxy. In this manner the source, as viewed on Earth, might appear distorted in various ways. The figure represents (not to scale) two light rays in an ideally symmetric lensing situation, in which the source S would appear to an observer as a ring, called an Einstein ring E, with the lensing (bending) object L at the ring's center. To an observer on Earth, the two light rays appear as two images I of the source, which form part of the Einstein ring.

gravitational wave A WAVE in the gravitational FIELD is termed a gravitational wave. ALBERT EINSTEIN's general theory of relativity shows that time-varying MASS configurations can produce waves in the gravitational field. Gravitational waves travel at the SPEED OF LIGHT and have other similarities to electromagnetic waves. They should be detectable through the time-varying forces they cause on bodies. There is an ongoing effort to construct gravitational-wave detectors of sufficient sensitivity to detect such waves. The most intense gravitational waves are candidates for earliest detection, when detectors become sensitive enough. They are expected to be generated by large masses undergoing high ACCELERATIONS. Such might include supernovas, massive stars falling into BLACK HOLES, and even the BIG BANG. (*See* ELECTROMAGNETISM; GRAVITATION; RELATIVITY, GENERAL THEORY OF.)

ground state A system's state of lowest ENERGY is termed the ground state of the system. Because systems tend to reduce their energy spontaneously, a ground state is a stable state. Any state of higher energy than the ground state is called an excited state. The terms are from QUANTUM MECHANICS, according to which the existence of ground states assures the continued existence of matter. As an example, consider an ATOM, which is a bound system of ELECTRONS and a NUCLEUS. According to CLASSICAL PHYSICS, the electrons are orbiting the nucleus (much as the planets orbit the sun) and are thus undergoing continual ACCELERATION. The result should be that the electrons radiate away their energy and fall into the nucleus within a very short time. In this manner, all ordinary matter should be highly unstable. Quantum mechanics determines that every atom possesses a ground state, from which no radiation is possible. An atom in an excited state can emit a PHOTON and undergo transition to a lower-energy state. But when the atom's ground state is reached, no further energy reduction is possible, and the atom is stable in that state. (*See* RADIATION.)

See also EQUILIBRIUM.

group speed Also called group velocity, group speed has to do with WAVES and is contrasted with PHASE SPEED, also called phase velocity. A periodic wave is characterized by its FREQUENCY, f, WAVELENGTH, λ, and propagation SPEED, v, where:

$$f\lambda = v$$

Here f is in hertz (Hz), λ is in meters (m), and v in meters per second (m/s). We are concerned here with sinusoidal *waves,* which form a special case of periodic waves. These are waves that exhibit a sinusoidal time dependence at any fixed point in space and, accordingly, a sinusoidal waveform, or spatial pattern, along the direction of propagation at any fixed time. The term *phase speed* refers to the propagation speed, v, of a sinusoidal wave. Overlapping sinusoidal waves of different frequencies interfere, resulting in a propagating INTERFERENCE pattern, such as a pulse. The speed of propagation of an interference pattern is called its group speed. If sinusoidal waves of different frequencies possess equal phase speeds, the interference pattern will propagate at the same speed, so that the group speed will equal the phase speed.

In a dispersive medium, the phase speed depends on the frequency/wavelength of the wave. In this case, the propagation speed of an interference pattern produced by overlapping sinusoidal waves, the pattern's group speed, will depend on the relation between the phase speed and the frequency/wavelength. A mathematical expression of such a relation is called a DISPERSION relation. Let the dispersion relation of the medium be expressed in terms of the dependence of the phase speed on the wavelength, $v(\lambda)$. For a range of useful situations, an expression for the group speed, v_g, can be derived. Then it can be shown that:

$$v_g = v - \lambda\, dv(\lambda)/d\lambda$$

where the expression is evaluated for λ and v equaling, respectively, the average wavelength for the set of overlapping waves and the corresponding phase speed.

As a trivial example, all sinusoidal electromagnetic waves in VACUUM travel at the SPEED OF LIGHT, c (i.e., the vacuum is nondispersive). So the wavelength dependence (actually, independence) of the phase speed in this case is:

$$v = c$$

The derivative of constant v vanishes, of course, which leaves for the group speed:

$$v_g = c$$

So, as stated above, when the phase speed is the same for all frequencies, the group speed equals the phase speed. (*See* ELECTROMAGNETISM.)

In a more general case, let the dispersion relation have the form:

$$v = v_0 + \alpha(\lambda - \lambda_0)$$

with α constant, as a linear approximation over some range of wavelengths centered around λ_0. Here, v_0 is the phase speed of a sinusoidal wave of wavelength λ_0. Then the expression for the group velocity is:

$$v_g = v_0 - \alpha\lambda_0$$

GUT *See* GRAND UNIFIED THEORY.

gyroscope A gyroscope is a rotating object that, when given a high ANGULAR MOMENTUM, tends to maintain its spatial orientation. When a TORQUE acts on a gyroscope, its axis of rotation tends to swing, or precess, around a fixed direction. More complex behavior is possible. When a gyroscope is rotating at a sufficiently high angular SPEED ω, in radians per second (rad/s), and has its off-vertical shaft supported at some point that is not the gyroscope's CENTER OF MASS, the gyroscope's WEIGHT produces a torque that causes the axis to precess at angular speed Ω, also in radians per second (rad/s), where:

A demonstration gyroscope. Note that the mass of the rotor is concentrated around the circumference to increase the moment of inertia. For scale, the outer diameter of the support ring is 8.9 centimeters (3.5 inches). ***(Photo by Frost-Rosen)***

$$\Omega = \frac{mgr}{I\omega}$$

Here m denotes the gyroscope's MASS in kilograms (kg); g is the ACCELERATION due to gravity, whose nominal value at the surface of the Earth is 9.8 meters per second per second (m/s^2); r is the distance in meters (m) from the gyroscope's center of mass to the point of support; and I represents the gyroscope's MOMENT OF INERTIA about its axis in kilogram·meter^2 ($kg \cdot m^2$). (*See* GRAVITATION; ROTATION.)

Gyroscopes lie at the heart of inertial guidance systems, the name of which derives from the orientation-maintaining, or inertial, property of a gyroscope. In such systems, the gyroscopes are mounted in such a manner that they are practically free from torque. They are initially set spinning, and the directions of their rotation axes do not vary from then on. As the vehicle or rocket travels, the gyroscopes of its inertial guidance system serve as a reference for determining direction. (*See* INERTIA.)

H

hadron The term *hadron* is the designation for any ELEMENTARY PARTICLE that is composed of QUARKs and antiquarks in various combinations (as well as of the GLUONs that bind the quarks and antiquarks together) and thus participates in the STRONG INTERACTION. The hadrons are a family of particles that comprises the BARYONs, which are matter particles, and the MESONs, which are force particles. The baryons all consist of three quarks each. Among the baryons, the most common are the PROTON (p) and the NEUTRON (n), which are the building blocks of atomic NUCLEI. The proton consists of two up quarks and one down quark (uud), while the neutron is made of one up quark and two downs (udd). Both particles have MASS and possess a half unit of SPIN. The proton carries electric CHARGE +1, and the neutron is electrically neutral. The baryons include additional particles, such as those named and denoted Λ^0 (called lambda-zero), Ω^- (omega-minus), and various versions of Δ, Σ, and Ξ. The Λ^0, for example, consists of an up, a down, and a strange quark (uds), while the Ω^- is made of three strange quarks (sss). (*See* ANTIMATTER; ATOM; ELECTRICITY.)

The mesons are made of quark-antiquark pairs. Among the mesons, the pions (π^+, π^-, π^0) are most noted for their role in intermediating the strong interaction among protons and neutrons that holds nuclei together. The π^+ consists of an up quark and an antidown quark ($u\bar{d}$), the π^- is made of an antiup and a down quark ($\bar{u}d$), while the π^0 comprises a quantum mixture of up-antiup ($u\bar{u}$) and down-antidown ($d\bar{d}$). The pions π^+, π^-, π^0 possess mass and have spin 0. Their electric charge is +1, −1, and 0, respectively. There are additional kinds of mesons as well, some of which are named and denoted η, η', J/ψ, Y, and various forms of K. The η', for instance, is composed of a strange-antistrange pair ($s\bar{s}$), and the J/ψ (jay-psi) comprises a charm-anticharm pair ($c\bar{c}$). (*See* QUANTUM PHYSICS.)

Among the hadrons, only the proton seems to be stable (although its long-term stability is put into question by certain elementary-particle theories). The neutron is basically unstable and decays via BETA DECAY. However, in nuclei it can be stable, which fact allows the existence of stable nuclei.

To every species of hadron there corresponds an antiparticle. Its quark composition matches that of the corresponding particle, with every quark replaced by its antiquark and vice versa. The antiproton ($\bar{p}$), for instance, consists of two antiups and an antidown ($\bar{u}\bar{u}\bar{d}$), the π^0 is its own antiparticle, and the π^+ and π^- are each other's antiparticles.

half-life *See* DECAY.

Hall, Edwin Herbert (1855–1938) American *Physicist* Known today as the discoverer of the effect that bears his name, Edwin Hall discovered the HALL EFFECT in 1879. He was then a graduate student at Johns Hopkins University in Baltimore, Maryland, where he was investigating electromagnetic effects. After receiving his doctorate in physics, Hall joined the physics department

Edwin Hall was a 19th/20th-century experimental physicist who discovered the Hall effect in electromagnetism while still a graduate student. ***(Photo by Bachrach)***

of Harvard University in Cambridge, Massachusetts, in 1881, where he remained until his retirement in 1921. He continued working in his laboratory as emeritus professor until shortly before his death.

See also ELECTROMAGNETISM.

Hall effect Named for the American physicist EDWIN HERBERT HALL, the Hall effect is the development of a VOLTAGE, or potential difference, called the Hall voltage, between opposite surfaces of a CONDUCTOR carrying an electric CURRENT in a perpendicular magnetic FIELD. The mechanism of the Hall effect is as follows. Let the conductor and the magnetic field both be horizontal, with the field perpendicular to the conductor. The moving free ELECTRONs that form the current are

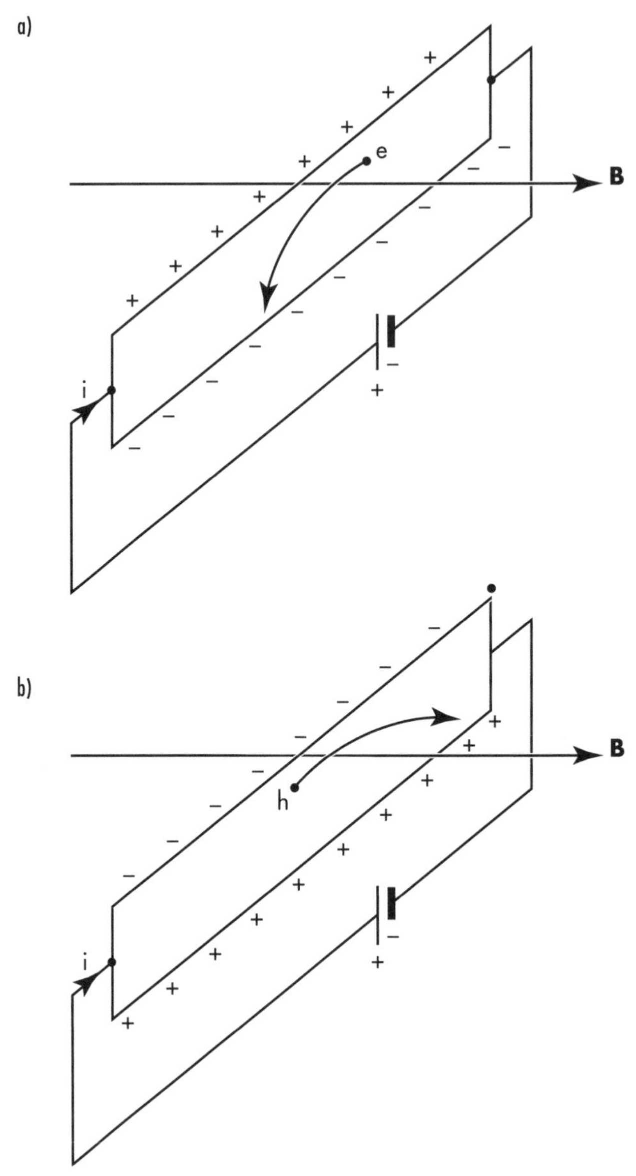

The Hall effect is illustrated for a right-pointing magnetic field, B, and an electric current, i, flowing into the page through a thin, vertical, rectangular conductor. As a result, the upper and lower edges of the conductor become oppositely charged, creating a voltage between the edges, called the Hall voltage. (a) The trajectory of negative current carrier e (electron) is shown. When the charge carriers are negative (or predominantly negative), this brings about a net negative charge on the lower edge of the conductor and a net positive charge on the upper edge, so the upper edge has a positive Hall voltage with respect to the lower. (b) Here the figure shows the trajectory of positive charge carrier h (hole). When conduction is by (or predominantly by) positive charge carriers, the upper edge gains a net negative charge and the lower edge a net positive charge. Thus the Hall voltage of the upper edge with respect to the lower is negative.

deflected by the magnetic FORCE either up or down, depending on the direction of the field; let us say down, for the discussion. As a result, there develops a higher-than-average DENSITY of free electrons on the lower surface of the conductor and a lower-than-average density on the upper surface. The upper and lower surfaces become equally and oppositely charged, producing a potential difference between the surfaces and a downward electric field inside the conductor. The effect of the electric field is to drive free electrons upward, in opposition to the effect of the magnetic field. In this manner, an EQUILIBRIUM is reached. The potential difference at equilibrium is the Hall voltage. (*See* CONDUCTION; ELECTRICITY; MAGNETISM; POTENTIAL, ELECTRIC.)

For given current, magnetic field, and conductor thickness, the magnitude and polarity of the Hall voltage allow calculation of the density of current carriers in the conductor and determination of the sign of their charge. Hall-effect experiments, together with their interpretation based on QUANTUM PHYSICS, help physicists to better understand electric conduction in METALs and SEMICONDUCTORs and the nature of electric RESISTANCE.

harmonic motion Commonly called simple harmonic motion, harmonic motion is one-dimensional oscillatory MOTION in which the DISPLACEMENT from the center of oscillation is described by a sinusoidal function of time. If we denote the displacement by x, then the most general harmonic motion is represented by:

$$x = A \cos(2\pi ft + \alpha)$$

where A denotes the (positive) amplitude of the oscillation, f its (positive) FREQUENCY (number of cycles per unit time), t the time, and α is a constant, the phase constant, or phase shift. Note that PHASE, in this context, refers to the stage in the cycle. Thus the motion takes place between the extremes of $x = A$ and $x = -A$. As for UNITs, x and A are in meters (m), f is in hertz (Hz), t in seconds (s), and α in radians (rad). Successive derivatives of x give the VELOCITY, v, in meters per second (m/s) and ACCELERATION, a, in meters per second per second (m/s^2) of the motion as follows:

$$v = dx/dt = -2\pi fA \sin(2\pi ft + \alpha)$$
$$a = dv/dt = -(2\pi f)^2 A \cos(2\pi ft + \alpha)$$

(*See* OSCILLATION.)

Note the proportionality, with negative coefficient, between the displacement and the acceleration that is characteristic of simple harmonic motion:

$$a = -(2\pi f)^2 x$$

It is easily seen that at the points of maximal displacement ($x = \pm A$), the SPEED is zero and the acceleration is maximal, is equal to $(2\pi f)^2 A$, and is directed toward the center. At the center ($x = 0$), the speed is maximal and equals $2\pi fA$, while the acceleration is zero.

In the particular case of motion starting at the center ($x = 0$) at time $t = 0$ and initially heading for increasing values of x, the expressions for displacement, velocity, and acceleration are:

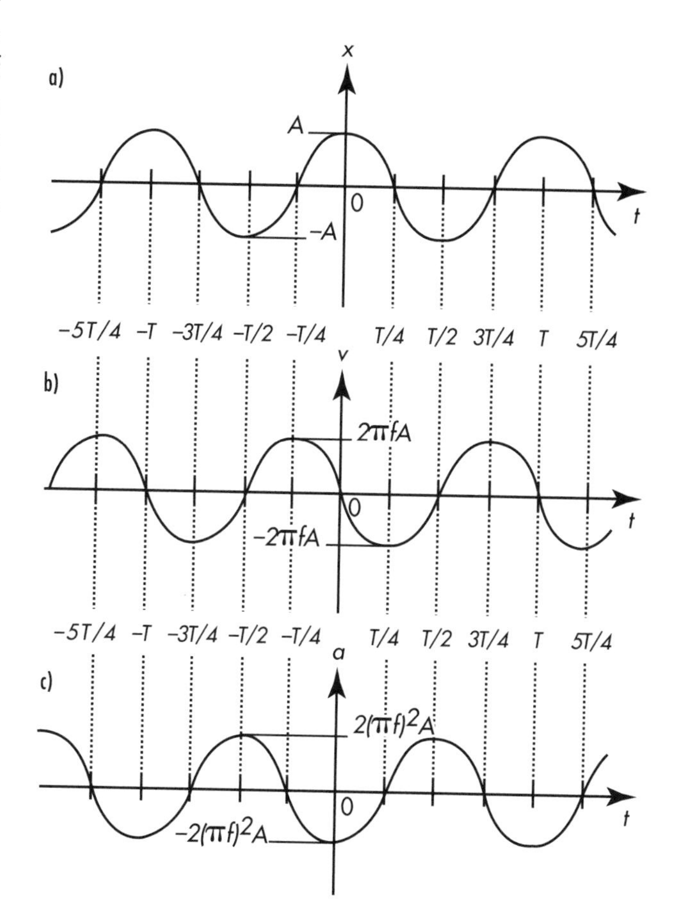

Harmonic motion is described by a sinusoidal function of time. (a) The figure depicts $x = A \cos 2\pi ft$, where x denotes the displacement from equilibrium, A the amplitude, f the frequency, and t the time. The period is $T = 1/f$. (b) The velocity of the motion as a function of time, $v = -2\pi fA \sin 2\pi ft$. (c) The acceleration as a function of time, $a = -(2\pi f)^2 A \cos 2\pi ft$.

$$x = A \sin 2\pi ft$$
$$v = 2\pi fA \cos 2\pi ft$$
$$a = -(2\pi f)^2 A \sin 2\pi ft$$

For motion starting at maximal positive displacement, $x = A$, at time $t = 0$, the corresponding expressions are:

$$x = A \cos 2\pi ft$$
$$v = -2\pi fA \sin 2\pi ft$$
$$a = -(2\pi f)^2 A \cos 2\pi ft$$

Harmonic motion is seen, for example, when the position of a point in CIRCULAR MOTION at constant angular speed around the origin is projected onto the x- or y-axis.

In physical systems, simple harmonic motion results when a system in stable EQUILIBRIUM and obeying HOOKE'S LAW (that the restoring FORCE is proportional to the displacement) is displaced from stable equilibrium and allowed to evolve from that state. Let x denote the displacement from equilibrium for such a system and F the restoring force in newtons (N). Hooke's law is then:

$$F = -kx$$

where k is the proportionality factor (called the spring constant in the case of a spring) in newtons per meter (N/m), and the minus sign indicates that the restoring force is in the opposite direction from that of the displacement. By ISAAC NEWTON's second law of motion:

$$F = ma$$

where m denotes the MASS of the moving part of the system in kilograms (kg) and a its acceleration. From the last two equations, we obtain that the acceleration is proportional, with negative coefficient, to the displacement:

$$a = -(k/m)x$$

As we saw above, this relation is characteristic of simple harmonic motion, with:

$$(2\pi f)^2 = k/m$$

Thus, such a system can undergo simple harmonic motion at the frequency:

$$f = \frac{1}{2\pi}\sqrt{\frac{k}{m}}$$

(*See* NEWTON'S LAWS.)

The mechanical ENERGY of such a system remains constant (in the ideal case) and is continually undergoing conversion from elastic POTENTIAL ENERGY, $kx^2/2$, to KINETIC ENERGY, $mv^2/2$, and back, so that:

$$\text{Mechanical energy} = kx^2/2 + mv^2/2 = kA^2/2$$

where energies are in joules (J).

In general, almost any system in stable equilibrium, when displaced from equilibrium, responds with a restoring force that, for sufficiently small displacements, is approximately proportional to the displacement. The proportionality coefficient might differ for different ways of displacing the system from equilibrium, such as in different directions. So systems in stable equilibrium can, in general, oscillate at a number of frequencies. (*See* FREQUENCY, NATURAL.)

All of the above discussion is applicable as well to nonmechanical systems, in which motion, displacement, velocity, acceleration, etc. refer to change in the value of some—not necessarily spatial—physical quantity, and force is generalized appropriately. Simple harmonic variation can be undergone, for instance, by VOLTAGEs and CURRENTs in electric circuits, by reagent concentrations in chemical reactions, and by air PRESSURE in musical instruments. (*See* ELECTRICITY.)

See also ELASTICITY.

harmonic oscillator Any device that is designed to oscillate in HARMONIC MOTION, either literally or figuratively, is called a harmonic oscillator. As an example, a weight suspended from a spring is a harmonic oscillator: when displaced from equilibrium, it oscillates in literal harmonic motion, with its vertical displacement varying as a sinusoidal function of time. For a figurative example, an electric circuit can be designed so that it produces a VOLTAGE that varies sinusoidally in time. Such a circuit is thus a harmonic oscillator. The essential physical property of a harmonic oscillator is that it possess a state of stable EQUILIBRIUM, such that when it is displaced from that state, there arises a restoring FORCE (or suitable generalization of force) whose magnitude is proportional to the magnitude of the displacement (or of its appropriate generalization). (*See* ELECTRICITY; OSCILLATION.)

Whereas, according to CLASSICAL PHYSICS, a harmonic oscillator may possess ENERGY in a continuous range of values, the energy of a quantum harmonic oscillator is constrained to a sequence of discrete values:

$$E_n = (n + 1/2)hf \text{ for } n = 0, 1, 2, \ldots$$

Here E_n denotes the value of the energy in joules (J) of the harmonic oscillator's nth energy level, f is the classical FREQUENCY of the oscillator in hertz (Hz), and h represents the PLANCK CONSTANT, whose value is $6.62606876 \times 10^{-34}$ joule·second (J·s). Note that the quantum harmonic oscillator cannot possess zero energy. That fact is related to the HEISENBERG UNCERTAINTY PRINCIPLE. (*See* QUANTUM PHYSICS.)

See also FREQUENCY, NATURAL.

harmonics The term *harmonics* refers to an infinite sequence of FREQUENCIES that are all integer multiples of the lowest among them:

$$f_n = nf_1 \text{ for } n = 1, 2, \ldots$$

where f_n denotes the nth frequency of the sequence, called the nth harmonic. Such a sequence is called a harmonic sequence or harmonic series. The lowest frequency of the series, f_1, the first harmonic, is called the fundamental frequency, or simply the fundamental. The higher frequencies of the series can also be referred to as the harmonics of the fundamental.

The name *harmonic* comes from the fact that the frequencies of musical tones that are generally judged to sound pleasant, or harmonious, when heard together are members of such a sequence. Because of this, musical instruments that produce definite pitches were developed so that their natural frequencies are members of a harmonic sequence. The acoustic pitch of such an instrument is determined by the fundamental of the sequence. For example, the natural OSCILLATION frequencies of a stretched string are (for small displacements from EQUILIBRIUM):

$$f_n = \frac{n}{2L}\sqrt{\frac{F}{\mu}} \text{ for } n = 1, 2, \ldots$$

where F denotes the tension of the string in newtons (N), L is the length of the string in meters (m), and μ is the string's linear DENSITY in kilograms per meter (kg/m). The fundamental frequency, which determines the pitch, is:

$$f_1 = \frac{1}{2L}\sqrt{\frac{F}{\mu}}$$

This is the acoustic foundation of all stringed instruments, such as the violin, guitar, and piano. (*See* ACOUSTICS; FREQUENCY, NATURAL.)

What happens when a pitched musical instrument is played at a pitch of frequency f_1 is that all or some of the harmonics of f_1 sound simultaneously. In the parlance of musical acoustics, the higher members of the harmonic sequence of f_1 are called overtones. The first overtone is the second harmonic, the second overtone the third harmonic, and so on. The relative INTENSITIES of the simultaneously sounding harmonics determine the tonal character, or timbre, of the instrument. In double-reed instruments, such as the oboe and the bassoon, for example, there is more acoustic ENERGY in the first overtone (the second harmonic) than in the fundamental. The clarinet, on the other hand, lacks the odd overtones (even harmonics) altogether and possesses a rather strong fundamental. The flute is very weak in all its overtones compared with its fundamental and thus has a very strong fundamental. A tone that consists of the fundamental alone and is completely lacking in overtones is referred to as a pure tone. On the other hand, there is a psychoacoustic effect that when the fundamental is lacking from a tone, while higher harmonics are present, a listener will nevertheless perceive the pitch of the fundamental.

Hawking, Stephen William (1942–) British *Physicist* Having written the best-seller *A Brief History of Time,* among other popular books, Stephen Hawking is famous as a cosmologist. He studied the general theory of relativity and COSMOLOGY at the University of Cambridge, England, received his Ph.D. in physics there in 1966, and has remained at Cambridge ever since. Hawking is well known also for remaining professionally active in spite of his severe disability due to motor neurone disease and amyotrophic lateral sclerosis, which have rendered him almost totally paralyzed and allow him to speak only via computer and voice synthesizer. Hawking's work deals mostly with the implications of the unification of the general theory of relativity and QUANTUM PHYSICS. One of his results is that BLACK HOLES are not completely black, but radiate their MASS away and eventually "evaporate."

See also RELATIVITY, GENERAL THEORY OF.

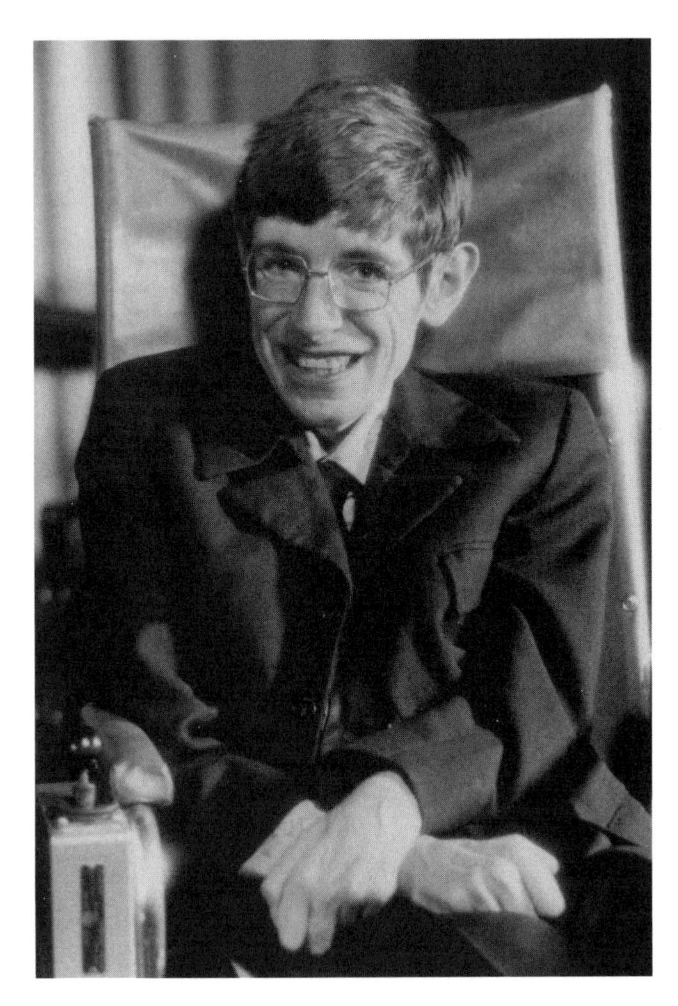

Stephen Hawking is a cosmologist who is investigating the implications of the unification of the general theory of relativity and quantum physics. ***(AIP Emilio Segrè Visual Archives, Physics Today Collection)***

heat Also called thermal energy, heat is the internal energy of matter, the ENERGY of the random motion of matter's microscopic constituents (its ATOMs, MOLECULEs, and IONs). In a GAS, the molecules or atoms are confined only by the walls of the container. The form of heat for a gas is then the random, relatively free-flying about of its molecules or atoms, occasionally bouncing off the walls and off each other. The microscopic constituents of a LIQUID are confined to the VOLUME of the liquid. So for a liquid, heat is the random dance of its constituents as they collide often with each other and occasionally with the liquid surface. In a SOLID, the constituents are confined to more or less fixed positions. Thus heat in solids is the random OSCILLATION of its constituents about their positions. (*See* KINETIC THEORY.)

Of all forms of energy, heat is unique in its random nature and in its not being fully convertible to WORK and, through work, to other forms of energy. THERMODYNAMICS is the field of physics that deals with heat and its transformation. It is the second law of thermodynamics that expresses the impossibility of total conversion of heat to work.

TEMPERATURE is intimately related to heat and is a measure of the average random energy of a single molecule (or atom, etc.) in the substance. Heat spontaneously flows from a region of higher temperature to one of lower temperature; never the opposite.

The transfer of heat occurs in one or more of three ways: CONDUCTION, convection, and RADIATION. In the former, heat flows through matter without concomitant transfer of the matter. In convection the transfer is accomplished through the motion of the material itself, such as when heated air from a furnace is blown through ducts into the rooms of a house. Radiation is the transfer of heat by electromagnetic radiation. (*See* ELECTROMAGNETISM.)

heat capacity Also called thermal capacity, heat capacity is the amount of HEAT, or thermal energy, required to raise the TEMPERATURE of a body by a single unit of temperature. Its SI UNIT is joule per kelvin (J/K). The heat capacity of a body depends both on the material the body is made of and the amount of material in the body. Related to heat capacity are specific heat capacity and molar heat capacity. They both are properties of materials and are independent of the amount of material.

Specific heat capacity (also specific thermal capacity or specific heat) is the amount of heat required to raise the temperature of one unit of mass by one unit of temperature. Its SI unit is joule per kelvin per kilogram [J/(kg·K)].

Molar heat capacity (also molar thermal capacity) is the amount of heat required to raise the temperature of one MOLE of a substance by one unit of temperature. Its SI unit is joules per kelvin per mole [J/(mol·K)].

The heat capacity of a body equals the product of its mass and the specific heat capacity of the substance the body is made of. As an example, the specific heat capacity of SOLID iron is 4.7×10^2 J/(kg·K), while the heat capacity of a five-kilogram iron dumbbell is $5 \times (4.7 \times 10^2) = 2.4 \times 10^3$ J/K. A body's heat capacity also equals the product of the number of moles of substance contained in the body and the substance's molar heat capacity.

heat engine Also called a thermal engine, a heat engine is any device that converts HEAT, or thermal ENERGY, to WORK and thence to other forms of energy. The second law of THERMODYNAMICS limits the efficiency of heat engines in that not all the heat taken in by a heat engine can be converted to work; some must be discharged. The maximal possible efficiency of heat engines that take in heat at a definite TEMPERATURE and discharge waste heat at a lower temperature is the efficiency of a CARNOT CYCLE operating between the same temperatures.

heat of fusion The term *fusion* is synonymous with MELTING, which is the PHASE TRANSITION from a SOLID to a LIQUID. It is the inverse process to FREEZING. As a sample of solid material is heated, its temperature rises until it reaches the melting point for that material. Further heating does not raise the temperature, but rather gives some of the particles that constitute the material, its ATOMS or MOLECULES, sufficient ENERGY to overcome the FORCES constraining them to fixed positions. They then form the liquid PHASE. When the solid phase is fully depleted and converted to liquid, additional heating raises the temperature. The amount of heat (i.e., thermal energy) needed to completely convert the solid phase to liquid at the melting point is the heat of fusion of the sample. (*See* STATE OF MATTER.)

In the freezing of a liquid as a result of heat being extracted from it, the inverse process takes place. When the temperature falls to the melting point, further cooling does not reduce the temperature, but rather causes solidification. The amount of heat that needs to be removed to fully convert the liquid phase to a solid at the melting point, the heat of freezing, equals the heat of fusion for that sample.

heat of transformation The quantity of HEAT, or thermal ENERGY, that is absorbed or released when a substance undergoes a PHASE TRANSITION is referred to as the heat of transformation. As an example, the HEAT OF FUSION of an amount of material is the quantity of heat required to melt it in its SOLID state without changing its TEMPERATURE (which is then at the material's melting point). That equals the quantity of heat that needs to be removed from the same amount of the same material in LIQUID state in order to freeze it at constant temperature (which is again the material's melting point). For another example, the HEAT OF VAPORIZATION of an amount of matter is the quantity of heat needed to convert it at its boiling point from its liquid state to a GAS. In the inverse process of CONDENSATION, the same quantity of heat must be removed from the gas to liquify it at the boiling point. There are also solid-solid phase transitions, such as from one crystalline structure to another, and they involve heat of transformations as well. (*See* BOILING; CRYSTAL; CRYSTALLOGRAPHY; FREEZING; MELTING; STATE OF MATTER.)

The heat of transformation of a substance depends both on the nature and the amount of material and is, in fact, proportional to the amount of material. Quantities that characterize the material itself and do not depend on its amount are specific heat of transformation and molar heat of transformation. The former is the heat of transformation of a single unit of MASS of the material. Its SI UNIT is joule per kilogram (J/kg). The latter is the heat of transformation of one MOLE of the material, with SI unit of joule per mole (J/mol).

heat of vaporization The quantity of HEAT, or thermal ENERGY, needed to convert a substance from its LIQUID state to a GAS at its boiling point is called the heat of vaporization. In the inverse process of CONDENSATION, the same quantity of heat must be removed from the same amount of the same gas to liquify it at the boiling point. The heat of vaporization of a substance depends both on the nature and the amount of material. It is proportional to the amount of material. Quantities that characterize the material itself and do not depend on its amount are specific heat of vaporization and molar heat of vaporization. The former is the heat of vaporization of a single unit of MASS of the material. Its SI UNIT is joule per kilogram (J/kg). The latter is the heat of vaporization of one MOLE of the material, with SI unit of joule per mole (J/mol). (*See* BOILING; STATE OF MATTER.)

Heisenberg, Werner Karl (1901–1976) German *Physicist* One of the pioneers of QUANTUM PHYSICS, Werner Heisenberg is famous for his uncertainty principle. He received his doctorate in physics in 1923 from the University of Munich, Germany. Until 1927, Heisenberg held various visiting positions and worked with other pioneers of quantum physics, such as NIELS HENRIK DAVID BOHR and MAX BORN. During that period, he progressed in the development of QUANTUM MECHANICS and applied it to ATOMs and MOLECULES. Heisenberg discovered theoretically that hydrogen molecules can exist in two different forms (allotropy), and in 1927, he discovered the uncertainty principle. In 1927, Heisenberg was appointed to a chair at the University of Leipzig, Germany, where he remained until 1941, when he became director of the Kaiser Wilhelm Institute for Physics in Berlin, Germany. During World War II, Heisenberg headed the German nuclear weapons project, which did not succeed in producing a nuclear weapon. After the war, he was interned, along with other leading German scientists, in England, but he returned to Germany in 1946 as director of the Max Planck Institute for Physics and Astrophysics in Göttingen, later in Munich, and remained as director until 1970. Heisenberg received the 1932 Nobel Prize in physics "for the creation of quantum mechanics, the application of which has, inter alia, led to the discovery of the allotropic forms of hydrogen."

See also HEISENBERG UNCERTAINTY PRINCIPLE.

Werner Heisenberg was a pioneer of quantum physics and is famous for his uncertainty principle. He received the 1932 Nobel Prize in physics. ***(AIP Emilio Segrè Visual Archives, W. F. Meggers Collection)***

Heisenberg uncertainty principle Named for its proposer, the German physicist WERNER KARL HEISENBERG, the Heisenberg uncertainty principle states that certain pairs of physical quantities cannot simultaneously possess sharp values, that greater sharpness of one is at the expense of less sharpness of the other. This principle is an essential ingredient of QUANTUM PHYSICS and QUANTUM MECHANICS.

The principle is expressed in terms of the uncertainty of the physical quantities, where uncertainty and sharpness are inversely related: greater uncertainty means less sharpness and less uncertainty signifies greater sharpness. Zero uncertainty is synonymous with perfect sharpness. Let us denote a pair of physical quantities by A and B and their respective uncertainties by ΔA and ΔB. (*See* QUANTUM MECHANICS for more detail about uncertainty.) Then the Heisenberg uncertainty principle states that the product of ΔA and ΔB cannot be less than a certain amount, specifically:

$$\Delta A \Delta B \geq \frac{h}{4\pi}$$

where h denotes the PLANCK CONSTANT, whose value is $6.62606876 \times 10^{-34}$ joule·second (J·s). Thus, given the uncertainty of one of such a pair, say ΔA, the least possible uncertainty of the other is determined by:

$$\Delta B \geq \frac{h}{4\pi \Delta A}$$

When one quantity of such a pair is sharp (i.e., its uncertainty is zero), the uncertainty of the other is infinite. That means its range of values is unlimited; it is as unsharp as possible.

The Heisenberg uncertainty relation is understood to concern the measured values of the physical quantities. It is not a matter of the system always possessing a sharp value and the uncertainty being in our knowledge of it. Rather, according to quantum physics, the system does not possess any value until a measurement

Physics in the Hierarchical Structure of Science

Science is our attempt to understand nature. As science developed over the past few centuries, it became apparent from very early on that nature does not at all present a uniform face. There are aspects of nature that seem largely independent of other aspects and accordingly deserve, and possess, a particular branch of science for each. That is why we often refer to the *sciences,* in the plural.

One aspect of nature is that of living organisms of all kinds. To study it, there are the life sciences, which we will refer to collectively as biology. Another apparently independent aspect of nature is that of MATTER, including its properties, composition, and structure. This aspect of nature is studied by the chemical sciences, which we will group together under the heading of chemistry. Then there is the aspect of nature that presents itself in the form of elementary entities, principles, and laws. The various fields of physics are concerned with this. Physics, chemistry, and biology make up what are called the *natural sciences.*

Beyond those three aspects, nature reveals the apparently independent aspect of mind. The science that seeks to understand mind is psychology. This science is conventionally not grouped with the natural sciences. Some organisms live social lives, and that aspect of nature seems largely independent of others. We will lump the social sciences together as sociology, also not considered a natural science.

Now, what do we mean by "independent" with regard to the five aspects of nature we have collected here? The idea is that it is possible to achieve a reasonable understanding of each aspect in terms of concepts that are particular to that aspect, with no, or little, reference to other aspects. This means that the corresponding science is, or can be, practically self-contained. As an example, biology, the science that studies living organisms, can deal with its subject almost wholly in terms of organisms and components of organisms. A better understanding is gained when concepts from chemistry and physics are introduced. However, that is not necessary for biology to operate as a science. In fact, the sciences developed very independently, with but little reference to each other, if any. Until the end of the 19th century, chemistry was doing very well with no knowledge of the atomic nature of matter. And even fields within a science were considerably isolated from each other. Well into the 20th century, biology was largely separated into botany (the study of plants) and zoology (the science of animals). It has become increasingly recognized that there is much unity within each science and that the various sciences can benefit from each other. (*See* ATOM.)

This benefit turns out to be mainly a one-way street, and strictly so in the natural sciences. Take chemistry, as an example. Modern chemists do not make do with merely describing chemical processes and properties and searching for empirical laws to help them organize their observations. They need to know the reasons for their findings. And the reasons invariably come from physics. Physics underlies chemistry. In principle, all of chemistry can be derived from physics (not that it would be possible in practice to do so). Much in chemistry depends on the FORCES between atoms. These forces are electromagnetic in nature, and at the atomic level the quantum character of nature is dominant. Both ELECTROMAGNETISM and QUANTUM MECHANICS, as well as their unification to QUANTUM ELECTRODYNAMICS, are fields of physics. Take the PERIODIC TABLE of the chemical ELEMENTS, a cornerstone of chemistry. It too is fully explained by physics, through the internal structure of atoms. Physics, on the other hand, has no need of chemistry, biology, psychology, or sociology to explain anything within its purview. It is truly an independent science and thus the most fundamental science.

Return to biology. Where do biologists turn to for extrabiological reasons for biological phenomena that do not find an explanation within biology? They seek their explanations from chemistry—largely from biochemistry—and from physics. In principle, all of biology should be reducible to chemistry and physics, and thus ultimately to physics, at least according to many people. There are, however, also many dissenters who hold that biology is, in some way, more than chemistry and physics. It remains true that extrabiological explanations in biology come from chemistry and from physics; never vice versa. It is not only physics that has no need of biology, but chemistry as well.

Among the natural sciences there exists a clear hierarchy, based on the directions of explanations among them. Physics is the most fundamental science, so we might picture it at the bottom of the hierarchy, in foundation position. All explanations from one science to another eventually reach physics. Then comes chemistry, one level above physics. Its extrachemical explanations are derived from physics alone. Above chemistry in the hierarchy lies biology. Extrabiological explanations come solely from chemistry and from physics.

Psychology, the science of mind, has long operated with little or no regard for the underlying mechanism of mind, which seems to be the operation of the brain, a biological organ. There is increasing interest among psychologists in the biological basis of mind. Whether mind is, at least in principle, completely explainable in biological terms

(continues)

Physics in the Hierarchical Structure of Science *(continued)*

is an open question at present. There are those who adamantly insist it is, those who equally adamantly insist it cannot be, and those who are waiting to see what develops. But the position of psychology in the hierarchy of science seems clear: right above biology. The most likely source of any extrapsychological explanation of mind appears to be biology. On the other hand, biology needs nothing from psychology with regard to explanation.

Sociology, the conglomeration of sciences that study society, should then find its hierarchy position above psychology. Its extrasociological explanations must surely derive from the nature of the members of a society and the interactions among them. That leads to the minds and bodies of the members, and thus to psychology and biology. Biology has no explanatory need for sociology. Psychology, however, just might, since real minds do not operate in a social vacuum. So the positions of psychology and sociology in the hierarchy of science might possess some degree of parallelism.

We see that science is structured in a rather clear-cut hierarchy, a hierarchy that is based on the directions of explanations among the various sciences. All the natural sciences are more fundamental than the others. And among the natural sciences, chemistry underlies biology, while physics holds the position of the most fundamental science of them all.

"forces" it to take one. Sharpness means that repeated measurements on the system in the same state (or measurements on many identical systems in the same state) always result in the same value for the quantity. In a situation of nonzero uncertainty, however, such a set of measurements produces a range of values. That range is what *uncertainty* means.

The pairs of physical quantities that are subject to the uncertainty principle include every generalized coordinate of the system together with its corresponding generalized momentum. The generalized coordinates are the set of physical quantities needed to specify a state of the system. The system's wave function is a function of all the generalized coordinates and of time. In the case of a system of particles, for instance, the generalized coordinates are simply the coordinates of the positions of all the particles, and the corresponding generalized momenta are the corresponding components of the particles' LINEAR MOMENTUM. As an example, let us consider the x-coordinate of a particle's position, x, and the x-component of its linear momentum, p_x, as a pair of quantities for which the Heisenberg uncertainty relation holds. Their respective uncertainties are denoted Δx and Δp_x and are subject to the constraint:

$$\Delta x \Delta p_x \geq \frac{h}{4\pi}$$

where Δx is in meters (m), and Δp_x is in kilogram·meters per second (kg·m/s). Given the uncertainty of Δx, say, the least possible uncertainty of Δp_x is determined by:

$$\Delta p_x \geq \frac{h}{4\pi \Delta x}$$

When one of these two quantities is sharp, the uncertainty of the other is infinite. If the particle's position is precisely known, the particle's momentum is maximally uncertain. And conversely, if the momentum is sharp, the particle has no location and can be anywhere.

An uncertainty relation also holds for ENERGY and TIME, although time is not considered a physical quantity, but rather a PARAMETER of evolution. The form of this uncertainty relation is:

$$\Delta t \Delta E \geq \frac{h}{4\pi}$$

where ΔE denotes the uncertainty of energy in joules (J), and Δt is in seconds (s). The significance of Δt is the duration, or lifetime, of the state of the system whose energy is E. The longer the state lives, the sharper its energy can be, and the briefer the state's lifetime, the more uncertain its energy. So stable states of ATOMs and NUCLEI, for example, and other states with long lifetimes possess sharp, well-defined values of energy. On the other hand, the energies of very unstable atomic and nuclear states can be quite indefinite and become less and less sharp for shorter and shorter lifetimes.

See also QUANTIZATION.

Helmholtz, Hermann Ludwig Ferdinand von (1821–1894) German *Physicist, Physiologist* Hermann Helmholtz was a versatile scientist, best known among physicists for his work in THERMODYNAMICS

Hermann Helmholtz was a 19th-century physiologist and physicist who worked in many fields, including thermodynamics and acoustics. His approach to science required that natural phenomena be explained wholly by physical mechanisms. *(AIP Emilio Segrè Visual Archives)*

and in musical ACOUSTICS. He graduated from the Royal Friedrich-Wilhelm Institute of Medicine and Surgery in Berlin, Germany, in 1843 and was immediately assigned to serve as a physician in an army unit. During his service, Helmholtz continued his research into basing physiology completely on physics and chemistry, with no recourse to nonphysical effects such as "vital force." (Although this is taken for granted today, at that time it was not.) In 1847, he published his first paper, on the mathematical principles underlying the conservation of ENERGY. As a result, Helmholtz was released from his military obligation in 1849 and until 1871 held professorships in physiology at a number of German universities. During that period, his interests and research shifted toward physics, and in 1871 Helmholtz became professor of physics at the University of Berlin. A fundamental theme of his approach to science was that natural phenomena must be explainable in terms of wholly physical mechanisms. And although he was involved in experimental research, his theoretical work was very mathematical in its approach. His fields of interest included, among others, the conservation of energy, thermodynamics (Helmholtz FREE ENERGY is named for him), musical acoustics and tone perception, and ELECTROMAGNETISM.

See also CONSERVATION LAW; EXPERIMENTAL PHYSICS; THEORETICAL PHYSICS.

Henry, Joseph (1797–1878) American *Physicist* A scientist of wide-ranging interests, Joseph Henry was one of the most eminent American scientists of his time. Physicists best know him for his work in ELECTROMAGNETISM. Henry graduated from the Albany Academy in Albany, New York, in 1822, and during 1826–32 served as professor there. He was appointed professor of physics at the College of New Jersey (later Princeton University) in Princeton in 1832 and stayed there until his resignation in 1848. In 1846, Henry became secretary of the Smithsonian Institution in

Joseph Henry was a 19th-century scientist, known among physicists for his work in electromagnetism. He directed the Smithsonian Institution for over 30 years. *(AIP Emilio Segrè Visual Archives, E. Scott Barr Collection)*

Washington, D.C., in which capacity he served until his death. Early in his career, Henry carried out research in electromagnetism, which laid a foundation for the telegraph, electric motor, and telephone. The SI UNIT of INDUCTANCE, the henry (H), is named for him. As secretary of the Smithsonian Institution, Henry promoted basic research, anthropology, meteorology, and education, among other fields.

Hertz, Heinrich Rudolf (1857–1894) German *Physicist* Best known among physicists for his work with electromagnetic WAVEs, Heinrich Hertz was the first to transmit and receive radio waves. From 1885 until his death, Hertz served as professor of physics in Germany, first at the technical school in Karlsruhe and then at the University of Bonn. His research in ELECTROMAGNETISM led to his achievement with radio waves and to his demonstration that LIGHT is a form of electromagnetic RADIATION, thus confirming JAMES CLERK MAXWELL's prediction. Hertz's experiments laid a foundation for wireless broadcasting and communication. The SI UNIT of FREQUENCY, the hertz (Hz), is named for him.

See also MAXWELL'S EQUATIONS.

A 19th-century physicist, Heinrich Hertz was the first person to transmit and receive radio waves. ***(Deutsches Museum München, courtesy AIP Emilio Segrè Visual Archives, Physics Today Collection)***

hole A missing ELECTRON in the VALENCE BAND of a SEMICONDUCTOR is called a hole. It behaves as a mobile carrier of positive electric CHARGE and thus contributes to the CONDUCTION of electricity by the semiconductor. Holes are produced by exciting electrons from the valence band to the CONDUCTION BAND. In a semiconductor, relatively little ENERGY is needed for that. A permanent supply of holes is introduced into a semiconductor by adding a tiny quantity of an appropriate impurity, called a dopant, to the basic, elemental semiconductor, such as silicon or germanium, during manufacture. The appropriate dopant for creating holes must possess less than four valence electrons per ATOM. (Gallium, whose atom has three valence electrons, is often used.) The dopant atoms replace semiconductor atoms, which have four valence electrons each, in the crystal lattice. That leaves a deficiency of valence electrons in the lattice. The missing electrons are the holes. Since a hole is a lack of a negative charge in a "sea" of negative charges, it behaves as a positive charge. It is mobile because an electron from an adjacent semiconductor atom can easily jump over and fill it, thus creating a hole at the next lattice site, and so on. Such a semiconductor is known as a p-type semiconductor. (*See* CRYSTAL; ELECTRICITY; LATTICE.)

holography In the most general sense of the term, holography is the recording and storage of three-dimensional information in a two-dimensional medium and its reconstruction from the medium. A two-dimensional record of three-dimensional information is called a hologram, or a holographic image. In particular, holography most commonly refers to three-dimensional scenes recorded on photographic film in such a manner that the scenes can be viewed in their full three-dimensionality from the image on the film. The procedure for producing a hologram is to allow a beam of coherent LIGHT both to illuminate a scene and to fall directly on the film. The light reflected to the film from the scene and the light directly reaching the film interfere and the resulting interference pattern is recorded on the film, thus creating a hologram. The holographic image contains information about both

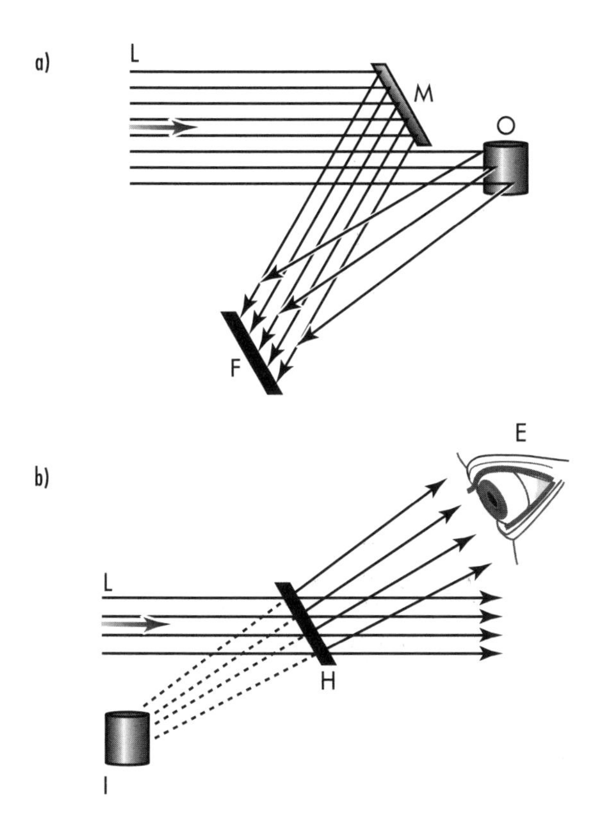

(a) A hologram is produced by having a coherent beam of light, L, reach photographic film, F, both directly (with the help of mirror, M, in this setup) and by reflection from an object, O. (b) A holographic virtual image, I, of the object is viewed by shining light, L, through the hologram, H, and looking back through it (represented by an eye, E). As the observer moves around, different sides of the image disappear from and come into view.

the amplitude, related to the INTENSITY, and the PHASE of the light from the scene, which allows a three-dimensional reconstruction of the scene. (In ordinary photography, on the other hand, it is only the intensity that is recorded in the image.) Reconstruction is achieved by shining light through or reflecting light from the hologram. The image thus generated can be viewed as if it were the original three-dimensional scene. As an observer moves about, different sides of the scene come into and disappear from view. (*See* COHERENCE; DIMENSION.)

Hooke, Robert (1635–1703) British *Scientist* One of Britain's greatest scientists of his time, along with ISAAC NEWTON, Robert Hooke is known among physicists for HOOKE'S LAW of ELASTICITY. A self-educated person, Hooke was involved in and made contributions to many areas of science throughout his lifetime. Examples include elasticity, LIGHT and color, GASes, microscopy, and astronomy. He was also an inventor, surveyor, and architect. Among his many accomplishments, Hooke invented what is essentially the modern air pump and designed an improved escapement mechanism for a spiral-spring-powered clock.

Hooke's law Named for the British scientist ROBERT HOOKE, Hooke's law of ELASTICITY states that when a material is deformed by an external deforming factor, called a STRESS, the STRAIN of the material—its response to the stress—is proportional to the stress. This is not a universal law, but merely describes the behavior of many elastic materials, i.e., materials that return to their initial configuration when the deforming stress is removed, at least if the stress is not too great. For such materials, the strain is typically proportional to the stress for stresses up to a proportionality limit. So materials with a significantly high proportionality limit can be said to obey Hooke's law (with the understanding that above the proportionality limit, Hooke's law becomes invalid for them). Although many elastic materials do obey Hooke's law, some, such as rubber, do not, or do so only approximately.

As an example, springs are normally constructed to obey Hooke's law. For a spring, the applied force, F, in newtons (N) and the spring's resulting displacement, x, in meters (m) are related by:

$$F = kx$$

for any force below the proportionality limit, where k is the spring constant in newtons per meter (N/m).

Materials that obey Hooke's law can be characterized by a number of quantities called moduli of elasticity, all of which are defined as the ratio of the stress to the resulting strain:

$$\text{Modulus} = \frac{\text{stress}}{\text{strain}}$$

In all cases, the stress is force per unit area, and its UNIT is the pascal (Pa), equivalent to newton per square meter (N/m^2). The strain is always a dimensionless ratio of a change in a quantity to the quantity. So the unit of the modulus of elasticity is the same as that of stress, the pascal. (*See* DIMENSION.)

For linear deformation caused by tension or compression, Young's modulus is the ratio of the longitudinal force per unit cross-section area to the relative change in length. If a longitudinal force F causes a length change ΔL in a length L, then Young's modulus, Y, is defined as:

$$Y = \frac{F/A}{\Delta L/L}$$

where A denotes the cross-section area. Here F is in newtons (N), A is in square meters (m^2), and ΔL and L are in the same units of length. F and ΔL are taken to have the same algebraic sign.

For volume change under pressure, the bulk modulus, B, is defined as the ratio of the PRESSURE to the relative change of volume:

$$B = -\frac{p}{\Delta V/V}$$

Here p denotes the pressure in pascals (Pa), V the volume, and ΔV the volume change, where ΔV and V are in the same units of volume. A minus sign is introduced to make B positive, since a positive pressure causes a decrease of volume (i.e., a negative value for ΔV).

An additional modulus is the shear modulus, G, for SHEAR deformation. It is defined as the ratio of the force per unit parallel area to the angle of deformation:

$$G = \frac{F/A}{\gamma}$$

Here F denotes the shear force in newtons (N); A is the area parallel to the force over which the force is acting, in square meters (m^2); and γ is the deformation angle in radians (rad).

Hubble effect Named for the American astronomer Edwin Hubble (1889–1953), the Hubble effect is the observation that, except for nearby GALAXIES, all observed galaxies are moving away from us. Moreover, on the average and to a good approximation, the speed of recession of a galaxy is proportional to its distance from us. That is expressed as Hubble's law:

$$\text{Recession speed} = H_0 \times \text{distance}$$

where, as is common in astronomy, the distance is in megaparsecs (Mpc) and the speed is in kilometers per second (km/s). The proportionality coefficient, H_0, is the Hubble constant and has a value in the range 50–80 kilometers per second per megaparsec [km/(s·Mpc)], based on the measured speeds and distances of many galaxies. The uncertainty in the value of H_0 has mostly to do with methodological problems in the determination of large astronomical distances. Recession speeds are deduced from the measured RED SHIFT of a galaxy's SPECTRUM.

The common interpretation of the Hubble effect is that the UNIVERSE is expanding. The observed red shifts are understood to be mainly due to the "stretching" of SPACE (and hence are called cosmological red shifts), according to ALBERT EINSTEIN's general theory of relativity. If the universe evolved from a BIG BANG and if the rate of expansion has been constant, Hubble's law gives an estimate for the age of the universe, since $1/H_0$ would then be the time taken by any galaxy, assuming it has existed for so long, to reach its measured distance from us while traveling at its measured speed. That comes to about 15 billion years. (*See* COSMOLOGY; RELATIVITY, GENERAL THEORY OF.)

Since the SPEED OF LIGHT is finite, astronomical observations do not reveal the present situation, but show what was happening as far back in the past is the time it took the observed light to reach us. In this way we can gather clues about the universe's past. One such clue, a deviation from Hubble's law for very distant galaxies as indicated by recent measurements, seems to show that in the past, the rate of expansion of the universe was lower than it is at present. In other words, the expansion of the universe appears to be accelerating. This result came as a surprise, since it had been generally assumed that the gravitational attraction among all the matter in the universe would cause the expansion to slow down. One proposed explanation is that the universe is pervaded by a uniform distribution of ENERGY, called dark energy, whose effect, according to the general theory of relativity, is repulsive. (*See* GRAVITATION.)

Huygens's principle Named for the Dutch physicist Christiaan Huygens (1629–95), Huygens's principle deals with the propagation of WAVES. It states that for any given WAVEFRONT, successive wavefronts can be constructed by considering every point on the initial wavefront as serving as a source of secondary waves emanating from it in all directions. All those waves undergo INTERFERENCE, resulting in the new wavefronts. Any new wavefront is constructed as a surface tangential to simultaneous secondary wavefronts from all the points of the initial wavefront. Huygens's principle gives correct results in many cases, although there are limitations to its validity.

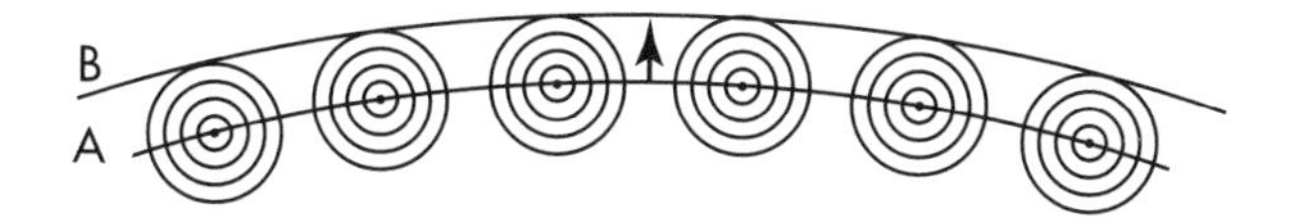

According to Huygens's principle, every point on a wavefront of a propagating wave is to be considered a source of secondary waves. The secondary waves from all the points on a wavefront interfere with each other, resulting in new wavefronts, which are constructed as surfaces that are tangent to the simultaneous secondary wavefronts. This cross-sectional view shows the secondary wavefronts emanating from a number of points on wavefront A. Successive wavefront B is shown as tangent to simultaneous secondary wavefronts. The direction of propagation is indicated by an arrow.

hydraulics Hydraulics is the field of physics and of engineering that deals with the mechanical properties of FLUIDS (i.e., LIQUIDS and GASes) at rest and in motion and the practical utilization of those properties. Examples of applications are hydraulic presses and lifts, pneumatic wrenches and drills, and hydraulic controls of machinery. (*See* MECHANICS.)

See also HYDRODYNAMICS; HYDROSTATICS.

hydrodynamics The field of physics that studies the flow of FLUIDS (i.e., LIQUIDS and GASes) is called hydrodynamics. Fluid flow is generally either laminar or turbulent. Laminar flow, also called streamline flow or steady flow, is a smooth flow in which the path of any particle of the fluid is a smooth line, a streamline. Turbulent flow, on the other hand, is chaotic and rapidly changing, with shifting eddies and tightly curled lines of flow. Observe the smoke rising from a smoking ember, for example. Close to the ember it ascends in smooth, laminar flow; as it rises higher, it breaks up into eddies and continues ascending in turbulent flow. (*See* TURBULENCE.)

The flow of fluids is, in general and even in laminar flow, a complex affair. VISCOSITY plays its role, acting in the manner of FRICTION by removing KINETIC ENERGY from the system and converting it to HEAT. The DENSITY of the fluid can change along the flow. Moreover, the particles of the fluid might attain rotational kinetic energy about their CENTERS OF MASS. And the flow, even if laminar and steady, might vary in time as the causes of the flow vary. For the following discussion, we make a number of very simplifying assumptions, which nevertheless allow the treatment and understanding of various useful phenomena. First, we assume that we are dealing with flow that is laminar and stationary: the flow is smooth and steady and, moreover, at each point in the fluid the VELOCITY of the flow does not change in time. Next, we assume that the fluid is incompressible, i.e., that its density is the same at all points and at all times. Further, we assume that the effect of viscosity is negligible, i.e., that the flow is nonviscous. Finally, we assume irrotational flow, i.e., that the fluid particles do not rotate around their centers of mass. (*See* ROTATION.)

With those assumptions, consider any flow tube, an imaginary tubelike surface whose walls are parallel to streamlines. The streamlines that are inside a flow tube stay inside it and those that are outside it do not penetrate and enter it. It is as if the fluid is flowing through a tube, except there is no actual tube. Let A denote the cross-section AREA in square meters (m^2) at any position along a flow tube and v the flow SPEED there in meters per second (m/s). The product Av is the volume flux or volume flow rate, the VOLUME of fluid flowing through the cross-section per unit time. Its SI UNIT is cubic meter per second (m^3/s). Under the assumptions, the volume flux is constant along a flow tube: the volume of fluid flowing past any location along a flow tube during some time interval is the same as that passing any other location along the same flow tube during the same time interval. This is expressed by the continuity equation:

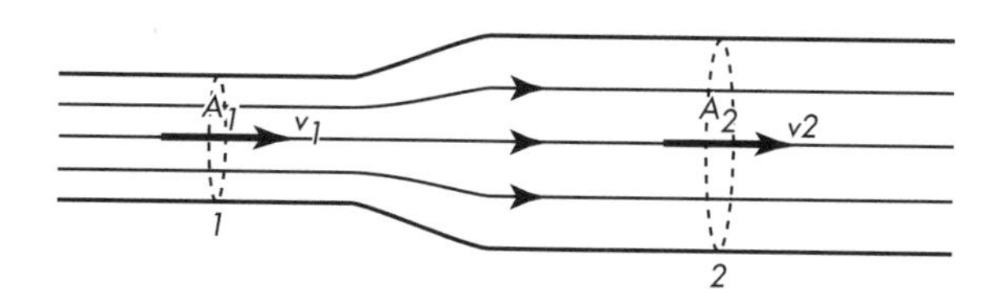

The continuity equation for an incompressible fluid is an expression of the fact that in a confined flow, or in an equivalent situation, the volume of fluid entering the flow during any period of time equals the volume exiting it during that time. The equation is $A_1v_1 = A_2v_2$, where A_1 and A_2 denote the cross-section areas of the flow at positions 1 and 2, respectively, while v_1 and v_2 are the respective flow speeds. It follows that the fluid flows faster in narrower regions than in wider ones.

$$A_1v_1 = A_2v_2$$

where the subscripts indicate any two locations along the flow tube. The continuity equation tells us, for instance, that where the flow narrows (and the streamlines bunch together), the flow speed increases, and where the flow broadens, it slows. One use of this effect is to increase the range of the water stream from a garden hose by narrowing the end of the hose, either with a finger or by means of a nozzle. (*See* FLUX.)

Consider a small volume of fluid that fills the cross-section of a flow tube and flows along the tube, changing in size and in speed as it flows, in accord with the continuity equation. The WORK-ENERGY THEOREM states that the WORK done on such a volume by the PRESSURE of the fluid, as the volume flows from one location along the flow tube to another, equals the sum of its gains in kinetic energy and POTENTIAL ENERGY. That results in BERNOULLI'S EQUATION, named for the Swiss mathematician-physicist DANIEL BERNOULLI:

$$p_1 + \rho v_1^2/2 + \rho g y_1 = p_2 + \rho v_2^2/2 + \rho g y_2$$

where p denotes the pressure of the fluid in pascals (Pa), equivalent to newtons per square meter (N/m²); ρ is the density of the fluid in kilograms per cubic meter (kg/m³); g is the acceleration due to gravity whose nominal value is 9.8 meters per second per second (m/s²); y denotes the height of the fluid volume in meters (m); and the subscripts denote any two locations along a streamline. (*See* ACCELERATION; GRAVITATION.)

An example of application of Bernoulli's equation is this. Let a fluid flow through a horizontal pipe that is narrower in some region than in the rest of the pipe.

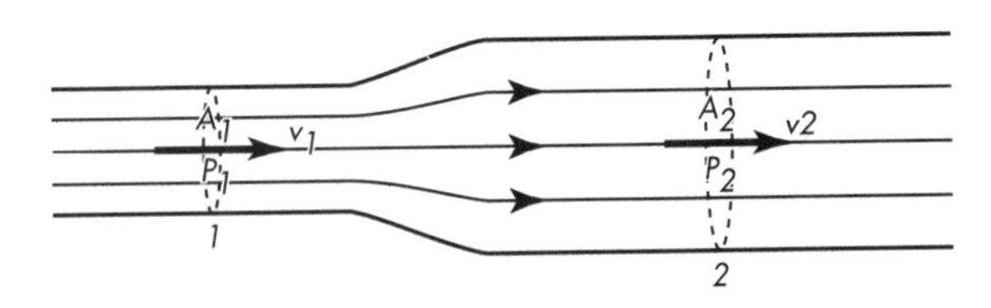

The Venturi effect in fluid flow, which follows from Bernoulli's equation, is the reduction of pressure in regions of higher flow speed. These are regions of smaller cross-section area, according to the continuity equation. A_1, v_1, and p_1 denote the cross-section area, flow speed, and pressure, respectively, at position 1 along the flow, while A_2, v_2, and p_2 denote the corresponding quantities at position 2. Since in this example $A_1 < A_2$, then $v_1 > v_2$, by the continuity equation, and $p_1 < p_2$, by Bernoulli's equation.

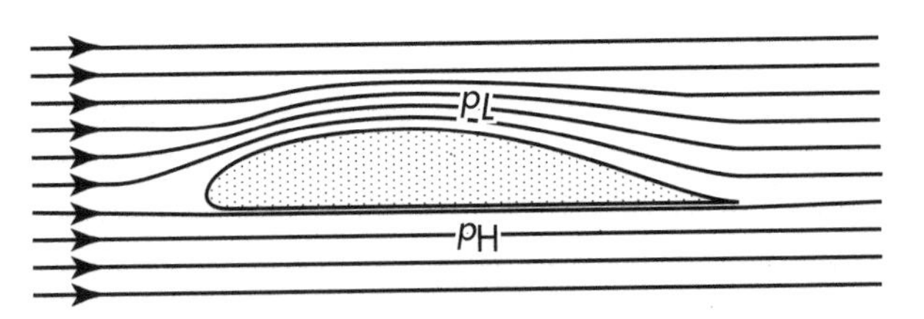

When air flows over and under a horizontal aircraft wing (or when the wing moves through air), the continuity equation and Bernoulli's equation show that the pressure beneath the wing is greater than the pressure above it. That results in a net upward force on the wing, called hydrodynamic lift. The symbols p_H and p_L denote the higher and lower pressures, respectively.

We saw—as an example of use of the continuity equation—that the fluid flows faster in the narrower region than in the wider sections. Putting $y_1 = y_2$ and $v_1 > v_2$ in Bernoulli's equation, we obtain $p_1 < p_2$. In other words, the pressure of the fluid in the narrower region, where it is flowing faster, is less than in the wider region, where its speed is lower. This is referred to as the Venturi effect.

As an additional example, the air flow around a horizontal airplane wing can be treated with the continuity equation and Bernoulli's equation similarly to the derivation of the Venturi effect. The result is that the air pressure beneath the wing is greater than the pressure above it. This is one source of the upward FORCE, called LIFT, that keeps airplanes in the air.

hydrostatics The field of physics that studies the behavior of FLUIDS (i.e., LIQUIDS and GASes) at rest is called hydrostatics. Some of the principles of hydrostatics are:

Direction of force. The FORCE exerted by a fluid at rest upon a SOLID surface is perpendicular to the surface.

Hydrostatic pressure. The PRESSURE, p, at depth h beneath the surface of a liquid at rest is given by:

$$p = p_0 + \rho g h$$

where p_0 is the pressure on the surface of the liquid (often atmospheric pressure), ρ is the DENSITY of the liquid, and g is the acceleration due to gravity. Here the pressures are in pascals (Pa), equivalent to newtons per square meter (N/m²); density is in kilograms per cubic meter (kg/m³); g is in meters per second per second (m/s^2) with value of about 9.8 m/s^2; and h is in meters (m). (*See* ACCELERATION; GRAVITATION.)

Transmission of pressure. PASCAL'S PRINCIPLE states that an external pressure applied to a fluid that is confined in a closed container is transmitted at the same value throughout the entire volume of the fluid.

Buoyancy. ARCHIMEDES' PRINCIPLE states that a body immersed in a FLUID, whether wholly or partially immersed, is buoyed up by a FORCE whose magnitude equals that of the WEIGHT of the fluid that the body displaces. (*See* BUOYANCY.)

hysteresis Hysteresis is the phenomenon that an effect lags behind the application of its cause and, moreover, can depend on the system's history. Hysteresis is seen in the response of materials to the application and removal of FORCES and of electric and magnetic FIELDS. (*See* ELECTRICITY; MAGNETISM.)

In elastic hysteresis, the STRAIN, the deformation, is not a function exclusively of the STRESS, the applied cause of deformation, but depends also on past states of the material. When a force is applied and removed, for instance, the material might retain some deformation and not return to its initial state. As a result, when the stress goes through a cycle, returning to its initial value, WORK is performed on the material, which causes heating. As an example, note the heating of a rubber band (which is more noticeable for a thick band) after it has been stretched and released a few times. (*See* ELASTICITY; ENERGY; HEAT; WORK-ENERGY THEOREM.)

In magnetic hysteresis, a material's magnetic POLARIZATION, or magnetization, caused by an external magnetic field might, similarly to the elastic case, depend also on past states of the material. When a magnetic field is applied to and removed from an initially unmagnetized ferromagnetic sample, for example, the material might retain some magnetic polarization, thereby becoming a permanent MAGNET. Again as in the elastic case, when a cycle is completed by the external magnetic field, work is done on the material, resulting in heating. This is one cause of heating in electric transformers, where the alternating electric CURRENTs (AC) produce cycling magnetic fields. The hysteresis of the transformer core material brings about undesired heating of the core, with consequent waste of electric energy.

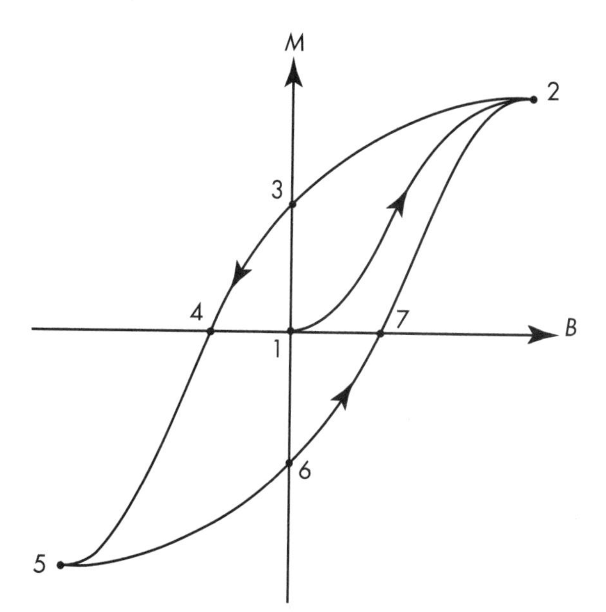

Example of a magnetic hysteresis curve for a ferromagnetic material. The axes represent the external magnetic field in some direction, *B*, and the magnetic polarization, or magnetization, of a sample, *M*. The example starts at 1, with initially unmagnetized sample and zero field. The field is increased and the sample polarizes until it becomes maximally magnetized, a state called saturation, at 2. When the field is reduced to zero, the sample retains some of its magnetization and has become a permanent magnet at 3. The field is reversed and increased in magnitude, causing the sample to lose magnetization at 4 and then reverse its polarization and achieve saturation at 5. The field is again reduced to zero and again the sample retains magnetization at 6. The field is reversed again and increased in magnitude until the polarization disappears at 7, then reverses and saturates again at 2.

Electric hysteresis, similar to magnetic hysteresis, can occur when a material responds to an applied electric field by becoming electrically polarized.

See also DIAMAGNETISM; DIELECTRIC; DIELECTRIC CONSTANT; FERROMAGNETISM; PARAMAGNETISM; PERMEABILITY; PERMITTIVITY.

I

ideal gas *See* GAS, IDEAL.

ideal gas law *See* GAS, IDEAL.

impedance Impedance is a quantity that plays the role for alternating CURRENT (AC) circuits that RESISTANCE plays for direct current (DC) circuits. Impedance appears analogously in the AC version of OHM'S LAW:

$$V = IZ$$

where Z denotes the impedance, whose SI unit is the ohm (Ω), and V and I represent the AC VOLTAGE and current, respectively, in volts (V) and amperes (A), both expressed as either effective values or maximal values. Impedance measures the extent to which a circuit opposes the passage of an alternating electric current. (*See* ELECTRICITY.)

The impedance of a RESISTOR is simply its RESISTANCE, R, in ohms (Ω). The instantaneous AC voltage across a resistor has the same PHASE as the instantaneous current through it.

For a CAPACITOR, the impedance is called capacitive reactance and has the value:

$$X_C = \frac{1}{2\pi f C}$$

where X_C is in ohms (Ω), f is the FREQUENCY in hertz (Hz) of the alternating current, and C is the CAPACITANCE in farads (F). The instantaneous voltage across a capacitor lags behind the instantaneous current through it by a quarter cycle ($\pi/2$ radians, or 90°). (*See* REACTANCE.)

The impedance of an INDUCTOR is referred to as inductive reactance, with the value

$$X_L = 2\pi f L$$

where X_L is in ohms (Ω), and L is the INDUCTANCE in henrys (H). The instantaneous voltage across an inductor leads the instantaneous current through it by a quarter cycle ($\pi/2$ radians, or 90°).

The total impedance of combinations of electric components of various kinds must take into account the AC phase differences that occur. One method of doing that is based on complex numbers. Another method, the phasor method, represents AC quantities by two-dimensional VECTORs, with the time-varying quantities—the instantaneous voltages and currents—appearing as rotating vectors. In the simple case of a resistor, an inductor, and a capacitor connected in series, referred to as a series RLC connection, the reactance is:

$$X = X_L - X_C$$

and the impedance is given by:

$$Z = \sqrt{R^2 + X^2}$$

When one AC circuit is producing electric POWER and is connected to another circuit that is receiving the power, maximal power transfer is achieved when the respective impedances of the two circuits are equal.

That is called IMPEDANCE MATCHING. An example is matching the impedance of an audio amplifier and the impedance of each speaker connected to it, in order to optimize power delivery from the amplifier to the speakers. The concept of impedance is more general that that described above. It is applied to the opposition of any system, such as a mechanical system, to external sinusoidal excitation. Also, in the general case, optimal power transfer between systems is achieved by impedance matching.

impedance matching To maximize the transfer of POWER between a power-generating system and a power-receiving one, the systems' respective IMPEDANCES are made equal. This is termed impedance matching, which is most familiar for the case of electric circuits, where one circuit delivers electric power to another, such as an audio amplifier to a speaker. Making the impedances of the circuits equal ensures that the greatest amount of power is delivered from one to the other. In the audio example, on the back of an amplifier that is not part of a system of components designed to work together, there is always an indication of the proper impedance (such as 8 Ω) for each of the speakers that are to be connected to it. Some amplifiers offer separate terminals for speakers of different impedances. (*See* CURRENT; ELECTRICITY.)

The concept of impedance is more general than just electric impedance. It is also applied to the opposition of any system, such as mechanical, to external sinusoidal excitation. Also, in the general case, optimal power transfer between systems is achieved by impedance matching.

impulse The effect of a FORCE, acting over TIME, in changing MOMENTUM is termed impulse. More precisely, the impulse of a force is the integral of the force over time:

$$\mathbf{I} = \int \mathbf{F}\,dt$$

where **I** represents the impulse, a VECTOR, in newton·seconds (N·s); **F** is the (possibly time varying) force, also a vector, in newtons (N); and t is the time in seconds (s).

According to ISAAC NEWTON's second law of MOTION in its most general form, the instantaneous effect of a force **F** acting on a body of MASS m is:

$$\mathbf{F} = d(m\mathbf{v})/dt$$

where **v** denotes the VELOCITY of the body is in meters per second (m/s) and m is in kilograms (kg). In other words, the force equals the rate of change in time of the momentum, $m\mathbf{v}$. To find the impulse, or cumulative effect over time, of the force for the time interval from t_1 to t_2, integrate this equation over the time interval:

$$\mathbf{I} = \int_{t_1}^{t_2} \mathbf{F}\,dt = (m\mathbf{v})_2 - (m\mathbf{v})_1$$

The subscripts on the right-hand side indicate the corresponding values of the momentum at the two times. So the impulse equals the change in momentum caused by the force during the time interval under consideration. (*See* NEWTON'S LAWS.)

For a constant force, **F**, acting over the time interval from t_1 to t_2, the impulse is simply:

$$\mathbf{I} = (t_2 - t_1)\,\mathbf{F}$$

The time average of a force over the time interval from time t_1 to t_2, $\mathbf{F}_{av}$, is defined as the constant force that, acting over the same time interval, would produce the same impulse as the given force:

$$(t_2 - t_1)\,\mathbf{F}_{av} = \int_{t_1}^{t_2} \mathbf{F}\,dt$$

Hence:

$$\mathbf{F}_{av} = \frac{1}{t_2 - t_1}\int_{t_1}^{t_2} \mathbf{F}\,dt$$

index of refraction The ratio of the SPEED OF LIGHT to the propagation SPEED of a sinusoidal electromagnetic wave in a material is the index of refraction of that material for the wave's frequency. This is expressed in the following formula:

$$n = c/v$$

where n denotes the index of refraction of a material for the wave's frequency, v is the propagation speed in the material of a sinusoidal electromagnetic wave in meters per second (m/s), and c represents the speed of light, whose value is 2.99792458×10^8 m/s. As a ratio of two speeds, the index of refraction is a dimensionless quantity. (*See* DIMENSION; ELECTROMAGNETISM.)

For visible LIGHT the propagation speed in matter is often, but not always, less than the SPEED OF LIGHT, giving an index of refraction greater than 1. On the other hand, an index of refraction less than 1 indicates a propagation speed that exceeds the speed of light. Such a situation does not violate the special theory of relativity, since no information or signal, and certainly no matter, is being transmitted superluminally. For the transmission of information, a sinusoidal wave must be modulated (i.e., a signal must be imposed upon it), and a modulation propagates at the GROUP SPEED, which is always less that the speed of light. (*See* RELATIVITY, SPECIAL THEORY OF.)

The name of this quantity derives from its importance in REFRACTION through SNELL'S LAW. This law shows how the refraction (i.e., bending) of a light ray that passes from one transmission medium into another depends on the indexes of refraction of the two media.

See also BIREFRINGENCE; DISPERSION.

inductance Also called self-inductance, the inductance of an electric circuit component, usually a coil, is a measure of the magnitude of the ELECTROMOTIVE FORCE (emf) that is induced in the component by a time-varying electric CURRENT flowing through it. More precisely, if L denotes the inductance of a component and i the current through it, then the induced emf, E, is given by:

$$E = -L\, di/dt$$

where t denotes the TIME. In words, the induced emf is proportional to the negative of the time rate of change of the current through the component, and the proportionality coefficient is the inductance. The SI UNIT of inductance is the henry (H), E is in volts (V), i is in amperes (A), and t is in seconds (s). This relation follows from AMPÈRE'S LAW and from the BIOT-SAVART LAW. The former determines that the induced emf is proportional to the time rate of change of the magnetic FLUX through the component, while from the latter it follows that the magnetic flux is proportional to the current. The negative sign in this equation indicates that the polarity of the induced emf is such as to reduce the cause of the emf, according to LENTZ'S LAW. (*See* ELECTRICITY; MAGNETISM.)

Self-inductance can equivalently be viewed as a measure of the magnetic flux created in a coil by a current in the coil, compensated for the number of turns in the coil. The relation is:

$$\Phi_m = Li/N$$

where Φ_m denotes the magnetic flux through the coil in webers (Wb), and N is the number of turns. From this, the self-inductance of a coil can be defined as:

$$L = N\Phi_m/i$$

See also IMPEDANCE; INDUCTION; MUTUAL INDUCTION; SELF-INDUCTION.

induction The term *induction* describes the causing of an effect by means of a FIELD. The term is most commonly used in any of three ways:

Electric (or *electrostatic*) *induction.* An applied electric field causes electric POLARIZATION of a material by aligning its atomic or molecular DIPOLES, by bringing about the separation of positive and negative CHARGES, or by both means together. That creates an internal electric field that is oppositely directed to the applied field and partially cancels it inside the material. (*See* ATOM; ELECTRICITY; MOLECULE.)

See also CAPACITOR; DIELECTRIC; DIELECTRIC CONSTANT; PERMITTIVITY.

Magnetic (or magnetostatic) induction. An applied magnetic field causes magnetic polarization of a material by creating atomic MAGNETIC DIPOLES, by aligning existing atomic magnetic dipoles, by increasing the size of the MAGNETIC DOMAINS that are aligned with the applied field at the expense of the other domains, by aligning misaligned magnetic domains, or by some combination of those. Whereas in DIAMAGNETISM the internal magnetic field that is produced opposes the applied field and weakens it in inside the material, in both PARAMAGNETISM and FERROMAGNETISM the internal field reinforces the applied field, weakly in the former case and strongly in the latter. (*See* MAGNETIC INDUCTION; MAGNETISM.)

See also MAGNET; PERMEABILITY.

Electromagnetic induction. Here magnetism creates electric effects and vice versa. (*See* ELECTROMAGNETISM.)

Two kinds of electromagnetic induction are the most familiar. They are:

Electromagnetic radiation. These WAVES propagate due to an induction effect described by MAXWELL'S EQUATIONS: a time-varying electric field creates a magnetic field, while a time-varying magnetic field produces an electric field. (*See* RADIATION.)

See also LIGHT.

Induced emf. According to FARADAY'S LAW, a time-varying magnetic FLUX through a conducting coil creates an induced ELECTROMOTIVE FORCE (emf) in the coil.

See also IMPEDANCE; INDUCTANCE; MUTUAL INDUCTION; SELF-INDUCTION.

inductor An inductor is any electric circuit component that is designed to possess a particular value of INDUCTANCE. It is also called a coil or a choke. In an AC circuit, the IMPEDANCE of an inductor is referred to as inductive REACTANCE and has the value:

$$X_L = 2\pi fL$$

where X_L is in ohms (Ω), f is the AC FREQUENCY in hertz (Hz), and L is the inductor's inductance in henrys (H). The instantaneous voltage across an inductor leads the instantaneous current through it by a quarter cycle ($\pi/2$ radians, or 90°). (*See* ELECTRICITY.)

inertia The tendency of a physical system to maintain its state under the action of an external influence is known as inertia. The best-known type of inertia is mechanical inertia. According to ISAAC NEWTON's first law of MOTION, inertial motion—motion in the absence of FORCES—is motion at constant VELOCITY, i.e., motion in a straight line at constant SPEED (or rest, which is but the special case of constant zero speed). The action of a force on a body is to change its velocity (i.e., to change its speed or direction of motion or both). A body's resistance to such a change is its inertia. According to Newton's second law of motion, a body's MASS is a measure of its inertia, since for a given force, the magnitude of the body's ACCELERATION—the time rate of change of its velocity—is inversely proportional to its mass. This inertia can also be termed translational inertia, to distinguish it from rotational inertia, which plays an analogous role in rotational motion. There, inertial motion is that of constant ANGULAR MOMENTUM, and a body's MOMENT OF INERTIA serves as a measure of its resistance to changes in its angular momentum due to the effect of TORQUES. (*See* NEWTON'S LAWS; ROTATION.)

The question of the origin of mechanical inertia has not yet been satisfactorily resolved. ERNST MACH proposed that inertia's origin is in the presence of all the matter of the UNIVERSE, an idea that is known as MACH'S PRINCIPLE. ALBERT EINSTEIN was greatly influenced by Mach's idea and attempted to incorporate it into his general theory of relativity. Other ideas have been put forth. But it seems that a full understanding of the origin of mechanical inertia has not yet been achieved. (*See* RELATIVITY, GENERAL THEORY OF.)

As an example of a different type of inertia, electromagnetic inertia is the tendency of an electric circuit to maintain the value of magnetic FLUX passing through it. According to FARADAY'S LAW and LENZ'S LAW, any change in flux induces an ELECTROMOTIVE FORCE (emf) in the circuit that causes a CURRENT that produces a magnetic FIELD so as to retard the flux change. Through the same mechanism, an INDUCTOR tends to maintain a constant current through itself. Any change in current induces an emf that causes an additional current such that the change is retarded. If the current is decreasing, the polarity of the emf will be such that the induced current will be in the same direction as the decreasing current, thus reducing the decrease. If the current is increasing, the induced current will oppose the increasing current, thus moderating the increase. (*See* BIOT-SAVART LAW; ELECTRICITY; ELECTROMAGNETISM; INDUCTION; MAGNETISM.)

inertial reference frame *See* REFERENCE FRAME, INERTIAL.

initial conditions The term *initial conditions* refers to a set of values for the variables that specify the state of a system and for their first, and possibly higher, TIME derivatives at any given time. A description of how a system evolves into and from the state specified by initial conditions is obtained by solving the system's EQUATIONS OF MOTION for a solution, in the form of the variables as functions of time, that fulfills the initial conditions. Such a solution describes both the evolution of the system that leads to the specified state at the given time and the system's subsequent evolution from that state.

A collection of inductors, also called coils or chokes. The two on the far left and right are adjustable: their inductance can be changed by screwing a ferrite core in and out. To set the scale of the figure, the full length of the body of the inductor on the far left is 5.2 centimeters (just over 2 inches). ***(Photo by Frost-Rosen)***

In the particular case of a mechanical system consisting of a number of point particles, the variables specifying the state of a system are the positions of the particles, say the x-, y-, z- coordinates of all the particles. The equations of motion are provided by ISAAC NEWTON's second law of MOTION and are second-order ordinary differential equations for the particle coordinates. Initial conditions for such a system are the values, for any given time, of all the particle coordinates and their first time derivatives, i.e., a specification of the positions and VELOCITIES (or, equivalently, LINEAR MOMENTA) of all the particles at the given time. The equations of motion can then be solved to obtain a solution that fulfills the initial conditions and thus describes the motions of the particles that lead to the specified state at the given time and the particles' subsequent motions for all later times. (*See* NEWTON'S LAWS.)

More fundamentally, initial conditions are part of an analysis of what is going on in the world in terms of developmental laws of nature and initial conditions. In that approach, the time evolution of physical systems is conceptually split into two aspects. One is the initial state, or initial conditions. This is simply any state of the system whose subsequent development over time is determined solely by nature. Otherwise initial conditions are arbitrary—within whatever constraints nature imposes through its existential laws: for instance, we can have particles moving at any SPEED that is less than the SPEED OF LIGHT. As experimenters, we set the initial conditions of a physical system as we please (again, within natural constraints), at least in principle. What then happens to a system whose initial conditions have been specified is solely up to nature; we, as experimenters, have no further hand in it. That is the second aspect of the time evolution of physical systems. We call it the developmental laws of nature. So initial conditions are the aspect of the evolution of systems that nature allows us to have some control over, while the developmental laws of nature are the aspect of the evolution that we have no control over. Note that initial conditions completely determine, through the action of the developmental laws of nature, the future state of the system at any time. (*See* LAWS OF NATURE.)

This approach to analyzing nature—conceptually splitting it into initial conditions and developmental laws—has been very successful in attaining an understanding of nature at scales from that of ELEMENTARY PARTICLES up to astronomical scales. However, the approach becomes problematic, if not altogether meaningless, in dealing with the UNIVERSE as a whole. That is because there is only one universe, or at least we have access to but a single universe, so there is no sense in which we can set initial conditions.

See also COSMOLOGY.

insulator A material of high RESISTIVITY, or equivalently, of low CONDUCTIVITY, is called an insulator. So an insulator is a material that presents a hindrance to the passage of electric CURRENT. (*See* ELECTRICITY.)

See also CONDUCTION; CONDUCTION BAND.

intensity The term *intensity* is used in physics in a number of ways, some loose and others precise. In loose usage, as in everyday speech, intensity is the strength or amount of a quantity. One might refer to the intensity of the electric or magnetic FIELD, for example, meaning the value of the field. Similar usage is applied to many other quantities and phenomena, such as electric CURRENT and RADIOACTIVITY. (*See* ELECTRICITY; MAGNETISM.)

In precise usage, "intensity" often involves POWER (i.e., ENERGY per unit TIME). In this sense we have:

Wave intensity is the power transported by any WAVE in the direction of propagation per unit area perpendicular to that direction. Its SI UNIT is watt per square meter (W/m^2).

To be more specific:

Radiant intensity characterizes a source or the propagation of electromagnetic RADIATION. It is the power emitted or propagated in a particular direction per unit solid angle that is subtended from the source of radiation. Its SI unit is watt per steradian (W/sr). (*See* ELECTROMAGNETISM.)

Radiation intensity can be used synonymously with radiant intensity. It also serves for particle radiation, where it is the number of particles passing by a location per unit time per unit area perpendicular to the direction of particle motion. In the latter use, its SI unit is the inverse second·meter^2 $[1/(s \cdot m^2)]$.

Luminous intensity, or *light intensity*, is particular for LIGHT, and furthermore, for the human perception of light. It is the power of light radiation at some location per unit solid angle as subtended from the light source, where the power is corrected and evaluated for the light's ability to cause visual sensation in humans. The difference between luminous intensity and radiant intensity is that, while the latter involves all radiation energy, the former weighs the energy according to the strength of its visual effect. Humans are insensitive to infrared and ultraviolet radiation. For visible light, we are most sensitive to a wavelength of about 560 nm, which produces the sensation of yellow-green. (Note that the radiant intensity of the Sun is maximal for this wavelength.) Luminous intensity serves as a BASE QUANTITY for the INTERNATIONAL SYSTEM OF UNITS (SI). Its SI unit is the candela (cd).

See also FLUX for radiant flux, radiant flux density, luminous flux, and luminous flux density.

Acoustic intensity, or *sound intensity,* is the wave intensity of a SOUND wave. It is the acoustic power transported by a sound wave in the direction of propagation per unit area perpendicular to that direction. Its SI unit is watt per square meter (W/m^2). (*See* ACOUSTICS.)

Intensity level is useful for any kind of intensity. It is the base-10 logarithm of the ratio of two intensities (or, for that matter, two powers, energies, voltages, etc.). It shows by how many powers of 10 one intensity is greater than another. Since intensity level is derived from a ratio, it is dimensionless. Nevertheless, it is usually expressed in the unit bel (B). If the logarithm is multiplied by 10, as is common, the intensity level is expressed in decibels (dB). In a formula:

$$\text{Intensity level (in decibels [dB])} = 10 \log_{10} (I/I_0)$$

where I and I_0 denote the two intensities involved, expressed in the same unit. So amplification, such as of voltages, for instance, can be expressed in decibels. For an amplification factor $A = I/I_0$, the corresponding amplification expressed in decibels is:

$$\text{Amplification (in decibels [dB])} = 10 \log_{10} A$$

(*See* DIMENSION.)

Intensity level is very useful when I_0 is taken as some reference intensity. Then the intensity level corresponding to I gives the relation of I to I_0 expressed logarithmically. This is common practice in ACOUSTICS, where sound intensities are expressed in decibels with respect to the sound intensity at the threshold of human hearing, which is assigned the nominal value $I_0 = 1.0 \times 10^{-12}$ W/m^2. The idea behind this is that the sensation of loudness in response to an acoustic stimulus is found to relate in an approximately logarithmic manner to the sound intensity. A similar situation holds for the relation between the visual sensation of brightness and the light intensity of the stimulus.

interaction In its most general sense, interaction is the effect of systems on each other. Two bodies might interact by colliding with each other, for example. Or a system might interact with its environment (which is also a system) by exchanging HEAT with it. More specifically, an interaction is any one of four fundamental ways in which ELEMENTARY PARTICLES affect each other. These interactions are understood as being mediated by FIELDs, whereby a matter particle affects the field and is affected by it through the emission and absorption of force particles. Matter particles then interact by exchanging force particles. (*See* COLLISION.)

In order of decreasing strength, the interactions are:

STRONG INTERACTION. For comparison, this interaction is assigned the strength 1. It operates among QUARKS, and its corresponding force particles are the GLUONS. It affects all HADRONS (whose fundamental constituents are quarks). It is the interaction that holds together the PROTONS and NEUTRONS that make up atomic NUCLEI, for example. The range of the strong interaction is around 10^{-15} meter (m), about the size of nuclei. (*See* ATOM.)

ELECTROMAGNETIC INTERACTION. The strength of this interaction is about 0.01. Its range is infinite. It affects all electrically charged elementary particles, which interact by exchange of PHOTONS. The electromagnetic interaction is what holds atoms and MOLECULES together and determines to a very large extent the properties of our everyday world. (*See* CHARGE; ELECTRICITY; FINE STRUCTURE CONSTANT.)

WEAK INTERACTION. This interaction has the strength 10^{-5}. It affects all hadrons and LEPTONS. Its force particles are the intermediate vector BOSONS, W^+, W^-, and Z^0. Among its other effects, it is responsible for the process of BETA DECAY, the decay of a neutron to a proton. This is the way that the radioactive ISOTOPE carbon-14 transmutes to the stable isotope nitrogen-14 and is used for dating archeological finds. (*See* RADIOACTIVITY.)

GRAVITATION. The strength of gravitation is about 10^{-38}, and its range is infinite. Its corresponding force particle is the graviton, and it affects all MASS and ENERGY (i.e., all elementary particles of all kinds). It appears to be the dominant player in cosmological and astronomical phenomena. (*See* COSMOLOGY.)

All four fundamental interactions are described by GAUGE THEORIES, and attempts are underway to unify them to fewer interactions, hopefully even to a single one. The electromagnetic interaction is already a result of the 19th-century unification of ELECTRICITY and MAGNETISM to ELECTROMAGNETISM. The electromagnetic and weak interactions were unified in the 20th century to the ELECTROWEAK INTERACTION. This unification is only in the sense that at sufficiently high energies (i.e., at high TEMPERATURES), the two interactions should merge into one. At early times in the life of the UNIVERSE, according to BIG-BANG-type cosmological models, the temperature was so high that the electromagnetic and weak interactions that we know today were indeed merged into the electroweak interaction. As the universe cooled over time, the two interactions separated and evolved into what we find today.

Further unification is expected to result in a merging of the electroweak and strong interactions to a single interaction described by a GRAND UNIFIED THEORY (GUT). It should attain full significance at energies (and temperatures) even higher than those required for the validity of the electroweak interaction and even earlier in the life of the universe. Beyond that, the unification of a future GUT with gravitation, although highly desired and expected eventually to be achieved, is problematic. The theories of the other three interactions can easily be quantized (i.e., made consistent with the quantum character of nature). Gravitation is not so amenable. This seems to be intimately related to the fact that gravitation, as described by the general theory of relativity, involves SPACE-TIME in an essential way, while the other three interactions "merely" appear to play their roles on the space-time stage, so to speak, without affecting or being affected by space-time. So a QUANTIZATION of gravitation is tantamount to a quantization of space-time itself. Clearly, new approaches are called for, and new approaches are indeed being developed. (*See* QUANTUM PHYSICS; RELATIVITY, GENERAL THEORY OF.)

interference The term interference describes the combined effect of two or more WAVES passing through

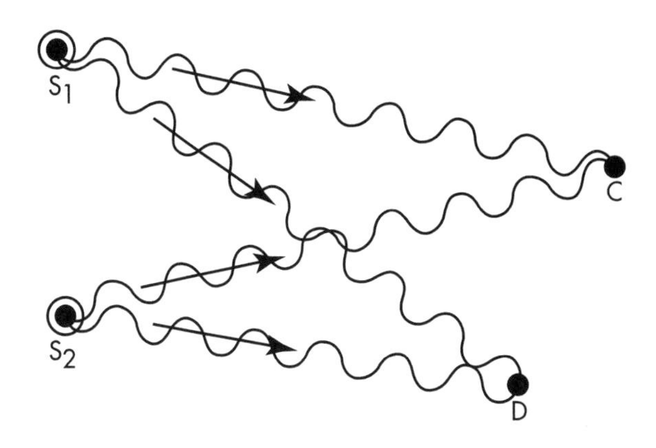

Constructive and destructive interference from two wave sources. Sources S_1 and S_2 emit sinusoidal waves at the same frequency and phase (i.e., the sources are coherent). Waves arrive at C in phase, resulting in maximal constructive interference. At D waves arrive in antiphase (maximally out of phase), which produces maximal destructive interference.

the same location at the same time. Under very special conditions, the waves might (1) reinforce each other, a situation called constructive interference, (2) partially or completely cancel each other, which is destructive interference, or (3) produce BEATS. In the general case, however, interference results in time-varying effects of no particular significance.

We confine our considerations to interference between two sinusoidal waves. For maximal constructive interference to occur at some location, the waves must have the same FREQUENCY and reach the location in phase, meaning that the effects of the individual waves at the location are simultaneously at the same stage of their respective cycles (i.e., their phase difference is zero). Both waves disturb the medium upward at the same time, then downward at the same time, and so on, as an example. The individual effects then reinforce each other maximally, and the result is a local oscillation that is stronger than that produced by each wave individually. In a linear medium, in which the combined effect of several influences is the sum of their individual effects, the amplitude of oscillation equals the sum of the amplitudes of the two waves. (*See* PHASE; SUPERPOSITION PRINCIPLE.)

For maximal destructive interference, the waves must possess equal frequencies, as before, but must reach the location in antiphase (i.e., maximally out of phase), whereby the effects of the individual waves differ in time by half a cycle. In our example, when one wave disturbs the medium maximally upward, the other disturbs it maximally downward. Then the individual effects weaken each other. If they are of equal amplitude, they will totally cancel each other, and there will be no oscillation at that location. For unequal amplitudes in a linear medium, the amplitude of oscillation equals the difference of the two amplitudes.

Intermediate cases of interference occur when the wave frequencies are equal and the phase difference of the two waves reaching the location is between zero and half a cycle. When the frequencies are not equal but are sufficiently close, beats will occur.

See also DIFFRACTION; YOUNG'S EXPERIMENT.

International Bureau of Weights and Measures (BIPM) Located in Sèvres, France, the International Bureau of Weights and Measures (Bureau International des Poids et Mesures, hence the initials BIPM) provides the basis for a single, coherent system of measurements throughout the world, traceable to the INTERNATIONAL SYSTEM OF UNITS (SI). With an international staff of about 70, the bureau carries out its mission in diverse ways, from direct dissemination of UNITS (as in the case of MASS and TIME) to coordination through international comparisons of national measurement standards (as in length, ELECTRICITY, radiometry, and ionizing RADIATION). (*See* IONIZATION.)

BIPM operates under the authority of the Convention of the Meter, which is a diplomatic treaty among 51 nations, including the United States, originally signed in 1875. The bureau performs measurement-related research, takes part in and organizes international comparisons of national measurement standards, and carries out calibrations for member states of the convention. Its work comprises the combined efforts of the national metrology laboratories of the member states (such as the NATIONAL INSTITUTE OF STANDARDS AND TECHNOLOGY of the United States) and BIPM's own laboratories.

The URL of the International Bureau of Weights and Measures website is http://www.bipm.org/. Its mailing address is Bureau International des Poids et Mesures, Pavillon de Breteuil, F-92312 Sèvres Cedex, France.

Steam cleaning of a 1-kilogram mass standard before a mass comparison at the International Bureau of Weights and Measures (BIPM). ***(Courtesy of the International Bureau of Weights and Measures [BIPM])***

International System of Units (SI) The International System of Units, abbreviated as SI (for Système International d'Unités, in French), is a unified, international system of UNITS for physical quantities, which allows scientific communication that is free of the need for unit conversion. Most countries have adopted some variation of the SI as their national standard. The United States, however, holds the distinction of being the only country of major importance that has not done so, still generally using, for example, miles rather than kilometers, pounds rather than kilograms, and Fahrenheit degrees rather than Celsius.

At the foundation of the SI lies a set of seven BASE QUANTITIES, which are physical quantities that serve to define the seven base units. The base quantities are length, MASS, TIME, electric CURRENT, TEMPERATURE, amount of substance, and luminous INTENSITY. The corresponding base units, which are defined directly from measurement and are not derived solely from other units, and their symbols are:

Base Quantity	Base Unit	Symbol
length	meter	m
mass	kilogram	kg
time	second	s
electric current	ampere	A
temperature	kelvin	K
amount of substance	molemol	
luminous intensity	candela	cd

(*See* BASE QUANTITY; ELECTRICITY; LIGHT.)

All other SI units are derived from the seven base units together with two supplementary units, based on the supplementary quantities of plane angle and solid angle. The corresponding supplementary units and their symbols are:

Supplementary Quantity	Supplementary Unit	Symbol
plane angle	radian	rad
solid angle	steradian	sr

Here is a list of some derived units that have names of their own, together with the corresponding physical quantities they are units of and their symbols:

Derived Quantity	Unit	Symbol
frequency	hertz	Hz
force	newton	N
pressure	pascal	Pa
energy	joule	J
power	watt	W
electric charge	coulomb	C
electric potential	volt	V
electric resistance	ohm	Ω
electric conductance	siemens	S
electric capacitance	farad	F
magnetic flux	weber	Wb
magnetic flux density	tesla	T
inductance	henry	H
luminous flux	lumen	lm
illuminance	lux	lx

The SI units can be used with multipliers (positive integer powers of 10, such as 100 or a 1,000) and dividers (negative integer powers of 10, such as a tenth or a millionth). Multiples and submultiples are indicated by a prefix to the name of the unit and a symbol prefixed to the unit's symbol. As examples, a kilovolt, denoted kV, is 1,000 (10^3) volts, while a millisecond, ms, is one thousandth (10^{-3}) of a second. (Note the exception: that the base unit for mass, the kilogram, is itself a multiple, 1,000 grams. So a millionth [10^{-6}] of a kilogram, for instance, is called a milligram, denoted mg.) Multipliers and dividers that are used with SI units, listed with their corresponding prefixes and symbols, are:

Multiplier	Prefix	Symbol	Divider	Prefix	Symbol
10^1	deca	da	10^{-1}	deci	d
10^2	hecto	h	10^{-2}	centi	c
10^3	kilo	k	10^{-3}	milli	m
10^6	mega	M	10^{-6}	micro	μ
10^9	giga	G	10^{-9}	nano	n
10^{12}	tera	T	10^{-12}	pico	p
10^{15}	peta	P	10^{-15}	femto	f
10^{18}	exa	E	10^{-18}	atto	a
10^{21}	zetta	Z	10^{-21}	zepto	z
10^{24}	yotta	Y	10^{-24}	yocto	y

International Union of Pure and Applied Physics (IUPAP) Established in Brussels in 1922, the International Union of Pure and Applied Physics is an association of national physics communities whose present membership numbers 48. IUPAP is itself a member of the International Council for Science and collaborates with similar unions from other disciplines. The mission of IUPAP is to assist in the worldwide development of physics, to foster international cooperation in physics, and to help in the application of physics toward solving problems of concern to humanity. IUPAP carries out its mission by sponsoring suitable international meetings and assisting meeting organizing committees; promoting international agreements on symbols, UNITS, nomenclature, and standards; fostering the free movement of physicists; and encouraging research and education, including sponsoring awards.

The work of the Union is carried out with the help of 20 subdisciplinary commissions, such as the Commission on Physics Education and the Commission on Computational Physics, and five working groups, including the International Committee for Future Accelerators and the Working Group on Women in Physics. (*See* ACCELERATOR, PARTICLE.)

The website URL for the International Union of Pure and Applied Physics is http://www.iupap.org/. The mailing address of the Union's secretariat in the United States is IUPAP, c/o The American Physical Society, One Physics Ellipse, College Park, MD 20740-3844.

invariance principle A statement that the laws of physics are invariant under a certain change is an invariance principle. In other words, if a certain change were imposed, the laws of physics would nevertheless remain the same. An invariance principle is a SYMMETRY of nature. Some of the best-known invariance principles are these:

Spatial-displacement invariance. The laws of physics are the same at all locations. The same experiment performed here or there gives the same result.

Temporal-displacement invariance. The laws of physics are the same at all times. The same experiment performed now or then gives the same result.

Rotation invariance. The laws of physics are the same in all directions. The same experiment aimed one way or another gives the same result.

Boost invariance. (A boost is a change of VELOCITY.) The laws of physics are the same at all velocities. The same experiment performed at this or that velocity gives the same result.

Those invariance principles all deal with changes that involve SPACE and TIME. They can also be expressed in the following way. No experiment carried out in an isolated laboratory can detect the lab's location or orientation in space, the reading of any clock outside the lab, the speed of the lab's straight-line motion, or the direction of that motion.

Two more invariance principles are obeyed by nature as long as the WEAK INTERACTION is not involved, since it violates them. They are:

Reflection invariance. The laws of physics, except for the weak interaction, do not change under mirror REFLECTION. Every experiment that does not involve the weak interaction gives the same result as its mirror-image experiment. The operation of reflection (or more precisely, the reversal of all three spatial directions) is conventionally denoted P (for PARITY).

Particle-antiparticle conjugation invariance. This is also called charge conjugation invariance. The laws of physics, except for the weak interaction, are the same for ELEMENTARY PARTICLES as for their respective antiparticles. Every experiment that does not involve the weak interaction gives the same result when all particles are replaced by their respective antiparticles. The change involved in this invariance principle is denoted C (for CHARGE). (*See* ANTIMATTER.)

The weak interaction *is* invariant under the combined operation of reflection and particle-antiparticle conjugation, denoted CP, even if not under each one separately. This invariance principle is nevertheless violated by a little-understood effect involving the elementary particle of type neutral kaon. An additional invariance principle, this one having to do with reversal of temporal ordering, is similarly valid for all of physics except the neutral kaon. The two invariance principles are:

CP invariance. The laws of physics are the same for elementary particles, except neutral kaons, as for their respective reflected antiparticles. Every experiment that does not involve neutral kaons gives the same result as its mirror-image experiment with all particles replaced by their respective antiparticles.

Time-reversal invariance. The name of this invariance principle is somewhat misleading, since time cannot be reversed. What is reversed is the temporal order of events making up a process. We compare any process with the process running in reverse order, from the final state to the initial state. Such pairs of processes are the time-reversal images of each other. The invariance principle is that, except for processes involving neutral kaons, the time-reversal image of any process that is allowed by nature is also a process allowed by nature. Except for experiments involving neutral kaons, if the result of any experiment is turned around and used as the input to an experiment, then the result of the latter experiment, when turned around, will be the same as the input to the original experiment. Time reversal is denoted by T.

In spite of the fact that P, C, CP, and T invariances are all violated in one way or another, there are compelling theoretical reasons to believe that the triple combined operation of CPT—particle-antiparticle conjugation, reflection, and time reversal together—is one that all laws of physics are invariant under. No experimental evidence has as yet contradicted this invariance.

CPT invariance. The laws of physics are the same for all elementary particle processes as for their respective particle-antiparticle conjugated, reflected, and time-reversed image processes. For every experiment, if the particles in the result are replaced by their respective antiparticles, are reflected, and are turned around and used as input to an experiment, then the result of the latter experiment—when turned around, reflected, and with all its particles replaced by their respective antiparticles—will be the same as the input to the original experiment.

Additional invariance principles are found to hold, including gauge symmetries. (*See* GAUGE THEORY.)

ion An electrically charged ATOM, MOLECULE, or radical (a bound group of atoms that cannot exist independently for long) is called an ion. It has lost or gained one or more ELECTRONs in relation to its neutral state. In terms of its structure, the total number of electrons in an ion does not equal the total number of PROTONs in the ion's NUCLEUS or nuclei. Ions are designated notationally by the chemical symbol for the corresponding neutral entity (element, molecule, or radical), with the addition of a right-hand superscript showing the charge in units of the fundamental charge. (*See* CHARGE; ELECTRICITY.)

As an example, sulfuric acid, H_2SO_4, breaks up in water, and each molecule separates into two singly charged positive hydrogen ions, H^+, and one doubly charged negative ion of the SO_4 radical, SO_4^{2-}. The full process is:

$$H_2SO_4 \rightarrow 2H^+ + SO_4^{2-}$$

Each hydrogen atom in the molecule gives up its electron to the SO_4 radical. As another example, sodium chloride (common table salt, NaCl) exists in ionic condition, $Na^+ + Cl^-$, both as a solid and dissolved in water. Each chlorine atom takes an electron from a sodium atom.

Oppositely charged ions have a strong attraction to each other due to the Coulomb FORCE between them. Because of their electric charge, ions are much more easily manipulated by means of electric and magnetic FIELDs than are neutral atoms and molecules. They can be accelerated in ACCELERATORs, for example, and their trajectories can be redirected by MAGNETs. (*See* COULOMB'S LAW; MAGNETISM.)

A positively charged ion is called a cation, while a negative ion is an anion.

See also IONIZATION.

ionization *Ionization* is the term describing the creation of an ION. This is any process by which an electrically neutral ATOM, MOLECULE, or radical (a bound group of atoms that cannot exist independently for long) has one or more ELECTRONs removed from it or added to it and thus becomes electrically charged. As an example, when sodium atoms, Na, and chlorine atoms, Cl, combine to form molecules of sodium chloride (ordinary table salt), NaCl, they undergo ionization. The affinity of a chlorine atom for an additional electron is so strong that it pulls an electron from the sodium atom. Thus both atoms are ionized in the process, with the sodium atom becoming a positive ion, Na^+, and the chlorine atom a negative ion, Cl^-. Similarly, the hydrogen atoms, H, and the sulfate radical, SO_4, undergo ionization in the sulfuric acid molecule, H_2SO_4. The sulfate radical takes an electron from each hydrogen atom and turns into a doubly charged negative ion, SO_4^{2-}, while each hydrogen atom becomes a singly charged positive ion, H^+. (*See* CHARGE; ELECTRICITY.)

Ionization can occur also as a result of RADIATION when a sufficiently energetic electron or PHOTON, as examples, collides with an atom or molecule and causes the ejection of one or more electrons from it. Ionization by radiation produces positive ions. Radiation that causes ionization is termed ionizing radiation. Among the various kinds of electromagnetic radiation, ultraviolet radiation and electromagnetic WAVEs of shorter WAVELENGTHs, such as X-RAYs and GAMMA RAYs, are capable of ionizing. In that range of wavelengths, the photons possess sufficient energy to knock out electrons from atoms and molecules. The ultraviolet component of sunlight, for instance, ionizes molecules in our skin and thus causes premature skin aging and even cancer. As an example of ionization by energetic particles, the AURORA is due to ionization of air molecules and atoms by charged particles arriving from the Sun. (*See* ELECTROMAGNETISM; ENERGY.)

An additional source of ionization is strong electric FIELDs. The oppositely directed FORCEs that an electric field causes on NUCLEI and on electrons can, if the field is sufficiently strong, tear one or more electrons from an atom or molecule. Such ionization occurs, for example, in CORONA DISCHARGE.

ionosphere The ionosphere is that layer of the Earth's atmosphere in which a considerable fraction of the air MOLECULEs and ATOMs are in ionic state as a result of their IONIZATION by the ultraviolet component of sunlight. The ionosphere contains positively charged ions and free ELECTRONs and extends from an altitude of about 50 kilometers (km) up to around 1,000 km. An important property of the ionosphere is that it can affect the propagation of electromagnetic WAVEs passing through it. It can reflect back to Earth radio waves reaching it from the Earth's surface, which allows long-distance radio transmission that would otherwise not occur due to the curvature of the Earth's surface.

See also CHARGE; ELECTRICITY; ELECTROMAGNETISM; ION; LIGHT.

isobaric process An isobaric process is a process that takes place at constant PRESSURE. The MELTING of ice to LIQUID water in an open drinking glass is an example of an isobaric process, since it is exposed to the atmosphere and occurs at the ambient atmospheric pressure. A graphic curve representing the relation between two variables of a system during an isobaric process, such as between VOLUME and TEMPERATURE or between pressure and volume, is called an isobar. Isobar curves are

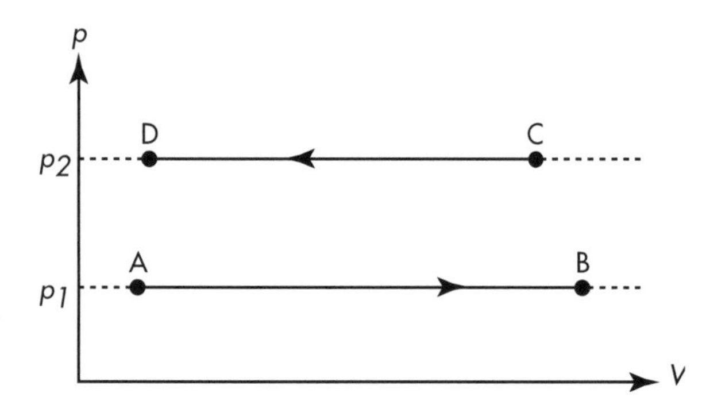

Two isobars are shown in a pressure-volume *(p-V)* diagram. An isobar is a representation of an isobaric process, which is a process taking place at constant pressure. Processes A→B and C→D are isobaric processes involving change of volume at pressures p_1 and p_2, respectively.

also used to indicate the geographic variation of atmospheric pressure on weather maps. In this context, an isobar is a line connecting locations at which the atmospheric pressure has equal values at the same time.

isochoric process An isochoric process is a process that occurs at constant VOLUME. Heating or cooling a GAS in a closed container is an example of an isochoric process. A graphic curve representing the relation between two variables of a system during an isochoric process, such as between PRESSURE and TEMPERATURE or between pressure and volume, is called an isochor.

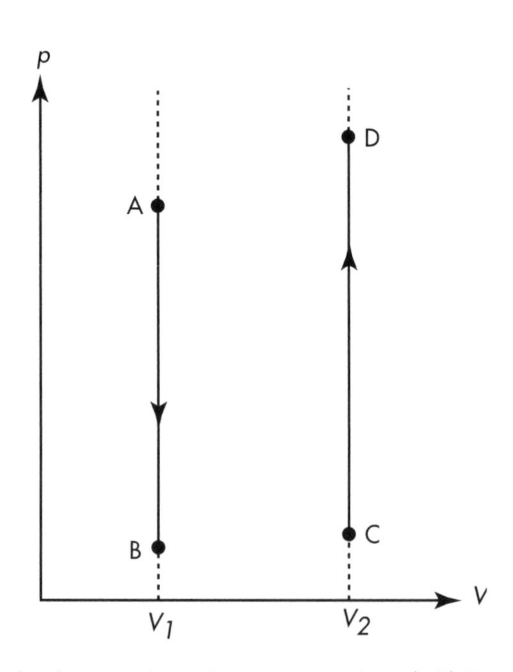

Two isochors are shown in a pressure-volume *(p-V)* diagram. An isochor is a representation of an isochoric process, which is a process taking place at constant volume. Processes A→B and C→D are isochoric processes involving change of pressure at volumes V_1 and V_2, respectively.

isothermal process An isothermal process is a process that happens at constant TEMPERATURE. In some cases, such a process is carried out by putting the system in thermal contact with a HEAT reservoir, which keeps the system at the temperature of the reservoir. As an example, the CARNOT CYCLE, which is of great importance in THERMODYNAMICS and gives the theoretical upper limit on the efficiency of real heat engines, includes two isothermal processes involving heat reservoirs. Another kind of isothermal process is a process of PHASE TRANSITION, or change of state of matter. BOILING, for instance, is an isothermal process, since the heat added to the LIQUID does not raise its temperature, but rather causes it to vaporize—to transform into a gas—at constant temperature. A graphic curve showing the relation between two variables of a system during an isothermal process, such as between PRESSURE and

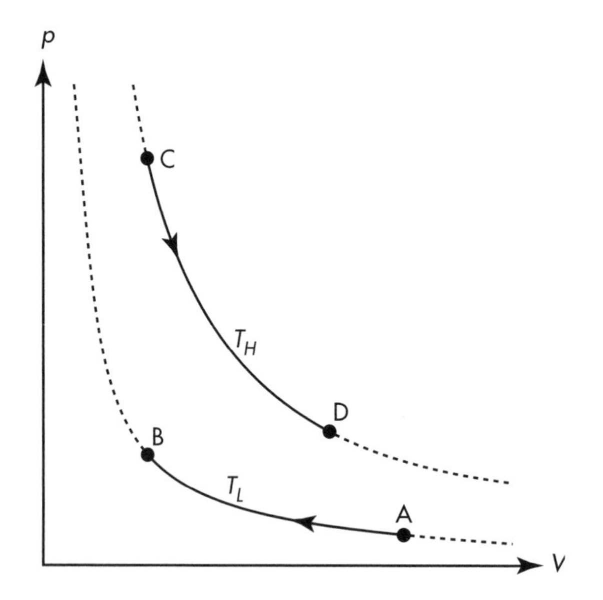

Two isotherms are shown for an ideal gas in a pressure-volume *(p-V)* diagram. An isotherm is a representation of an isothermal process, which is a process taking place at constant temperature. For an isothermal process in an ideal gas, $pV = nRT =$ constant, where *n* denotes the number of moles of the gas, *R* is the gas constant, and *T* is the absolute temperature (which is constant in the process). Processes A→B and C→D are such isothermal processes, the former at lower temperature T_L and the latter at higher temperature T_H.

VOLUME, is called an isotherm. Isotherm curves are also used to exhibit the geographic variation of temperature on weather maps or to otherwise show spatial variation of temperature. In this context, an isotherm is a line connecting locations at which the temperature has equal values at the same time.

isotope An isotope is a version of an ELEMENT that has both a definite ATOMIC NUMBER and a definite ATOMIC MASS. All NUCLEI of any given isotope possess the same number of PROTONs and the same number of NEUTRONs. Different isotopes of the same element possess the same ELECTRON structure and the same chemical properties, although their physical properties may differ as a result of their different MASSes, since their nuclei contain different numbers of neutrons. Most naturally occurring elements are mixtures of isotopes. Natural carbon, for instance, with atomic number 6, contains a mixture of three isotopes having atomic masses 12, 13, and 14. The nuclei of each isotope contain six protons as well as six, seven, or eight neutrons, respectively. The first, carbon-12, is by far the most abundant isotope. Both carbon-12 and carbon-13 are stable isotopes, while carbon-14 is radioactive and is used for dating archeological specimens. (*See* RADIOACTIVITY.)

J

Josephson, Brian David (1940–) British *Physicist* A theoretical physicist, Brian Josephson is famous for his discovery of the JOSEPHSON EFFECT. He received his Ph.D. in physics in 1964 from the University of Cambridge, England. Josephson has been at Cambridge continuously since 1967, and in 1974 he was appointed professor of physics there. Josephson's field of research has been CONDENSED-MATTER PHYSICS, in particular SOLID-STATE PHYSICS. He studied quantum effects in SOLIDs, especially SUPERCONDUCTIVITY. That led to his discovery of the Josephson effect, including the Josephson junction and Josephson TUNNELING. The effect named for him, among its various uses, allows high-speed switching for supercomputers and plays a role in the measurement of fundamental constants. In 1973, Josephson was awarded half the Nobel Prize in physics "for his theoretical predictions of the properties of a supercurrent through a tunnel barrier, in particular those phenomena which are generally known as the Josephson effects."

Brian Josephson is a theoretical condensed-matter physicist who, in his investigation of superconductivity, discovered the Josephson effect. For that, he shared the 1973 Nobel Prize in physics. ***(AIP Emilio Segrè Visual Archives, Physics Today Collection)***

See also CONSTANTS, FUNDAMENTAL; QUANTUM PHYSICS; THEORETICAL PHYSICS.

Josephson effect Predicted by the British theoretical physicist BRIAN JOSEPHSON, the Josephson effect is the flow of an electric CURRENT consisting of COOPER PAIRS, called a supercurrent, between superconductors that are weakly connected. A weak connection, called a Josephson junction, refers to a small area of contact, a thin insulating barrier, or a thin, normally conducting metallic interface. In the insulator case, TUNNELING is involved, called Josephson tunneling. Josephson tunneling, like all tunneling, is a quantum mechanical effect, whereby particles can pass into regions that are considered inaccessible by CLASSICAL PHYSICS. As long as the supercurrent does not exceed a certain maximal value, no VOLTAGE develops across the barrier. This is the DC Josephson effect. (*See* ELECTRICITY; QUANTUM MECHANICS; SUPERCONDUCTIVITY; THEORETICAL PHYSICS.)

When a constant voltage is maintained across the insulating barrier, an alternating supercurrent flows between the superconductors. Its FREQUENCY is proportional to the imposed voltage and is given by:

$$f = 2eV/h$$

where f denotes the frequency in hertz (Hz), V is the voltage in volts (V), e is the magnitude of the ELECTRON's electric CHARGE with the value 1.602176462 $\times 10^{-19}$ coulomb (C), and h represents the PLANK CONSTANT, whose value is 6.62606876 $\times 10^{-34}$ joule·second (J·s). This effect is called the AC Josephson effect. Additional effects occur when the imposed voltage is an alternating (AC) voltage and when the junction is immersed in a magnetic FIELD. When the Josephson junction is not an insulator and tunneling is not involved (i.e., when the DC Josephson effect is not in play), effects appear that are similar to the above. (*See* MAGNETISM.)

The Josephson effect, among its various uses, allows high-speed switching for supercomputers and plays a role in the measurement of fundamental constants. (*See* CONSTANTS, FUNDAMENTAL.)

James Joule, a 19th-century physicist, sought to discover unity among energy, heat, and work. ***(AIP Emilio Segrè Visual Archives, Physics Today Collection)***

Joule, James Prescott (1818–1889) British *Physicist* Best known among physicists for his investigations of HEAT, James Joule has the SI UNIT of ENERGY, the joule (J), named for him. Joule was educated at home and at the University of Cambridge, England, and led a life of physics research and experimentation. His research was mainly in the areas of ENERGY, heat, and WORK, and he sought to discover unity among them. Among his results were the measured conversion of work to heat, indications of conservation of energy, and the measurement of heat produced by an electric CURRENT through a RESISTOR. The heat generated in the latter effect is known as Joule heat, and its relation to current and RESISTANCE is known as Joule's law. Joule's experiments showed that heat is a form of internal MOTION.

See also CONSERVATION LAW; ELECTRICITY.

K

Kamerlingh Onnes, Heike (1853–1926) Dutch *Physicist* Famous for his investigations in low-TEMPERATURE physics, also known as CRYOGENICS, Heike Kamerlingh Onnes was the first to discover the effect of SUPERCONDUCTIVITY. He received his doctorate in physics in 1879 from the University of Groningen, in the Netherlands. In 1882, Kamerlingh Onnes was appointed professor of physics at the University of Leyden, the Netherlands. He reorganized the laboratory there—now known as the Kamerlingh Onnes Laboratory—to suit his research goals, and the laboratory became a world-famous center for low-temperature research. Among the laboratory's achievements, in 1908 he succeeded in liquefying helium by lowering its temperature to 0.9 kelvins (K), the lowest temperature ever achieved at that time. He discovered superconductivity, the complete disappearance of electric RESISTANCE, in 1911. Kamerlingh Onnes was awarded the Nobel Prize in physics in 1913 "for his investigations on the properties of matter at low temperatures which led, inter alia, to the production of liquid helium."

See also ELECTRICITY; LIQUID.

Shown here in his laboratory, Heike Kamerlingh Onnes investigated low-temperature phenomena, was the first to liquify helium, and discovered superconductivity. He was awarded the 1913 Nobel Prize in physics. ***(Massachusetts Institute of Technology Burndy Library, courtesy AIP Emilio Segrè Visual Archives)***

Kapitsa (Kapitza), Pyotr Leonidovich (1894–1984) Russian *Physicist* Famous for his work in low-TEMPERATURE physics, also known as CRYOGENICS, Pyotr Leonidovich Kapitsa is also well known for his achievements in PLASMA physics. In 1918, he graduated from the Petrograd Polytechnical Institute in Russia. Kapitsa spent the years 1923–34 at the University of

Pyotr Leonidovich Kapitsa worked in low-temperature physics and in plasma physics and discovered superfluidity in helium. He shared the 1978 Nobel Prize in physics. ***(AIP Emilio Segrè Visual Archives, Segrè collection)***

Cambridge, England, where he first worked on NUCLEAR PHYSICS with ERNEST RUTHERFORD, later turning to low-temperature physics. During 1930–34, he served as director of the Royal Society Mond Laboratory, which was devoted to low-temperature research. While on a visit to the Soviet Union in 1934, Kapitsa was detained there by order of Soviet dictator Joseph Stalin, and in 1935, he was appointed director of the Institute of Physical Problems of the Soviet Academy of Sciences in Moscow, where he continued work in low-temperature physics. One outcome of his research was the discovery of a superfluid state of helium. Apparently due to his refusing to take part in the development of nuclear weapons and consequently running afoul of Stalin, Kapitsa was relieved of his position in 1946. After Stalin's death, Kapitsa was reinstated in 1955 to his former position, where he remained until his death. During the latter part of his life, Kapitsa worked in the fields of plasma physics and controlled nuclear fusion. In 1978, he was awarded half of the Nobel Prize in physics "for his basic inventions and discoveries in the area of low-temperature physics."

See also FUSION, NUCLEAR; SUPERFLUIDITY.

Kelvin, Lord (William Thomson) (1824–1907) British *Physicist* A theoretical physicist of broad interests, William Thomson is best known among physicists for his work on the nature of HEAT. The SI UNIT of TEMPERATURE, the kelvin (K), is named for him. He graduated from the University of Cambridge, England, in 1845. In 1846, Thomson was appointed professor of

Active mainly in the 19th century, Lord Kelvin (William Thomson) investigated the nature of heat and attempted to describe all physical phenomena in a unified way. ***(AIP Emilio Segrè Visual Archives, Zeleny Collection)***

physics (then called "natural philosophy") at the University of Glasgow, Scotland, where he remained. His physics investigations included the fields of heat and ELECTROMAGNETISM. Thomson's approach to physics was mathematical, and he attempted to find a unified description of all physical phenomena.

See also THEORETICAL PHYSICS.

Kepler, Johannes (1571–1630) German *Astronomer, Mathematician* A versatile scientist, Johannes Kepler is most famous for his laws of planetary MOTION. He studied astronomy and mathematics, among other subjects, at the University of Tübingen in Germany. Subsequently, Kepler held a number of successive positions, including that of assistant to the Danish astronomer Tycho Brahe (1546–1601) and the position of imperial mathematician. In his capacity as Brahe's assistant, he had access to Brahe's meticulous observations of the motions of the planets. Kepler used those data to formulate his three laws of planetary motion, which had all the planets, including the Earth, revolving around the Sun. In doing so, he ran afoul of the religious authorities, since a heliocentric scheme of the heavens contradicted geocentric religious dogma. The latter had the Earth holding a central position in the UNIVERSE, with all astronomical objects revolving around it. Kepler also did work in OPTICS and mathematics and developed telescopes.

See also KEPLER'S LAWS.

Johannes Kepler was a 16th/17th-century astronomer and mathematician who is famous for his laws of planetary motion. ***(AIP Emilio Segrè Visual Archives)***

Kepler's laws In the 17th century, German astronomer and mathematician JOHANNES KEPLER discovered three laws that describe the MOTION of the planets that were known at the time: Mercury, Venus, Earth, Mars, Jupiter, and Saturn. The later discovery of the planets Uranus, Neptune, and Pluto added further confirmation. Although the laws were based on observations of the Sun's planets, they are valid as well for any astronomical situation in which satellites revolve around a much more massive central body under the sole influence of its gravitational attraction. Such situations might be moons around a planet (the moons of Jupiter, for instance), planets around a star (as in the SOLAR SYSTEM, studied by Kepler), or stars of a GALAXY around a massive galactic core. Kepler's laws follow from ISAAC NEWTON's laws of MOTION and law of GRAVITATION, which were developed mainly to explain Kepler's laws. (*See* MASS; NEWTON'S LAWS.)

These are Kepler's laws of planetary motion:

First law. The orbit of each planet lies wholly in a plane and has the form of an ellipse, with the Sun at one focus.

Except for Pluto, the orbits of all the planets of the solar system lie very much in the same plane. Their elliptical orbits are close to being circular, with the orbit of Pluto the most elongated.

Second law. As each planet moves in its orbit, its radius-vector (i.e., the imaginary line connecting the planet and the Sun) sweeps out equal areas in equal time intervals.

To accomplish that, a planet must move faster when it is closer to the Sun and slower when it is farther away.

Third law. This law relates the orbital PERIODs of the planets (i.e., the durations of the planets' "years") to their respective distances from the Sun. It states that for any planet, the square of the orbital period is proportional to the cube of the orbit's semimajor axis (or of the planet's average distance from the Sun). The formula is:

$$T^2/a^3 = \text{constant}$$

or:

$$T_1^2/a_1^3 = T_2^2/a_2^3$$

where T denotes the orbital period (in any unit of time), a is the orbit's semimajor axis (or the planet's average distance from the Sun, in any length unit), and the subscripts indicate any pair of planets.

So the inner planets, those closer to the Sun than the Earth, must have shorter years than ours, while the years of the outer planets should lengthen progressively the farther out the planets are. That is indeed the case. Double the distance, for example, and the year increases by a factor of 1.59. Halve the distance, and the year is shorter by a factor of 0.63.

The value of the constant in the formula depends on the units used for time and length. If the Earth year is taken as the unit of time and the Earth's average distance from the Sun (called an astronomical unit) as the unit of length, then when a represents a planet's average distance from the Sun, the constant's value will be 1. That is for the solar system. For other systems, the constant will have different values, depending on the mass of the central body.

Kepler's laws of planetary motion challenged ISAAC NEWTON to find a theory (i.e., an explanation) for them. The result was Newton's laws of motion and law of gravitation.

Kerr effect Named for Scottish physicist John Kerr (1824–1907), the Kerr effect is the phenomenon of BIREFRINGENCE appearing in certain LIQUIDs when an electric FIELD is applied. Birefringence is the property of a material that it possesses two INDEXES OF REFRACTION, one for each of two mutually perpendicular directions of linear POLARIZATION of LIGHT. The Kerr effect lies at the heart of the Kerr cell, which is a device that can change its transparency very rapidly under electric control and can thus serve as a very fast electric switch for light. (*See* ELECTRICITY.)

kinematics The study of the MOTION of bodies without regard to the causes of the motion is known as kinematics. In general, a body's motion consists of a combination of translational motion and ROTATION. The former has to do with a body's position as a function of time. This can be expressed in terms of a body's coordinates (i.e., the coordinates of its CENTER OF MASS or its actual coordinates, if it is a point particle) with respect to some REFERENCE FRAME or COORDINATE SYSTEM, say $(x(t), y(t), z(t))$, or in terms of its position VECTOR, $\mathbf{r}(t)$, which is a vector pointing from the origin to the body's position and whose x-, y-, and z-components are just the coordinates, $(x(t), y(t), z(t))$.

Translational Motion

A body's instantaneous VELOCITY, $\mathbf{v}(t)$, a vector, is the time rate of change of its position (i.e., the derivative of its position vector with respect to time):

$$\mathbf{v}(t) = d\mathbf{r}(t)/dt$$

The x-, y-, and z-components of the velocity vector, $(v_x(t), v_y(t), v_z(t))$, are the time derivatives of the body's x-, y-, and z-coordinates, respectively:

$$v_x(t) = dx(t)/dt$$
$$v_y(t) = dy(t)/dt$$
$$v_z(t) = dz(t)/dt$$

The body's instantaneous SPEED, $v(t)$, a SCALAR, is the magnitude of its instantaneous velocity:

$$v(t) = |\mathbf{v}(t)| = \sqrt{v_x^2 + v_y^2 + v_z^2}$$

The average velocity for the interval between times t_1 and t_2 is defined as the DISPLACEMENT of the body during the interval divided by the time duration:

$$\mathbf{v}_{\text{av}} = \frac{\mathbf{r}(t_2) - \mathbf{r}(t_1)}{t_2 - t_1}$$

The instantaneous ACCELERATION, $\mathbf{a}(t)$, is the derivative of the instantaneous velocity with respect to time:

$$\mathbf{a}(t) = d\mathbf{v}(t)/dt$$

Its x-, y-, and z-components, $(a_x(t), a_y(t), a_z(t))$, are the time derivatives of the respective velocity components:

$$a_x(t) = dv_x(t)/dt$$
$$a_y(t) = dv_y(t)/dt$$
$$a_z(t) = dv_z(t)/dt$$

Accordingly, the instantaneous acceleration is the second time derivative of the position vector:

$$\mathbf{a}(t) = d^2\mathbf{r}(t)/dt^2$$

and its components are the second time derivatives of the respective coordinates:

$$a_x(t) = d^2x(t)/dt^2$$
$$a_y(t) = d^2y(t)/dt^2$$
$$a_z(t) = d^2z(t)/dt^2$$

The average acceleration for the interval between times t_1 and t_2 is defined as:

$$\mathbf{a}_{av} = \frac{\mathbf{v}(t_2) - \mathbf{v}(t_1)}{t_2 - t_1}$$

The SI UNITS for all those: time in seconds (s), coordinates in meters (m), velocity and speed in meters per second (m/s), and acceleration in meters per second per second (m/s^2). Higher-order derivatives can be useful in certain circumstances.

Simple Translational Motions

Although translational motion can, in general, involve an arbitrary dependence of the body's position on time, there are some particularly simple motions that both have importance and are easy to treat mathematically. The simplest is rest, the situation of constant, time-independent position:

$$\mathbf{r}(t) = \mathbf{r}_0$$

where $\mathbf{r}_0$ denotes the body's constant position vector, whose components are (x_0, y_0, z_0). In terms of coordinates:

$$x(t) = x_0$$
$$y(t) = y_0$$
$$z(t) = z_0$$

This is motion at constant zero velocity:

$$\mathbf{v}(t) = 0$$

The next simplest motion is motion at constant nonzero velocity:

$$\mathbf{v}(t) = \mathbf{v}_0$$

where $\mathbf{v}_0 \neq 0$ represents the body's constant velocity, whose components are (v_{x0}, v_{y0}, v_{z0}). This is straight-line motion at constant speed, motion with constant zero acceleration:

$$\mathbf{a}(t) = 0$$

The position as a function of time is then given by:

$$\mathbf{r}(t) = \mathbf{v}_0 t + \mathbf{r}_0$$

where $\mathbf{r}_0$ now denotes the position at time $t = 0$. In terms of components and coordinates, we have:

$$v_x(t) = v_{x0}$$
$$v_y(t) = v_{y0}$$
$$v_z(t) = v_{z0}$$

and:

$$x(t) = v_{x0}t + x_0$$
$$y(t) = v_{y0} + y_0$$
$$z(t) = v_{z0}t + z_0$$

The graph of any coordinate against time has the form of a straight line whose slope equals the respective component of the velocity. Note that the state of rest is a special case of constant-velocity motion, when the velocity is constantly zero. Constant-velocity motion is, according to Newton's second law of motion, inertial motion (i.e., motion in the absence of FORCES). (*See* INERTIA; NEWTON'S LAWS.)

The next simplest motion, after constant-velocity motion, is motion at constant nonzero acceleration:

$$\mathbf{a}(t) = \mathbf{a}_0$$

where $\mathbf{a}_0 \neq 0$ is the constant acceleration, whose components are (a_{x0}, a_{y0}, a_{z0}). It follows that the time dependence of the velocity is:

$$\mathbf{v}(t) = \mathbf{a}_0 t + \mathbf{v}_0$$

where $\mathbf{v}_0$ denotes the instantaneous velocity at time $t = 0$. The position is given as a function of time by:

$$\mathbf{r}(t) = \mathbf{a}_0 t^2/2 + \mathbf{v}_0 t + \mathbf{r}_0$$

In terms of components and coordinates, these relations take the form:

$$a_x(t) = a_{x0}$$
$$a_y(t) = a_{y0}$$
$$a_z(t) = a_{z0}$$

for the acceleration:

$$v_x(t) = a_{x0}t + v_{x0}$$
$$v_y(t) = a_{y0}t + v_{y0}$$
$$v_z(t) = a_{z0}t + v_{z0}$$

for the velocity, and for the coordinates:

$$x(t) = a_{x0}t^2/2 + v_{x0}t + x_0$$
$$y(t) = a_{y0}t^2/2 + v_{y0}t + y_0$$
$$z(t) = a_{z0}t^2/2 + v_{z0}t + z_0$$

The graph of any velocity component against time shows a straight line whose slope is the respective component of the acceleration, while the graph of any coordinate against time has the form of a parabola. Motion at constant velocity is a special case of constant-acceleration motion, that with constant zero acceleration. Motion at constant acceleration is, according to Newton's second law of motion, the result of a constant force acting on a body. Such is the situation, say, near the surface of the Earth, where the acceleration of FREE FALL is, indeed, constant (and has the same value for all bodies falling in VACUUM, about 9.8 meters per second [m/s²]).

Rotational Motion

Considerations similar to the above hold for the rotational motion of a body. In the case of rotation about an axis (through the body's center of mass) of fixed direction, the body's orientation is specified by a single angle in radians (rad) that is a function of time, $\theta(t)$. The time derivative of this function gives the angular speed in radians per second (rad/s):

$$\omega(t) = d\theta(t)/dt$$

This can also be expressed as a vector, the angular velocity vector, $\boldsymbol{\omega}(t)$. The magnitude of $\boldsymbol{\omega}(t)$ is $\omega(t)$ and its direction is the direction of the rotation axis, so that if the right thumb points in its direction, then the curled fingers of that hand indicate the sense of rotation. The average angular speed for the interval between times t_1 and t_2 is defined as the angular displacement of the body during the interval divided by the time duration:

$$\omega_{av} = \frac{\theta(t_2) - \theta(t_1)}{t_2 - t_1}$$

The angular acceleration in radians per second per second (rad/s²) is the time derivative of the angular speed and, accordingly, the second time derivative of the orientation angle:

$$\alpha(t) = d\omega(t)/dt = d^2\theta(t)/dt^2$$

The average angular acceleration for the interval between times t_1 and t_2 is defined as:

$$\alpha_{av} = \frac{\omega(t_2) - \omega(t_1)}{t_2 - t_1}$$

Simple Rotational Motions

The simplest rotational motion about a fixed axis is the state of rotational rest, a state of constant, time-independent orientation. Then:

$$\theta(t) = \theta_0$$

where θ_0 denotes the body's constant orientation angle. This is rotation at constant zero angular speed:

$$\omega(t) = 0$$

The next simplest rotational motion about a fixed axis is one with constant nonzero angular speed:

$$\omega(t) = \omega_0$$

where $\omega_0 \neq 0$ represents the body's constant angular speed. Then the orientation angle as a function of time is given by:

$$\theta(t) = \omega_0 t + \theta_0$$

where θ_0 denotes the orientation angle at time $t = 0$. This is rotation at constant zero angular acceleration:

$$\alpha(t) = 0$$

The next simplest rotation after that is rotation at constant nonzero angular acceleration:

$$\alpha(t) = \alpha_0$$

with $\alpha_0 \neq 0$ denoting the value of the constant angular acceleration. It follows that the time dependence of the angular speed is then:

$$\omega(t) = \alpha_0 t + \omega_0$$

where ω_0 now denotes the value of the angular speed at time $t = 0$. The orientation angle is given as a function of time by:

$$\theta(t) = \alpha_0 t^2/2 + \omega_0 t + \theta_0$$

where θ_0 now denotes the orientation angle at time $t = 0$.

See also DYNAMICS.

kinetic energy ENERGY due to MOTION is called kinetic energy. The kinetic energy in joules (J) of a point particle possessing MASS m in kilograms (kg) and SPEED v in meters per second (m/s) is:

$$\text{Kinetic energy} = mv^2/2$$

Note that kinetic energy is relative, in that observers in different states of motion assign different values to the kinetic energy of the same particle, since they observe it as moving at different speeds.

Consider now a system with structure, such as a bowling ball. Its external kinetic energy is the sum of its translational kinetic energy and its rotational kinetic energy. The former is the system's kinetic energy due to the motion of its CENTER OF MASS, whereby the system is equivalent to a point particle of the same mass located at its center of mass. So for a system with total mass m in kilograms (kg), whose center of mass is moving with speed v in meters per second (m/s), the translational kinetic energy in joules (J) is:

$$\text{Translational kinetic energy} = mv^2/2$$

(*See* ROTATION.)

According to ALBERT EINSTEIN's special theory of relativity, this expression is valid only for speeds that are sufficiently small compared with the SPEED OF LIGHT. In general, the translational kinetic energy of a body is obtained by subtracting the body's rest energy from its total energy:

$$\text{Translational kinetic energy} = mc^2\left(\frac{1}{\sqrt{1-v^2/c^2}}-1\right)$$

where m denotes the body's REST MASS and c represents the speed of light, with value 2.99792458×10^8 meters per second (m/s). At low speed, this expression takes the form:

$$\text{Translational kinetic energy} \approx mv^2/2 + \text{negligible terms}$$

(*See* MASS ENERGY; RELATIVITY, SPECIAL THEORY OF.)

If the system is rotating about its center of mass, it possesses rotational kinetic energy, given by

$$\text{Rotational kinetic energy} = I\omega^2/2$$

where ω denotes the magnitude of the system's angular velocity in radians per second (rad/s) and I is the system's MOMENT OF INERTIA in kilogram·meter² (kg·m²) with respect to the axis of rotation.

Although total energy is always conserved in COLLISIONs in isolated systems, the *kinetic* energy of the collision results might not equal that of the colliding objects. In a head-on car crash, for example, almost all the initial kinetic energy of the colliding cars is converted to energy in the form of HEAT, SOUND, and energy of metal deformation. Those collisions in which kinetic energy *is* conserved are called elastic collisions. An example is a spacecraft performing a planetary flyby. If the spacecraft gains speed, and thus kinetic energy, from such an encounter, it is at the expense of the kinetic energy of the planet. Collisions in which kinetic energy is not conserved are called inelastic collisions. When the colliding objects stick together and do not separate after collision, there is maximal loss of kinetic energy. That is termed completely inelastic collision. In the extreme, for instance, when two bodies of equal mass and speed collide head on and stick together, the bodies stop dead, and all their initial kinetic energy is lost.

A system with structure can also possess internal kinetic energy, which appears as thermal energy, or heat, of the system. It is the kinetic energy of the random motion of the system's microscopic constituents.

See also POTENTIAL ENERGY.

kinetic theory The explanation of properties of bulk MATTER in terms of the MOTION of its constituent particles (ATOMs, IONs, MOLECULEs) is called kinetic theory. In particular, the kinetic theory of GASes derives the properties of PRESSURE, TEMPERATURE, and quantity from the particle nature of gases. Pressure is understood as the average perpendicular FORCE on a unit AREA of container wall due to collisions of the gas particles with the wall. Temperature is related to the average KINETIC ENERGY of a particle of gas. And quantity of gas is basically the number of particles in the container.

For an ideal gas consisting of a single species of particle, the kinetic theory straightforwardly gives some simple results. Note that the EQUATION OF STATE of an ideal gas is

$$pV = nRT$$

where p denotes the pressure in pascals (Pa), equivalent to newtons per square meter (N/m²); V is the VOLUME of the gas in cubic meters (m³); T is the absolute temperature in kelvins (K), n is the amount of gas in MOLEs (mol); and R is the gas constant, whose value is 8.314472 joules per mole per kelvin [J/(mol·K)]. (*See* GAS, IDEAL.)

The pressure can be expressed as:

$$p = \frac{2N(E_{\text{kin}})_{\text{av}}}{3V}$$

where N denotes the number of gas particles and $(E_{\text{kin}})_{\text{av}}$ represents the average kinetic energy of a gas particle in joules (J). The latter is given as:

$$(E_{\text{kin}})_{\text{av}} = m(v^2)_{\text{av}}/2$$

where m is the mass of a single gas particle in kilograms (kg), and $(v^2)_{\text{av}}$ denotes the average squared speed of a gas particle, with speed in meters per second (m/s). The absolute temperature, T, is related to $(E_{\text{kin}})_{\text{av}}$ by:

$$(E_{kin})_{av} = (3/2)kT$$

where k denotes the BOLTZMANN constant, whose value is $1.3806503 \times 10^{-23}$ joule per kelvin (J/K). The number of moles, n, and the number of particles, N, are related by:

$$N = N_A n$$

where N_A is AVOGADRO'S NUMBER, which is the number of particles per mole and has the value $6.02214199 \times 10^{23}$ per mole (mol^{-1}). In words, the number of particles equals the number of moles multiplied by the number of particles per mole. Note that the gas constant, R, the Boltzmann constant, k, and Avogadro's number, N_A, are related by:

$$R = N_A k$$

See also MAXWELL DISTRIBUTION.

Kirchhoff's rules Kirchhoff's two rules for the analysis of electric circuits, or networks, allow one to find the CURRENTs in all branches of any circuit. Although in practice other methods of analysis might be used, they are equivalent to Kirchhoff's rules. The rules are named for the German physicist Gustav Robert Kirchhoff (1824–87). (*See* ELECTRICITY.)

First, here is some foundation. A branch of a circuit is a single conducting path through which a definite current flows. Branches join at junctions, or nodes. A loop in a circuit is a closed conducting path. We consider the application of Kirchhoff's rules to direct current (DC) circuits, consisting of RESISTORS (or other components possessing electric RESISTANCE) and sources of ELECTROMOTIVE FORCE (emf), such as BATTERIES. (*See* CONDUCTION.)

Next, prepare the given circuit diagram (with given emf sources and resistances) for the application of Kirchhoff's rules as follows. Identify the branches of the circuit, arbitrarily choose a direction of current in each branch, and assign a symbol to each current. (If you choose the wrong direction, that is of no concern, since the algebra will straighten things out at the end.) Now, apply the junction rule at each junction.

Kirchhoff's junction rule. The algebraic sum of all currents at a junction—with entering currents taken positive and exiting currents negative—equals zero. Equivalently, the sum of currents entering a junction equals the sum of currents leaving it.

The significance of this rule is basically conservation of electric CHARGE, which is neither created nor destroyed at a junction; what goes in must come out. (*See* CONSERVATION LAW.)

The application of the junction rule for all junctions gives a number of equations for the unknown currents. Use the equations to algebraically eliminate as many unknown currents as possible. You are now left with some irreducible set of unknown currents. The equations obtained from the junction rule express all the other currents in terms of those in the irreducible set.

At this point, arbitrarily choose a number of independent loops in the circuit. (See below for more about "independent.") The number of loops should equal the number of unknown currents in the irreducible set. Apply the loop rule to every loop you chose.

Kirchhoff's loop rule. The algebraic sum of the potential rises and drops that are encountered as the loop is completely traversed equals zero.

The loop rule is none other than conservation of ENERGY together with the conservative nature of the electric FORCE. It tells us that if a test charge is carried around a loop, it will end up with the same energy it had initially. (*See* CONSERVATIVE FORCE.)

In detail, apply the loop rule in this manner. On each loop choose a point of departure. Traverse the loop in whichever direction you like. As you encounter circuit components, algebraically sum their potential changes. Crossing a battery from negative terminal to positive is a potential rise (positive change); crossing from positive to negative is a drop (negative change). The emf of each battery is given, so no unknowns are involved in this. However, in crossing a resistor or the internal resistance of a battery, the magnitude of potential change equals the product of the unknown current and the known resistance, by OHM'S LAW. Crossing in the direction of the current gives a potential drop (negative change); crossing against the current gives a rise (positive change). (That is because a current flows from higher to lower potential.) When your traversal takes you back to your point of departure, equate the sum you have accumulated to zero.

If you chose independent loops, the equations you have are algebraically independent, and you have as many equations for the unknown currents in the irreducible set as there are currents in the set. Solve the equations for those currents. If you cannot solve for all of them, your equations are not independent, and you should modify your choice of loops and apply the loop rule again. When you have solved for the currents in the irreducible set, use the junction-rule equations to find all the other currents. A negative value for a current simply indicates that your guess of its direction was wrong; reverse the corresponding arrow in the circuit diagram and note a positive value instead. Thus the magnitudes and directions of the currents in all the branches of the circuit are now known.

Kirchhoff's rules are applicable to alternative current (AC) circuits as well.

L

Landau, Lev Davidovich (1908–1968) Azerbaijani/Russian *Physicist* Famous especially for his work in CONDENSED-MATTER PHYSICS, Lev Davidovich Landau contributed to nearly all fields of THEORETICAL PHYSICS. He received his education in physics at Baku University in Azerbaijan and at the University of Leningrad and the Leningrad Institute of Physics and Technology in Russia. During 1929–31, Landau traveled to several institutes in Europe and spent some time at NIELS HENRIK DAVID BOHR's Institute for Theoretical Physics in Copenhagen, Denmark. He headed the theoretical department of the Ukrainian Physico-Technical Institute at Kharkov, Ukraine, during 1932–39, and from 1937 until his death he served as head of the theoretical department of the Institute for Physical Problems of the Academy of Sciences of the U.S.S.R. in Moscow, Russia. In the field of condensed-matter physics, Landau's investigations included FERROMAGNETISM, PHASE TRANSITIONS, SUPERCONDUCTIVITY, and SUPERFLUIDITY, in particular the superfluidity of LIQUID helium. He also worked in atomic physics, NUCLEAR PHYSICS, PLASMA physics, HYDRODYNAMICS, and QUANTUM FIELD THEORY, among other areas. In 1962, Landau was awarded the Nobel Prize in physics "for his pioneering theories for condensed matter, especially liquid helium."

See also ATOM.

Lev Davidovich Landau was a versatile theoretical physicist who is known especially for his work in condensed-matter physics. He was awarded the 1962 Nobel Prize in physics. *(AIP Emilio Segrè Visual Archives, Physics Today Collection)*

Laplace, Pierre-Simon (1749–1827) French *Mathematician, Physicist* A great and influential scientist, Pierre-Simon Laplace is known for his work in mathematics, including probability, in mathematical

The mathematician and physicist Pierre-Simon Laplace, working in the 18th and 19th centuries, carried out theoretical investigations of, among other phenomena, the solar system and its planets. ***(AIP Emilio Segrè Visual Archives)***

astronomy, and in physics, especially on the theory of HEAT. He studied mathematics at the University of Caen, France. Then he moved to Paris to continue his studies, while supporting himself by teaching. As his reputation grew, he gained better-paying positions. In 1770, Laplace commenced production of a steady flow of papers. His work in mathematical astronomy dealt with such subjects as the orbits and shapes of planets and the stability and formation of the SOLAR SYSTEM. Laplace's studies in physics included, in addition to heat, also CAPILLARY FLOW, REFRACTION, the SPEED of SOUND, and other subjects.

laser This is an acronym of "light amplification by stimulated emission of radiation." A laser is a device that produces LIGHT that is monochromatic (i.e., is characterized by a single WAVELENGTH, or FREQUENCY), highly coherent, and relatively intense. The emitted light beam is very parallel, propagating with but little spreading. Laser light can be focused down to extremely small regions, giving tremendous radiant-power intensities on the illuminated areas. Those properties are exploited in lasers' many and diverse applications, such as in medicine, communications, industry, and science research. The light used to "read" a compact disk in a CD player or a CD-ROM drive, for example, is produced by a laser. And in surgery, laser light serves both to sever tissue, in place of a scalpel, and to join tissue, as in attaching a detached retina. (*See* COHERENCE; ELECTROMAGNETISM; FLUX; INTENSITY; POWER; RADIATION.)

The principle of a laser's operation is as follows. Light generation takes place within the volume of an appropriate material, which might be a SOLID, LIQUID, or GAS. The material's ATOMs or MOLECULEs must possess a fairly stable ENERGY state above its GROUND STATE and an unstable state of even higher energy. We denote the energy difference between the ground state and the state of next-higher energy by ΔE. An external energy source excites many of the particles of the material from their ground state to their next-higher-energy state via the unstable state, producing a population inversion, a situation in which there are more particles in the next-higher-energy state than in the ground state (whereas under EQUILIBRIUM conditions, the ground state is more populated than any higher-energy state.) Some of the excited particles spontaneously DECAY to their ground state, thereby emitting PHOTONs of energy ΔE. The passing photons of energy ΔE cause additional excited particles to decay to their ground state too and, in doing so, emit additional photons of the same energy. This process is called stimulated emission. The photons thus emitted join the stimulating photons in a coherent fashion to stimulate the emission of even more photons, and so on. In this manner, a light beam builds up. The wavelength of this light, λ, is determined by the energy difference, ΔE, through the relation:

$$\lambda = hc/\Delta E$$

where h denotes the PLANCK CONSTANT, whose value is $6.62606876 \times 10^{-34}$ joule·second (J·s), and c is the SPEED OF LIGHT, which has the value 2.99792458×10^8 meters per second (m/s). Here λ is in meters (m) and ΔE in joules (J). The frequency, f, in hertz (Hz) is:

$$f = \Delta E/h$$

A pair of carefully aligned and positioned mirrors at opposite sides of the material causes the light beam to be reflected back and forth many times and to undergo constructive INTERFERENCE with itself as it increases in intensity through stimulated emission. That tends to deplete the next-higher-energy state, but the external energy source repopulates it and maintains the population inversion, which permits further increase of intensity. One of the mirrors is only partially reflecting, allowing some of the light to escape as a monochromatic, coherent, intense, nearly parallel laser beam.

The choice of material determines the wavelength of the laser light through the material's ΔE, according to the above relation. Lasers have been constructed for wavelengths throughout the range of visible light and in the infrared and ultraviolet. Higher-power lasers tend to be of the pulsed variety, while lower-power lasers can run continuously. A device that operates much like a laser but produces electromagnetic radiation in the microwave range is called a maser, an acronym for "microwave amplification by stimulated emission of radiation."

lattice A regularly repeating spatial arrangement of geometric points in two or three DIMENSIONs is termed a lattice. In a CRYSTAL, the constituent ATOMs, MOLECULEs, or IONs are located at the points of a lattice. Thus, the structure of a crystal is determined to a large extent by its lattice. Crystal lattices are the subject of investigation of theoretical, or mathematical, CRYSTALLOGRAPHY. They are classified according to their SYMMETRY properties, of which there are 230 different types, called space groups. Those can be grouped into 32 crystal classes, or symmetry classes. These in turn are further divided into seven crystal systems, which are characterized by the relations among three VECTORs that represent the repeat distances of the lattice in three independent directions. The seven crystal systems are:

- *Cubic,* or *isometric, system.* Three equal, mutually perpendicular vectors
- *Tetragonal system.* Two equal vectors and one not equal to them, where all three are mutually perpendicular
- *Orthorhombic system.* Three unequal, mutually perpendicular vectors
- *Hexagonal system.* Two equal vectors at 60° to each other and a third vector, which is unequal to the others and perpendicular to them
- *Monoclinic system.* Three unequal vectors, of which only two pairs are perpendicular
- *Trigonal system.* Three equal, nonperpendicular vectors, with equal angles between all pairs of them
- *Triclinic system.* Three unequal, nonperpendicular vectors, with unequal angles between pairs of them

Lawrence, Ernest Orlando (1901–1958) American *Physicist* Famous as the inventor of the cyclotron, Ernest Lawrence was an experimental physicist. He received his Ph.D. in physics from Yale University in New Haven, Connecticut, in 1925. In 1928, Lawrence joined the physics department of the University of California, Berkeley, and in 1936 was appointed director of the university's Radiation Laboratory. He held both positions until his death, upon which the laboratory was renamed the Lawrence Radiation Laboratory. In 1929,

Ernest Lawrence was an experimental nuclear physicist and the inventor of the cyclotron, a particle accelerator. He was awarded the 1939 Nobel Prize in physics. ***(AIP Emilio Segrè Visual Archives)***

Lawrence invented the cyclotron, a particle accelerator that imparted high ENERGIES to ELEMENTARY PARTICLES and IONs without using high VOLTAGEs for the purpose. Lawrence's research focused on NUCLEAR PHYSICS, and he used the cyclotron to bombard NUCLEI and obtain new ELEMENTs and radioactive ISOTOPEs of known elements. He was also involved in other areas of experimental physics. During World War II, Lawrence contributed to the development of nuclear-FISSION ("atomic") bombs. In 1939, he was awarded the Nobel Prize in physics "for the invention and development of the cyclotron and for results obtained with it, especially with regard to artificial radioactive elements."

See also ACCELERATOR, PARTICLE; EXPERIMENTAL PHYSICS; RADIOACTIVITY.

laws of nature The laws of nature are expressions of the orderly and predictable aspects of nature. They fall into two categories: (1) existential laws, which tell us what may and may not be, and (2) developmental laws, telling us how systems evolve in time. An example of an existential law is the ideal gas law. It tells us, for instance, that if we take a certain amount of gas, place it in a container of a certain VOLUME, and give it a certain TEMPERATURE, its PRESSURE will be such and such. The existence of that amount of gas, occupying that volume, having that temperature, but possessing a different pressure, is precluded by nature. For another example, the existence of a type of ELEMENTARY PARTICLE called a PROTON is a law of this kind. Conversely, the nonexistence of a particle with half the proton's

Laws: There Is Order in the World

In science generally, and in physics particularly, a law is an expression of a pattern, or some ORDER, that exists among natural phenomena. Consider CHARLES AUGUSTIN DE COULOMB's law as an example. This law states that the magnitude of the electrostatic FORCE of attraction or repulsion between two point electric CHARGES is proportional to the magnitude of each charge and inversely proportional to the square of the distance between them. The pattern and order expressed by the law are, first of all, that the magnitude of the force is not random, but depends on physical quantities that can be measured and controlled. In this case, it depends on three quantities: the magnitude of each charge and the distance between the charges. Second, the dependence on these quantities is not wild and useless, but tame and serviceable. It can be expressed mathematically, which makes it possible to apply the law and make use of it. In the example, the dependence involves the mathematical relations of proportionality, inverse proportionality, and second power. (*See* COULOMB'S LAW; ELECTRICITY; LAWS OF NATURE.)

Laws are discovered by searching for and finding patterns in experimental and observational data. Take Coulomb's law again. In order to discover the law, Coulomb performed many measurements of the force with which two small, charged spheres attract or repel each other. He repeated his experiments, using different magnitudes of charges and different separation distances. He might have noticed, for instance, that whenever he doubled the charge on one sphere, the magnitude of the force doubled as well. He noticed that when he decreased the separation to a half of its value, the magnitude of force quadrupled. In such a manner he arrived at his law for the electrostatic force:

$$F = Kq_1q_2/r^2$$

Here F denotes the magnitude of the electrostatic force, q_1 and q_2 are the magnitudes of the charges, and r represents the distance between the point charges. The value of the proportionality constant, K, depends on the UNITs used for the various quantities. This relation summarized in a succinct mathematical expression all the data that Coulomb had collected.

Laws are based on a finite amount of data. The amount might be very large, but it is always limited. The mathematical relation that expresses a law forms a summary of the relevant data. But it offers more than that. It allows the prediction of the results of new experiments or observations. Thus the law can be tested and confirmed, or tested and contradicted. In the example of Coulomb's law, one can take a set of values for q_1, q_2, and r that Coulomb never tried, calculate the resulting force, run an experiment with those values, measure the force, and test the law.

When a law has been confirmed by overwhelming evidence for its validity, i.e., by many successful predictions, it becomes accepted as correct and can serve as a useful, predictive tool. Such is the case now with Coulomb's law. It is so well confirmed that no physicist feels the need to reconfirm it, and it serves as a standard tool in the toolbox of physics. At least in principle, there always exists the chance, however tiny, that the law might fail for some as-yet untried combination of q_1, q_2, and r. Every law, through its use, is continually being retested and reconfirmed.

MASS but with all its other properties the same as those of the proton is also a law of nature. Existential laws allow prediction. In the gas example, for a given amount, volume, and temperature, the ideal gas law allows prediction of the gas's pressure. (*See* GAS, IDEAL.)

Developmental laws are part of an analysis of what is going on in nature by conceptually splitting the time evolution of physical systems into two aspects: INITIAL CONDITIONS and developmental laws, which are laws of evolution. In this approach, initial conditions are any state of the system that is consistent with nature's existential laws and whose subsequent development over time is determined solely by nature. Otherwise, initial conditions are arbitrary. For example, we can have particles moving at any speed that is less than the SPEED OF LIGHT. As experimenters, we set the initial conditions of a physical system as we please (again, within natural constraints), at least in principle. What then happens to a system whose initial conditions have been specified is solely up to nature; we, as experimenters, have no further hand in it. That is the second aspect of the time evolution of physical systems. We call it the developmental laws of nature. So initial conditions are the aspect of the evolution of systems that nature allows us to have some control over, while developmental laws of nature are the aspect of the evolution that we have no control over. Note that initial conditions completely determine, through the action of developmental laws, the future state of the system at any time. So the developmental laws of nature allow prediction.

Every law possesses a limited known range of validity. What this means is that every law is known to be correct only for certain ranges of values of the physical quantities that enter into it. For Coulomb's law, the known range of validity covers charges that are not too large and distances that are neither too large nor too small. Within its known range of validity, a law serves as a tool. Under conditions that exceed its known range of validity, a law is being tested; its envelope is being stretched, so to speak.

Every known law of physics is expected to fail under sufficiently extreme conditions. For some laws, the actual range of validity has been found. For others, it is surmised. Here are some examples. Due to quantum effects, Coulomb's law is expected to fail for sufficiently large charges and for sufficiently small separations. For large distances, its range of validity is not known. The ideal GAS law fails for gas DENSITIES that are not sufficiently low. ISAAC NEWTON's law of GRAVITATION fails for strong gravitational FIELDS and for high SPEEDS. ALBERT EINSTEIN's general theory of relativity fixes that, but it is expected to fail at very small distances, since it is incompatible with QUANTUM PHYSICS. (*See* GAS, IDEAL; NEWTON'S LAWS; RELATIVITY, GENERAL THEORY OF.)

Let us take another example of the discovery of a law. GALILEO GALILEI performed many experiments on the behavior of uniform spheres rolling down straight, inclined tracks. He released spheres from rest and measured the distances they roll from rest position, d, during various elapsed times from release, t. In this way Galileo collected a lot of experimental data in the form of d-t pairs. He studied the numbers, performed calculations on them, and perhaps plotted them in various ways, all in search of pattern and order. What he found was that, for a fixed angle of incline, the distance a rolling sphere covers from rest position is proportional to the square of the elapsed time from its release:

$$d = bt^2$$

The value of the proportionality coefficient, b, depends on the incline angle and on the units used. It does not depend on the size, MASS, or material of the sphere.

This mathematical relation summarizes Galileo's experimental results and forms a law. Although it is based on only a finite number of d-t pairs, the relation is able to predict the distance covered for *any* elapsed time, whether Galileo actually had that d-t pair among his data or not. Thus it predicts the results of new experiments. Perhaps Galileo never measured the distance for an elapsed time of exactly 2.5903 seconds, for instance. This relation predicts what that will be. The predictive power of the law allows it to be tested, and it has been tested and confirmed sufficiently to be considered correct. It is valid as long as the mass of the rolling sphere is not too small nor its speed too high (i.e., as long as the force of gravity acting on the sphere is sufficiently greater than the retarding force on it due to the VISCOSITY of air).

Although physicists are always happy to discover laws, laws themselves do not fulfill the need of physicists to understand nature. Laws can be useful for physics applications. They might serve engineers and chemists for their purposes. But for the understanding that physicists are striving for, laws are only a step in the right direction. The next step is to develop theories to explain the laws. That, however, is the subject of other essays.

This approach to analyzing nature—conceptually splitting it into initial conditions and developmental laws—has been very successful in attaining an understanding of nature at scales from that of ELEMENTARY PARTICLES up to the astronomical. However, the approach becomes problematic, if not altogether meaningless, in dealing with the UNIVERSE as a whole. (*See* COSMOLOGY.) That is because there is only one universe, or at least we have access to but a single universe, so there is no sense in which we can set initial conditions.

In physics, the existential laws of nature are often expressed as EQUATIONS OF STATE for the systems involved. The ideal gas law is one example. The developmental laws are very often expressed as equations for the variables determining the state of the system of interest, the system's EQUATIONS OF MOTION. When those equations are solved for any set of initial conditions for the system, their solution, in the form of the variables as functions of time, describes the system's evolution from the initial conditions into the future. They might also describe the evolution that leads to the initial state.

LED *See* LIGHT-EMITTING DIODE.

Leidenfrost effect Named after the German physician Johann Gottlieb Leidenfrost (1715–94), the Leidenfrost effect is the dancing about of drops of water on a sufficiently hot surface, such as the bottom of a well-heated pan. The effect can serve as a rough indication of the pan's TEMPERATURE. The mechanism of the effect is based on BOILING at the drop's lower surface due to heat radiated from the pan's surface and absorbed by the drop at its lower surface. The water vapor thus produced acts as a low-friction cushion between the drop and the pan, insulating the drop from rapid heat transfer from the pan. Thus, the drop can survive for considerable time before finally boiling away, although the temperature of the pan might be well above the boiling point of water.

See also ABSORPTION; GAS; HEAT; INSULATOR; RADIATION.

length contraction Also called the Lorentz-FitzGerald contraction, after the Dutch physicist HENDRIK ANTOON LORENTZ and the Irish physicist George Francis FitzGerald (1851–1901), relativistic length contraction is the contraction of the lengths of moving objects.

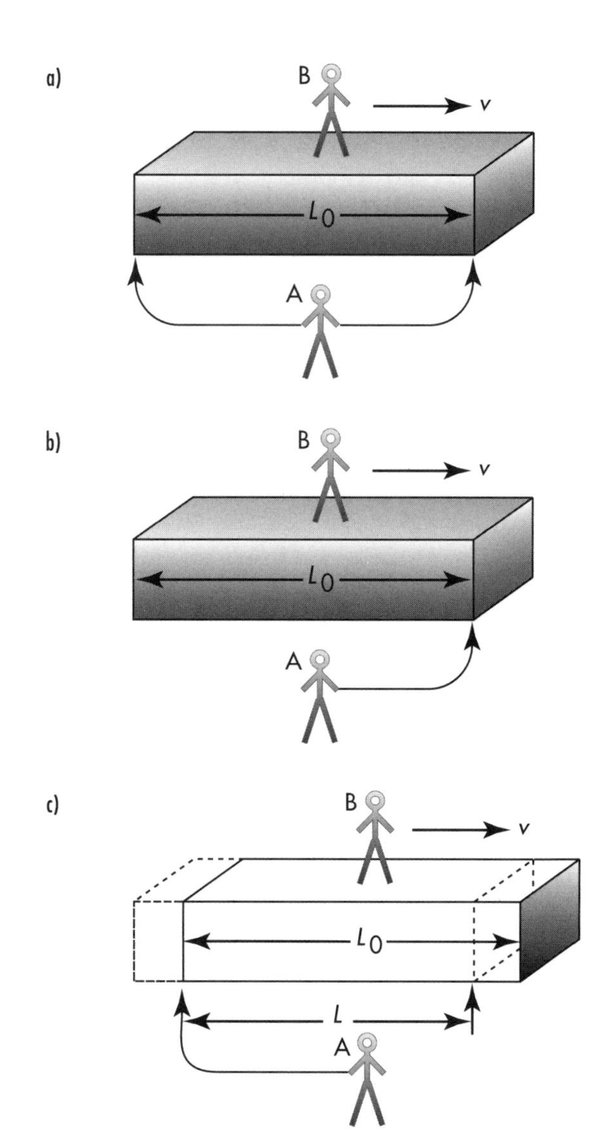

The effect of length contraction: an object of rest length L_0 that is moving at speed v in the direction of its length is measured as having length $L = L_0\sqrt{1 - v^2/c^2}$, with c the speed of light. For nonzero v, L is less than L_0. (a) Observer A measures the length of the moving object, whose rest length is L_0, by simultaneously determining the positions of its ends and finds L. Observer B moves with the body. (b) For observer B, A's measurements are not simultaneous. B sees A first measure the position of the body's front. (c) Then B sees A measure the position of the body's rear. In the interval between the measurements, B sees the body moving with respect to A, so A's length measurement, L, is less than L_0. Thus B agrees that A should find a shorter length than the rest length, as in fact A finds.

More specifically, the length of a moving object in the direction of its motion, measured as L by an observer with respect to whom the object is moving at SPEED v, is found to be:

$$L = L_0\sqrt{1 - v^2/c^2}$$

where L_0 denotes the body's rest length (i.e., its length as measured when it is at rest), and c is the SPEED OF LIGHT, whose value is 2.99792458×10^8 meters per second (m/s). For nonzero speed v, L is less than L_0, so there seems to have occurred a contraction of the length. Here v is in meters per second (m/s), while L and L_0 are expressed in the same units of length. Note that at everyday speeds, length contraction is undetectable, while at 10 percent of the speed of light, the contraction is about a half percent. In order to achieve 50 percent contraction (to half the rest length), the speed must reach approximately 87 percent of the speed of light. As the object's speed approaches the limit of the speed of light, its length contracts toward zero. No object can achieve the speed of light, however, so no length can contract to zero. The formula for length contraction follows from the LORENTZ TRANSFORMATION. (*See* RELATIVITY, SPECIAL THEORY OF.)

Length contraction is a mutual effect: each of two observers in relative MOTION measures the objects moving with the other as contracted. The effect is easily understood as following from the relativity of SIMULTANEITY: measurements that are simultaneous for one observer are not simultaneous for an observer moving with respect to the first. Thus, the simultaneous measurements of the positions of a moving object's front and rear made by observer A (and from which she calculates the object's length) are viewed as nonsimultaneous measurements by observer B, for whom the object is at rest. B observes A determining the position of the *front* of the object *before* she measures the position of its *rear*. During the time interval between the measurements, B observes the rear of the object approaching the location that A previously measured for its front, so that by the time A measures the position of the rear, the rear is closer to where the front was than would be the case if the measurements were made simultaneously. In this way, B agrees that A should find a shorter length than the rest length, as she in fact finds. And the situation is simply reversed when B measures the length of an object that is at rest with respect to A. So each observer measures a shorter length than the rest length for objects that are at rest with respect to the other observer.

lens A lens is an optical device that modifies, by means of REFRACTION and in a controlled manner, the direction of LIGHT rays passing through it. (The concept is suitably generalized for other types of WAVE, such as other electromagnetic RADIATION and SOUND waves.) A lens is made of a transparent material, often glass or plastic, and typically causes light rays either to converge or to diverge. (*See* ELECTROMAGNETISM; OPTICS.)

Here we deal with the simplest situation and assume that:

- The lens material is homogeneous. It is characterized by its INDEX OF REFRACTION.
- The lens is immersed in air. We take the index of refraction of air to equal 1, to a sufficient approximation.
- Both lens surfaces have the form of spherical surfaces. So each lens surface possesses a radius of curvature, which is the radius of the sphere of whose surface it is a part. A plane surface is a special case of a spherical surface, one with infinite radius of curvature.

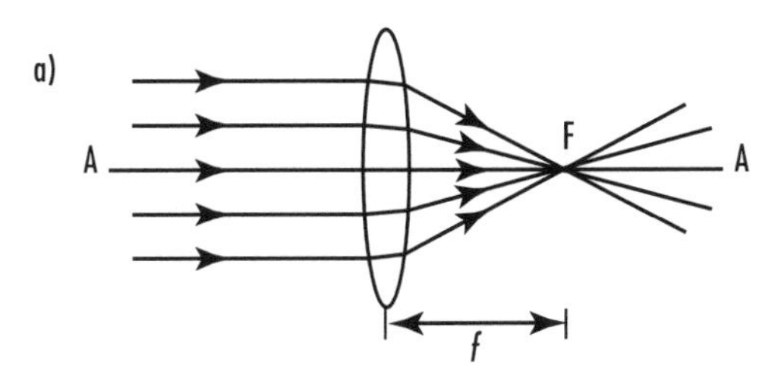

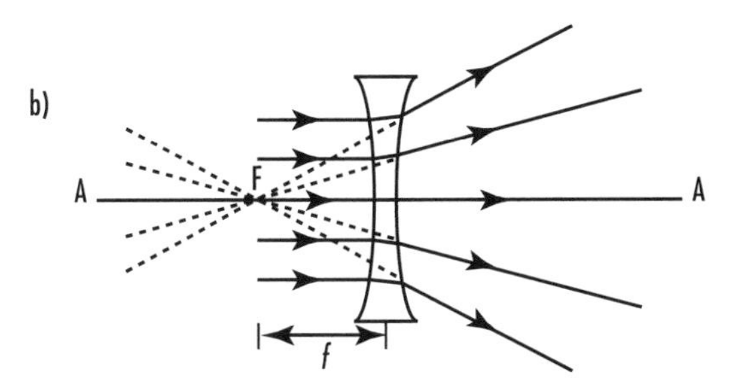

The focal length of a lens. (a) When rays strike a converging lens parallel to its axis A-A, they are made to converge to a point F on the axis, a focal point of the lens. The distance of this point from the center of the lens is the lens's focal length *f*, a positive number. The lens possesses another focal point, at the same distance from the lens on its other side. (b) Similar rays that are incident on a diverging lens are made to diverge as if they emanate from a point F on the lens's axis A-A, a virtual focal point. This point's distance from the center of the lens is the lens's focal length *f*, taken as a negative number. The lens's other virtual focal point is located on its other side at the same distance from the lens.

- The lens possesses an axis. This is an imaginary line through both surfaces, such that the surfaces are perpendicular to the line at their respective points of intersection with it.
- The lens is thin. This means that its thickness is much smaller than its other dimensions. The center of the lens is at the midpoint of the segment of the lens's axis that is between the lens surfaces.
- The light rays are close to the axis and almost parallel with it. Such rays are called paraxial rays.

Focal Length

Consider the effect of a lens on rays that are parallel with its axis. Such rays are either made to converge to a point, termed a focal point, or made to diverge as if emanating from a point on the other side of the lens, called a virtual focal point. One says that the rays are brought, or focused, to a real or virtual focus. In the former case, the lens is called a converging lens, while in the latter the lens is a diverging lens, as the effect is independent of which direction the light passes through the lens. A converging or diverging lens is always thicker or thinner, respectively, at its center than at its edge. The distance between the center of the lens and its real or virtual focal point is the lens's focal length, and is denoted f. The focal length is the same for either direction of light passage. A positive value for the focal length, f, indicates a converging lens, while a diverging lens has a negative value. Since the focal length is a length, the SI UNIT of f is the meter (m).

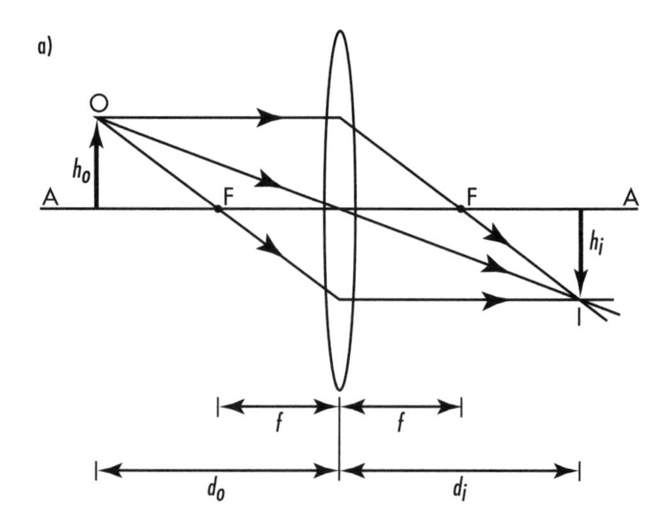

An object O and its real image I as formed by a converging lens of focal length f(> 0). The lens's focal points are denoted F and its axis A-A. The object and image distances are denoted d_o and d_i, respectively. Both are positive. The heights of the object and image are denoted, respectively, h_o and h_i. The former is positive and the latter negative. Three rays from the object to the image are shown. A real image is formed whenever the object is farther from the lens than the focal point, as in the figure.

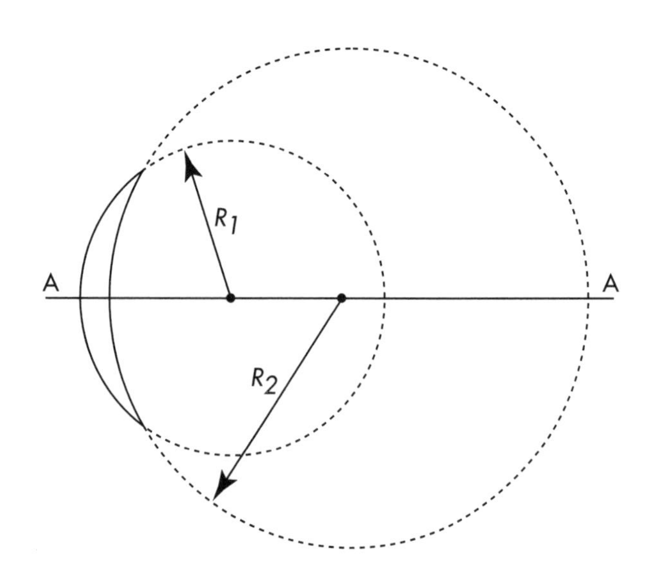

An example of radii of curvature of lens surfaces. A-A denotes the lens axis. The radius of curvature of the surface on the left is R_1, and that of the right surface is R_2. According to the sign convention, both R_1 and R_2 are positive in this example.

A mathematical relation exists that relates the focal length of a lens to its material and shape. Consider any one of the lens's surfaces that is not planar. Denote its radius of curvature by R_1 in meters (m). If the surface bulges away from the lens (i.e., if the surface is convex), make R_1 a positive number. If the surface is concave, bulging in toward the lens, make R_1 negative. Denote the radius of curvature of the other surface by R_2, also in meters (m). If this surface is convex, assign a negative value to R_2. If the surface is concave, make R_2 positive. (Note that the sign convention for R_2 is opposite to that for R_1.) If the second surface is planar, then $R_2 = \infty$ and the sign is not important. For a lens material possessing index of refraction n, the lens maker's formula is:

$$\frac{1}{f} = (n-1)\left(\frac{1}{R_1} - \frac{1}{R_2}\right)$$

The inverse of the focal length is called the optical power of the lens:

$$D = 1/f.$$

Its SI unit is the inverse meter (1/m), also referred to as diopter (D). The significance of this is that the closer the focal point is to the center of the lens, and the smaller the focal length in absolute value, the greater is the convergent or divergent effect of the lens on the light passing through it. The optical power expresses that numerically. As it is for focal length, a positive or negative value for D indicates a converging or diverging lens, respectively.

Combinations of lenses can be formed by having them touching at their centers with a common axis. The focal length, f, of such a combination of lenses having focal lengths $f_1, f_2, \ldots$ is given by:

$$\frac{1}{f} = \frac{1}{f_1} + \frac{1}{f_2} + \ldots$$

Thus, the optical power of the combination, D, equals the sum of the optical powers of the components:

$$D = D_1 + D_2 \ldots$$

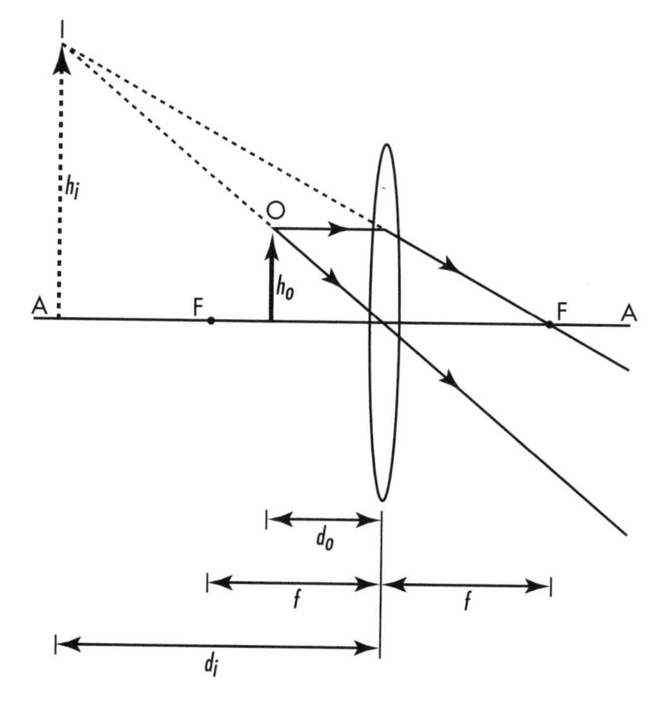

The formation of a virtual image I of an object O by a converging lens of focal length f (> 0), which is acting as a magnifier. The lens's focal points are denoted F and its axis A-A. The object and image distances are denoted d_o and d_i, respectively, where d_o is positive and d_i negative. The heights of the object and virtual image are denoted, respectively, h_o and h_i. Both are positive. Two rays from the object to the viewer are shown. As in the figure, whenever the object is located between the lens and the focal point, the image is virtual.

Lens Equation

For the focusing effect of a lens on nonparallel rays, we assume, for the purpose of presentation, that the light source, or object, is located to the left of the lens, so the light diverges from the source and passes through the lens from left to right. Denote the distance of the object from the lens by d_o in meters (m). The light rays from the object might converge as they emerge from the lens, in which case they converge to an image, called a real image, on the right side of the lens. Denote the distance of the image from the lens by d_i in meters (m) and assign it a positive value. If the emerging light rays diverge, they will diverge as if they come from a virtual image located on the left side of the lens. The distance of a virtual image from the lens is denoted by d_i as well, but it is given a negative sign. In brief, a positive or negative value for d_i indicates a real image (on the right) or a virtual image (on the left), respectively. If the object is located at the focal point, the rays emerge from the lens parallel to the axis, neither converging nor diverging. In this case, the image distance is infinite and its sign is not important.

A real image can be caught by placing a screen, such as a movie screen, at its location. A virtual image cannot be caught in that way. Rather, a virtual image can be seen or photographed by aiming the eye or camera toward it through the lens, possibly using a magnifying glass or binoculars.

It might happen that the rays entering the lens from the left are converging rather than diverging. This occurs when the lens intercepts the converging rays emerging from, say, another lens. In the absence of our lens, the rays would converge to a real image, located to the right of our lens. That real image serves as a virtual object for our lens. Denote the distance of the virtual object from our lens by d_o and give it a negative sign. So a positive or negative sign for d_o indicates a real object (on the left) or a virtual object (on the right), respectively. If a virtual object is located at the focal point, the rays emerging from a diverging lens are parallel to the axis, and the image distance is infinite.

The object and image distances and the focal length of a lens are related by the lens equation:

$$\frac{1}{d_o} + \frac{1}{d_i} = \frac{1}{f}$$

This relation implies the following behavior of d_i as a function of d_o for fixed f and real object (positive d_o):

Converging Lens ($f>0$)

Object Distance (d_o)	Image Distance (d_i)	Image Type
∞	f	Real
Decreasing from ∞ to $2f$	Increasing from f to $2f$	Real
$2f$	$2f$	Real
Decreasing from $2f$ to f	Increasing from $2f$ to ∞	Real
f	$\pm\infty$	—
Decreasing from f to $f/2$	Increasing from $-\infty$ to $-f$ (while decreasing in absolute value)	Virtual
$f/2$	$-f$	Virtual
Decreasing from $f/2$ to 0	Increasing from $-f$ to 0 (while decreasing in absolute value)	Virtual

Diverging Lens ($f<0$)

Object Distance (d_o)	Image Distance (d_i)—All Images Are Virtual
∞	$-\lvert f\rvert$
Decreasing from ∞ to f	Increasing to from $-\lvert f\rvert$ to $-\lvert f\rvert/2$ (while decreasing in absolute value)
f	$-\lvert f\rvert/2$
Decreasing from f to 0	Increasing from $-\lvert f\rvert/2$ to 0 (while decreasing in absolute value)

Magnification

Denote the size, or height, of an object in a direction perpendicular to the axis by h_o and that of the corresponding image by h_i. The SI unit for both quantities is the meter (m). A positive value for h_o indicates that the object is upright in relation to some standard for orientation, while a negative value shows that it is inverted. Similarly, positive h_i indicates an upright image, and a negative value indicates an inverted image. The object and image sizes are proportional to the object and image distances, respectively, through the relation:

$$h_i/h_0 = -d_i/d_o.$$

It follows immediately that the real image of a real object is always oriented oppositely to the object, while its virtual image always has the same orientation. On the other hand, a virtual object and its real image both have the same orientation, and a virtual image of a virtual object is oppositely oriented. For a given object size, the farther the image is from the lens, the larger it is.

The MAGNIFICATION is defined as:

$$M = h_i/h_0$$

which, by the above relation, is:

$$M = -d_i/d_o$$

Magnification is dimensionless. An absolute value of M greater than 1 ($\lvert d_i\rvert > \lvert d_o\rvert$) indicates strict magnification, i.e., the image is larger than the object. When $\lvert M\rvert$ equals 1 ($\lvert d_i\rvert = \lvert d_o\rvert$), the image is the same size as the object, while for $\lvert M\rvert$ less than 1 ($\lvert d_i\rvert < \lvert d_o\rvert$) we have reduction. Positive M (d_i, d_o of opposite signs) shows

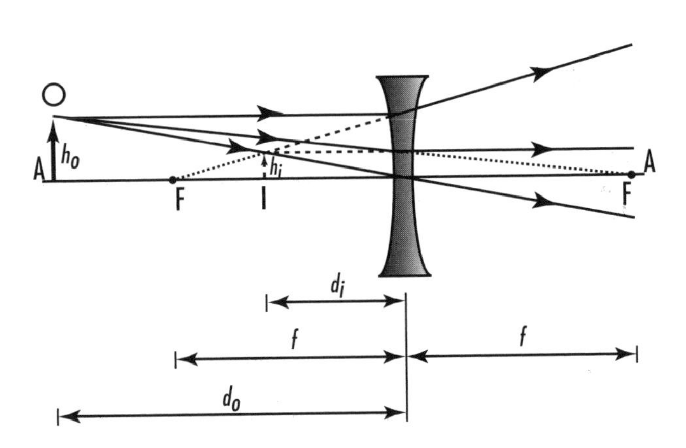

A diverging lens always forms a virtual image of an object, labeled I and O, respectively. The lens's focal length is $f(<0)$, its virtual focal points are denoted by F, and its axis is A-A. Object and image distances are denoted d_o and d_i, respectively, where d_o is positive and d_i negative. The heights of the object and virtual image are denoted, respectively, h_o and h_i. Both are positive. Three rays from the object to the viewer are shown.

that the image and the object have the same orientation, while negative M (d_i, d_o of the same sign) indicates they are oriented oppositely. For a series of magnifications—where the image produced by the first lens (or other optical device) serves as the object for the second, the image formed by the second serves as the object for the third, and so on—the total magnification is the product of the individual magnifications:

$$M = M_1 M_2 \ldots$$

(*See* DIMENSION.)

The following is a summary of the sign conventions used previously:

Sign Conventions

Quantity	Positive	Negative
R_1	First surface is convex	First surface is concave
R_2	Second surface is concave	Second surface is convex
f	Converging lens	Diverging lens
d_o	Real object	Virtual object
d_i	Real image	Virtual image
h_o	Upright object	Inverted object
h_i	Upright image	Inverted image
M	Object and image in same orientation	Object and image oppositely oriented

Real lenses in real situations are subject to various kinds of ABERRATION.

See also MIRROR.

Lenz's law Named for the German/Russian physicist Heinrich Friedrich Emil Lenz (1804–1865), Lenz's law states that the polarity of the ELECTROMOTIVE FORCE (emf) induced in a conducting coil by a changing magnetic FLUX through the coil is such as to produce an electric CURRENT whose magnetic effect tends to cancel the cause of the emf. If the flux were increasing in one direction, the current would be such as to produce a magnetic FIELD in the opposite direction, which would reduce the rate of increase. If, on the other hand, the flux were decreasing, the current would create a magnetic field in the same direction as that of the decreasing flux, which would reinforce the decreasing flux and retard its decrease. (*See* BIOT-SAVART LAW; CONDUCTION; ELECTRICITY; FARADAY'S LAW; INDUCTION; MAGNETISM.)

Lenz's law extends to actual FORCEs that act to prevent change of magnetic flux. Let a MAGNET be brought toward a CONDUCTOR with its north pole first, for instance. The resulting increase of flux into the conductor induces currents in the conductor that produce a magnetic field that, in the vicinity of the magnet, points away from the conductor. Thus the magnet effectively "sees" a north magnetic pole in front of it and is repelled. A person pushing the magnetic might actually feel an opposing force. Now, let the magnet, with its north pole facing the conductor, be withdrawn from the conductor. As a result, the flux into the conductor diminishes. That induces currents that create a magnetic field in the region of the magnet that points toward the conductor. The moving magnet now "sees" a south magnetic pole behind it, which pulls it back toward the conductor and opposes its motion.

lepton The leptons are a family of ELEMENTARY PARTICLEs that consist of the ELECTRON (e^-), the muon (μ), the tau (τ), their corresponding NEUTRINOs—the electron-type neutrino (ν_e), the muon-type neutrino (ν_μ), and the tau-type neutrino (ν_τ)—and the antiparticles of all six. (The antiparticles are denoted by adding a bar on top of the symbol of the corresponding particle, such as $\bar{\mu}$, $\bar{\nu}_e$, etc. An exception to this rule is that the electron's antiparticle, the POSITRON, is often denoted e^+.) The leptons are matter particles and are FERMIONS, possessing a half unit of SPIN. They participate in all the fundamental INTERACTIONs except the STRONG INTERACTION. (Only the electrically charged leptons participate in the ELECTROMAGNETIC INTERACTION.) As far as is presently understood, the leptons are structureless and pointlike. (*See* ANTIMATTER; CHARGE; ELECTRICITY.)

The electron, muon, and tau each carry one elementary unit of negative electric charge (i.e., $-e = -1.602176462 \times 10^{-19}$ coulomb [C]). Their MASSes are 0.5110, 105.7, and 1784 MeV/c^2, respectively. Only the electron is stable, while the muon and tau DECAY with lifetimes of about 2×10^{-6} and 5×10^{-13} second (s), respectively. The corresponding antiparticles have the same masses, respectively, and carry one elementary unit of positive electric charge.

The neutrinos are electrically neutral. They possess masses that are considerably smaller than those of the other leptons but have not yet been measured to more than very coarse accuracy. The neutrinos do not decay, but they do undergo the process of mixing, whereby, as they travel, they continuously convert into varying quantum mixtures of all three neutrino types. (*See* QUANTUM PHYSICS.)

Leptons appear to be organized into generations, where the higher generations are "exotic" and more massive versions of the first generation. The first generation consists of the electron and the electron-type neutrino. The muon and the muon-type neutrino form the second generation, and the tau and tau-type neutrino make up the third. A similar organization appears among the QUARKs, hinting at the possibility that the leptons and the quarks are related in some deep manner. (*See* GRAND UNIFIED THEORY [GUT].)

Leptons carry additional CHARGEs, which are various kinds of lepton number that are conserved in all interactions. All leptons have a lepton number: +1 for leptons and –1 for antileptons. In addition, each lepton generation possesses its own number, which is separately conserved. The electron-type lepton number is +1 for the electron and the electron-type neutrino, –1 for their antiparticles, and 0 for all other leptons. The same applies to the muon-type lepton number and the tau-type lepton number. As an example of charge conservation, consider the decay process of the muon (through the WEAK INTERACTION):

$$\mu \rightarrow e^- + \nu_\mu + \bar{\nu}_e$$

whereby the muon decays into an electron, a muon-type neutrino, and an electron-type antineutrino. First consider the process with regard to conservation of electric charge:

$$(-1) \rightarrow (-1) + 0 + 0$$

With regard to lepton-number conservation, we have:

$$1 \rightarrow 1 + 1 + (-1)$$

Conservation of electron-type lepton number is:

$$0 \rightarrow 1 + 0 + (-1)$$

And muon-type lepton-number conservation appears as:

$$1 \rightarrow 0 + 1 + 0$$

See also CONSERVATION LAW.

lever A lever is a simple mechanical device for multiplying FORCE or MOTION. A lever is basically a rigid beam that is free to rotate about a fixed fulcrum, or pivot. A force applied to the beam at any place on it causes the beam to apply a force to a load at any other position on the beam. Considering only forces that are perpendicular to the beam, denote the magnitude of the applied force by F_a and that of the force on the load by F_l. Denote the distance from the point of application of the applied force to the fulcrum by d_a and the distance between the load and the fulcrum by d_l. Further, denote the arc through which the point of application moves by s_a and the arc through which the load moves by s_l. As for units, it is necessary only that the two forces, the

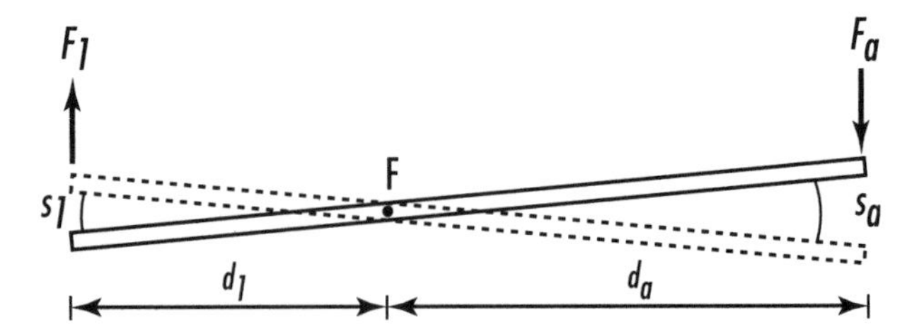

A lever, with fulcrum at F. The forces are assumed to be perpendicular to the lever. The magnitude of the applied force is F_a and that of the lever's force on the load is F_l. The distances from the fulcrum of the point of application of the applied force and of the load are d_a and d_l, respectively. The variables s_a and s_l denote, respectively, the arcs through which the point of application and the load move.

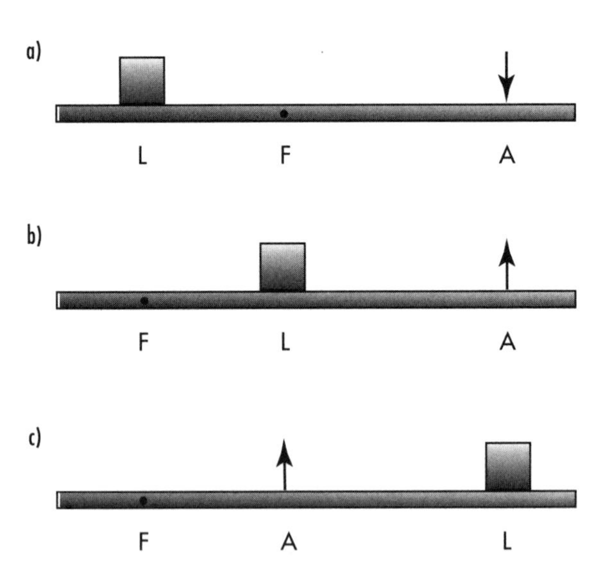

Three classes of lever. (a) The fulcrum F is between the point of application of the applied force A and the load L. (b) The load is between the fulcrum and the point of application of the force. (c) The point of application of the force is between the fulcrum and the load.

two distances, and the two arcs be expressed, respectively, in the same units. Then the following relation holds in an ideal, frictionless, almost-static situation:

$$F_l/F_a = d_a/d_l$$

In any situation we also have the geometric relation:

$$d_a/d_l = s_a/s_l$$

(*See* FRICTION; ROTATION.)

In other words, the ratio of the force magnitudes equals the inverse ratio of the corresponding distances. The ratio of the motion arcs equals the ratio of the corresponding distances. So by choosing appropriate distances for the applied force and the load, the applied force can be multiplied as desired in its effect on the load. Alternatively, the applied motion can be magnified or reduced in its realization at the load.

Levers are conventionally divided into three classes, according to whether the fulcrum is between the application point and the load or not, and if not, whether the load or the application point is closer to the fulcrum.

levitation, magnetic (maglev) *See* MAGNETIC LEVITATION.

lift Also called aerodynamic lift or hydrodynamic lift, lift is a component of the FORCE acting on a body due to the flow of a FLUID (LIQUID or GAS) past the body. It is the component in the direction perpendicular to the direction of flow if the body were absent. It is just such lift due to moving air that keeps kites aloft, for instance. It is the relative MOTION that matters; the body might in fact be moving through a stationary fluid, such as in the case of an airplane. One source of lift is a PRESSURE difference above and below the body resulting from faster flow above than below, according to BERNOULLI'S EQUATION. Another source is fluid deflection by the body, when the body causes the flow to change direction. In such a situation, the force on the body is the reaction force, according to ISAAC NEWTON's second law of motion, to the force the body exerts on the fluid to change its direction of flow. The analysis of lift forces is complicated, especially due to TURBULENCE and VISCOSITY. (*See* NEWTON'S LAWS.)

Lift forces caused by moving air keep aloft heavier-than-air aircraft, as well as birds and flying insects, for example. Submarines are designed to exploit a combination of lift and BUOYANCY forces to control their depth in water.

light Light is electromagnetic RADIATION, or electromagnetic WAVEs, that humans can see, i.e., in the WAVELENGTH range of about 400 nanometers (nm) to 700 nm, corresponding to the FREQUENCY range of around 750 to 430 terahertz (THz). As are all electromagnetic waves, light is a transverse wave and can thus be polarized. And as does all electromagnetic radiation, light travels in VACUUM at the SPEED OF LIGHT, conventionally denoted *c*, with the value 2.99792458×10^8 meters per second (m/s). The speed of light is found to have the same value no matter what the state of MOTION of the observer. This seemingly paradoxical effect led ALBERT EINSTEIN to the special theory of relativity and serves as one of the theory's cornerstones. (*See* ELECTROMAGNETISM; POLARIZATION; RELATIVITY, SPECIAL THEORY OF.)

Light is produced by any object at a TEMPERATURE of at least about 900 kelvins (K), such as a light bulb or a star, called an incandescent source. Other sources of light include LASERs, fluorescent tubes, LIGHT-EMITTING DIODEs (LEDs), and chemical reactions. In the perception of light by humans, different wavelengths, combinations of wavelengths, and continuous ranges of wavelengths arouse sensations of various colors. Light of a single wavelength is called monochromatic. As an example, one of the important characteristics of laser light is its monochromaticity. (*See* FLUORESCENCE.)

The colors that are perceived as a result of visual stimulation by monochromatic light fall roughly into the following ranges:

Color	Frequency Range (THz)	Wavelength Range (nm)
Red	430–460	700–650
Orange	460–500	650–600
Yellow	500–550	600–550
Green	550–600	550–500
Blue	600–670	500–450
Indigo	670–710	450–420
Violet	710–750	420–400

Two or more distinct wavelengths together are termed a discrete spectrum. An example of this is the light from a discharge tube, such as are used for outdoor advertising and are generally referred to as neon lights (although they might contain GASes other than neon). Continuous range(s) of wavelengths are called a continuous spectrum. Such spectra are typically produced by incandescent sources. (*See* SPECTRUM.)

According to QUANTUM PHYSICS, light is viewed as a flow of particles, called PHOTONs. So light is endowed with a particle aspect as well as a wave aspect. The connection between the wave and particle aspects is that the ENERGY, E, of each individual photon among those that form a light wave of frequency f is given by:

$$E = hf$$

where E is in joules (J), f is in hertz (Hz), and h is the PLANCK CONSTANT, with value $6.62606876 \times 10^{-34}$ joule·second (J·s). (*See* WAVE-PARTICLE DUALITY.)

See also LUMINESCENCE; OPTICS; PHOSPHORESCENCE; PHOTOELECTRIC EFFECT; YOUNG'S EXPERIMENT.

light, speed of *See* SPEED OF LIGHT.

light-emitting diode (LED) A light-emitting diode is a SEMICONDUCTOR device that emits LIGHT or infrared RADIATION when electric CURRENT passes through it. A diode, in general, is any electric device with two terminals that presents low RESISTANCE to an electric current in one direction and high resistance to a current in the opposite direction. In a semiconductor, the electric current is carried by both ELECTRONs (negative) and HOLES (positive). In an LED, when current is flowing in the "easy" direction, electron-hole pairs combine with a consequent release of energy as PHOTONs, one photon from each combination. The materials forming the diode set the energy, E, of the photon emitted by every pair combination. That determines the FREQUENCY of the emitted light, f, through the relation:

$$f = E/h$$

where E is in joules (J), f is in hertz (Hz), and h is the PLANCK CONSTANT, whose value is $6.62606876 \times 10^{-34}$ joule·second (J·s). Although LED light is monochromatic, as is LASER light, it is neither coherent, narrow-beamed, nor relatively intense, although continuing development is bringing about improvement in INTENSITY. (*See* ELECTRICITY.)

linear momentum The linear momentum, $\mathbf{p}$, of a particle of MASS m and VELOCITY $\mathbf{v}$ is

$$\mathbf{p} = m\mathbf{v}$$

where $\mathbf{p}$ is in kilogram·meters per second (kg·m/s), m in kilograms (kg), and $\mathbf{v}$ in meters per second (m/s). Linear momentum is a VECTOR quantity. The adjective *linear* is used to distinguish linear momentum from ANGULAR MOMENTUM and can be dropped when the situation is clear. The momentum of a collection of particles is the vector sum of the individual momenta.

ISAAC NEWTON's second law of MOTION is often formulated as "FORCE equals mass times ACCELERATION":

$$\mathbf{F} = m\mathbf{a} = m\, d\mathbf{v}/dt$$

where $\mathbf{F}$ denotes the force in newtons (N), $\mathbf{a}$ the acceleration in meters per second per second (m/s^2), and t the time in seconds (s). It is more correctly expressed as "force equals time-rate-of-change of the linear momentum":

$$\mathbf{F} = d\mathbf{p}/dt$$

It is in the latter form that ALBERT EINSTEIN's special theory of relativity uses the law. The former expression is an approximation to the latter for SPEEDs that are sufficiently small compared with the SPEED OF LIGHT. (*See* NEWTON'S LAWS; RELATIVITY, SPECIAL THEORY OF.)

Linear momentum obeys a CONSERVATION LAW, which states that in the absence of external forces acting on a system, the system's total linear momentum remains constant in time.

liquid One of the ordinary STATES OF MATTER, a liquid is characterized by having fixed VOLUME, to a good approximation (i.e., it has low compressibility), but no fixed shape: under the influence of gravity, a liquid pools at the bottom of its container and takes the shape of the container there, forming a free surface at its top. The particles constituting a liquid—its ATOMS, MOLECULES, or IONs—move about randomly within the liquid, colliding with each other and with the container walls, yet they possess a degree of short-range order (as contrasted to the long-range order in a SOLID). This means that in the immediate vicinity of each particle, other particles tend to maintain, on the average, constant distances from it and from each other. Water, which we drink, wash in, swim in, and consist of to a large extent, is a very common example of a liquid. (*See* GRAVITATION.)

See also GAS; LIQUID CRYSTAL; SURFACE TENSION.

liquid crystal The term *liquid crystal* describes a STATE OF MATTER that combines properties of a LIQUID and of a SOLID. While a liquid crystal flows like a liquid, it possesses long-range order and structure as in a solid and exhibits anisotropy (i.e., its physical properties differ in different directions). It can be birefringent, for example. Liquid crystals typically rotate the plane of polarized LIGHT passing through them, and that effect might be affected by an applied electric FIELD. This phenomenon allows certain liquid crystals to be used in displays—liquid crystal displays (LCDs), such as in calculators and digital watches—that are electrically controlled. (*See* BIREFRINGENCE; ELECTRICITY; POLARIZATION.)

See also CRYSTAL.

Lorentz, Hendrik Antoon (1853–1928) Dutch *Physicist* Greatly admired by his contemporaries, Hendrik Lorentz was a theoretical physicist who made major contributions to the foundations of ALBERT EINSTEIN's special theory of relativity. Lorentz obtained his doctorate in physics in 1875 from the University of Leyden in the Netherlands. Three years later, he was appointed to the position of professor of physics at the same university, where he remained until his retirement in 1912 to take up the directorship of the Teyler Institute in Haarlem, the Netherlands. Lorentz strove to elaborate on JAMES CLERK MAXWELL's theory of ELECTROMAGNETISM, as expressed by MAXWELL'S EQUATIONS, and investigated the interaction of LIGHT with MATTER, specifically with the electric CHARGEs in matter. In that way, he effectively predicted the existence of ELECTRONs before the structure of ATOMs was known. In this connection, he proposed what is now known as the LORENTZ FORCE. Lorentz's study of the effects of high-speed MOTION led to the Lorentz-FitzGerald contraction and the LORENTZ TRANSFORMATION, both of which form components of the special theory of relativity. In 1902, Lorentz shared the Nobel Prize in physics with Pieter Zeeman "in recognition of the extraordinary service they rendered by their researches into the influence of magnetism upon radiation phenomena."

See also ELECTRICITY; LENGTH CONTRACTION; RELATIVITY, SPECIAL THEORY OF; SPEED; THEORETICAL PHYSICS.

A theoretical physicist, Hendrik Antoon Lorentz made major contributions to the foundations of the special theory of relativity. He shared the 1902 Nobel Prize in physics. ***(AIP Emilio Segrè Visual Archives, Physics Today Collection)***

Lorentz force Named for the Dutch physicist HENDRIK ANTOON LORENTZ, the lorentz force is the FORCE acting on an electrically charged particle in an electromagnetic FIELD. This force is the resultant of the electric force and the magnetic force acting on the particle due to the electric and magnetic fields, respectively. The vectorial expression for the Lorentz force is:

$$\mathbf{F} = q(\mathbf{E} + \mathbf{v}\times\mathbf{B})$$

where **F** denotes the force in newtons (N), q is the electric charge of the particle in coulombs (C), **E** is the value of the electric field in newtons per coulomb (N/C) (equivalent to volts per meter [V/m]) at the particle's location, **v** is the particle's VELOCITY in meters per second (m/s), and **B** represents the value of the magnetic field in teslas (T) at the particle's location. (*See* CHARGE; ELECTRICITY; ELECTROMAGNETISM; MAGNETISM; VECTOR.)

Lorentz transformation Named for the Dutch physicist HENDRIK ANTOON LORENTZ, Lorentz trans-

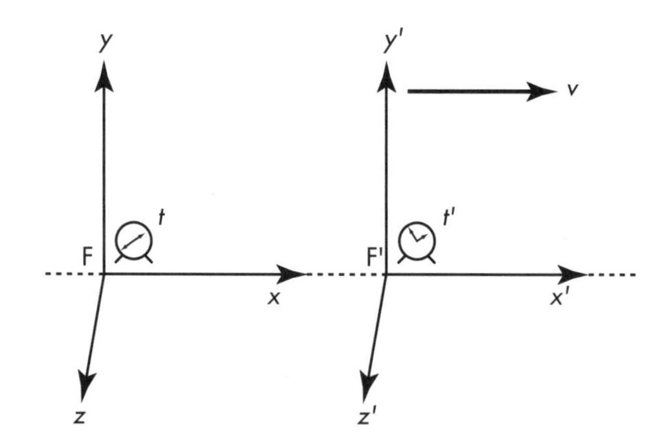

Lorentz transformations relate the coordinates and time variables of two inertial reference frames. The figure shows two such reference frames, F and F′, with coordinates (x, y, z) and (x', y', z') and time variables t and t', respectively. F′ is moving at constant speed v, which may approach the speed of light, in the positive x direction with respect to F, and their coordinate axes coincide at some time, which is taken to be $t = t' = 0$.

formations are mathematical relations that might hold between the respective SPACE coordinates and TIME variables used by any two observers in relative constant-VELOCITY motion (i.e., in relative motion that is at constant SPEED and in a straight line), according to ALBERT EINSTEIN's special theory of relativity. The most general such mathematical relations are called POINCARÉ TRANSFORMATIONS. A Poincaré transformation between the coordinates and time variables of two such observers is a Lorentz transformation in the special case that the respective coordinate origins coincide at some instant, and at that instant both their clocks read zero. (*See* COORDINATE SYSTEM; RELATIVITY, SPECIAL THEORY OF.)

As an example of a Lorentz transformation, let coordinates (x, y, z) and time variable t be those used by observer O to designate locations and times in her REFERENCE FRAME, while (x', y', z') and t' are used by observer O′ in his. Let the reference frame of observer O′ be moving at constant speed v with respect to the reference frame of observer O, in O's positive x direction. To fulfill the special condition for a Lorentz transformation, let the respective coordinate axes of O and O′ instantaneously coincide when the observers' clocks at their coordinate origins both read zero. Physically, this is no limitation, since whenever one reference frame is in constant-speed rectilinear motion with respect to another, their coordinate axes can always be displaced and rotated and their clocks can always be reset so as to make this the case. Then, if an event is observed to occur at position (x, y, z) and at time t with respect to the reference frame of O, and if the same event is observed by O′ at position (x', y', z') and time t' with respect to his reference frame, the Lorentz transformation relating the variables is:

$$x' = \gamma(x - vt)$$
$$y' = y$$
$$z' = z$$
$$t' = \gamma(t - vx/c^2)$$

or inversely:

$$x = \gamma(x' + vt')$$
$$y = y'$$
$$z = z'$$
$$t = \gamma(t' + vx'/c^2)$$

The factor γ is defined as:

$$\gamma = \frac{1}{\sqrt{1 - v^2/c^2}}$$

and c denotes the SPEED OF LIGHT, whose value is 2.99792458×10^8 meters per second (m/s). All spatial coordinates are in meters (m), all times in seconds (s), and v is in meters per second (m/s).

Note in particular that the observers' clock readings for the same event are related in a way that depends on the location of the event and on the speed v. Note also that as v/c goes to 0, both γ goes to 1 and the x- and x'-dependence drops out of the time transformation. So for speeds that are sufficiently low compared with the speed of light, the Lorentz transformation is well approximated by the corresponding GALILEI TRANSFORMATION. The formulas for LENGTH CONTRACTION and TIME DILATION follow from the Lorentz transformation.

Space and time are merged to form four-dimensional SPACE-TIME, in which events are represented by sets of four coordinates (x, y, z, t). In analogy to the distance between points, an interval, D, is defined for every pair of events in space-time and is given by:

$$D^2 = (x_2 - x_1)^2 + (y_2 - y_1)^2 + (z_2 - z_1)^2 - c^2(t_2 - (t_1)^2$$

where (x_1, y_1, z_1, t_1) and (x_2, y_2, z_2, t_2) are the space-time coordinates of the two events. Although the coordinates of the same event have different values with respect to different reference frames (i.e., in different coordinate systems) in space-time, the interval between any two events is invariant for reference frames in relative constant-velocity motion. In other words, the interval between a pair of events does not change under Lorentz (or Poincaré) transformations. If unprimed and primed symbols denote coordinates with respect to two such reference frames, the invariance is expressed as:

$$(x_2' - x_1')^2 + (y_2' - y_1')^2 + (z_2' - z_1')^2 - c^2 (t_2' - t_1')^2$$
$$= (x_2 - x_1)^2 + (y_2 - y_1)^2 + (z_2 - z_1)^2 - c^2 (t_2 - t_1)^2$$

This invariance assures that the speed of light has the same value with respect to both reference frames. (*See* DIMENSION.)

Los Alamos National Laboratory (LANL) Founded in 1943 at Los Alamos, New Mexico, Los Alamos National Laboratory was established as part of the Manhattan Project, the U.S. government's World War II effort to develop a nuclear-FISSION ("atomic") bomb. It is now a multiprogram, multidisciplinary research laboratory, operating under the National Nuclear Security Administration of the U.S. Department of Energy. LANL employs over 12,000 people, of whom about 7,900 are regular personnel, 900 are students, and 3,300 are contractors. (*See* NUCLEAR PHYSICS.)

The laboratory's principal missions are:

- Ensuring the safety and reliability of America's nuclear weapons
- Reducing threats to U.S. security, especially with regard to weapons of mass destruction
- Contributing technical solutions to U.S. security problems in ENERGY, environment, infrastructure, and health

The programs being carried out at LANL include:

- *Nuclear stockpile stewardship*. LANL maintains the nuclear stockpile of the United States without actu-

Aerial view of Los Alamos National Laboratory. ***(Courtesy of Los Alamos National Laboratory)***

ally detonating nuclear devices, through a program of computations, simulations, experiments, and manufacturing.

- *High-performance computing.* Very powerful computers perform numerical modeling, which allows the visualization and prediction of diverse phenomena, such as the progress of explosions in nuclear weapons, the course of wildfires, global weather patterns, and epidemics. The program helps ensure the effectiveness of the U.S. nuclear arsenal.
- *Advanced materials.* The Laboratory carries out research on the behavior of materials, which is crucial to predicting the performance of nuclear weapons and developing high-tech products. The latter include such as improved batteries, more efficient fuel cells, and stronger composite materials.
- *Bioscience and biotechnology.* LANL seeks to understand the dangers of nuclear, biological, and chemical weapons and to search for ways to protect people from them. The laboratory also researches bioscience and biotechnology more generally and is involved in developing the human genome map.
- *Earth and environmental science.* The laboratory seeks to better understand the complex geophysical systems that drive the earth, oceans, and atmosphere and analyzes how man-made hazards might harm human health and the environment.
- *Physics.* LANL pursues research in EXPERIMENTAL PHYSICS and THEORETICAL PHYSICS to gain understanding of the physical world.
- *User facilities.* Hundreds of scientists and researchers from around the world use LANL's more than 50 cross-disciplinary facilities. The latter include, for example, the National High Magnetic Field Laboratory. (*See* FIELD; MAGNETISM.)

The URL of the Los Alamos National Laboratory's website is http://www.lanl.gov/. Its mailing address is P.O. Box 1663, Los Alamos, NM 87545.

luminescence The emission of LIGHT or of other electromagnetic RADIATION by a material due to the de-excitation of its constituent particles (ATOMs, MOLECULEs, IONs) following previous excitation is called *luminescence.* After having been excited to ENERGY states above their GROUND STATE, the particles spontaneously make transitions to lower-energy states, emitting PHOTONS in the process. When emission ceases with, or immediately following, cessation of excitation, the luminescence is called FLUORESCENCE. Otherwise, if emission is significantly extended after excitation has stopped, the effect is called PHOSPHORESCENCE. (*See* ELECTROMAGNETISM.)

There are different terms for luminescence according to its method of excitation. Photoluminescence is excitation by photons (i.e., by electromagnetic radiation). An example is fluorescent yellow and pink traffic signs. They are excited by ultraviolet radiation and emit visible light. In chemiluminescence, chemical reactions are the source of excitation. An example of that is the glowing tubes used for emergency lighting and for amusement at parties. Light emission is initiated by causing chemicals to mix and interact. Bioluminescence is chemiluminescence that occurs in living organisms, such as the distinctive glow of fireflies. Electroluminescence is luminescence resulting from excitation by energetic ELECTRONs. This is what occurs in cathode ray tubes (CRTs), such as television picture tubes and computer monitor tubes. There, accelerated electrons strike the coating on the inside of the picture surface, which consequently emits visible light. Another effect is triboluminescence, which is emission resulting from friction. This effect is easily seen by briskly pulling adhesive tape, such as insulating or cellophane tape, off its roll in the dark.

M

Mach, Ernst (1838–1916) Austrian *Physicist, Philosopher* Famous for MACH'S PRINCIPLE, Ernst Mach is also known for the MACH NUMBER. He was a skilled experimental physicist who also performed theoretical investigations and studied the philosophical foundations of physics. Mach adhered to and argued for the empiricist view that what is inherently unobservable is meaningless and has no place in physics. His research included ELECTROMAGNETISM, the DOPPLER EFFECT, OPTICS, the physiology of the sense organs, and ACOUSTICS, including the acoustic effects of a projectile in supersonic MOTION (which involve the Mach number). Mach's ideas and principle, involving the origin of mechanical INERTIA, had a profound effect on ALBERT EINSTEIN and influenced Einstein's development of the general theory of relativity. Mach studied at the University of Vienna, Austria, and held positions at the University of Graz, Austria, at Charles University of Prague, in what is now the Czech Republic, and at the University of Vienna.

See also EXPERIMENTAL PHYSICS; MECHANICS; RELATIVITY, GENERAL THEORY OF; THEORETICAL PHYSICS.

The 19th/20th-century physicist Ernst Mach performed experimental and theoretical investigations in various fields of physics, including supersonic motion. His principle involving the origin of mechanical inertia had a deep effect on Albert Einstein's development of the general theory of relativity. ***(AIP Emilio Segrè Visual Archives, Physics Today Collection)***

Mach number Named for the Austrian physicist ERNST MACH, the mach number is the ratio of the SPEED of a moving object to the speed of SOUND in the medium through which the object is moving. The term is most familiar in connection with the speed of aircraft. An airplane traveling at Mach 1.2, for instance, is traveling 20 percent faster than the speed of sound in air at the same conditions. (At ordinary temperatures on the ground, the speed of sound in air is roughly 340 meters per second [m/s].) Speeds that are less than the speed of sound, with Mach numbers smaller than 1, are called subsonic, while Mach numbers greater than 1 indicate supersonic speeds.

Mach 1, of course, designates travel at the speed of sound, also referred to as the "sound barrier." What happens at this speed is that the sound WAVES the aircraft is emitting in the forward direction do not succeed in leaving it, since the aircraft is moving at the same speed as they are. Thus the aircraft is continually immersed in those waves, which attain high INTENSITY, causing VIBRATION of the aircraft's structure. At Mach numbers greater than 1, the aircraft emits SHOCK WAVES that propagate through the air and can sometimes be heard on the ground and might even cause damage there. Such shock waves are called "sonic boom."

Mach's principle Named after the Austrian physicist ERNST MACH, Mach's principle deals with the origin of mechanical INERTIA. Mechanical inertia is expressed by ISAAC NEWTON's first and second laws of MOTION. According to the first law, in the absence of forces acting on a body (or if the forces on it cancel out), the body remains at rest or moves with constant VELOCITY (i.e., at constant SPEED in a straight line). The second law states how a body resists changes in its force-free motion, or inertial motion. Such a change is an ACCELERATION, and according to the second law, the acceleration resulting from a force is inversely proportional to a body's MASS: the greater the mass, the less change a force can effect. (Thus mass serves as a measure of a body's inertia.) *See* NEWTON'S LAWS.

The question immediately arises: in what way do inertial motion and accelerated motion differ? Since in an appropriately moving REFERENCE FRAME any motion can appear to be inertial, the question can be restated as: what distinguishes inertial reference frames—those in which Newton's laws of motion are valid—from noninertial ones? According to Newton, the difference lies in the state of motion relative to absolute SPACE. Newton assumed the existence of an absolute space (as well as an absolute TIME) that is unobservable except for its inertial effects. Inertial and accelerated motions are with respect to absolute space. Accordingly, inertial reference frames are reference frames moving at constant velocity (or at rest) with respect to absolute space. (*See* REFERENCE FRAME, INERTIAL.)

Mach's philosophical approach to science, however, did not allow entities that are inherently unobservable, including absolute space. But if there is no absolute space with respect to what is inertial motion or inertial reference frames, are the latter different from all other motions or reference frames? Considerations, such as of the fundamental meaning of motion for a single body, two bodies, and three bodies in an otherwise empty UNIVERSE, led Mach to his principle: the origin of inertia lies in all the MATTER of the universe. Since the matter of the universe consists overwhelmingly of distant stars, GALAXIES, etc., Mach's principle can also be given the formulation that the origin of inertia lies in the distant stars.

So, when the motions of the distant stars are averaged out, they define a reference frame for rest. Stated in more detail, this reference frame—let us call it the universal reference frame—is that in which the average LINEAR MOMENTUM of all the matter in the universe is zero. Then, according to Mach, inertial reference frames are those, and only those, that are at rest or moving at constant velocity relative to the just-defined reference frame. All other reference frames, which are in relative acceleration or ROTATION, are noninertial. In terms of motions of bodies, the rest, velocity, and acceleration that Newton's first two laws of motion are concerned with are relative to the universal rest frame defined by the matter in the universe.

ALBERT EINSTEIN, in his development of the general theory of relativity, was greatly influenced by Mach's thinking and principle. Nevertheless, the theory does not succeed in fully implementing the principle, in showing just *how* distant matter engenders inertial effects. Neither has Mach's principle yet been successfully implemented by other theories in a way that is generally accepted. So at present, Mach's principle can best be viewed as a guiding principle. (*See* RELATIVITY, GENERAL THEORY OF.)

maglev *See* MAGNETIC LEVITATION.

magnet Any object that, when at rest, can produce and be affected by a magnetic FIELD is a magnet. Every magnet possesses two poles, a north pole (N) and a south pole (S), and is thus a MAGNETIC DIPOLE, characterized by its magnetic dipole moment. Same-type poles (N-N, S-S) of different magnets repel each other, while opposite-type poles (N-S) mutually attract. The Earth forms a magnet whose poles wander geographically but are in the general vicinity of the geographic poles. The north pole of any magnet is defined as the pole that is attracted to the Earth's magnetic pole that is near its north geographic pole. So the Earth has a *south* magnetic pole at its *north* geographic pole and vice versa. A magnetic pole cannot be isolated; poles always appear in north-south pairs. If a magnet is cut in two in an attempt to isolate its poles, each part is found to possess a pair of poles. (*See* GEOMAGNETISM; MAGNETISM; MOMENT, MAGNETIC DIPOLE.)

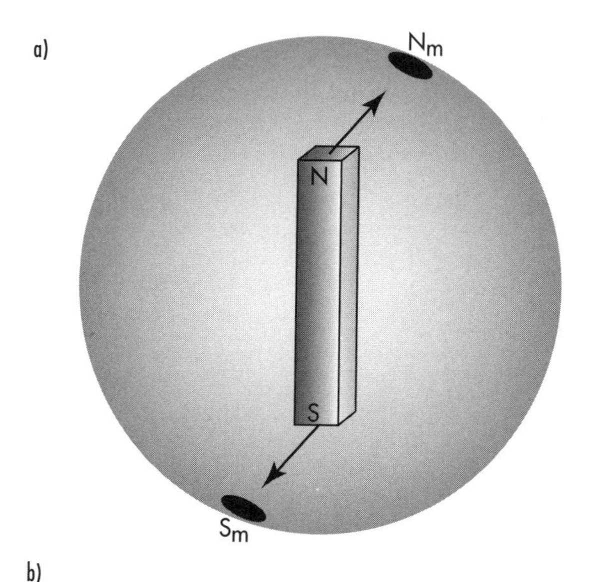

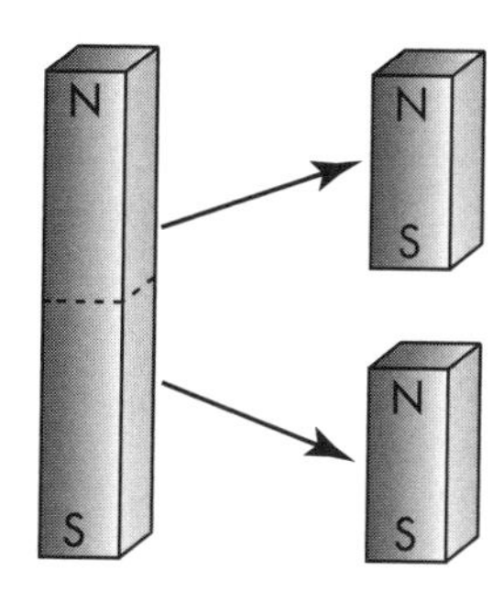

(a) A magnet possesses two poles, a north pole, N, and a south pole, S. The magnet's north pole is attracted to the Earth's north geomagnetic pole N_m (near the north geographic pole) and its south pole to the Earth's south geomagnetic pole S_m (near the south geographic pole). (b) A magnet's poles cannot be separated. When a magnet is cut into two parts, each part is found to have a pair of north and south poles.

The source of a magnet's magnetism is ultimately either an electric CURRENT, the intrinsic magnetism of ELECTRONs in ATOMs, or a combination of both. In the first case, a current in a coil becomes a magnet, an ELECTROMAGNET. The second case is represented by a permanent magnet, in which the aligned magnetic dipole moments of the material's atoms, which are due to the atoms' possessing unpaired electrons, combine to produce the magnetic field of the magnet in an effect that is an aspect of FERROMAGNETISM. The two sources of magnetism come into combined play when ferromagnetic material is used for the core of an electromagnet. Then the magnetic field of the current induces the core to become a magnet. (*See* ELECTRICITY; INDUCTION.)

magnetic dipole A pair of equal-magnitude and opposite magnetic poles at some distance from each other, or any situation that is equivalent to that, is known as a magnetic dipole. A bar magnet is a straightforward example of a magnetic dipole. The term includes all magnets, as well as electric CURRENT configurations—typically loops and coils—that behave like magnets. ATOMs might also be magnetic dipoles. That is brought about by the net ANGULAR MOMENTUM of their ELECTRONs or by their possession of unpaired electrons (or by both). In the former case, electrons in the atom act like electric current loops. In the latter, an unpaired electron itself—like many other kinds of ELEMENTARY PARTICLE—behaves like a magnetic dipole. A magnetic dipole is characterized by its magnetic dipole moment. (*See* ELECTRICITY; ELECTROMAGNET; MAGNET; MAGNETISM; MOMENT, MAGNETIC DIPOLE.)

Note that there are no real magnetic poles. So a magnet, or any other magnetic dipole, is always the result of an electric CURRENT or currents, either macroscopic or atomic, or of the intrinsic magnetic dipoles of elementary particles.

A uniform magnetic FIELD exerts no net force on a magnetic dipole. It does exert a TORQUE, whose magnitude depends on the orientation of the dipole in the field. In a nonuniform magnetic field, however, a net force can act on a magnetic dipole in addition to a torque.

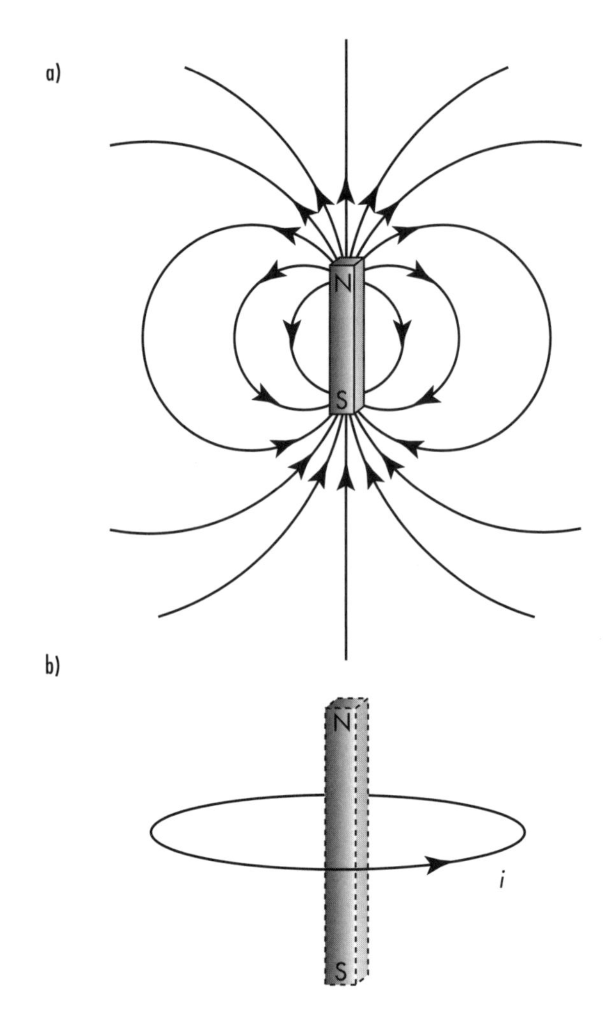

A magnetic dipole is a pair of separated equal-magnitude and opposite magnetic poles. (a) A bar magnet serves as a typical example of a magnetic dipole. Its north and south poles are labeled N and S, respectively. A number of magnetic field lines are shown for the dipole. (b) An electric current loop *i* is equivalent to a magnetic dipole, shown in dashed lines. The dipole's polarity relates to the direction of the current in the loop as shown in the figure.

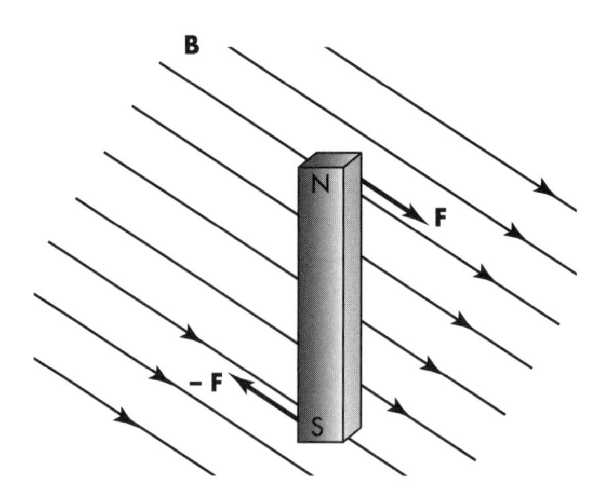

A uniform magnetic field exerts no net force on a magnetic dipole. Rather, the dipole's poles are affected by equal-magnitude and opposite forces **F** and –**F**, with the north pole N pulled in the direction of the field **B** and the south pole S pulled in the opposite direction. That results in a torque whose magnitude depends on the dipole's orientation in the field. Some of the field lines are indicated.

magnetic dipole moment *See* MOMENT, MAGNETIC DIPOLE.

magnetic domain A region of a ferromagnetic material in which all the atomic magnetic dipole moments are aligned with each other is called a magnetic domain. Their volumes fall in the range 10^{-12} to 10^{-8} cubic meter (m^3), and each contains some 10^{17} to 10^{21} ATOMs. The boundaries between adjacent domains are termed domain walls. Ferromagnetic materials below their Curie TEMPERATURE spontaneously form magnetic domains throughout all their volume. The mechanism of domain formation is as follows. Each atom of a ferromagnetic material has a relatively large number of unpaired ELECTRONs, on the average, whose MAGNETIC DIPOLES (related to the electron's SPIN) endow it with a considerable magnetic dipole moment. There is strong coupling among the atoms' magnetic dipole moments, which causes the atoms to spontaneously form local clusters in each of which all the magnetic dipoles are aligned. Those clusters are the magnetic domains, which are observable under a microscope. A domain serves as an elementary MAGNET. With no application of an external magnetic FIELD, the domains are oriented randomly, so their contributions to the magnetic field cancel, and the bulk material is not magnetized. A magnetic field affects domains by bringing about a movement of domain walls in such a way that those domains whose magnetic dipole moments are oriented in the direction of the field increase in volume at the expense of the other domains. A sufficiently strong field can rotate the magnetic dipoles of unfavorably oriented domains to align them with the field. (*See* CURIE, PIERRE; FERROMAGNETISM; MAGNETISM; MOMENT, MAGNETIC DIPOLE.)

a)

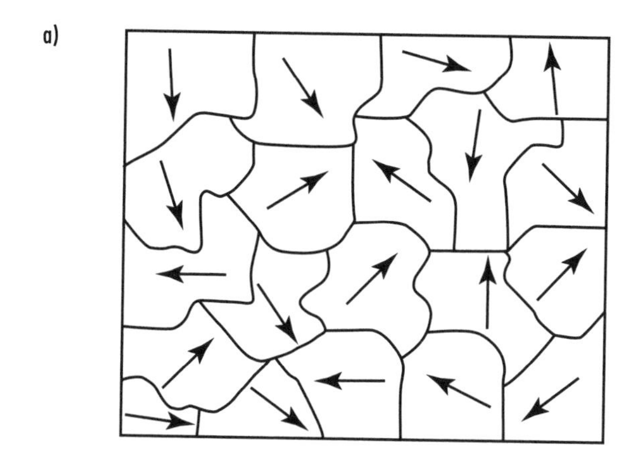

b)

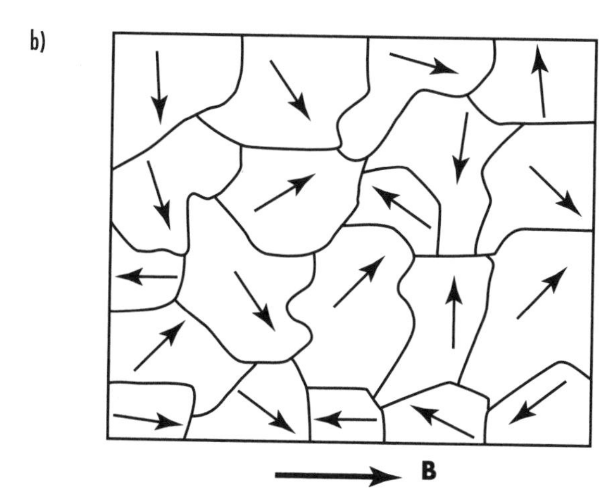

c)

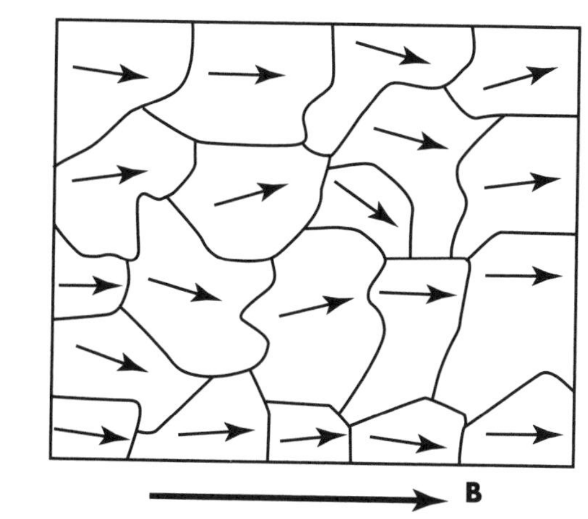

A magnetic domain is a volume in a ferromagnetic material in which all the atomic magnetic dipoles are aligned and which serves as an elementary magnet. (a) In an unmagnetized sample, the domains are randomly oriented. (b) The effect of an applied magnetic field B is to cause the domains whose magnetic dipole moment is oriented in the general direction of the field to grow at the expense of the other domains. (c) Sufficiently strong fields reorient domains to align their magnetic dipole moments with the applied field.

magnetic drag Retarding FORCEs that arise when a MAGNET and an electric CONDUCTOR are in relative motion are called magnetic DRAG forces. Examples are a flat sheet of metal being pulled between the poles of a horseshoe magnet and a bar magnet dropped down a vertical metal tube. According to FARADAY'S LAW, LENZ'S LAW, and the BIOT-SAVART LAW, the relative motion induces electric CURRENTs in the conductor, which produce magnetic FIELDs that repel the magnet. The magnet "sees" another magnet of such polarity that like poles repel each other. The WORK required to maintain motion against magnetic drag forces, or the KINETIC ENERGY that disappears when such forces are used for magnetic braking, appears as HEAT in the conductor. The induced currents cause heating due to the RESISTANCE of the conductor. (*See* ELECTRICITY; INDUCTION; MAGNETISM.)

magnetic induction Also called magnetostatic induction, magnetic induction is the production of magnetic POLARIZATION in materials by an applied magnetic FIELD. The effect is viewed as a combination of three distinct effects:

- In DIAMAGNETISM, the applied field creates atomic MAGNETIC DIPOLEs, which in turn produce an internal magnetic field that weakly opposes the applied field.
- The phenomenon of PARAMAGNETISM is the aligning of existing atomic magnetic dipoles by an applied field. In this case, the internal magnetic field weakly reinforces the applied field.
- FERROMAGNETISM is more complex. The effect involves MAGNETIC DOMAINs, which are regions of the material in which existing atomic magnetic dipoles have spontaneously aligned themselves. The applied field increases the size of those magnetic domains that are aligned with the applied field at the expense of the other domains and might also align misaligned domains. In this case, the internal magnetic field strongly reinforces the applied field.

See also ATOM; MAGNET; MAGNETISM; PERMEABILITY.

magnetic levitation (maglev) The use of magnetic FORCEs to overcome gravity is called magnetic levitation. The term is commonly used in connection with transportation, such as maglev trains, but the effect can be seen in other circumstances. The use of an ELECTROMAGNET to lift and move steel scrap is an example of magnetic levitation. MAGNETIC INDUCTION causes the steel to become a magnet and be attracted to the electromagnet. A more exotic example is the floating of a small magnet above a superconductor. This is an extreme case of MAGNETIC DRAG. As the magnet falls, or is carried, toward the superconductor, electric CURRENTs are induced in the superconductor, which produce magnetic FIELDs that repel the magnet. In an ordinary CONDUCTOR, due to its electric RESISTANCE, such induced currents disappear when the motion of the magnet ceases. In a superconductor, however, the induced currents continue to flow even when the magnet comes to rest. Thus, there exists an equilibrium height above the surface of the superconductor, where the magnet can oscillate vertically or remain at rest. (*See* BIOT-SAVART LAW; ELECTRICITY; FARADAY'S LAW; INDUCTION; LENZ'S LAW; MAGNETISM; MEISSNER EFFECT; OSCILLATION; SUPERCONDUCTIVITY.)

In the application of magnetic levitation to vehicles, there are numerous and various proposed designs. They involve combinations from among permanent magnets, electromagnets, superconducting electromagnets (in which the magnet coils are superconducting), and conductors in various configurations. The design might involve one or more among a diverse range of guideways, lower rails, and overhead rails. In addition to levitation, maglev systems can also provide noncontact magnetic guidance and propulsion. Maglev can totally eliminate the FRICTION that is involved in conventional propulsion and guidance, as well as the rolling friction of wheeled travel. There still remain the DRAG forces caused by air, however. These can, at least in principle, be greatly reduced by operating the vehicle in an evacuated tube. Propulsion might then be achieved by introducing a small air pressure behind the rear of the vehicle. But maglev inherently creates magnetic drag, which is absent in conventional modes of transportation. Magnetic levitation offers the potential for achieving considerably higher vehicular SPEEDs and significantly greater ENERGY efficiency than in conventional transportation. (*See* VACUUM.)

magnetic monopole A hypothetical entity that possesses the characteristics of a single pole of a MAGNET (i.e. possesses magnetic CHARGE) is called a magnetic monopole. Magnetic charge would play the role in MAGNETISM that is analogous to the role of electric charge in ELECTRICITY. It would come in two varieties, north (N) and south (S). Like monopoles (N-N, S-S) would repel each other, while unlike monopoles (N-S) would attract. Monopoles would affect and be affected by the magnetic FIELD. Moving monopoles would produce electric fields and would be affected by them. As far as is known, however, magnetic monopoles do not exist in nature. Searches for them have come up negative. The nonexistence of magnetic charge might well seem odd in light of the existence of electric charge and the fact that MAXWELL'S EQUATIONS, which describe ELECTROMAGNETISM, treat electricity and magnetism equivalently in all other ways.

magnetic resonance imaging (MRI) Magnetic resonance imaging is a noninvasive imaging technique, widely used in medicine. It is based on nuclear magnetic resonance. This effect occurs for nuclei that possess SPIN (and accordingly a magnetic dipole moment) and are located in a magnetic FIELD. In that situation, the magnetic dipole moment VECTOR tends to align itself with the magnetic field, but it can precess about the field's direction, much as a GYROSCOPE precesses when a torque acts upon it. The FREQUENCY of precession, or resonant frequency, is determined by the magnitudes of the magnetic dipole moment and the magnetic field. Such a system absorbs energy from radio WAVEs at its resonant frequency that impinge upon it and also emits such waves. (*See* ELECTROMAGNETISM; MAGNETISM; MOMENT, MAGNETIC DIPOLE; NUCLEUS; RESONANCE.)

For imaging, the object being imaged, such as part of a human body, is immersed in a very strong magnetic field. That necessitates using a very strong ELECTROMAGNET, which might be a superconducting magnet (an electromagnet with superconducting coils). A magnetic pulse "kicks" the magnetic nuclei of the object into precession, and the radio waves they emit are detected. From the frequency of the waves, the identity and environment of the emitting nuclei can be inferred. Careful control of the magnetic field allows the location of the nuclei to be determined as well. In this way, a three-dimensional image of the object can be constructed. A computer performs the mathematical analysis required for that. (*See* DIMENSION; SUPERCONDUCTIVITY.)

magnetism At its most fundamental, magnetism is the effect that electric CHARGEs in relative motion have on each other over and above their mutual electric attraction or repulsion. In other words, magnetism is that component of the mutual influence of electric charges that depends on their relative VELOCITY. Magnetic FORCES are described by means of the magnetic FIELD, which is a VECTOR quantity that possesses a value at every location in SPACE and can vary over TIME. The magnetic field is considered to mediate the magnetic force, in the sense that any moving charge, as a *source charge,* contributes to the magnetic field, while any moving charge, as a *test charge,* is affected by the field in a manner that causes a force to act on the charge. (*See* ELECTRICITY.)

Force and Field

The force F in newtons (N) that acts on an electric charge q in coulombs (C) as test charge moving with velocity v in meters per second (m/s) at a location where the magnetic field has the value B in teslas (T) is given by:

$$\mathbf{F} = q\mathbf{v} \times \mathbf{B}$$

This formula gives the magnitude and the direction of the force. Its direction is perpendicular to both the velocity and the magnetic field. Since it is perpendicular to the velocity (i.e., to the DISPLACEMENT), the magnetic force performs no work, and the ACCELERATION caused by the magnetic force changes only the direction

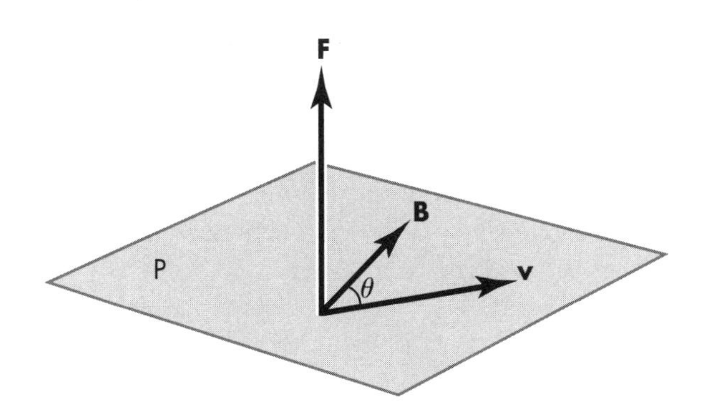

The force, F, on a moving point charge due to a magnetic field is given by F = *q*v × B. Here *q*, v, and B denote, respectively, the amount of charge, the velocity of the charge, and the magnetic field at the location of the charge. Plane P contains v and B. For positive charge, F is perpendicular to P in the sense shown. If *q* is negative, the force is in the opposite direction. The smaller angle between v and B is denoted *θ*. The magnitude of the force equals |*q*|*vB* sin *θ*, where *v* and *B* denote the magnitudes of v and B, respectively.

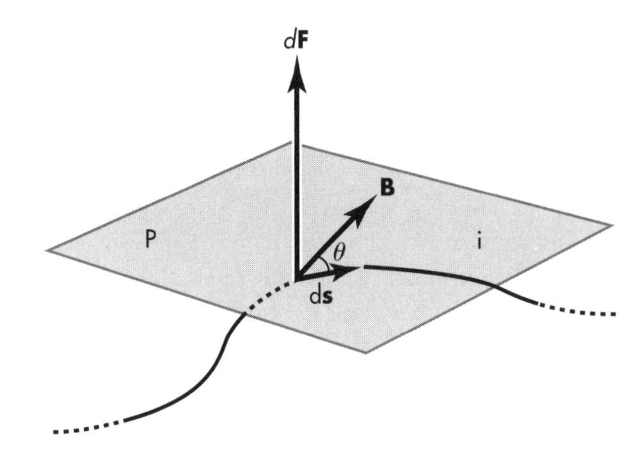

The infinitesimal force, *d*F, on an infinitesimal length of current-carrying conductor due to a magnetic field is given by *d*F = *i d*s × B, where *i*, *d*s, and B denote, respectively, the current, the infinitesimal directed element of conductor length (pointing in the sense of the current), and the magnetic field at the location of the conductor element. Plane P contains *d*s and B. *d*F is perpendicular to P in the sense shown. The smaller angle between *d*s and B is denoted *θ*. The magnitude of the force equals *i ds B* sin *θ*, where *ds* and *B* denote the magnitudes of *d*s and B, respectively.

of the velocity but not the SPEED. The magnitude of the force, F, is:

$$F = |q|vB \sin \theta$$

where v and B denote the magnitudes of the velocity and the magnetic field, respectively, and θ is the smaller angle between v and B.

For an electric CURRENT in a magnetic field, the infinitesimal force $d\mathbf{F}$ acting on an infinitesimal directed element of CONDUCTOR length $d\mathbf{s}$ in meters (m), carrying current i in amperes (A) and pointing in the sense of the current (i.e., the force acting on an infinitesimal current element $i\,d\mathbf{s}$), is:

$$d\mathbf{F} = i\,d\mathbf{s} \times \mathbf{B}$$

The magnitude of this force, dF, is:

$$dF = i\,ds\,B \sin \theta$$

where ds denotes the magnitude of $d\mathbf{s}$ (i.e., the infinitesimal length of the conductor) and θ is the smaller angle between the direction of the current (the direction of $d\mathbf{s}$) and B. Accordingly, the force F on a finite-length straight conductor, represented by the vector L in meters (m), is given by:

$$\mathbf{F} = i\mathbf{L} \times \mathbf{B}$$

with the current flowing in the direction of **L**. The magnitude of the force is:

$$F = iLB \sin \theta$$

where L denotes the length of the conductor.

Field Production

As mentioned previously, the source of the magnetic field is moving charges. The BIOT-SAVART LAW gives the infinitesimal magnetic field, $d\mathbf{B}$, produced by the infinitesimal element of current $i\, d\mathbf{s}$:

$$d\mathbf{B} = \frac{\mu_0}{4\pi} \frac{i\, d\mathbf{s} \times \mathbf{r}}{r^3}$$

Here **r** is the vector pointing from the current element to the point where the magnetic field is being calculated (the field point) whose magnitude, r, is the distance between the two points in meters (m), and μ_0 is the magnetic PERMEABILITY of the VACUUM, whose value is $4\pi \times 10^{-7}$ T·m/A. The magnitude of $d\mathbf{B}$, dB, is given by:

$$dB = \frac{\mu_0}{4\pi} \frac{i\, ds \sin \theta}{r^2}$$

where θ denotes the angle between the current element (i.e., between $d\mathbf{s}$) and **r**.

An example of the use of the Biot-Savart law is finding the magnitude of the magnetic field at perpendicular distance r from a long straight wire carrying current i. Integration gives the result:

$$B = \frac{\mu_0}{4\pi} \frac{i}{r}$$

The Biot-Savart law can also be expressed in terms of the magnetic field produced by an electric charge q moving with velocity **v**, giving:

$$B = \frac{\mu_0}{4\pi} \frac{q\, d\mathbf{v} \times \mathbf{r}}{r^3}$$

where **r** is the distance vector from the charge to the field point.

Magnetic Dipole and Field Lines

An electric current flowing in a loop is a MAGNETIC DIPOLE and behaves like a MAGNET, effectively possessing a pair of north and south magnetic poles. It is characterized by a magnetic dipole moment, $\boldsymbol{\mu}$, whose SI UNIT is ampere·meter² (A·m²). A TORQUE, $\boldsymbol{\tau}$, acts on a magnetic dipole in a magnetic field and is given by:

$$\boldsymbol{\tau} = \boldsymbol{\mu} \times \mathbf{B}$$

where torque is in units of newton·meters (N·m). It tends to align the magnetic dipole moment with the field. The POTENTIAL ENERGY, E_p, in joules (J), of a magnetic dipole in a magnetic field is:

$$E_P = -\boldsymbol{\mu} \cdot \mathbf{B}$$

It is lowest when the dipole moment is aligned with the field and highest when the dipole moment points in the opposite direction. The potential-energy difference between the two orientations equals $2\mu B$, where μ denotes the magnitude of $\boldsymbol{\mu}$. (*See* MOMENT, MAGNETIC DIPOLE.)

A magnetic field line is a directed line in space whose direction at every point on it is the direction of the magnetic field at that point. Magnetic field lines make a very useful device for describing the spatial configurations of magnetic fields. For example, the field lines for the magnetic field produced by a single positive point charge moving in a straight line are circles centered on the line of motion, whose planes are perpendicular to that line. The sense of the circles is determined by the right-hand rule, whereby if the extended thumb of the right hand lies along the line of motion and points in the direction of motion, the other fingers curve in the sense of the circles. The field lines associated with a magnetic dipole, including a magnet, emanate from the north magnetic pole and enter the south pole. Only a single magnetic field line can pass through any point in space.

Energy is stored in the magnetic field. The energy DENSITY (i.e., the energy per unit volume), in joules per cubic meter (J/m³) at a location where the magnitude of the magnetic field is B, is given by:

$$\text{Energy density of magnetic field} = \frac{B^2}{2\mu_0}$$

Magnetism and electricity are two aspects of ELECTROMAGNETISM, which subsumes the two and, moreover, has them affecting each other.

See also DIAMAGNETISM; FERROMAGNETISM; PARAMAGNETISM.

magnetohydrodynamics The study of systems consisting of electrically conducting FLUIDs (i.e., a conducting LIQUID or GAS) and magnetic FIELDs is called magnetohydrodynamics. The conducting fluid might be a PLASMA (an ionized gas), a liquid METAL, or an ionic solution, as examples. (*See* CONDUCTION; ELECTRICITY; ION; IONIZATION; MAGNETISM.)

An example of a plasma is an extremely hot gas in which the collisions among its ATOMs or MOLECULES are sufficiently energetic to knock ELECTRONs out of them and thus to ionize them. The electrons and the positively charged ions can form electric currents, so a plasma is a CONDUCTOR. A metal, which is an intrinsic conductor, can be turned into a conducting liquid by MELTING it. The chemical element mercury is unique among metals in that it is a liquid at ordinary TEMPERATUREs. A common example of an ionic solution is seawater. Its solutes consist mostly of table salt, NaCl, which dissolves in water as $Na^+ + Cl^-$. The ions, being charged, can form electric currents, so such a solution is a conductor. (*See* ENERGY.)

magnification The ratio of the size of an object's image to the size of the object is termed magnification. Sometimes an algebraic sign is attached to the magnification, whereby a positive number indicates that the image has the same orientation as the object, while a negative magnification shows that the image is oriented oppositely. When the absolute value of the magnification is greater than, equal to, or less than one, there is, respectively, an increase in size (which is magnification in the strict sense), no change in size, or reduction. The following table summarizes the meaning of the ranges of values of magnification.

Meaning of Magnification, *M*

If	then
$\lvert M\rvert > 1$	the image is larger than the object
$\lvert M\rvert = 1$	the image is the same size as the object
$\lvert M\rvert < 1$	the image is smaller than the object
$M > 0$	the image has the same orientation as the object
$M < 0$	the image has the opposite orientation to the object

For an imaging system with more than a single stage, such as an optical system involving two or more LENSes, the overall magnification equals the product of the individual magnifications of the separate stages. In a two-stage optical system, for example, let the first-stage magnification be $M_1 = 60$. Thus, the first stage creates an image of the object that is 60 times larger and in the same orientation. Suppose the second-stage magnification is $M_2 = -0.5$. That indicates the second stage treats the image from the first stage as an object and produces an image of it that is half the size and oppositely oriented. The total magnification of the system is:

$$M = M_1M_2 = 60 \times (-0.5) = -30$$

meaning that the final image is larger than the object by a factor of 30 and is oriented oppositely to the object. (*See* OPTICS.)

See also LENS; MIRROR.

Malus's law Named for the French physicist Étienne-Louis Malus (1775–1812), Malus's law deals with the reduction in INTENSITY of plane-polarized light upon passing through a second polarizer. (Although any device that polarizes light is called a polarizer, sometimes a polarizer that acts on already-polarized light is called an analyzer, while the device that polarizes the beam is called a polarizer; the polarizer polarizes, while the analyzer tests the light as to its polarization.) Let a plane-polarized beam of light of intensity I_0 impinge on a polarizer (analyzer) whose transmission axis is oriented at angle θ with respect to the transmission axis of the

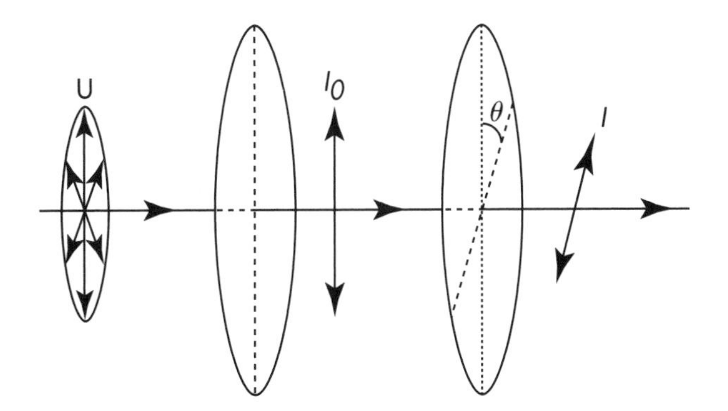

Malus's law deals with the reduction in intensity of a plane-polarized beam of light upon passing through an additional polarizer, called an analyzer in this capacity, whose transmission axis is rotated with respect to the beam's direction of polarization. A beam of unpolarized light U is propagating from left to right. It passes through a vertically oriented polarizer P and becomes vertically polarized with intensity I_0. The beam then passes through an analyzer A whose transmission axis forms angle θ with that of the polarizer and thus with the polarization direction of the beam impinging on it. The light leaving the analyzer has intensity I. Malus's law states that $I = I_0 \cos^2 \theta$.

polarizer that is polarizing the beam (i.e., with respect to the direction of polarization of the beam). Malus's law states that the intensity of the transmitted beam, I, is given by

$$I = I_0 \cos^2 \theta$$

When the transmission axes are parallel ($\theta = 0$), there is no decrease in intensity. When the axes are perpendicular ($\theta = 90°$), called "crossed," no light is transmitted. The formula is valid strictly only for an ideal polarizer (analyzer), one that is perfectly transparent and does not absorb light. (*See* ABSORPTION; POLARIZATION.)

many-body theory The MECHANICS of an arbitrary number of bodies in mutual INTERACTION is called many-body theory. In classical mechanics, exact solutions of the EQUATIONS OF MOTION can be obtained only for two bodies. For three bodies or more, approximate solutions can be obtained only for situations that allow simplification of the mathematics. As an example, if one of the bodies is very much more massive than all the others combined, the bodies are not too close together, and they interact via GRAVITATION, then the situation is similar to the SOLAR SYSTEM. In such a case, KEPLER'S LAWS can be shown to hold. In general, however, the motion of many-body systems must be solved by numerical methods. Computers are used for such calculations. The constant increase over time in available computing speed allows the treatment of systems of greater and greater numbers of bodies. The evolution of GALAXIES and COLLISIONs between galaxies, for example, are being studied by computer modeling that involves computing the motions of the many stars constituting the models. (*See* CLASSICAL PHYSICS.)

In QUANTUM MECHANICS, many-particle systems are described by many-particle wave functions, which allow the calculation of the probabilities of the various possible configurations of the system. Systems of identical particles obey BOSE-EINSTEIN STATISTICS or FERMI-DIRAC STATISTICS, depending on whether the particles are BOSONs or FERMIONs, respectively.

See also MAXWELL-BOLTZMANN STATISTICS; MAXWELL DISTRIBUTION; STATISTICAL MECHANICS.

mass The notion of mass is a complex one. That is because mass has a number of aspects, which seem to be quite different in concept, yet are apparently interconnected in ways that are still not fully understood. The SI UNIT of mass is the kilogram (kg). Here are various aspects of mass.

Amount of matter. Mass serves as a measure of the amount of MATTER. For instance, a 6-kilogram (kg) block of iron contains twice as many iron ATOMs as does a 3-kg block.

Inertial mass. This is mass as a measure of the INERTIA of a body, as a measure of a body's resistance to changes in its VELOCITY (or LINEAR MOMENTUM) (i.e., its resistance to ACCELERATION). ISAAC NEWTON's second law of MOTION deals with this aspect of mass. (*See* NEWTON'S LAWS.)

Gravitational mass. Mass measures participation in the gravitational INTERACTION. In this regard, it is further subcatagorized as active or passive. Active gravitational mass is a measure of the capability of whatever possesses it to serve as a source of the gravitational FIELD. Passive gravitational mass measures the degree to which the gravitational field affects its possessor. ALBERT EINSTEIN's general theory of relativity is a theory of GRAVITATION and thus deals with gravitational mass. (*See* RELATIVITY, GENERAL THEORY OF.)

Energy. According to Einstein's special theory of relativity, mass and ENERGY are related and can even be considered to reveal two faces of the same essence. The well-known formula relating mass and energy is:

$$E = mc^2$$

where m denotes the mass in kilograms (kg), E the energy in joules (J), and c the SPEED OF LIGHT, whose value is 2.99792458×10^8 meters per second (m/s). What this means is that every system possesses an intrinsic internal energy due to its mass, its MASS ENERGY. Some of this energy might be convertible to other forms of energy, such as through nuclear processes. All of the mass energy might be converted to a nonmaterial form of energy in the process of mutual annihilation of an ELEMENTARY PARTICLE and its antiparticle when they collide. An ELECTRON and a POSITRON, for instance, can undergo annihilation that results in a pair of GAMMA RAYs, which are particlelike manifestations of electromagnetic energy. Alternatively, nonmaterial forms of energy might be converted to matter. A gamma ray can transform into an electron-positron pair. In all such cases, Einstein's mass-energy formula is obeyed. Conversely, all energy possesses mass. So nonmaterial forms of energy, such as photons, exhibit inertia and affect and are affected by the gravitational field. As an example, light passing close to the Sun deviates from a straight path in the Sun's gravitational field. (*See* ANTIMATTER; COLLISION; ELECTROMAGNETISM; PAIR PRODUCTION; PHOTON; NUCLEAR PHYSICS; NUCLEUS; RELATIVITY, SPECIAL THEORY OF.)

REST MASS is an intrinsic property of an elementary particle, or of any body, and is defined as the particle's mass when the particle is at rest. A massless particle, such as a photon, is one that moves at the speed of light, and only at that speed, and is never at rest. The special theory of relativity nominally assigns zero rest mass to such a particle, although its energy still possesses a mass equivalent according to the above mass-energy relation applied in reverse. Usually when mass is referred to, it is rest mass that is intended.

For the mass equivalent, m, of the total energy of a body, due both to the body's rest mass and its motion, the special theory of relativity gives the expression:

$$m = \frac{m_0}{\sqrt{1 - v^2/c^2}}$$

where m_0 denotes the body's rest mass and v is its SPEED in meters per second (m/s). For speeds that are sufficiently small compared with the speed of light, this relation takes the form:

$$m = m_0 + \frac{1}{c^2}\frac{m_0 v^2}{2} + \text{negligible terms}$$

which shows the mass equivalent of the total energy as a sum of the rest mass and the mass equivalent of the body's kinetic energy (in its low-speed form).

Masses of elementary particles are often given in terms of their energy equivalents, according to Einstein's mass-energy relation. The most common unit is the mass equivalent of one mega-electron-volt (MeV), which is 1 MeV/c^2, related to the kilogram by 1 MeV/c^2 = $1.78266173 \times 10^{-30}$ kg.

mass energy The ENERGY equivalent of MASS, according to ALBERT EINSTEIN's special theory of relativity, is known as mass energy. This equivalence is expressed by the well-known formula:

$$E = mc^2$$

where m denotes the mass in kilograms (kg), E the energy equivalent of the mass in joules (J), and c the SPEED OF LIGHT, whose value is 2.99792458×10^8 meters per second (m/s). The idea is that every system possesses an intrinsic energy due to its mass. Some of this energy might be convertible to other forms of energy, such as through nuclear processes. That is the source of nuclear energy, which is the energy equivalent of the mass difference between the interacting nuclei and the final products. This energy appears as KINETIC ENERGY of the products. (*See* NUCLEAR PHYSICS; NUCLEUS; RELATIVITY, SPECIAL THEORY OF.)

All of the mass energy might be converted to a nonmaterial form of energy in the process of mutual annihilation of an ELEMENTARY PARTICLE and its antiparticle when they collide. An ELECTRON and a POSITRON, for instance, can undergo annihilation that results in a pair of GAMMA RAYs, which are particlelike manifestations of electromagnetic energy. Alternatively, nonmaterial forms of energy might be converted to matter. A gamma ray can transform into an electron-positron pair. In all such cases, Einstein's mass-energy formula is obeyed. As an example, the mass energy of one gram of matter is 9×10^{13} joules, which is the amount of energy that a 100-megawatt electric power plant supplies in about ten days. Note, however, that this energy could be obtained and used in full only through the mutual annihilation of a half gram of MATTER with a half gram of corresponding antimatter, which is not feasible in practice. (*See* ANTIMATTER; COLLISION; ELECTROMAGNETISM; PHOTON.)

The above mass-energy relation can also be applied in reverse to give the mass equivalent of any energy.

mass spectroscopy Mass spectroscopy is a laboratory technique for the analysis of MATTER into its atomic and molecular constituents, determination of the MASSes—and thus the identities—of the constituents, and measurement of their relative abundances. The instrument used for mass spectroscopy is the mass spectrometer, whose principle of operation is as follows. The whole procedure takes place in a continuously pumped VACUUM. The sample to be analyzed is vaporized, if necessary, and ionized to positive IONS by COLLISIONS with ELECTRONs in an electron beam. The ions are accelerated through an electric potential difference, or VOLTAGE, and enter a region with a magnetic FIELD that is perpendicular to their direction of motion. The magnetic field causes their trajectories to bend by an amount that depends on, among other

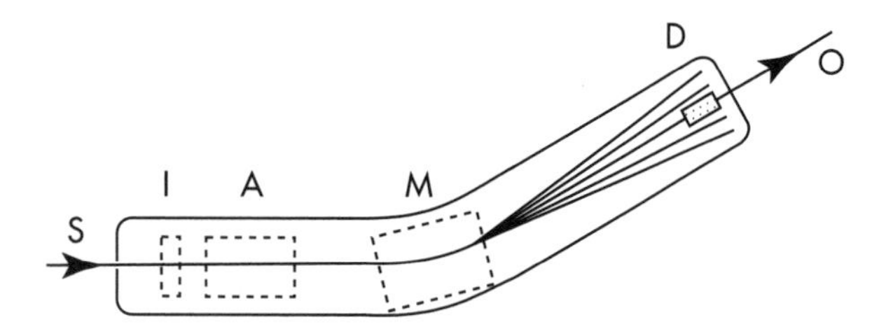

The basic structure of a mass spectrometer. The gaseous sample to be analyzed enters at S and is ionized at I. The ions are accelerated and the ion beam narrowed at A. The beam enters a magnetic field at M, where the ions are deflected by an amount that depends on their masses and the magnitude of the magnetic field. The deflected ions continue toward the detector, D, which produces an output signal O according to the flux of ions reaching it. By correlating the output signal with the magnitude of the varying magnetic field, the masses and abundances of the sample's atomic and molecular constituents can be determined. The process is carried out in vacuum.

factors, the ions' masses and the strength of the field. The field can thus be adjusted to cause only ions of a certain mass to strike a fixed detector. (The magnetic field, together with an electric field, focuses the ion beam on the detector.) The detector produces a signal indicating the INTENSITY of the beam reaching it. (*See* ATOM; ELECTRICITY; IONIZATION; MAGNETISM; MOLECULE; POTENTIAL, ELECTRIC.)

Assuming singly charged ions, the KINETIC ENERGY that an ion of mass m gains upon acceleration through a voltage V is given by:

$$mv^2/2 = eV$$

where v denotes the SPEED of the ion and e is the magnitude of the electric CHARGE of the electron, whose value is $1.602176462 \times 10^{-19}$ coulomb (C). Here m is in kilograms (kg), v is in meters per second (m/s), and V is in volts (V). The magnetic field causes the ion to move along a circular arc whose radius of curvature, r, in meters (m) relates to the other quantities through the identity of the magnetic force, evB, as CENTRIPETAL FORCE, mv^2/r, giving:

$$evB = mv^2/r$$

where B is the magnitude of the magnetic field in teslas (T). Eliminating v from the two equations, we obtain:

$$m = \frac{er^2}{2V} B^2$$

Since r is fixed by the structure of the mass spectrometer, then for a given accelerating voltage, the mass of the ions striking the detector is proportional to the square of the magnitude of the magnetic field that is bending the ion beam. By varying the magnetic field and correlating it with the signals from the detector, the masses and relative abundances of the ions making up the beam can be found.

Mass spectroscopy, in this and in other variations, is used for such tasks as measuring the mass of NUCLEI, analyzing the composition of materials, and separating ISOTOPEs and analyzing their abundances.

matter At its most general, matter is whatever consists of, though not necessarily only of, matter particles. Matter particles are the FERMIONs among the ELEMENTARY PARTICLEs, carrying half-integer values of SPIN, i.e., 1/2, 3/2, etc. Ordinary matter is in the form of ATOMs, MOLECULEs, and IONs. Most of the MASS of matter resides in NUCLEI, which are usually considered as being formed from NUCLEONs (i.e., from PROTONs and NEUTRONs). MESONs and PHOTONs, too, are present in nuclei, as carriers of the STRONG INTERACTION and ELECTROMAGNETIC INTERACTION, respectively, which act among the nucleons and affect the structure and stability of nuclei. ELECTRONs are an additional component of ordinary matter. They are not very localized. In atoms and ions, they are bound to particular nuclei. In molecules, they might be part of particular atoms or bound to the molecule as a whole. In SOLIDs, electrons might be bound only to the volume of the material. In PLASMAs, there are electrons that are not bound at all.

The term *exotic matter* refers to other types of matter from ordinary matter, such as might exist under extreme conditions. As an example of exotic matter, the final stage of evolution of a star with a mass of 8–25 times the mass of the Sun is a NEUTRON STAR, which is a single gigantic nucleus held together by GRAVITATION. More generally, it is widely thought that much of the matter in the UNIVERSE is not in the form of ordinary matter, although, if that is true, it is not clear what kind of matter it might be. (*See* ASTROPHYSICS; COSMOLOGY.)

Matter is also used to distinguish from ANTIMATTER, which is a form of matter that is a kind of "mirror image" of ordinary forms of matter.

See also STATE OF MATTER.

Physics and Mathematics: A Remarkable Partnership

It is mostly taken for granted that the laws and theories of physics are expressed in mathematical terms, that physics—and science in general—is couched in the language of mathematics. For an example, we might reach back to JOHANNES KEPLER's laws of planetary MOTION. They are indeed formulated in terms of mathematics (including geometry), which include ellipses, areas, ratios, powers, and equalities. Then we might consider ISAAC NEWTON's laws of motion and law of GRAVITATION, which form a theory—i.e., an explanation—of Kepler's laws. They, too, are stated in mathematical language, as they speak of zero, straight lines, constant values, proportionality and inverse proportionality, magnitudes, directions, and more. Similar examples go on and on, through all fields of physics and through all times. (*See* KEPLER'S LAWS; NEWTON'S LAWS.)

At its most fundamental, what is going on might be embodied in this archetypal example. Physicists study some phenomenon, say an object in FREE FALL, and make a finite number of measurements. The number of measurements might be large, but it is still finite. The measurements that are involved possess limited precision, which might be high, but it is still limited. After careful study, the physicists might notice that the numbers more or less match a very simple mathematical relation. In the case of free fall from initial rest, for example, the distance, y, that the falling object falls and the time, t, during which the object covers that distance seem more or less to obey the mathematical relation

$$y = (1/2)gt^2$$

where g is a constant that has the same value for all drops of all the bodies that were dropped.

Consider what has happened so far. The researchers performed measurements that gave numerical results. Instead, they might have recorded their thoughts about the falling objects, or they could have preferred to record their emotional reactions. They might have considered the comments of the bystanders or the "state of the union" during the objects' falls. But no, they aimed for numbers. Why? Then, having numbers to work with, they looked for simple mathematical relations among them. More complicated relations might well have better matched their numbers, passing through all points on their graphs rather than only passing near them. Yet, they preferred to match to a simple relation, even if the match was less than perfect. Why?

But that is only the start. What the physicists next do requires terrific nerve. They take the simple mathematical relation they have found—which is based on an approximate matching to a finite number of measurements of limited precision—and propose it as a law that should precisely hold for all falls of all objects. What justifies their belief that what they are doing makes sense? Moreover, not only does it make sense, but it turns out to be correct! How is it that nature is describable by mathematical relations at all, and furthermore, by *simple* ones?

We are seeing here what EUGENE PAUL WIGNER called "the unreasonable effectiveness of mathematics in the natural sciences." Why does mathematics work so well in understanding the world? People have given thought to this issue, but no decisive and generally accepted answer has been forthcoming. Some say, "God is a mathematician." But that is no answer. It simply pushes the question into another domain. The obvious response is, "*Why* is God a mathematician?" Others suggest that scientists have *chosen* to treat nature quantitatively and mathematically, while alternative ways of approaching nature might be—some say, are—equally, if not more, successful. Perhaps. However, while the mathematical approach can evaluate its clear success by definite, objective criteria, other approaches do not seem to be able to offer anything close to definite, objective criteria.

The following could serve as a not unreasonable, very partial attempt to answer the question. Start with the assertion that scientists in general, and physicists in particular, use mathematics because that is the only tool they have, that is how their brains work. This statement is quite reasonable. Just ask a physicist if he or she can suggest any other way of dealing with nature that might lead to meaningful understanding. More than likely, your question will evoke a very perplexed expression and no answer. Let us assume that, indeed, this is the way our brains work. Now use an argument based on evolution. The way our brains work cannot be very much out of tune with reality. Otherwise our ancestors could not have survived to pass on their genes. Furthermore, evolution refines the way human brains work, thereby improving humans' evolutionary advantage. Presumably, that means becoming more and more in tune with reality. This leads to the conclusion that the way we do physics—mathematically—is the best way of dealing with nature, or at least very close to being the best way.

So it is plausible to think we now understand that nature is best dealt with mathematically, that there is no point in even considering alternative modes of approach. If that is correct, it is a small step in the right direction. Yet, there still remains the gaping chasm to cross: *why* is nature best dealt with mathematically?

(continues)

Physics and Mathematics: A Remarkable Partnership *(continued)*

Here is another line of reasoning. It is not based on the way our brains work or on evolution, but rather on the character of science, on the fact that science is the human endeavor of attempting to achieve an *objective* understanding of nature. Our demand of objectivity leads immediately to the necessity of a rational approach to nature, which means relying on the methods and tools of logic. Any other approach would involve subjectivity, which must be avoided in science to the extent possible. Objectivity also leads us to the quantification of—i.e., assigning numerical values to—physical effects. Any other way of considering, recording, reporting, or otherwise treating physical effects would involve subjectivity. Mathematics is the branch of logic that deals with numerical quantities and with relations among them. The conclusion of this line of reasoning is that we use mathematics in science—and in physics, in particular—because of the objectivity of science.

If we accept this argument, then, since science works so well, our question reduces to: why does the physical world possess objective aspects? We are thus led to a very fundamental question of philosophy and will leave it there.

The main point is that, while the enormous importance of mathematics in physics is indisputable, the reason for its importance is not at all obvious. Is it due, at least in part, to the properties of the human mind? Or could it result from the objective nature of science? And what is it telling us about nature itself and about reality? Clearly, there is still much to learn.

matter wave The wave corresponding to the WAVE aspect of MATTER, according to QUANTUM THEORY, is called a matter wave. WAVE-PARTICLE DUALITY states that not only do waves possess a particlelike aspect (such as PHOTONS being the particles corresponding to electromagnetic RADIATION), but matter particles have a wave character, described by a wave function. (*See* ELECTROMAGNETISM; QUANTUM MECHANICS.)

Quantum mechanics gives relations between the ENERGY and LINEAR MOMENTUM of the particles, on the one hand, and the FREQUENCY and WAVELENGTH of the waves associated with the particles, on the other. For a beam of particles of individual energy E in joules (J), the frequency, f, in hertz (Hz) of the associated wave is given by:

$$f = E/h$$

where h is the PLANCK CONSTANT and has the value $6.62606876 \times 10^{-34}$ joule·second (J·s). If the magnitude of the linear momentum of each particle is p in kilogram·meters per second (kg·m/s), the wavelength of the matter wave, λ, in meters (m) is:

$$\lambda = h/p$$

The propagation SPEED, or PHASE SPEED, v_p, of the matter wave associated with a beam of particles of individual energy E and magnitude of momentum p is given by:

$$v_p = f\lambda = E/p$$

Using the appropriate relativistic relations, we obtain:

$$v_p = c^2/v$$

where v denotes the speed of the beam particles and c is the SPEED OF LIGHT, whose value is 2.99792458×10^8 meters per second (m/s). Since a particle's speed is always less than the speed of light, the phase speed of matter waves is accordingly always greater than c. That is not in violation of the special theory of relativity, since neither matter nor information is moving faster than light. (*See* RELATIVITY, SPECIAL THEORY OF.)

It can be shown that:

$$v_p = c\sqrt{1+(mc\lambda/h)^2}$$

where m denotes the MASS of an individual beam particle in kilograms (kg). Note that the phase speed depends on the wavelength, so matter waves exhibit DISPERSION. A localized wave effect, such as an individual particle, is described by a wave packet, which is an INTERFERENCE pattern of limited spatial extent produced by overlapping waves. The propagation speed of a wave packet is the GROUP SPEED, v_g, given by:

$$v_g = v_p - \lambda\frac{dv(\lambda)}{d\lambda} = \frac{c}{\sqrt{1+(mc\lambda/h)^2}} = c^2/v_p = v$$

The speed of the matter wave packet, v_g, indeed equals the particle speed, v.

One use of matter waves is in ELECTRON MICROSCOPY.

Maxwell, James Clerk (1831–1879) British *Physicist* One of the greatest theoretical physicists of the 19th century, James Clerk Maxwell is best known today for his theory of ELECTROMAGNETISM and his work in the KINETIC THEORY of GASes. He graduated from Cambridge University, England, in 1854. Subsequently, Maxwell held positions in Scotland and in England. His last position, starting 1871, was that of professor of physics at Cambridge University, where he remained until his early death. Maxwell's investigations of electromagnetism resulted in a theory in the form of a set of equations, called MAXWELL'S EQUATIONS, couched in terms of the electric and magnetic FIELDs. Among other effects, the equations predicted the existence of WAVEs that travel at the SPEED OF LIGHT. Accordingly, Maxwell correctly proposed that light is indeed an electromagnetic wave. His work in STATISTICAL MECHANICS and the kinetic theory of gases resulted in MAXWELL-BOLTZMANN STATISTICS for gas MOLECULEs, the MAXWELL DISTRIBUTION of molecular speeds, and clarification of the nature of HEAT. Maxwell's other fields of research included color vision and the structure of the rings of Saturn.

See also ELECTRICITY; MAGNETISM; THEORETICAL PHYSICS.

James Clerk Maxwell was a 19th-century theoretical physicist who discovered the fundamental laws of classical electromagnetism and furthered understanding in the kinetic theory of gases. ***(Original photograph in the possession of Sir Henry Roscoe, courtesy AIP Emilio Segrè Visual Archives)***

Maxwell-Boltzmann statistics Named for the British physicist JAMES CLERK MAXWELL and Austrian physicist LUDWIG BOLTZMANN, Maxwell-Boltzmann statistics deals with collections of distinguishable particles in thermal EQUILIBRIUM. One result is that in a GAS that is in thermal equilibrium at absolute TEMPERATURE *T*, the number of MOLECULES per unit VOLUME that have SPEEDs in the small range from v to $v + dv$ is given by $n(v)\,dv$, where $n(v)$, the speed distribution function, is:

$$n(v) = \frac{4\pi N}{V}\left(\frac{m}{2\pi kT}\right)^{3/2} v^2 e^{-\frac{mv^2}{2kT}}$$

Here v is in meters per second (m/s), T is in kelvins (K), N denotes the total number of molecules, V is the volume of the gas in cubic meters (m^3), m is the MASS of a molecule in kilograms (kg), and k is the Boltzmann constant, whose value is $1.3806503 \times 10^{-23}$ joule per kelvin (J/K). (*See* HEAT.)

In terms of ENERGY, the number of molecules per unit volume with KINETIC ENERGY in the small range from E to $E + dE$ is $n(E)\,dE$, where the ENERGY DISTRIBUTION FUNCTION, $n(E)$, is given by:

$$n(E) = \frac{2\pi N}{V}\left(\frac{1}{\pi kT}\right)^{3/2} \sqrt{E}\, e^{-\frac{E}{kT}}$$

with E in joules (J).

The root-mean-square speed of the molecules, v_{rms}—the square root of the average of the molecules' squared speeds, $\sqrt{(v^2)_{av}}$—is given by:

$$v_{rms} = \sqrt{\frac{3kT}{m}}$$

The average kinetic energy of a molecule, $(E_k)_{av}$, is:

$$(E_k)_{av} = m(v^2)_{av}/2 = mv_{rms}^2/2 = 3kT/2$$

That is in accord with EQUIPARTITION OF ENERGY, whereby each degree of freedom—three in this case—possesses average energy $kT/2$.

See also BOSE-EINSTEIN STATISTICS; FERMI-DIRAC STATISTICS; FREE ENERGY; KINETIC THEORY; MAXWELL DISTRIBUTION; STATISTICAL MECHANICS.

Maxwell distribution Named for the British physicist JAMES CLERK MAXWELL, the Maxwell distribution is the distribution of SPEEDs among the MOLECULEs of a GAS in thermal EQUILIBRIUM, according to MAXWELL-BOLTZMANN STATISTICS. The number of molecules per unit VOLUME that have speeds in the small range from v to $v + dv$ is given by $n(v)\,dv$, where $n(v)$, the speed distribution function, is:

$$n(v) = \frac{4\pi N}{V}\left(\frac{m}{2\pi kT}\right)^{3/2} v^2 e^{-\frac{mv^2}{2kT}}$$

Here v is in meters per second (m/s), T is the absolute TEMPERATURE in kelvins (K), N denotes the total number of molecules, V is the volume of the gas in cubic meters (m^3), m is the MASS of a molecule in kilograms (kg), and k is the Boltzmann constant, whose value is $1.3806503 \times 10^{-23}$ joule per kelvin (J/K). It follows that the root-mean-square speed of the molecules, v_{rms}—the square root of the average of the molecules' squared speeds, $\sqrt{(v^2)_{av}}$—is given by:

$$v_{rms} = \sqrt{\frac{3kT}{m}}$$

See also HEAT; KINETIC THEORY.

Maxwell's equations Named for the British physicist JAMES CLERK MAXWELL, Maxwell's equations are to ELECTROMAGNETISM what NEWTON'S LAWS of motion are to MECHANICS. They provide a complete description of the behavior of the electromagnetic FIELD, where the latter comprises the electric and magnetic fields. Although Maxwell's equations were formulated well before ALBERT EINSTEIN proposed his special theory of relativity, they are fully consistent with that theory. (*See* ELECTRICITY; MAGNETISM; RELATIVITY, SPECIAL THEORY OF.)

There are four equations, in their common expression in terms of the electric and magnetic fields. In their integral form, the equations, as they apply in the VACUUM, are as follows. (The numbering of the equations is arbitrary and is introduced here merely for convenience.)

First Equation

$$\oint_{\text{closed surface}} \mathbf{E} \cdot d\mathbf{A} = (1/\varepsilon_0)\Sigma q$$

This is GAUSS'S LAW, from which COULOMB'S LAW can be derived. It states that the outward electric FLUX through an arbitrary closed surface equals the algebraic sum (i.e., taking signs into account) of electric CHARGEs enclosed by the surface, divided by the PERMITTIVITY of the vacuum. The electric flux, Φ_e, in UNITs of newton·meter² per coulomb ($N{\cdot}m^2/C$) through any surface (closed or not), is given by:

$$\Phi_e = \int_{\text{surface}} \mathbf{E} \cdot d\mathbf{A}$$

where the integration is performed over the surface. Here $\mathbf{E}$ denotes the electric field in newtons per coulomb (N/C) and $d\mathbf{A}$ is an infinitesimal VECTOR area element whose magnitude is the infinitesimal surface area in square meters (m^2) at a point on the surface and whose direction is perpendicular to the surface and toward the same side of the surface over the whole surface. (For a closed surface, $d\mathbf{A}$ points outward.) On the right-hand side, q represents an electric charge inside the closed surface in coulombs (C), and Σq is the algebraic sum of all such charges. The permittivity of the vacuum, ε_0, has the value $8.85418781762 \times 10^{-12}$ $C^2/(N{\cdot}m^2)$.

Second Equation

$$\oint_{\text{closed surface}} \mathbf{B} \cdot d\mathbf{A} = 0$$

This equation is the magnetic analog of the first equation, but the zero on the right-hand side expresses the fact that there are no magnetic charges in nature. It states that the magnetic flux, Φ_m, through an arbitrary closed surface is zero, where the magnetic flux through any surface in webers (Wb) is given by:

$$\Phi_m = \int_{\text{surface}} \mathbf{B} \cdot d\mathbf{A}$$

Here $\mathbf{B}$ represents the magnetic field in teslas (T). (*See* MAGNETIC MONOPOLE.)

Third Equation

$$\oint_{\text{closed path}} \mathbf{E} \cdot d\mathbf{s} = -d\Phi_m/dt$$

This equation states that the integral of the electric field around a closed path equals the negative of the time rate of change of the magnetic flux through the surface enclosed by the closed path. Here $d\mathbf{s}$ denotes an infinitesimal vector element of directed path length in meters (m), and time, t, is in seconds (s). The direction of positive flux is related to the sense of traversal of the closed path by the right-hand rule, as described for AMPÈRE'S LAW. This equation is the foundation of FARADAY'S LAW. It shows how changes in the magnetic field affect the electric field.

Fourth Equation

$$\oint_{\text{closed path}} \mathbf{B} \cdot d\mathbf{s} = \mu_0(i + \varepsilon_0 d\Phi_e/dt)$$

This equation shows how the integral of the magnetic field around a closed path is affected by both the electric currents flowing through the surface enclosed by the path and by the changing electric flux through that surface. Here i denotes the net current flowing through the surface in amperes (A), and μ_0 is the PERMEABILITY of the vacuum, whose value is $4\pi \times 10^{-7}$ T·m/A. This equation is the magnetic analog of the third equation, except that the third has no term corresponding to the electric-current term in the fourth equation. That is because, since there are no magnetic charges in nature, there are also no magnetic currents. The relation between the direction of positive flux and current, on the one hand, and the sense of traversal of the closed path, on the other, is given by the right-hand rule, as described for Ampère's law. The fourth equation shows how changes in the electric field affect the magnetic field. Without the flux term, the fourth equation is just Ampère's law.

Maxwell's equations show that disturbances in the electromagnetic field propagate as WAVEs, called electromagnetic waves. The propagation SPEED of such waves in vacuum is known as the SPEED OF LIGHT, is conventionally denoted by c, and has the value 2.99792458×10^8 meters per second (m/s). The speed of light is related to the electric and magnetic constants that appear in Maxwell's equations by:

$$c = \frac{1}{\sqrt{\varepsilon_0 \mu_0}}$$

mean free path The average distance a randomly moving particle travels in free MOTION between successive COLLISIONs is called the mean free path. In the case of a FLUID (i.e., a LIQUID or a GAS), the collisions are mainly with the other particles. In that case, the mean free path, λ, in meters (m) is approximately:

$$\lambda = \frac{1}{2\sqrt{2}\pi r^2 n}$$

where r denotes the radius of a particle in meters (m), and n is the particle DENSITY in inverse cubic meters ($1/m^3$) (i.e., the number of particles per cubic meter). In a CONDUCTOR, such as a METAL, the conduction ELECTRONs move freely between collisions with the ATOMs of the material. Thus, their mean free path might be expected to be on the order of the LATTICE spacing. Due to their WAVE nature, however, the electrons scatter from irregularities in the lattice rather than from the lattice itself, and their mean free path can be much larger, even thousands of times larger, than the lattice spacing. (*See* CONDUCTION; WAVE-PARTICLE DUALITY.)

The mean free time is the average time a particle is in free motion between successive collisions. It is given by:

$$\tau = \lambda / v_{\text{av}}$$

where τ is in seconds (s), and v_{av} denotes the average SPEED of the particles in meters per second (m/s).

mechanics The field of physics that deals with the MOTION of bodies and with the effect of FORCEs on the motion is called mechanics. One subfield of mechanics is KINEMATICS, which is the description of motion without regard to its causes. Another subfield is STATICS, the study of EQUILIBRIUM situations, when the forces and TORQUEs on a body cancel out. A further subfield of mechanics is DYNAMICS, which is the study of the causes of motion and of its change (i.e., the study of the effect of forces on motion). There are classical mechanics and QUANTUM MECHANICS. In the domain of SPEEDs that are small compared with the SPEED OF LIGHT, the former is governed by ISAAC NEWTON's laws of MOTION and is also called Newtonian mechanics. More accurately, classical mechanics is governed by relativistic mechanics, or Einsteinian mechanics, according to ALBERT EINSTEIN's special theory of relativity, which is valid for all speeds. And to an even higher degree of accuracy, classical mechanics is governed by Einstein's general theory of relativity, which is a theory of GRAVITATION and generalizes Newton's law of gravitation. Classical mechanics is typically valid in the macroscopic domain.

Quantum mechanics, however, offers the more fundamental description of nature, as its use is generally required in dealing with the microscopic world, i.e., the domain of MOLECULES, ATOMs, and ELEMENTARY PARTICLEs, from which macroscopic bodies are constituted. Systems of many bodies can be difficult, or even impossible, to analyze on an individual-body basis. STATISTICAL MECHANICS treats such systems by studying the statistical properties of a system's constituents and relating them to properties of the system as a whole. (*See* CLASSICAL PHYSICS; MANY-BODY THEORY; NEWTON'S LAWS; RELATIVITY, GENERAL THEORY OF; RELATIVITY, SPECIAL THEORY OF.)

Meissner effect Named for the German physicist Walther Hans Meissner (1882–1974), the Meissner effect is the expulsion of an existing magnetic FIELD from a material when the material is cooled to below its critical TEMPERATURE and becomes superconducting (i.e., completely lacking electric RESISTANCE). So a superconductor is also a perfect diamagnet, in that it completely cancels externally applied magnetic fields and maintains a zero-field state in its interior. That is accomplished by surface currents, which are electric CURRENTS flowing within a thin layer at the surface of the superconductor. It is to be expected, according to the BIOT-SAVART LAW, FARADAY'S LAW, and LENZ'S LAW, that a superconductor will maintain whatever magnetic field exists within it upon its becoming superconducting. The Meissner effect, however, shows that a superconductor goes beyond that, by completely canceling any interior magnetic field and then maintaining the field-free state. In this way, the floating of a small magnet above a superconductor can be viewed as resulting from a "repulsion" of the magnet's magnetic field by the superconductor. It is as if the field lines behave like springs beneath the magnet and support the magnet above the impenetrable surface of the superconductor. (*See* DIAMAGNETISM; ELECTRICITY; MAGNETIC LEVITATION; MAGNETISM; SUPERCONDUCTIVITY.)

melting Also called fusion, melting is a PHASE TRANSITION of a material, whereby its SOLID form, or solid PHASE, changes into a LIQUID. Melting is achieved by adding HEAT (i.e., thermal ENERGY) to the solid. The TEMPERATURE at which a solid melts at standard atmospheric pressure is called its melting point. The melting point of ice, for example is 0 degrees Celsius (°C), that for lead 328°C, and for oxygen –218°C. The melting point tends to decrease as the pressure on the solid is increased and rises with reduced pressure. An example of that is the gradual descent of a heavy object resting on ice, when the temperature is not too low. At the area of contact between the object and the ice, the pressure lowers the melting point of the ice to below the ambient temperature, the ice melts under the body, and the body sinks into the ice. (*See* STATE OF MATTER.)

As a solid is heated, its temperature increases until the melting point is reached. Additional heating no longer raises the temperature, but rather the increased thermal energy causes the conversion of the solid to a liquid. At the molecular scale, according to the KINETIC THEORY of matter, a rise in temperature means an increase in the average energy of the MOLECULEs of the solid, which are constantly oscillating about their fixed locations and possess a range of energies. At the melting point, the most energetic of the molecules have just enough energy to overcome the cohesive FORCEs holding the solid together, and they commence wandering, forming the liquid phase. Further heating of the solid produces more high-energy molecules that join the liquid phase, leaving unchanged the average energy of the remaining molecules and consequently also the temperature of the solid. The amount of heat required to fully convert a sample of solid to a liquid at the melting point is the HEAT OF FUSION of the sample. When the phase transition is complete for the sample, further heating raises the temperature. (*See* COHESION; OSCILLATION.)

Melting is an inverse process to FREEZING.

meson The mesons form a class of ELEMENTARY PARTICLEs that are both HADRONs and BOSONs. Every meson consists of a QUARK and an antiquark, where the antiquark possesses the anticolor of the quark's color and the natures of the constituent quark and antiquark determine the type of meson. For every meson, there is a corresponding antimeson, although in certain cases, a meson is its own antimeson. While at the most fundamental level the STRONG INTERACTION is mediated by GLUONs, at the nuclear level the action of the strong interaction is described by the exchange of mesons among BARYONs, especially among NUCLEONs (PROTONs and NEUTRONs). The family of mesons includes, among others, the PIONs (π^+, π^-, π^0), the kaons (K^+, K^-, K^0), eta (η), and eta-prime (η'). (*See* ANTIMATTER; NUCLEUS.)

Physics and Metaphysics: Where Lies the Border between Them?

Physics is the human attempt to understand the most fundamental aspects of nature. It also forms the most fundamental branch of science. Physics studies phenomena that underlie the rest of science, that make up the foundation that supports the structure of science. In their dealings with such matters, physicists occasionally find themselves involved in considerations that stretch the legitimate domain of science and possibly exceed it. What they are then getting into is metaphysics.

In general, metaphysics is a branch of philosophy that deals with being and reality. In a narrower sense that is more relevant to our discussion, metaphysics is the philosophical framework in which science operates. In this sense, metaphysics is concerned with what lies around, below, above, before, and beyond science. While science involves observation of nature, metaphysics might consider the significance of the observer-observed dichotomy. Or, while science searches for and finds pattern, ORDER, and LAWS OF NATURE, metaphysics might consider what constitutes evidence, confirmation, and proof and why there are order and laws of nature at all. Concepts such as world view, fundamentality, simplicity, causation, and beauty are often used in science, and in physics in particular. But the concepts themselves lie beyond the domain of science and belong to the domain of metaphysics.

Considerations of physics and considerations of metaphysics—although each kind is perfectly respectable in its own domain—must not be confused with each other. It behooves physicists to stick to physics considerations, avoiding metaphysics, for as wide a range of phenomena as possible without exceeding the domain of science. And when people choose to do metaphysics, they must be sure that the subject of their investigation lies beyond the domain of science so that they will not find themselves inappropriately treating matters that can be well treated through science.

The domain of science is delineated very clearly. In detail, it is as follows. Science studies the reproducible and predictable aspects of nature, where the term *nature* refers to the material universe with which we can, or can conceivably, interact. Any phenomenon, concept, law, theory, etc. that does not toe that line lies beyond the domain of science. Science is not concerned with phenomena, concepts, etc. that are not aspects of the material universe, such as God and possibly feeling. Neither is science concerned with aspects of the material universe with which we cannot, or cannot conceivably, interact, such as other universes; or which are irreproducible, such as extrasensory perception (ESP); or are unpredictable, such as ESP again. If such a phenomenon, concept, etc. is related to the philosophical framework in which science is done, it is then grist for the mill of metaphysics. Theories, for example, are scientific stuff, while their possible beauty is a metaphysical matter. (*See* INTERACTION.)

Physicists have a good track record of sticking to physics in their study of nature. It is true that in the earliest eras of physics, before the Age of Enlightenment at around the 18th century, religious considerations were routinely introduced into the domain of science. However, that is long past. On the other hand, even today we find nonscientists and pseudoscientists introducing metaphysical considerations into the domain of science. The most egregious examples in the United States are surely creationism and "intelligent design" as proposed "explanations" for the existence of life and, in particular, the existence of species. The point is that such explanations, as emotionally satisfying as they might be for some people, are no different, in essence, from explanations such as: life exists because Mars and Venus created it, or because of the "vital force" (whatever that might be). Emotional satisfaction or religious claims are not valid explanations in science. The issue is one of truth: objective truth that holds for everybody vs. subjective truth, such as religious truth, that holds only for its believers.

In science, there exist rather strict criteria for objective truth. Whatever one might think or whatever controversy might arise concerning nature, in the final analysis, experiments are performed, observations are made, and it is nature itself that is the supreme arbiter. In order not to mislead, it should be added that the matter is not quite as clear as it might appear. There are metaphysical questions and controversy among philosophers about just what scientific truth means (if anything), how it might be attained through science (if it is attainable at all), and so on. But scientists, and physicists in particular, have no problems with such issues. While the philosophers debate, scientists proceed to gain meaningful and useful understanding of nature.

The situation is quite different in metaphysics. Metaphysical considerations are unburdened by any need to be "true." They might, perhaps, reasonably be expected not to contradict scientific truths and to be logically self-consistent (i.e., not to contradict themselves). But even those expectations possess no substantial foundation, and one can find metaphysical positions that do contradict the truth as seen by science or that are inconsistent or both. Personal taste, preference, and belief reign supreme in metaphysics, whereas in science their influence, though present, is not of great importance.

(continues)

Physics and Metaphysics: Where Lies the Border between Them? *(continued)*

There are situations in which physicists, of necessity, find themselves operating at, or even beyond, the border between physics and metaphysics. That occurs in the most fundamental fields of physics, in COSMOLOGY and ELEMENTARY PARTICLE physics. Consider cosmology: this is the study of the UNIVERSE as a whole. There is only one universe. Or at least we can interact with only one universe. So the universe is a unique phenomenon. It is irreproducible. There is no way to run experiments on the universe as a whole as one might, say, repeatedly drop cannon balls from the Leaning Tower of Pisa. Physicists cannot perform experiments in universe evolution. By the strict definition of the domain of science, then, the universe as a whole lies outside it.

Yet, physicists still study the universe as a whole, and cosmology forms a meaningful endeavor. However, the character of the understanding that is achieved in cosmology differs from that in other, nonborder fields. In other fields, any proposed law of evolution can be tested by allowing systems to evolve and then observing the results. That cannot be done for the universe. A proposed evolution scenario for the universe can be checked for its consistency with the past, but due to the time scales involved, there is no way to test its predictions for the future. In order to make some sense of the universe and its origin, there are proposals that involve the existence of other universes. We cannot, even conceivably, interact with such universes. Thus they, too, lie outside the strict domain of science.

In the field of elementary particle physics, the physics-metaphysics border is being stretched in the opposite way, toward the very small. From QUARKS to extra DIMENSIONS to strings, much research effort is being devoted to entities that are inaccessible. Are they on the physics side of the border or beyond it? Better, perhaps, to think of this part of the border as blurred, even uncharted. (*See* STRING THEORY.)

We have seen how physics and metaphysics differ and where the border between their domains lies. Roughly, physics studies nature, while metaphysics studies physics (and science in general). The truths of physics are as objective as possible, while those of metaphysics suffer from a high degree of subjectivity. We have also seen that physicists, while almost totally restricting their endeavors to the domain of physics, must, in the most fundamental fields of physics, stretch the border somewhat or even venture beyond it.

metal The metals form a class of chemical ELEMENTS whose characteristic properties are determined by the readiness of their ATOMS to give up one or more ELECTRONS. Except for mercury, they are all SOLIDS at ordinary TEMPERATURES. They are good CONDUCTORS of HEAT and ELECTRICITY and, in their solid state, are relatively easily deformed (stretched, drawn into wires, rolled flat). Their atoms pack closely, so they possess comparatively high DENSITIES. They are opaque, but they are good reflectors of LIGHT and other electromagnetic RADIATION. They give off electrons from their surface when heated (THERMIONIC EMISSION) and when irradiated by light (PHOTOELECTRIC EFFECT). (*See* CONDUCTION; CONDUCTION BAND; ELECTROMAGNETISM; REFLECTION.)

Michelson, Albert Abraham (1852–1931) American *Physicist* Best known today for his leading role in the MICHELSON-MORLEY EXPERIMENT, whose negative result formed part of the foundation of ALBERT EINSTEIN's special theory of relativity, Albert Michelson was the first American to be awarded a Nobel Prize. He graduated from the United States Naval Academy at Annapolis, Maryland, in 1873. After two years of postgraduate studies in Europe, Michelson held several positions in the United States, the last being that of professor of physics at the University of Chicago, Illinois, where he served from 1892 until 1929. Michelson's expertise was in the design and construction of optical instruments and in carrying out precise optical measurements, including measurement of the SPEED OF LIGHT. The Michelson-Morley experiment was intended to measure the SPEED of Earth's MOTION through the ether, a hypothesized medium for the propagation of light WAVES. The consistently negative result of this experiment indicated the nonexistence of the ether. Michelson was awarded the 1907 Nobel Prize in physics "for his optical precision instruments and the spectroscopic and metrological investigations carried out with their aid."

See also LIGHT; OPTICS; RELATIVITY, SPECIAL THEORY OF.

Michelson-Morley experiment Named for the American physicist ALBERT ABRAHAM MICHELSON and the American physicist and chemist Edward William Morley (1838–1923), the Michelson-Morley experi-

An experimental physicist, Albert Michelson led in performing the Michelson-Morley experiment, whose negative result became part of the foundation of the special theory of relativity. He was awarded the 1907 Nobel Prize in physics. ***(Albert A. Michelson Collection, Special Collections & Archives Division, Nimitz Library, U.S. Naval Academy, courtesy AIP Emilio Segrè Visual Archives)***

ment was designed to detect the motion of the Earth through the ether. The ether was postulated as a material medium that carried the WAVES associated with LIGHT and other electromagnetic RADIATION. If there were an ether, the Earth would be moving through it. In the Michelson-Morley experiment, a beam of light was split by a partially reflecting MIRROR into two perpendicular beams that were reflected back to the beam splitter and recombined, resulting in an INTERFERENCE pattern of parallel stripes of light, called fringes, separated by dark stripes. If the interference pattern shifted when the experiment was rotated, that would indicate unequal changes in the SPEEDs of the light in the two beams, which would be evidence that the apparatus was moving with respect to the ether. The result of the experiment was negative, however, showing that the ether concept was erroneous and that electromagnetic waves do not require a medium for their propagation. The negative result of the Michelson-Morley experiment served as important evidence that ALBERT EINSTEIN took into account in formulating his special theory of relativity. (*See* ELECTROMAGNETISM; REFLECTION; RELATIVITY, SPECIAL THEORY OF; SPEED OF LIGHT.)

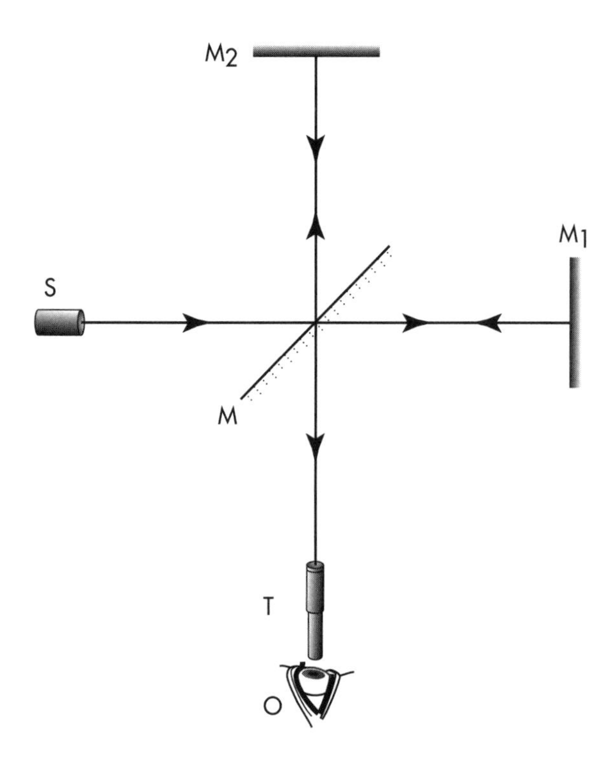

The Michelson-Morley experiment was performed in an attempt to detect the motion of the Earth through the ether. The ether was hypothesized as a medium that carries electromagnetic waves, including light. The results of the experiment were negative. A beam of light from a light source S is split by a partially reflecting mirror M into two beams, which are sent down the two perpendicular arms of the apparatus. The beams are reflected back by mirrors M_1 and M_2, recombine at M, and enter a telescope T. An observer O through the telescope sees an interference pattern of parallel stripes of light. The apparatus is aligned with one arm pointing in the direction of the assumed motion of the Earth through the ether. The interference pattern is observed and followed as the apparatus is then rotated so the other arm is pointing in that direction. A shift in the pattern would indicate Earth's motion, but no shift was ever observed.

In somewhat more detail, the experiment was typically performed by aiming one of the split beams in the supposed direction of the Earth's motion in the ether, then rotating the apparatus to bring the other beam to the supposed direction of motion. If the two beams are of equal length, L, in meters (m), the speed of light in the ether is c in meters per second (m/s), the supposed speed of the Earth in the ether is v in meters per second (m/s), and the WAVELENGTH of the light that was used is denoted by λ in meters (m), then the number of fringes the diffraction pattern should shift is given by:

$$\text{Number of fringes of shift} = \frac{2L}{\lambda}\left(\frac{v}{c}\right)^2$$

In order to maximize the shift, the effective length of the two beams was greatly increased by repeated reflections. Nevertheless, the experiment consistently gave a null result.

Millikan oil-drop experiment The oil-drop experiment, named for its designer, the American physicist Robert Andrews Millikan (1868–1953), confirmed that electric CHARGE appears in nature only in multiples of a fundamental unit of charge and measured the magnitude of that unit. What Millikan in fact measured was the charge of the ELECTRON. The procedure of the experiment was to follow and measure the SPEED of electrically charged tiny oil drops in air as they fell under the effect of gravity and as they rose in an electric FIELD. Due to the DRAG of the air on the drops, they quickly reach a constant TERMINAL SPEED in either direction of motion. If the FREE-FALL speed of a drop is denoted by v in meters per second (m/s), the upward speed in the electric field by u in meters per second (m/s), the MASS of the drop by m in kilograms (kg), the ACCELERATION due to gravity by g in meters per second per second (m/s^2), and the magnitude of the electric field by E in newtons per coulomb (N/C), then the magnitude of the drop's charge, q, in coulombs (C) is given by:

$$q = \frac{mg}{E}\,\frac{v+u}{v}$$

(*See* ELECTRICITY; GRAVITATION.)

It is not necessary to know m, g, and E in order to demonstrate that there is a fundamental unit of charge. As charged drops move upward under the influence of the electric field, their charge sometimes changes (by losing an electron, say, or attaching an ION), which causes an abrupt change of upward speed. If the new upward speed is denoted u', the corresponding charge is:

$$q' = \frac{mg}{E}\,\frac{v+u'}{v}$$

The ratio of the two equations gives:

$$\frac{q}{q'} = \frac{v+u}{v+u'}$$

So if the right-hand side, calculated for pairs of measurements on the same drop, consistently gives ratios of integers, the discrete nature of electric charge is demonstrated. That was indeed what Millikan found.

The acceleration due to gravity can be determined by PENDULUM measurements (as one possible method). The magnitude of the electric field is set by the experimenter. So in order to find the magnitudes of the drops' charges, and then the fundamental unit of charge, it remains to measure the masses of the observed drops. That is done by using STOKES'S LAW to find the radius of a drop, r, in meters (m) from its free-fall terminal speed v; the DENSITY of the oil, ρ, in kilograms per cubic meter (kg/m^3); and the VISCOSITY of the air, η, in kilograms per meter·second [kg/(m·s)]:

$$r = \sqrt{\frac{9\eta v}{2\rho g}}$$

Then the mass of a drop is obtained from the drop's VOLUME, $(4/3)\pi r^3$, and the oil's density:

$$m = (4/3)\pi r^3 \rho$$

In this way, Millikan was able to measure the magnitude of the electron's charge, whose present-day value is $1.602176462 \times 10^{-19}$ C.

mirror An optical device that modifies, by means of REFLECTION and in a controlled manner, the direction of LIGHT rays impinging on it is called a mirror. (The concept is suitably generalized for other types of WAVE, such as other electromagnetic RADIATION and SOUND waves.) A mirror for light is made of polished metal or of a metallic coating on a polished supporting substrate. (For waves other than light waves, a mirror might be constructed of other materials.) (*See* ELECTROMAGNETISM; OPTICS.)

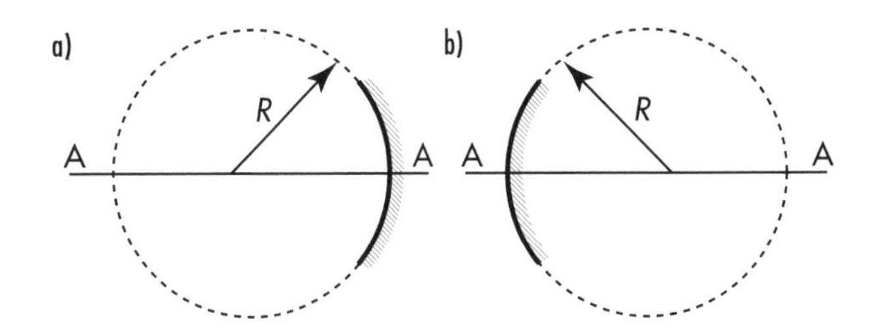

Radius of curvature *R*: (a) of a concave mirror and (b) of a convex mirror. *R* is taken negative for a concave mirror and positive for a convex one. The mirrors' axes are denoted A-A.

The most familiar type of mirror is a plane mirror. Its action is to reflect the light rays striking it so that they seem to be emanating from sources behind the mirror. The apparent sources are the virtual images of the actual sources, which are termed the objects. The designation *virtual* is used to indicate that the image appears to be located where no light rays are actually reaching. Although a virtual image can be viewed, it cannot be caught on a screen. In the case of a plane mirror, the location of an object's image appears to be the same distance behind the mirror as the object is in front. The image and object have the same size, and they both possess the same orientation: the image of an upright object is upright as well.

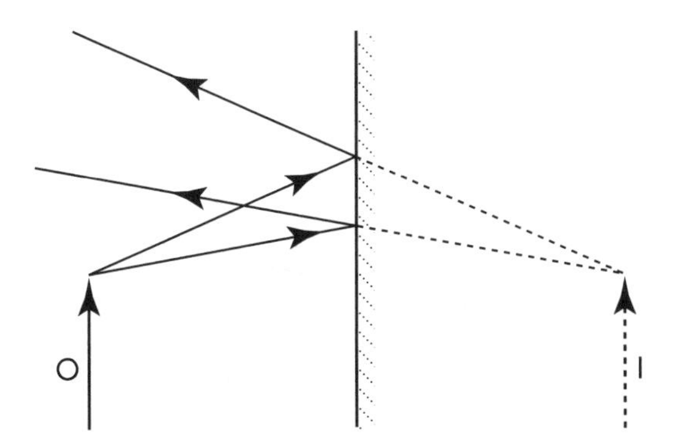

The action of a plane mirror. The virtual image I of object O appears to be located on the other side of the mirror from the object and at the same distance from the mirror as the object. It possesses the same size and orientation as does the object. Two rays from the object to the viewer are shown.

A mirror in the shape of a concave paraboloid has the useful property that all light rays that strike the mirror's surface parallel to the mirror's axis are reflected to the same point on the axis, called the *focal point* of the mirror. That property is used, for example, to enable a microphone to pick up sounds only from a particular distant source, ignoring other sounds. The microphone is placed at the focal point of a paraboloidal (also called "parabolic") reflector, which is aimed at the sound source.

Paraboloidal mirrors can often be reasonably approximated by spherical mirrors, mirrors whose shape is that of a surface of a sphere, as long as their depth is small compared with their diameter. Since paraboloidal mirrors for light are relatively expensive to manufacture, spherical mirrors are used when extreme precision is not necessary. In the following, we deal only with spherical mirrors. A spherical mirror is characterized by its radius of curvature, which is the radius of the sphere of which the surface forms a part. Note that a plane mirror is a limiting case of a spherical mirror, one of infinite radius of curvature. We also make the assumption that the light rays impinging on a mirror are almost parallel to each other and are almost perpendicular to the mirror's surface where they strike it (i.e., are almost parallel to the radius of the surface there).

Focal Length

Consider the effect of a mirror on rays that are parallel to each other and to the radius of the mirror's surface in the center of the area where they strike. After reflection, such rays are either made to converge to a point, the focal point, or made to diverge as if emanating from a point behind the mirror, called the virtual focal point. One says that the rays are brought, or focused, to a real or virtual focus. In the former case, the mirror is a converging mirror, while the latter is a diverging mirror. A converging mirror is always concave, and a diverging mirror is always convex. The distance between the mirror's surface and its real or virtual focal point is the lens's focal length and is denoted f. A positive value for the focal length, f, indicates a converging mirror, while a diverging mirror has a negative value of f. Since the focal length is a length the SI UNIT of f is the meter (m). Let us denote the radius of curvature of a mirror by R in meters (m), taking R positive for a convex mirror and negative for concave. Then this relation holds:

$$f = -R/2$$

The focal length of a spherical mirror equals half its radius of curvature and has the opposite sign.

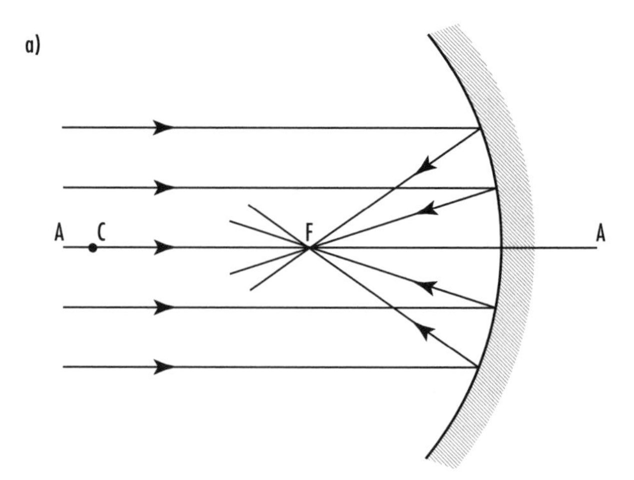

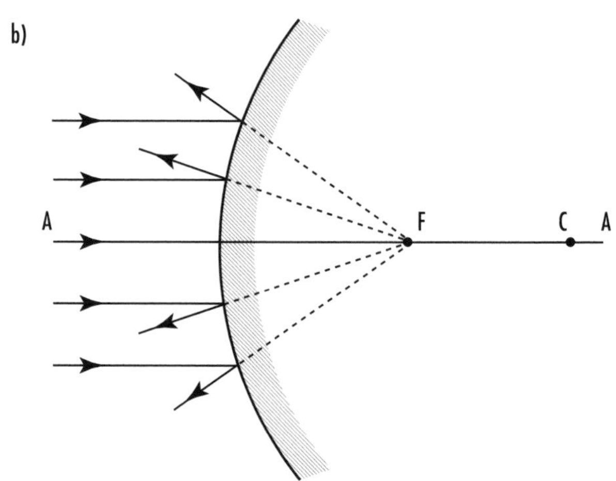

The focal length of a mirror. (a) When rays strike a converging mirror parallel to its axis A-A, they are made to converge to a point F on the axis, the focal point of the mirror. The distance of this point from the surface of the mirror along the axis is the mirror's focal length *f*, a positive number. (b) Similar rays that are incident on a diverging mirror are made to diverge as if they emanate from a point F on the mirror's axis A-A, a virtual focal point. This point's distance from the surface of the mirror along the axis is the lens's focal length *f*, taken as a negative number. In both cases, the focal point is located midway between the mirror's center of curvature, C, and its surface.

Mirror Equation

For the focusing effect of a mirror on nonparallel rays, we assume, for the purpose of presentation, that the mirror is positioned so that its central radius is horizontal, extending left-right in front of us, with the reflecting surface on the left. The light source, or object, is located to the left of the mirror, so the light diverges from the source and strikes the mirror on its reflecting surface. Denote the distance of the object from the mirror by d_o in meters (m) and give it a positive sign. The light rays from the object might converge as they are reflected from the mirror, in which case they converge to an image, called a real image, on the left side of the mirror. Denote the distance of the image from the mirror by d_i in meters (m) and assign it a positive value. If the reflected light rays diverge, they will diverge as if they come from a virtual image located on the right side of the mirror. The distance of a virtual image from the reflecting surface is denoted by d_i as well, but given a negative sign. In brief, a positive or negative value for d_i indicates a real image (on the left) or a virtual image (on the right), respectively. If the object is located at the focal point, the rays are reflected from the mirror parallel to the radius through the focal point, neither converging nor diverging. In that case, the image distance is infinite, and its sign is not important.

A real image can be caught by placing a screen, such as a movie screen, at its location. A virtual image cannot be caught in that way. Rather, a virtual image can be seen or photographed by aiming the eye or camera toward it, such as in the case of a bathroom or automobile mirror.

It might happen that the rays impinging on the mirror from the left are converging rather than diverging. This occurs when the mirror intercepts the converging rays emerging from, say, another mirror or from a LENS. In the absence of our mirror, the rays would converge to a real image, located to the right of our mirror. That real image serves as a virtual object for our mirror. Denote the distance of the virtual object from the surface of our mirror by d_o, and give it a negative sign. A positive or negative sign for d_o indicates a real object (on the left) or a virtual object (on the right), respectively. If a virtual object is located at the virtual focal point of a diverging mirror, the rays reflected from the mirror are parallel to the axis, and the image distance is infinite.

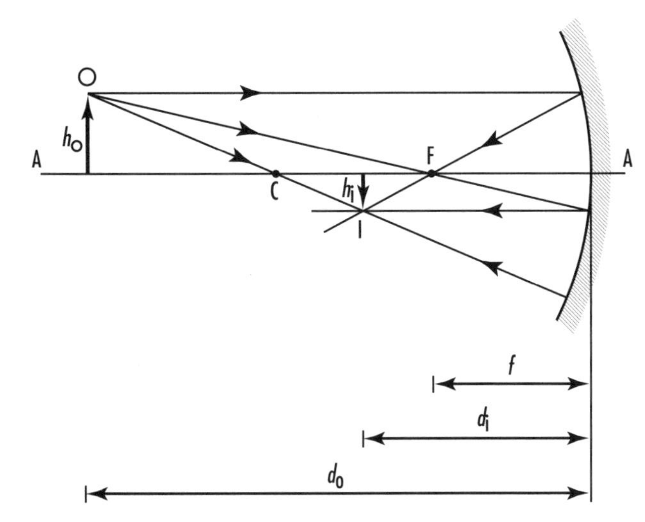

An object O and its real image I as formed by a converging mirror of focal length f (> 0). The mirror's focal point is denoted F, its center of curvature C, and its axis A-A. The object and image distances are denoted d_o and d_i, respectively. Both are positive. The heights of the object and image are denoted, respectively, h_o and h_i. The former is positive and the latter negative. Three rays from the object to the image are shown. A real image is formed whenever the object is farther from the mirror than the focal point, as in the figure.

The object and image distances and the focal length of a mirror are related by the mirror equation:

$$\frac{1}{d_o} + \frac{1}{d_i} = \frac{1}{f}$$

This relation implies the following behavior of d_i as a function of d_o for fixed f and real object (positive d_o):

Converging Mirror (f > 0)

Object Distance (d_o)	Image Distance (d_i)	Image Type
∞	f	Real
Decreasing from ∞ to 2f	Increasing from f to 2f	Real
2f	2f	Real
Decreasing from 2f to f	Increasing from 2f to ∞	Real
f	±∞	—
Decreasing from f to f/2	Increasing from −∞ to −f (while decreasing in absolute value)	Virtual
f/2	−f	Virtual
Decreasing from f/2 to 0	Increasing from −f to 0 (while decreasing in absolute value)	Virtual

Diverging Mirror (f < 0)

Object Distance (d_o)	Image Distance (d_i)—All Images Are Virtual
∞	$-\lvert f \rvert$
Decreasing from ∞ to f	Increasing to from $-\lvert f \rvert$ to $-\lvert f \rvert/2$ (while decreasing in absolute value)
f	$-\lvert f \rvert/2$
Decreasing from f to 0	Increasing from $-\lvert f \rvert/2$ to 0 (while decreasing in absolute value)

Magnification

Denote the size, or height, of an object in a direction perpendicular to the light rays by h_0 and that of the corresponding image by h_i. The SI unit for both quantities is the meter (m). A positive value for h_o indicates that the object is upright in relation to some standard for orientation, while a negative value shows that it is inverted. Similarly, positive h_i indicates an upright image, and a negative value indicates an inverted image. The object and image sizes are proportional to the object and image distances, respectively, through the relation:

$$h_i/h_O = -d_i/d_o$$

If follows immediately that the real image of a real object is always oriented oppositely to the object, while its virtual image always has the same orientation. On the other hand, a virtual object and its real image both have the same orientation, and a virtual image of a virtual object is oppositely oriented. For a given object size, the farther the image is from the mirror, the larger it is.

The MAGNIFICATION is defined as:

$$M = h_i/h_O$$

which, by the above relation, is:

$$M = -d_i/d_o$$

Magnification is dimensionless. An absolute value of M greater than 1 ($|d_i| > |d_o|$) indicates strict magnification, that the image is larger than the object. When $|M|$ equals 1 ($|d_i| = |d_o|$), the image is the same size as the object, while for $|M|$ less than 1 ($|d_i| < |d_o|$) we have reduction. Positive M (d_i, d_o of opposite signs) shows that the image and the object have the same orientation, while negative M (d_i, d_o of the same sign) indicates that they are oriented oppositely. For a series of magnifications—where the image produced by the first mirror (or other optical device) serves as the object for the second, the image formed by the second serves as the object for the third, and so on—the total magnification is the product of the individual magnifications:

$$M = M_1 M_2 \ldots.$$

(*See* DIMENSION.)

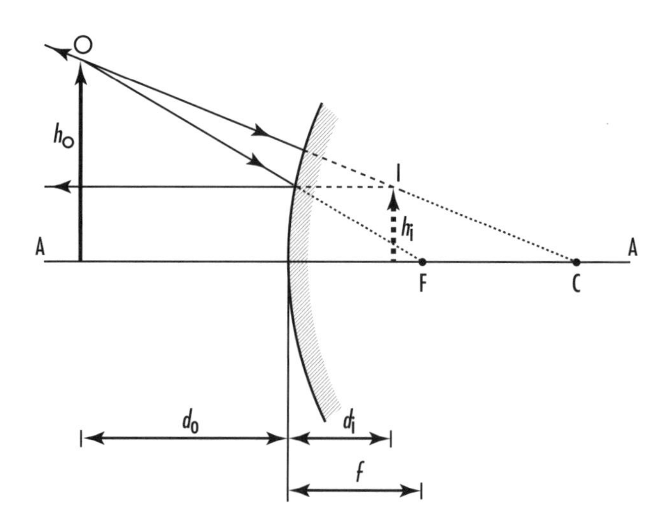

A diverging mirror always forms a virtual image of an object, labeled I and O, respectively. The mirror's focal length is f(< 0). Its virtual focal point is denoted by F, its center of curvature by C, and its axis by A-A. Object and image distances are denoted d_o and d_i, respectively, where d_o is positive and d_i negative. The heights of the object and virtual image are denoted, respectively, h_o and h_i. Both are positive. Two rays from the object to the viewer are shown.

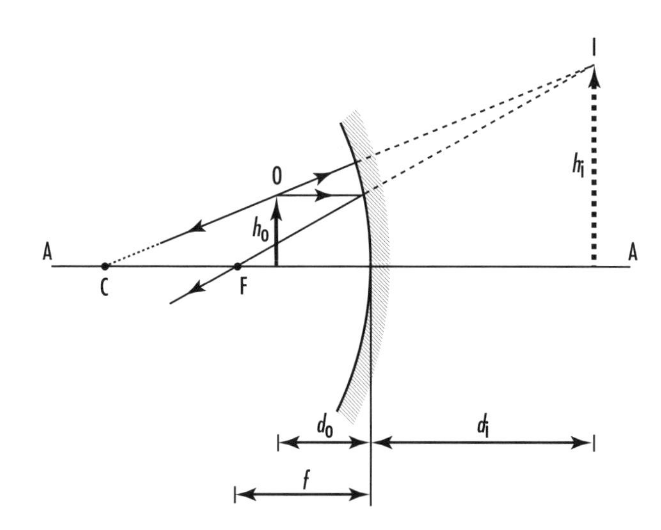

The formation of a virtual image I of an object O by a converging mirror of focal length f(> 0). The mirror's focal point is denoted F, its center of curvature C, and its axis A-A. The object and image distances are denoted d_o and d_i, respectively, where d_o is positive and d_i is negative. The heights of the object and virtual image are denoted, respectively, h_o and h_i. Both are positive. Two rays from the object to the viewer are shown. As in the figure, whenever the object is located between the mirror and the focal point, the image is virtual.

The following table is a summary of the sign conventions:

Sign Conventions

Quantity	Positive	Negative
R	Surface is convex	Surface is concave
f	Converging mirror	Diverging mirror
d_o	Real object	Virtual object
d_i	Real image	Virtual image
h_o	Upright object	Inverted object
h_i	Upright image	Inverted image
M	Object and image in same orientation	Object and image oppositely oriented

Real mirrors in real situations are subject to various kinds of ABERRATION.

See also LENS.

mole The SI base UNIT for amount of substance is the mole and is denoted mol. It is, by definition, the amount of substance that contains as many elementary entities (such as ATOMs, MOLECULEs, IONs, ELECTRONs, etc., which must be specified) as there are atoms in exactly 0.012 kilogram (12 grams) of the ISOTOPE carbon-12. That number, called AVOGADRO'S NUMBER, equals $6.02214199 \times 10^{23}$ per mole (mol^{-1}). (*See* BASE QUANTITY; INTERNATIONAL SYSTEM OF UNITS.)

molecular weight The molecular weight of a chemical compound is the number whose value equals the MASS, expressed in grams, of one MOLE of that compound. A mole consists of an AVOGADRO'S NUMBER of MOLECULEs (i.e., $6.02214199 \times 10^{23}$ molecules). The molecular weight can be calculated from the ATOMIC WEIGHTs of the elements composing the compound by adding them up, with each atomic weight multiplied by the multiplicity of that element in the compound. Consider water, for instance, whose chemical formula is H_2O. The molecular weight of water is the sum of twice the atomic weight of hydrogen and the atomic weight of oxygen (i.e., $2 \times 1.0079 + 15.9994 = 18.0152$). The atomic weights of the elements can be found in the PERIODIC TABLE of the elements that appears in Appendix III as well as in every introductory chemistry textbook and many introductory physics textbooks.

molecule A molecule is a chemically bound group of ATOMs. It is the smallest identifiable quantity of a chemical COMPOUND. As an example, the molecule of silicon dioxide, denoted SiO_2, consists of a silicon atom bound to two oxygen atoms. Or, a methane atom is formed from a central carbon atom connected to four hydrogen atoms. Accordingly, the chemical symbol for methane is CH_4. Even some chemical ELEMENTs have a molecular form. Both nitrogen and oxygen in their normal gaseous state, for instance, are in the form of diatomic molecules, N_2 and O_2, respectively. Oxygen even possesses a triatomic gaseous form, O_3, called ozone. (*See* GAS; STATE OF MATTER.)

Not all compounds possess a molecular form. As an example, common table salt, sodium chloride, denoted NaCl, in its solid, crystalline form, consists of a three-dimensional array of alternating sodium and chlorine IONs, Na^+ and Cl^-, respectively. Such a crystal might be thought of as a single gigantic molecule. In aqueous solution, the ions separate. So a salt solution is composed of sodium and chlorine ions wandering among water molecules. Neither as a solid nor as a solute does the conventionally denoted NaCl actually take the form of diatomic NaCl molecules. (*See* CRYSTAL.)

moment, electric dipole Electric dipole moment is a measure of the strength of an ELECTRIC DIPOLE. It is a VECTOR quantity. For a pair of equal-magnitude, opposite electric CHARGEs, both of magnitude q in coulombs (C) and separated by distance d in meters (m), the magnitude of the electric dipole moment, μ_e, in coulomb·meters is given by:

$$\mu_e = qd$$

The direction of the electric dipole moment vector is along the line connecting the centers of the charges and pointing from the negative to the positive charge. Neutral charge configurations for which the center of positive charge does not coincide with the center of negative charge are similarly characterized by an electric dipole moment. (*See* ELECTRICITY.)

An electric FIELD exerts a TORQUE on an electric dipole, since it pulls the positive and negative charges in opposite directions, the positive charge in the direction of the field and the negative in the opposite direction. The torque tends to align the electric dipole moment with the field. The vector relation of torque to electric dipole moment and electric field is:

$$\boldsymbol{\tau} = \boldsymbol{\mu}_e \times \mathbf{E}$$

where $\boldsymbol{\tau}$ denotes the torque vector in newton·meters (N·m), $\boldsymbol{\mu}_e$ is the electric dipole moment vector, and E is

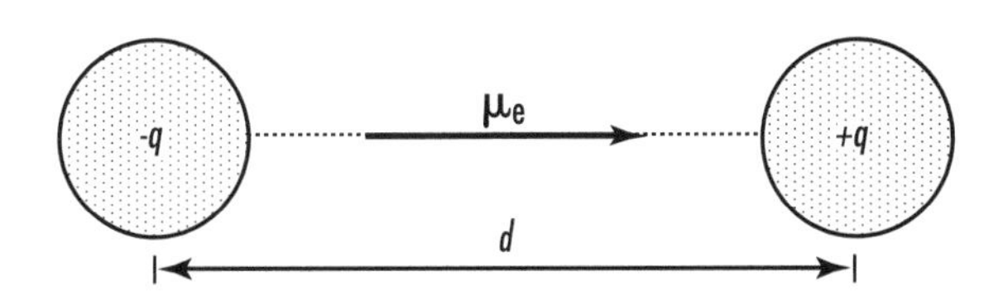

The electric dipole moment of an electric dipole, consisting of a pair of separated equal-magnitude and opposite electric charges, is a vector $\boldsymbol{\mu}_e$ that points from the negative charge $-q$ to the positive one $+q$. Its magnitude μ_e is given by $\mu_e = qd$, where q is the magnitude of either charge and d the distance between the centers of the charges.

the electric field at the location of the dipole in newtons per coulomb (N/C) or, equivalently, volts per meter (V/m). The magnitude of the torque, τ, is:

$$\tau = \mu_e E \sin \varphi$$

where E denotes the magnitude of the electric field and φ is the smaller angle (less than 180°) between $\boldsymbol{\mu}_e$ and **E**. The direction of $\boldsymbol{\tau}$ is perpendicular to both $\boldsymbol{\mu}_e$ and **E** and points such that, if the right thumb is aimed in the same direction, the curved fingers of the right hand will indicate a rotation from the direction of $\boldsymbol{\mu}_e$ to that of **E** through the smaller angle between them. Although the sole effect of a *uniform* electric field on an electric dipole is a torque, as just described, in a *nonuniform* field a net FORCE acts on it as well, since then the forces on the positive and negative charges are not equal and opposite.

A POTENTIAL ENERGY is associated with a dipole in an electric field, since WORK is involved in rotating the dipole. The potential energy in joules (J) is:

$$\text{Potential energy} = -\boldsymbol{\mu}_e \cdot \mathbf{E} = -\mu_e E \cos \varphi$$

It follows that the potential energy of the system is lowest when the electric dipole moment and the field are parallel ($\varphi = 0$, $\cos \varphi = 1$), which is thus the orientation of stable EQUILIBRIUM, and highest when they are antiparallel ($\varphi = 180°$, $\cos \varphi = -1$). The potential-energy difference between the two states is $2\mu_e E$.

moment, magnetic dipole Magnetic dipole moment is a measure of the strength of a MAGNETIC DIPOLE. It is a VECTOR quantity. The magnitude of the magnetic dipole moment in ampere·meters² (A·m²) produced by an electric CURRENT, i, in amperes (A) flowing in a flat loop is:

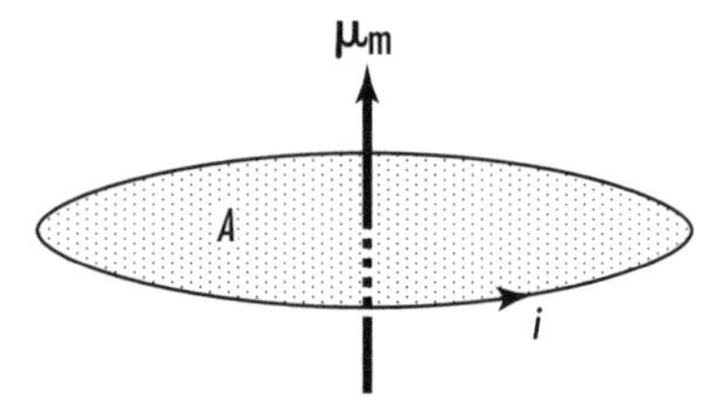

The magnetic dipole moment of a flat electric current loop is a vector $\boldsymbol{\mu}_m$ that is perpendicular to the plane of the current loop in the direction shown. Its magnitude is given by $\mu_m = iA$, where i denotes the current and A is the area of the loop. For a flat coil of N loops $\mu_m = NiA$.

$$\mu_m = iA$$

where A denotes the area in square meters (m²) enclosed by the loop. For a coil of N loops, the magnitude of the magnetic dipole moment is:

$$\mu_m = NiA$$

The direction of the magnetic dipole moment vector, $\boldsymbol{\mu}_m$, is perpendicular to the plane of the current loop(s) and pointing such that, if the right thumb is aimed in the same direction, the curved fingers of the right hand will indicate the direction of the electric current in the loop(s). (*See* ELECTRICITY.)

A TORQUE acts on a magnetic dipole in a magnetic FIELD. It tends to align the magnetic dipole moment with the field. The vector relation of torque to magnetic dipole moment and magnetic field is:

$$\boldsymbol{\tau} = \boldsymbol{\mu}_m \times \mathbf{B}$$

where $\boldsymbol{\tau}$ denotes the torque vector in newton·meters (N·m), and **B** is the magnetic field at the location of the dipole in teslas (T). The magnitude of the torque, τ, is:

$$\tau = \mu_m B \sin \varphi$$

where B denotes the magnitude of the magnetic field and φ is the smaller angle (less than 180°) between $\boldsymbol{\mu}_m$ and **B**. The direction of $\boldsymbol{\tau}$ is perpendicular to both $\boldsymbol{\mu}_m$ and **B** and points such that, if the right thumb is aimed in the same direction, the curved fingers of the right hand will indicate a rotation from the direction of $\boldsymbol{\mu}_m$ to that of **B** through the smaller angle between them. Although the sole effect of a *uniform* magnetic field on a magnetic dipole is a torque, as just described, in a *nonuniform* field a net FORCE acts on it as well. (*See* MAGNETISM.)

A POTENTIAL ENERGY is associated with a dipole in a magnetic field, since WORK is involved in rotating the dipole. The potential energy in joules (J) is:

$$\text{Potential energy} = -\boldsymbol{\mu}_m \cdot \mathbf{B} = -\mu_m B \cos \varphi$$

It follows that the potential energy of the system is lowest when the magnetic dipole moment and the field are parallel ($\varphi = 0$, $\cos \varphi = 1$), which is thus the orientation of stable EQUILIBRIUM, and highest when they are antiparallel ($\varphi = 180°$, $\cos \varphi = -1$). The potential-energy difference between the two states is $2\mu_m B$.

moment of force *See* TORQUE.

moment of inertia The moment of inertia, I, of a rigid body consisting of a number of point particles, with respect to any axis of ROTATION, is:

$$I = \Sigma m_i r_i^2$$

In this formula, m_i denotes the MASS of the ith particle in kilograms (kg), r_i is the particle's perpendicular distance from the axis in meters (m), and the summation is taken over all the particles constituting the body. For continuous bodies, the summation is replaced by integration. For a single point particle, the moment of inertia is simply:

$$I = mr^2$$

The SI UNIT of moment of inertia is kilogram·meter² (kg·m²).

Moment of inertia is the rotational analog of mass. We see that in the following two examples. Whereas the magnitude of LINEAR MOMENTUM, p, in kilogram·meters per second (kg·m/s) of a body of mass m in kilograms (kg) and SPEED v in meters per second (m/s) is given by:

$$p = mv$$

the magnitude of a body's ANGULAR MOMENTUM, L, is:

$$L = I\omega$$

where ω denotes the body's angular speed in radians per second (rad/s). In addition, the rotational version of ISAAC NEWTON's second law of MOTION is:

$$\tau = I\, d\omega/dt$$

or more generally:

$$\tau = dL/dt$$

where τ denotes the magnitude of the TORQUE acting on a rigid body in newton·meters (N·m), $d\omega/dt$ is the magnitude of the body's resulting angular acceleration in radians per second per second (rad/s²), and t denotes TIME in seconds (s). The linear version of Newton's second law of motion is:

$$F = m\, dv/dt$$

or more generally:

$$F = dp/dt$$

where F denotes the magnitude of the FORCE acting on a body in newtons (N), and dv/dt is the magnitude of the body's acceleration in meters per second per second (m/s²). The analogy here should be clear. In both cases—the linear and the angular—the mass and the moment of inertia measure the body's INERTIA, its resistance to changes in its VELOCITY and its angular velocity, respectively. (*See* ACCELERATION; NEWTON'S LAWS; VELOCITY.)

The radius of gyration of a rigid body rotating about an axis is the distance from the rotation axis at which a point particle, of the same mass as the body, would have the same moment of inertia as does the body. In other words, a point particle of the body's mass that is located at the radius of gyration is rotationally equivalent to the body itself. If we denote the radius of gyration by r_{gyr} in meters (m), we have:

$$r_{gyr} = \sqrt{\frac{I}{m}}$$

where I denotes the body's moment of inertia and m its mass.

Moment of inertia obeys the parallel-axis theorem, which relates a body's moment of inertia with respect to any axis to its moment of inertia with respect to a parallel axis through its CENTER OF MASS. If a body's moment of inertia with respect to an axis through its center of mass is I_0, then its moment of inertia, I, with respect to a parallel axis at perpendicular distance d in meters (m) from the center-of-mass axis is given by:

$$I = I_0 + md^2$$

where m denotes the body's mass.

momentum *See* LINEAR MOMENTUM.

Mössbauer effect Named for the German physicist Rudolph L. Mössbauer (1929–), the effect has to do with the FREQUENCIES of the PHOTONS (GAMMA RAYS) that are absorbed and emitted by NUCLEI as they make transitions between the same pair of ENERGY levels. For energy difference ΔE in joules (J) between two states of a nucleus, it might be expected, based on conservation of energy, that photons of exactly the frequency:

$$f = \Delta E/h$$

will excite such resting nuclei from the lower-energy state to the higher, where f is in hertz (Hz), and h is the PLANCK CONSTANT, whose value is $6.62606876 \times 10^{-34}$ joule·second (J·s). When such resting nuclei de-excite from the higher-energy state to the lower, it might similarly be expected that the emitted photons will possess the same frequency. Due to conservation of LINEAR MOMENTUM, however, the nuclei undergo recoil upon absorption and emission of photons. So in order to excite a resting nucleus, the photon must have a higher energy than ΔE, and consequently a higher frequency than the above, since some fraction of its energy subsequently appears as KINETIC ENERGY of the recoiling excited nucleus. In the inverse process of de-excitation, the amount of energy released by a resting nucleus, ΔE, is divided between the photon and the recoiling nucleus (as its kinetic energy), so that the photon's energy is less than ΔE and its frequency less than $\Delta E/h$. When the nuclei are initially in motion, the DOPPLER EFFECT, too, affects the photons' frequencies of absorption and emission. Therefore, such nuclei in a sample of material, due to their recoil and thermal motion, neither absorb nor emit photons at a single, well-defined frequency. (*See* CONSERVATION LAW; HEAT.)

Mössbauer discovered that the recoil can be eliminated by having the nuclei firmly attached to a CRYSTAL structure, so that it is the sample as a whole that recoils. Since the MASS of the sample is enormously greater than that of a single nucleus, the linear momentum transferred to the sample rather than to the nucleus results in negligible recoil motion. If, in addition, the sample is cooled considerably in order to reduce thermal motion, then the photon absorption and emission energies become very well defined and equal to each other. The Mössbauer effect allows production of photons of very precisely defined energy (and frequency) as well as high-resolution measurement of photon energy. As to the latter, if the photon energy does not quite match the nuclear excitation energy, the difference can be precisely measured by utilizing the Doppler effect and moving the sample toward or away from the arriving photons at such a speed that the photons' frequency relative to the sample is shifted to that required for excitation.

One important use of the Mössbauer effect was to confirm ALBERT EINSTEIN's prediction that photons lose or gain energy, with a corresponding decrease or increase of frequency, when they move upward or downward (i.e., against a gravitational FIELD or with it), respectively. (*See* GRAVITATION; RELATIVITY, GENERAL THEORY OF.)

See also RED SHIFT.

motion Change of spatial position or orientation taking place over TIME is what is called motion. An ideal point particle can only change position, while an object that possesses spatial extent can change both its position and its orientation. Change of position is linear motion, or translation, or translational motion, while change of orientation is ROTATION. For a body, it is convenient and common to consider separately the motion of its CENTER OF MASS and the body's rotation about its center of mass. If the center of mass moves, then the body is undergoing translational motion. Otherwise it is not. The most general motion of a body involves simultaneous motion of its center of mass and rotation about the center of mass. (*See* SPACE.)

As an example, consider the motion of the Moon. It is revolving around the Earth while keeping its same side facing the Earth. So as its center of mass performs nearly circular translational motion around the Earth with a PERIOD of about a month, the Moon is simultaneously rotating about its center of mass at exactly the same rate. (*See* CIRCULAR MOTION.)

The study of motion without regard to its causes is KINEMATICS. DYNAMICS studies the causes of motion and of changes in motion. Classical, nonrelativistic kinematics and dynamics are based on ISAAC NEWTON's laws of motion. ALBERT EINSTEIN generalized those laws in his special theory of relativity and further generalized them in his general theory of relativity to include GRAVITATION. (*See* CLASSICAL PHYSICS; NEWTON'S LAWS; RELATIVITY, GENERAL THEORY OF; RELATIVITY, SPECIAL THEORY OF.)

Free motion, or inertial motion, is motion in the absence of FORCEs. According to Newton's first law of motion, such motion is one of constant VELOCITY (i.e., motion at constant SPEED and in a straight line, which includes rest [zero speed]). In Einstein's general theory of relativity, free motion is geodesic motion, which is generally neither at constant speed nor in a straight line, but is the generalization of constant-velocity motion as appropriate to the theory. The path of a PHOTON passing near the Sun, for example, is not a straight line, but bends toward the Sun. According to the general theory of relativity, the photon is in free motion and is following the trajectory that is closest to a straight line—is a generalization of a straight line—in the SPACE-TIME geometry determined by the nearby presence of the Sun. (*See* INERTIA.)

mutual induction Mutual induction is the magnetic effect that one electric circuit or circuit element, usually a coil, has upon another. Mutual INDUCTION is indeed mutual, since if one circuit affects another, then the latter also affects the former. More precisely, let us denote by i_1, i_2 and by E_1, E_2 the electric CURRENTs in amperes (A) and the induced ELECTROMOTIVE FORCEs (emfs) in volts (V), respectively, in coils 1 and 2. Then these relations hold:

$$E_1 = -M\, di_2/dt$$

and:

$$E_2 = -M\, di_1/dt$$

where t is the TIME in seconds (s), and M denotes the mutual inductance of the two coils. The SI UNIT of mutual induction is the henry (H). Note that the same coefficient, M, appears in both equations. (*See* ELECTRICITY; MAGNETISM.)

A changing current in one coil induces an emf in the other coil to the same extent that a changing current in the second coil induces an emf in the first. These relations follow from AMPÈRE'S LAW and from the BIOT-SAVART LAW. The former determines that the induced emf in a coil is proportional to the time rate of change of the magnetic FLUX in that coil, while from the latter it follows that the magnetic flux in that coil is proportional to the current in the other coil. The negative sign in the equations indicates that the polarity of the induced emf is such as to reduce the cause of the emf, according to LENTZ'S LAW.

Mutual inductance can equivalently be viewed as a measure of the magnetic flux created in one coil by a current in the other coil, compensated for the number of turns in the first coil. Denote the number of turns in the coils by N_1, N_2. Also, denote the magnetic flux in coil 2 due to the current in coil 1 by Φ_{m21} and the magnetic flux in coil 1 due to the current in coil 2 by Φ_{m12}. The SI unit of magnetic flux is the weber (Wb). Then the relations are:

$$\Phi_{m21} = Mi_1/N_2$$

and:

$$\Phi_{m12} = Mi_2/N_1$$

Note again that the same coefficient, M, appears in both equations. So the mutual inductance of the pair of coils can be defined as:

$$M = N_2\Phi_{m21}/i_1 = N_1\Phi_{m12}/i_2$$

See also IMPEDANCE; SELF-INDUCTION.

N

National Institute of Standards and Technology (NIST) Founded in 1901 as the National Bureau of Standards, today's National Institute of Standards and Technology is a U.S. government agency under the Department of Commerce located at two sites: in Gaithersburg, Maryland, and in Boulder, Colorado. NIST's primary mission is to promote U.S. economic growth by working with industry to develop and apply measurements, standards, and technology. Its staff comprises about 3,000 scientists, engineers, technicians, and support personnel, together with around 1,600 visiting researchers.

This scanning electron micrograph shows a suspended microheater measuring about four micrometers (4×10^{-6} m) across at the National Institute of Standards and Technology (NIST). Arrays of such microheaters are used to test and calibrate infrared cameras and other imaging systems. ***(Courtesy of the National Institute of Standards and Technology)***

NIST boasts two Nobel Prizes in physics (1997 and 2001), both for experimental work at low TEMPERATURES.

The institute carries out its mission through four interwoven programs:

1. *Measurement and Standards Laboratories.* These include separate laboratories for each of physics, electronics and electrical engineering, manufacturing engineering, chemical science and technology, materials science and engineering, building and fire research, and information technology. The laboratories form a national resource for the standards, measurements, data, and calibrations that help U.S. industry succeed.
2. *Advanced Technology Program.* This program provides cost-shared awards to industry for the development of high-risk, enabling technologies with broad economic potential. It seeks to build bridges between basic research and product development for the purpose of accelerating technologies that otherwise are unlikely to be developed in time for rapidly changing markets.
3. *Manufacturing Extension Partnership.* Through a nationwide network of affiliated extension centers run by local, state, and nonprofit groups, NIST offers technological and business assistance to smaller manufacturers.
4. *Malcolm Baldrige National Quality Award Program.* This award and program recognize and promote achievements of U.S. companies in performance excellence and quality achievement.

The physics laboratory is involved in numerous projects, including:

- Standards and calibrations for diagnostic and therapeutic RADIATION in medicine.
- Calibrations of sensors used in monitoring the environment, such as remote temperature sensors in satellites and chemical sensors of atmospheric pollution and of chemical warfare agents.
- Maintenance and development of accurate clocks. Such TIME standards are required for the Global Positioning System and for telecommunications systems. The current atomic clock serving as a time standard for the United States is accurate to within a second in 6 million years. (*See* ATOM.)
- Research on ultrasmall structures, using special NIST-developed electron microscopes, to help the U.S. electronics industry develop smaller, thus faster, circuits and achieve higher densities of data storage. (*See* ELECTRON MICROSCOPY.)
- Research at the low-temperature frontier of science, including such as Bose-Einstein condensates. The resulting improved understanding about the behavior of atoms is being applied to improve the accuracy of atomic clocks and might have application in quantum computing. (*See* BOSE-EINSTEIN STATISTICS; QUANTUM PHYSICS.)

The National Institute of Standards and Technology website URL is http://www.nist.gov/, and that of the physics laboratory is http://physics.nist.gov/. The institute's mailing address is 100 Bureau Drive, Gaithersburg, MD 20899.

See also INTERNATIONAL BUREAU OF WEIGHTS AND MEASURES.

natural frequency *See* FREQUENCY, NATURAL.

neutrino A neutrino is any of a family of electrically neutral ELEMENTARY PARTICLEs that are the least massive of the LEPTONs. They are FERMIONs and have a half unit of SPIN. The masses of the neutrinos have not been measured to more than very coarse accuracy, but are considerably smaller than the mass of the ELECTRON. There are three types of neutrino, the electron-type neutrino (ν_e), the muon-type neutrino (ν_μ), and the tau-type neutrino (ν_τ), and their corresponding antiparticles. These, together with the electron (e^-), the muon (μ), and the tau (τ), constitute the leptons. All neutrino types interact very weakly with MATTER (via the WEAK INTERACTION and through GRAVITATION), so detecting them and investigating their properties are major undertakings. The Sun serves as a natural source of neutrinos that reach the Earth and mostly pass through it and through everything on it. Neutrinos do not DECAY, but they do undergo the process of mixing, whereby, as they travel, they continuously convert into varying quantum mixtures of all three neutrino types. As far as is presently understood, the neutrinos, like the other leptons, are structureless and pointlike. (*See* ANTIMATTER; CHARGE; ELECTRICITY; INTERACTION; MASS; QUANTUM PHYSICS.)

neutron This is a type of ELEMENTARY PARTICLE that belongs to the family of BARYONs and to the wider family of HADRONs. The conventional symbol for the neutron is n. Neutrons and PROTONs are the constituents of atomic NUCLEI and together are referred to as NUCLEONs. The neutron is electrically neutral and is a FERMION with a half unit of SPIN. Its MASS is 1.675×10^{-27} kilogram (kg) = 939.6 MeV/c^2. The neutron is not strictly elementary, as it is composed of three QUARKs: one up quark and two down quarks (udd). Free neutrons are unstable and DECAY, with a half-life of around 12 minutes, to a proton (p), an ELECTRON (e^-), and an electron-type antineutrino ($\bar{\nu}_e$):

$$n \rightarrow p + e^- + \bar{\nu}_e$$

This is the same process as that of nuclear BETA DECAY, in which the decay of a neutron in an unstable nucleus leaves a proton in the nucleus and emits an electron and an antineutrino. Neutrons are affected by all the fundamental INTERACTIONs. That includes the ELECTROMAGNETIC INTERACTION, even though neutrons are electrically neutral, since they do possess a magnetic dipole moment. (*See* ANTIMATTER; ATOM; CHARGE; ELECTRICITY; MOMENT, MAGNETIC DIPOLE; NEUTRINO.)

neutron star A neutron star is the final stage in the evolution of stars whose MASS is in the range of approximately eight to 25 times the mass of the Sun. It is an extremely dense object consisting of tightly packed NEUTRONs, mostly (hence its name), and PROTONs, as well as ELECTRONs. A neutron star contains around the mass of the Sun, but not more than three solar masses, within a diameter of about 10 kilometers. (The initial star ejects most of its mass as it collapses to form a neutron star.) Its DENSITY is that of nuclear matter, between 10^{17} and 10^{18} kilograms per cubic meter. Neutron stars normally rotate at very high angular SPEEDS. (*See* ASTROPHYSICS; DENSITY; NUCLEUS; ROTATION.)

Newton, Sir Isaac (1642–1727) British *Mathematician, Physicist* One of the most famous scientists of all time, Sir Isaac Newton is best known today for his laws of MOTION and GRAVITATION and his invention of calculus. The SI UNIT of FORCE, the newton (N), is named for him. Newton studied at Cambridge University, England, from 1661 to 1668, when he was appointed professor of mathematics there. He remained at Cambridge for almost 30 years and then took a government position in London. Newton's years at Cambridge were his most fruitful and productive ones for science research. In addition to the work leading to his laws of motion and gravitation, known as NEWTON'S LAWS, he also carried out investigations in OPTICS, chemistry, and mathematics, among other fields. It is generally accepted that one of Newton's greatest achievements was his demonstration that scientific principles are of universal application. This ushered in the Age of Reason, characterized by the expectation that the methods of science allow understanding of the UNIVERSE by revealing the fundamental laws that govern it.

See also LAWS OF NATURE.

Sir Isaac Newton, who lived in the 17th and 18th centuries, was one of the most famous scientists of all time and is best known for his laws of motion and gravitation and the invention of calculus. *(Massachusetts Institute of Technology Burndy Library, courtesy AIP Emilio Segrè Visual Archives)*

Newton's laws Named for their discoverer, the British physicist and mathematician ISAAC NEWTON, Newton's laws consist of his three laws of MOTION and his law of gravitation, sometimes called the law of universal GRAVITATION. These laws are found to govern classical MECHANICS (i.e., the mechanics of the macroscopic world) as long as the SPEEDs involved are not too great, the DENSITIES not too high, and the distances between bodies not too small, but less than cosmological distances. (*See* CLASSICAL PHYSICS; COSMOLOGY.)

In modern form, Newton's laws can be stated as follows:

First Law

First law of motion. In the absence of FORCEs acting on it, a body will remain at rest or in motion at constant speed in a straight line.

This law is also referred to as the law of inertia, since it describes a body's inertial motion (i.e., its motion in the absence of forces). The law states that the inertial motion of a body is motion with no ACCELERATION. The effect of a force, then, is to bring about noninertial motion, which is accelerated motion. Another view of this law is as a definition of inertial reference frames, in which the second law of motion is valid. One assumes that the presence of forces acting on a body is readily identifiable independently of their effect on the body, such as by contact with another body or by the presence of nearby bodies affecting it gravitationally or electromagnetically. When forces are known to be absent and a body's motion is unaccelerated, the reference frame in use is an inertial one. But if in the absence of forces a body's motion is accelerated, it is being observed in a noninertial reference frame. Such accelerations are sometimes "explained" by fictional forces, such as CENTRIFUGAL FORCE. (*See* CENTRIPETAL FORCE; ELECTRICITY; ELECTROMAGNETISM; INERTIA; MAGNETISM; REFERENCE FRAME, INERTIAL.)

Second Law

Second law of motion. The effect of a force on a body is to cause it to accelerate, where the direction of the acceleration is the same as the direction of the force, and the magnitude of the acceleration is proportional to the magnitude of the force and inversely proportional to the body's MASS. In a formula:

$$F = m\mathbf{a}$$

where F denotes the force, a VECTOR, in newtons (N); m is the body's mass in kilograms (kg); and **a** is the body's resulting acceleration, also a vector, in meters per second per second (m/s^2).

An alternative formulation of the second law of motion, one that is more generally valid, is as follows. The effect of a force on a body is a change in the body's LINEAR MOMENTUM, where the direction of the change is the same as the direction of the force, and the magnitude of the rate of change over time is proportional to the magnitude of the force. Expressed in a formula:

$$F = d\mathbf{p}/dt$$

where **p** denotes the body's momentum, a vector, in kilogram·meters per second (kg·m/s), and t is the time in seconds (s). The momentum is:

$$\mathbf{p} = m\mathbf{v}$$

with **v** the body's VELOCITY, also a vector, in meters per second (m/s).

The second law of motion is valid only in inertial reference frames (i.e., reference frames in which the first law of motion holds).

Third Law

Third law of motion. When a body exerts a force on another body, the second body exerts on the first a force of equal magnitude and of opposite direction.

This law is also referred to as the law of action and reaction. What it is telling us is that forces come in pairs, where one is the action force and the other the REACTION FORCE. Note that the forces of an action-reaction pair *act on different bodies.* Either of such a pair can be considered the action force, although one usually thinks of the force we have more control over as the action, with the other serving as its reaction. If you push on a wall, for instance, the wall pushes back on you with equal magnitude and in opposite direction. We usually think of your push as the action and the wall's push as its reaction. But the reaction force is no less real than the action force. If you are standing on a skateboard, for example, and push on a wall, you will be set into motion away from the wall by the wall's force on you, according to the second law of motion.

While the first and third laws of motion appear to be generally valid, the second law is correct only for speeds that are small compared with the SPEED OF LIGHT. The second law, in momentum form, was generalized for any speed by ALBERT EINSTEIN's special theory of relativity. Accordingly, Newtonian mechanics is an approximation of Einsteinian mechanics that is valid for sufficiently small speeds. (*See* RELATIVITY, SPECIAL THEORY OF.)

Gravitation

Law of gravitation. Every pair of particles in the UNIVERSE undergoes mutual attraction, where the force on each particle is directed toward the other, and the force's magnitude is proportional to the product of the particles' masses and inversely proportional to the square of the distance between the particles. In a formula, the law of gravitation for the magnitude of the mutual-attraction force, F, is:

$$F = Gm_1m_2/r^2$$

where F is in newtons (N), m_1 and m_2 denote the masses of the particles in kilograms (kg), r is the distance separating the particles in meters (m), and G is the gravitational constant, whose value is found by measurement to be 6.67259×10^{-11} N·m²/kg².

Newton's law of gravitation is valid for sufficiently low densities and large separations, as long as they are less than cosmological distances. Einstein's general theory of relativity generalizes Newtonian gravitation for all situations in the macroscopic domain. (*See* RELATIVITY, GENERAL THEORY OF.)

Newton developed his laws mainly as a theory of (i.e., an explanation for) KEPLER'S LAWS.

node Any location in an oscillating continuous system at which the amplitude of oscillation is zero is referred to as a node. Such oscillations are in reality a standing-WAVE phenomenon, so a node is, equivalently, any location along a standing wave at which the medium is at rest. The term *antinode* is used for any position in an oscillating continuous system where the oscillation amplitude is maximal. In one-dimensional systems, nodes and antinodes occur at alternating points

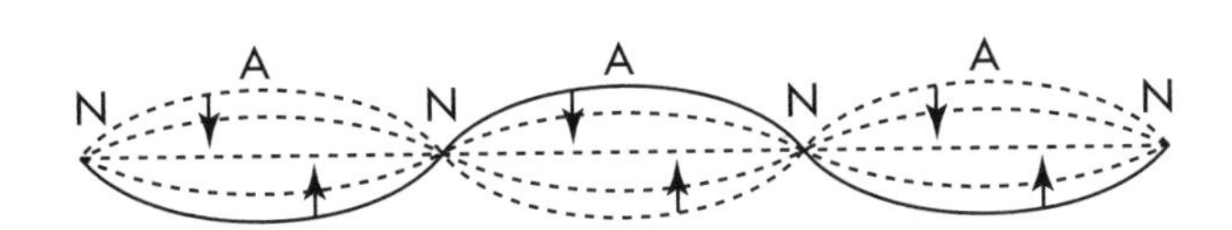

An example of nodes and antinodes in an oscillating stretched string, a one-dimensional continuous system. Nodes, denoted N, are the locations at which the amplitude of oscillation is zero. Locations of maximal amplitude are the antinodes, labeled A.

along the system. As an example, shake the end of a stretched rope transversely and find FREQUENCIES at which standing waves are generated (i.e., at which the rope oscillates without apparent wave propagation). Alternating nodes and antinodes will be obvious along the rope. Their number increases with increase of frequency. For standing waves in a uniform rope, the distance between adjacent nodes or adjacent antinodes equals half the WAVELENGTH of the transverse wave that is propagating in the rope, reflecting from the ends, and interfering with itself, thus producing the standing wave. In systems of two or three DIMENSIONS, such as a stretched membrane or an air-filled cavity, respectively, the nodes and antinodes might appear at points, along lines, or over surfaces and can take on various and complex configurations. (*See* FREQUENCY, NATURAL; INTERFERENCE; OSCILLATION; REFLECTION.)

nonconservative force *See* CONSERVATIVE FORCE.

nuclear physics The field of physics that deals with atomic nuclei is called nuclear physics. It studies the properties of nuclei, their structure, and their reactions, as well as applications of all those. While the STRONG INTERACTION is at its strongest when it binds together the triplets of QUARKs that form NUCLEONS (PROTONs and NEUTRONs, which compose nuclei), it "leaks" from the nucleons in a weakened version that holds nucleons together as nuclei, against the repulsive electric FORCE among the protons. So the study of nuclei is a way of gaining understanding of the strong interaction. A major tool for the study of nuclei is the particle accelerator. One of the important applications of nuclear physics is nuclear power, the production of ENERGY from nuclear reactions. Controlled production is performed exclusively in nuclear REACTORs, in which energy is obtained from controlled FISSION reactions. Uncontrolled nuclear fission is the basis of nuclear-fission bombs, also called ATOM bombs. Nuclear-fusion bombs, often referred to as hydrogen bombs, are based on uncontrolled FUSION reactions. Such applications are the domain of nuclear engineering. The goal of obtaining controlled nuclear power from fusion reactions is being actively pursued. Another important range of application is in medicine. Medical uses of nuclear physics form a field called nuclear medicine. (*See* ACCELERATOR, PARTICLE; ELECTRICITY; NUCLEUS.)

nucleon The nucleons are the family of ELEMENTARY PARTICLEs that compose atomic NUCLEI. The family comprises the PROTON and the NEUTRON. The nucleons form a subfamily of the BARYONs, which in turn make up a division of the HADRONs, the particles that are affected by the STRONG INTERACTION. Like the other baryons, the nucleons are composed of three QUARKs each, bound together by the strong interaction, which is mediated by GLUONs. The effect of the interquark interaction ranges beyond the volume of each nucleon to produce a weaker version that binds the nucleons together to form nuclei. In its internucleon version, the strong interaction is mediated by PIONs. In order to enable stable structures for nuclei, the internucleon strong interaction must overcome the mutual electric repulsion of the protons. (*See* ATOM; ELECTRICITY.)

nucleus The compact central component of an ATOM is called the nucleus. It contains practically all the MASS of the atom, while its size is on the order 10^{-14} meter (m), about one ten-thousandth the size of an atom. Nuclei are in general composed of both PROTONs, which carry one elementary unit of positive electric CHARGE each, and NEUTRONS, which are electrically neutral. (The sole exception is the hydrogen nucleus, consisting of only a single proton.) Both kinds of particle are called NUCLEONs. The nucleons in a nucleus are bound together by the STRONG INTERACTION, which in its internucleon version is mediated by PIONs. (*See* ELECTRICITY.)

The number of protons in a nucleus is its ATOMIC NUMBER, which determines the nucleus's electric charge and its atom's chemical properties and identity (i.e., to which chemical ELEMENT its atom belongs). The total number of nucleons in a nucleus is its atomic number (or atomic mass number or mass number). A form of

MATTER whose nuclei possess both a definite atomic number and a definite mass number is an ISOTOPE of the corresponding chemical element. Naturally occurring elements generally consist of a mixture of isotopes. The term *nuclide* is used for a type of nucleus characterized by both a particular atomic number and a particular mass number. Thus the nuclei of an isotope are all of the same nuclide.

The electric charge of a nucleus, q, is proportional to its atomic number, Z, and has the value:

$$q = Ze,$$

where q is in coulombs (C), and e is the magnitude of the elementary unit of electric charge, with the value $1.602176462 \times 10^{-19}$ C.

The nucleons in a nucleus are closely packed, so all nuclei possess practically the same DENSITY, which is uniform throughout most of a nucleus's VOLUME and has the value of around 2.3×10^{17} kilograms per cubic meter (kg/m^3). So the volume of a nucleus is proportional to its atomic mass, giving the relation between the radius, R, of a nucleus and its atomic mass, A:

$$R = R_0 A^{1/3}$$

Here R and R_0 are in meters (m), and R_0 has the value of about 1.2×10^{-15} m.

Unstable nuclides correspond to radioactive isotopes. They undergo DECAY to daughter nuclides in a number of possible ways, including ALPHA DECAY, BETA DECAY, inverse beta decay, and GAMMA DECAY. The specific DECAY PROCESS depends on the nuclide. (*See* RADIOACTIVITY.)

See also BINDING ENERGY; FISSION; FUSION, NUCLEAR; NUCLEAR PHYSICS.

Oak Ridge National Laboratory (ORNL) Established as Clinton Laboratories in 1943 at Oak Ridge, Tennessee, Oak Ridge National Laboratory (ORNL) was founded with the sole purpose of investigating the production and separation of plutonium for the Manhattan Project. This project formed the U.S. government's World War II effort to develop a nuclear-FISSION ("atomic") bomb. Since then, ORNL has evolved into a multiprogram science and technology laboratory under the U.S. Department of Energy. The laboratory employs more than 3,800 people, of whom 1,500 are scientists and engineers, and hosts annually about 3,000 guest researchers who spend at least two weeks there, including some 700 from industry. (*See* NUCLEAR PHYSICS.)

The laboratory's work addresses national and global ENERGY and environmental issues and, in particular, involves the development of new energy sources, technologies, and materials. In addition, ORNL carries out research projects in the biological, chemical, computational, engineering, environmental, physical, and social sciences. Its five major initiatives are:

- *Neutron sciences.* These include nuclear ASTROPHYSICS, NEUTRON scattering, radioisotope production, irradiation testing of materials, and neutron activation analysis. In this connection, ORNL is constructing the accelerator-based Spallation Neutron Source for neutron research, scheduled for completion in 2006. (*See* ACCELERATOR, PARTICLE; ISOTOPE.)
- *Complex biological systems.* Included here are functional genomics and proteomics, structural biology,

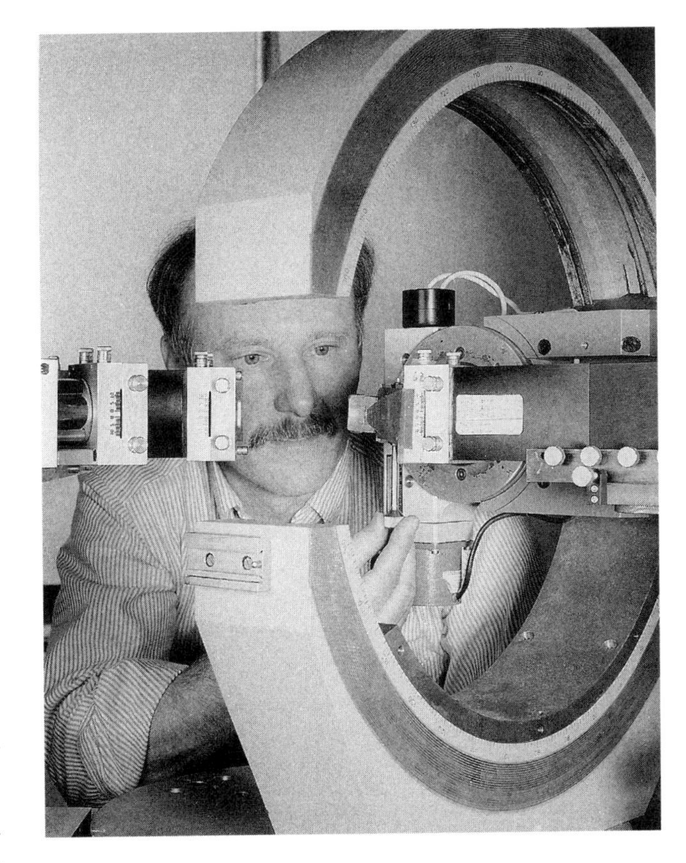

An X-ray diffractometer—an instrument for measuring the diffraction of X-rays—is used in the Solid State Division of Oak Ridge National Laboratory to analyze the structure of substrates and thin films for superconductors. It was found, for instance, that rolled pieces of silver and other metals can increase the superconductivity of an overlying film of yttrium-barium-copper-oxide, which is a high-temperature superconductor. ***(ORNL Photo by Tom Cerniglio. Courtesy of Oak Ridge National Laboratory)***

plant and microbial genomics, and computational biology and bioinformatics.

- *Terascale computing and simulation science.* ORNL is developing computational tools to address complex problems in science and engineering, including properties of materials, climate change, fuel combustion, and protein folding.
- *Energy and environmental systems of the future.*
- *Advanced materials.* This includes nanoscale science, engineering, and technology; materials characterization; and the synthesis, characterization, and processing of soft materials.

New facilities are under construction at ORNL, which include the Functional Genomics Center, the Center for Nanophase Materials Science, the Advanced Materials Characterization Laboratory, and the Joint Institute for Computational Science.

The laboratory's physics division performs research in the fields of theoretical and experimental nuclear and atomic physics. (*See* ATOM; EXPERIMENTAL PHYSICS; THEORETICAL PHYSICS.)

The URL of Oak Ridge National Laboratory website is http://www.ornl.gov/, and the website of the ORNL physics division is found at the URL http://www.phy.ornl.gov/. The laboratory's mailing address is P.O. Box 2008, Oak Ridge, TN 37831.

Ohm's law This law, named for the German physicist Georg Simon Ohm (1789–1854), states that the electric potential difference, or VOLTAGE, V, across an electric circuit component and the CURRENT, i, through the component are proportional to each other:

$$V = iR$$

where R denotes the component's RESISTANCE. Here V is in volts (V), i is in amperes (A), and R is in ohms (Ω). This relation is valid for direct current (DC) as well as for alternating current (AC). Ohm's law is not a law in the usual sense but, rather, a description of the behavior of a class of materials for a limited range of voltages. A material that obeys Ohm's law for some range of voltages is termed an ohmic conductor. A nonohmic conductor conducts electricity but does not exhibit proportionality between the current and the voltage. (*See* CONDUCTOR; ELECTRICITY; POTENTIAL, ELECTRIC.)

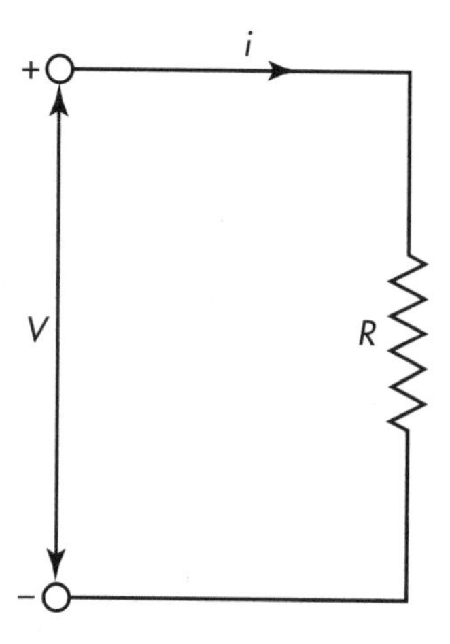

Ohm's law states that the voltage, *V*, across a circuit component, in this example a resistor, and the electric current, *i*, through the component are proportional to each other: *V* = *iR*, where *R* is the component's resistance. The current flows from higher potential (+) to lower (−).

optics The study of the behavior of LIGHT, in particular, and of electromagnetic and other WAVES and RADIATION, in general, is called optics. Wave optics, or physical optics, takes into account the wave nature of light. An important principle of wave optics is HUYGENS'S PRINCIPLE, which helps to construct the WAVEFRONTS that evolve by propagation from any given wavefront. Wave optics involves effects such as DIFFRACTION and INTERFERENCE. Ray optics, or geometric optics, deals with situations in which light propagates as rays in straight lines, and its wave nature can be ignored. Major topics in ray optics include REFLECTION and REFRACTION, which relate to MIRRORS and LENSES. (*See* ELECTROMAGNETISM.)

order The possession of differences or distinctions by a system is termed order. It is the opposite of DISORDER, which is the lack of differences and distinctions. As an example, consider a sample of some substance in its LIQUID form and in its CRYSTAL form. In the liquid form, the MOLECULES are moving about randomly and, over time, any molecule might be at any location within the sample's VOLUME. There is no distinction of one location from another or one direction from another within the material. In the crystal form, on the other hand, the molecules are fixed at locations on the crystal LATTICE. That endows the crystal with distinctions among different locations and among different directions. So the crystal form possesses more order than the liquid, or conversely, the liquid has more disorder than does the crystal.

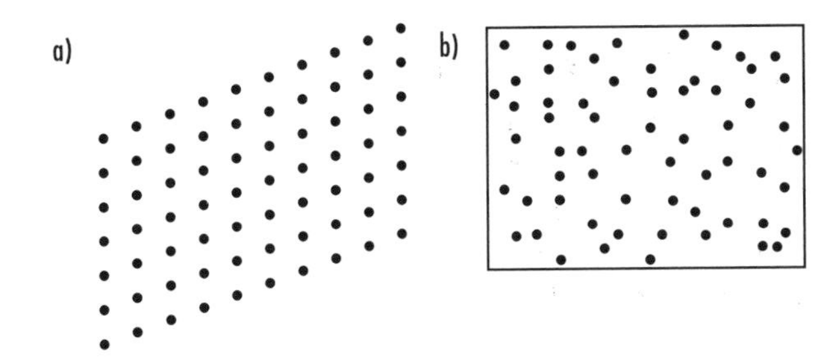

Order-disorder comparison. A substance in its crystal form (a) possesses a higher degree of order than it does in its liquid form (b). Conversely, the latter form has more disorder than the former.

Any situation can be described as lying somewhere along the order-disorder axis, where disorder correlates with homogeneity, randomness, and regularity, and order reflects heterogeneity, nonrandomness, and irregularity. ENTROPY is a measure of disorder. SYMMETRY, too, is correlated with disorder. (*See* STATISTICAL MECHANICS; THERMODYNAMICS.)

oscillation The repetitive variation in the value of a quantity is called oscillation. As an example, the position of a swinging PENDULUM undergoes oscillation, or oscillates, about the EQUILIBRIUM position. Also, the VOLTAGE in alternating CURRENT oscillates about the zero value. The minimal segment of oscillation that is repeated is a cycle. The time duration of a cycle is the PERIOD of the oscillation. The number of cycles that occur per unit of time is the FREQUENCY. The period, T, in seconds (s) and the frequency, f, in hertz (Hz) are mutual inverses:

$$f = 1/T$$

An important type of oscillation is simple harmonic oscillation, in which the variation is a sinusoidal function of time. The voltage variation in alternating current, for instance, is of this type:

$$V = V_0 \sin (2\pi ft + \alpha)$$

where V denotes the oscillating voltage in volts (V); V_0 is the maximal voltage, also in volts; f is the frequency in hertz (Hz); t represents the time in seconds (s); and α is the phase constant, or phase shift, in radians (rad). (*See* HARMONIC MOTION; PHASE.)

See also VIBRATION.

P

pair production The conversion of a PHOTON, of sufficient ENERGY to be designated a GAMMA RAY, into an ELECTRON and a POSITRON, the electron's antiparticle, is called pair production. The process cannot take place in isolation, since it cannot simultaneously conserve both energy and LINEAR MOMENTUM. Pair production can take place in the vicinity of an atomic NUCLEUS, which then absorbs some of the linear momentum. Since the energy equivalent of the sum of the MASSes of an electron and a positron is 1.022 mega-electron-volts (MeV), in order for a photon to be able to produce an electron-positron pair it must possess at least that amount of energy. Any extra energy appears as KINETIC ENERGY, mostly of the electron and positron, and also of the helper nucleus. With sufficient energy, a gamma ray might produce any particle-antiparticle pair, such as a PROTON and an antiproton. Among the pairs whose production is allowed by a photon's energy, the greater a pair's mass, the lower the probability of its production by the photon. (*See* ANTIMATTER; ATOM; CONSERVATION LAW; MASS ENERGY.)

The inverse process to pair production is particle-antiparticle annihilation. In this process, a particle and its antiparticle meet and, while undergoing mutual annihilation, convert to a pair of photons. The process can take place in isolation. The combined energy of the photons equals the total energy of the particle-antiparticle pair before annihilation, including its mass energy. In the CENTER-OF-MASS REFERENCE FRAME, the photons leave the annihilation site in opposite directions.

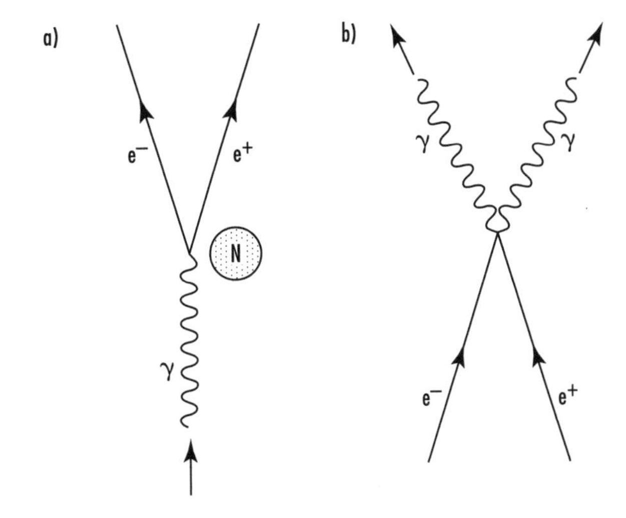

Pair production and annihilation. (a) A gamma ray (energetic photon), γ, converts to an electron-positron pair, e^- and e^+, in the vicinity of an atomic nucleus, N. (b) In the inverse process, an electron and a positron mutually annihilate and produce two gamma rays. Time runs upward in the figures.

paramagnetism The weak response of certain materials to an applied magnetic FIELD, by which the material becomes polarized as a MAGNET (i.e., magnetized) so that it is weakly attracted to the polarizing magnet, is referred to as paramagnetism. The effect is a result of the material's ATOMs each possessing one or a few unpaired ELECTRONs, which make the atom a MAGNETIC DIPOLE and endow it with a magnetic dipole moment. That occurs because each electron is itself a magnetic dipole, which is a property related to the electron's SPIN.

The applied magnetic field tends to align the atomic magnetic dipole moments against the disorienting effect of the atoms' random thermal motion. The result is a weak net orientation, which gives the material a weak net magnetic dipole moment in the direction of the applied field. When the applied field is removed, the material immediately loses its magnetization. A paramagnetic material tends to enhance the magnetic field acting on it, so its PERMEABILITY is greater than the permeability of the VACUUM, whose value is $4\pi \times 10^{-7}$ tesla·meter per ampere (T·m/A). The effect is weak, however, so the permeability of a paramagnetic material is not much more than the permeability of the vacuum. (*See* HEAT; MAGNETISM; MOMENT, MAGNETIC DIPOLE; POLARIZATION.)

See also DIAMAGNETISM; FERROMAGNETISM.

parameter This is any physical quantity that characterizes the state of a system and is constant in time or only slowly varying or is a quantity that specifies a physical process. INITIAL CONDITIONS are among the parameters of a process. As an example, consider the system of a block on an inclined plane and the process of its sliding down the plane. The parameters of the system are the MASS of the block, the slope of the plane, the coefficient of FRICTION between the block and the plane, and the value of the ACCELERATION due to gravity. To specify the sliding process, the additional parameters of the block's initial position and VELOCITY are needed. (*See* GRAVITATION.)

For another example, in the approximation of point bodies and ignoring everything but the Sun and the planets, the parameters characterizing the evolution of the SOLAR SYSTEM are the masses of the Sun and planets, the planets' positions and velocities at any time (as initial conditions), and the value of the gravitational constant. As a further example, for the isobaric expansion of a sample of GAS when HEAT is added to it, the parameters are the quantity of gas, its HEAT CAPACITY, the amount of heat added, and the gas's initial VOLUME and PRESSURE (or volume and TEMPERATURE, or pressure and temperature). (*See* ISOBARIC PROCESS.)

Note that in common usage, the word *parameter* is often taken to mean constraint or limit, such as in the expression "within these parameters." That is different from the way the term is used in science, as described above.

parity Parity has to do with the behavior of functions of position under the operation of SPACE inversion, which is inversion through the coordinate origin. When Cartesian coordinates are used, inversion is performed by replacing x, y, z with $-x$, $-y$, $-z$. If a function of position or positions remains unchanged under the transformation, the function is said to possess even, or positive, parity. If the function changes only its sign, it has odd, or negative, parity. If it changes in any other way, the function is not characterized by parity. While parity is of use in CLASSICAL PHYSICS, it is important in QUANTUM PHYSICS, where parity QUANTUM NUMBERS can be assigned to states and to the wave functions representing them: +1 for even (positive) and –1 for odd (negative). Then parity considerations can be invoked in SELECTION RULEs for transitions among quantum states. (*See* COORDINATE SYSTEM.)

Parity quantum numbers are assigned to the ELEMENTARY PARTICLEs and to collections of particles, whether bound, such as NUCLEI, or not. The parity of a group of particles equals the product of the parities of the individual constituents multiplied by a factor involving their ANGULAR MOMENTUM. Parity is conserved in processes governed by the STRONG INTERACTION, the ELECTROMAGNETIC INTERACTION, or, as far as is presently known, GRAVITATION. The WEAK INTERACTION, on the other hand, does not conserve parity. (*See* CONSERVATION LAW.)

particle physics Particle physics is the field of physics that studies the properties and INTERACTIONs of the ELEMENTARY PARTICLEs. Particle physics is also called elementary particle physics, high-energy physics, or physics of particles and fields. The field is divided into three subspecialties: experimental particle physics, particle phenomenology, and theoretical particle physics. This trifold division is unusual, since most other fields of physics are divided only into experimental and theoretical subspecialties, and some not even into those. The particle-physics experimentalists perform experiments and collect data. Their tools are mainly particle accelerators, detectors, and, it almost goes without saying, powerful computers. The experiments involve bringing about COLLISIONs among particles and detecting and analyzing the results of the collisions. The collisions might be between moving particles and particles at rest or between moving particles and particles moving in the opposite direction. The moving particles might be supplied by an accelerator—either directly or as particles produced when accelerated particles collide with stationary ones—or by nature, in the form of particles reaching Earth from beyond. The particles emanating from the collisions are detected by detecting devices whose signals are analyzed by computer to determine the identities and properties of the collision products. (*See* ACCELERATOR, PARTICLE; COSMIC RAY; EXPERIMENTAL PHYSICS.)

The subfield of particle phenomenology is the study and interpretation of experimental data. Phenomenologists plot the results of the experiments in various ways and attempt to find ORDER among them and to interpret them in light of theoretical ideas and models. In general, it is the results from the phenomenologists that are the grist for the mill of the theoreticians. The basic goal of theoretical particle physics is to understand, in fundamental terms, the properties and interactions of the elementary particles (i.e., to devise theories of the elementary particles). Currently accepted theories are constantly tested against experimental data. Deviations from expected results are studied as possible guides to modifications of theories or to new theories altogether. (*See* THEORETICAL PHYSICS.)

There appears to be an intimate relation between particle physics and COSMOLOGY. The types and properties of particles and particle interactions that exist at present presumably came about through cosmological processes. On the other hand, certain stages in the evolution of the UNIVERSE, such as immediately following the BIG BANG, according to current thinking, must surely have been affected by the properties of the elementary particles and their interactions.

Pascal's principle Formulated by the French physicist and mathematician Blaise Pascal (1623–62), the principle named for him states that a pressure applied to a FLUID (i.e., a GAS or a LIQUID) in a closed container is transmitted undiminished throughout the fluid as well as to the walls of the container. This principle is one of the principles of HYDRAULICS and of HYDROSTATICS. Pascal's principle underlies the operation of hydraulic car lifts, for example, in which a pump raises the pressure in the operating fluid (a liquid in this case), which is transmitted to the piston that pushes the car upward.

Pauli, Wolfgang (1900–1958) Austrian/Swiss *Physicist* Famous mostly for his quantum principle, called

Wolfgang Pauli contributed to the foundations of quantum field theory. He discovered the exclusion principle, named for him, for which he was awarded the 1945 Nobel Prize in physics. *(AIP Emilio Segrè Visual Archives, Goudsmit Collection)*

the PAULI EXCLUSION PRINCIPLE, Wolfgang Pauli was a theoretical physicist and an active researcher in QUANTUM PHYSICS and the theory of ELEMENTARY PARTICLES. Pauli received his doctorate in physics in 1922 from the University of Munich, Germany. He served for five years as a lecturer at the University of Hamburg, Germany. In 1928, Pauli was appointed professor of THEORETICAL PHYSICS at the Federal Institute of Technology in Zurich, Switzerland. He remained there for the rest of his life, except for the period of World War II, which he spent in the United States at the Institute for Advanced Study in Princeton, New Jersey. In his work in quantum physics, he discovered the extremely profound exclusion principle, named for him, which is essential for understanding the structure of ATOMs and the PERIODIC TABLE of the chemical ELEMENTs. Pauli also contributed to the foundations of QUANTUM FIELD THEORY. His study of BETA DECAY led to his proposal that an undetected particle is emitted together with the ELECTRON in this decay. Pauli's assumption was confirmed, and the particle is now known as an antineutrino of electron type. In 1945, Pauli was awarded the Nobel Prize in physics "for the discovery of the Exclusion Principle, also called the Pauli Principle."

See also ANTIMATTER; NEUTRINO.

Pauli exclusion principle Named for the Austrian/Swiss physicist, WOLFGANG PAULI, who discovered it, the Pauli exclusion principle states that no more than a single FERMION of the same type, such as an ELECTRON or a NEUTRON, can exist in the same quantum state. The principle is incorporated into FERMI-DIRAC STATISTICS, which governs collections of identical fermions. (*See* QUANTUM THEORY.)

One application of the Pauli principle is to the understanding of the operation of CONDUCTORS, SEMICONDUCTORS, and INSULATORS. There, the principle explains the manner in which electrons occupy the available ENERGY levels of the material. Whereas in the absence of the exclusion principle the lowest-energy states would be much more heavily populated than the higher-energy ones, in reality the states are populated two electrons per level from the lowest energy up. The reason for two electrons per level rather than one is that the SPIN of the electron can point in one of two opposite directions (say, up and down), and each direction determines a different state. What the Pauli principle forbids in this case is two or more electrons occupying the same energy level and having their spins in the same direction.

Another application of the Pauli exclusion principle is to atomic structure. The states available to electrons in an ATOM are ordered into shells and subshells and are characterized by a set of three QUANTUM NUMBERS. Without the principle, the electrons would tend to heavily populate the lowest-energy states. What occurs, however, is that the states are populated two electrons per state from the lowest-energy state, i.e., the GROUND STATE, up. Thus, each electron in an atom can be viewed as being characterized by a unique set of values for four quantum numbers, consisting of the three above-mentioned ones with the addition of the spin quantum number ($+^1/_2$ for "spin up," and $-^1/_2$ for "spin down").

pendulum Any object that is suspended from a pivot and can oscillate by swinging about its EQUILIBRIUM position is termed a pendulum. The equilibrium position of a pendulum is that in which the body's CENTER OF MASS is directly below the pivot. When the body is rotated from its equilibrium position and released, it swings back toward equilibrium, overshoots that position, comes to momentary rest, returns toward equilibrium, overshoots again, and so on. As a pendulum oscillates, it gradually loses ENERGY through FRICTION at the pivot and with the air and eventually comes to rest. (*See* INERTIA; OSCILLATION; ROTATION.)

From the energy point of view and ignoring friction, the swinging of a pendulum is a continuous, repetitive conversion and reconversion of gravitational POTENTIAL ENERGY to KINETIC ENERGY and back again, so that the sum of the two forms of energy is constant. At the highest point of its swing, a pendulum is momentarily at rest and all its energy is in potential form. As the pendulum descends, accelerating, toward equilibrium, its potential energy is being converted to kinetic energy, and as it passes through equilibrium position, its SPEED and kinetic energy are greatest and its potential energy least. As it continues and ascends from equilibrium position, the pendulum decelerates, thereby losing kinetic energy and gaining potential energy. When it reaches its highest point and is momentarily at rest again, it has lost all kinetic energy and possesses the same amount of potential energy it had on the other side of its swing. So it attains the same height on both sides of all swings. (*See* ACCELERATION; CONSERVATION LAW; GRAVITATION.)

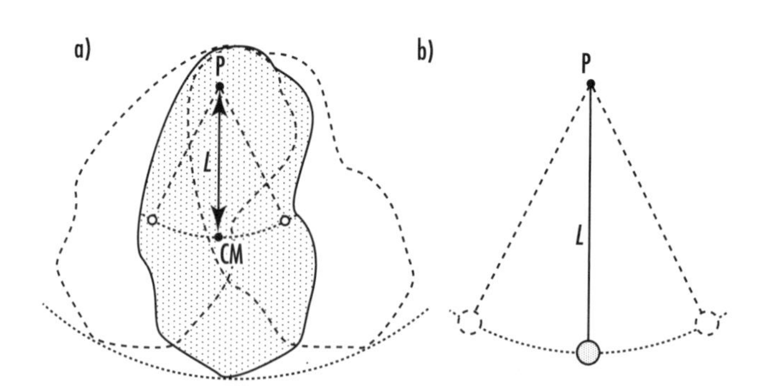

(a) A real (physical) pendulum. P denotes the pivot and CM the pendulum's center of mass. The distance between the two is *L*. (b) An ideal (mathematical) pendulum. *L* denotes the pendulum's length.

The PERIOD of a pendulum's oscillations depends on the amplitude of its swing. For small amplitudes, however, the period is independent of amplitude and is given by:

$$T = 2\pi\sqrt{\frac{I}{mgL}}$$

where T denotes the period in seconds (s), I is the MOMENT OF INERTIA of the pendulum with respect to the pivot in kilogram·meters² (kg·m²), m is the pendulum's mass in kilograms (kg), g is the acceleration due to gravity with nominal value 9.8 meters per second per second (m/s²), and L is the distance in meters (m) between the body's center of mass and the pivot. "Small amplitude" means that the pendulum's maximal linear DISPLACEMENT from equilibrium position is small compared with L, or that its angular displacement is no more than a few degrees. Since m appears explicitly in the formula for T, it might seem that the period depends on mass. That is not the case in fact. The mass of the pendulum enters into the moment of inertia and cancels in the above formula. What the period does depend on in this regard is the spatial distribution of the pendulum's mass with respect to the pivot.

The general pendulum is called a real pendulum, or physical pendulum. An ideal pendulum, also called a mathematical pendulum, is a point body suspended by a massless, inextensible string. An approximation to an ideal pendulum is a small body suspended by a low-mass string that is stretched but little by the WEIGHT of the body. A "small" body is one whose size is small compared to the length of the string, while the mass of a "low-mass" string is small in relation to the mass of the body. Similarly, "stretched but little" means that the string's extension is negligible compared to its length. For an ideal pendulum, the formula for the period of small-amplitude swings is:

$$T = 2\pi\sqrt{\frac{L}{g}}$$

where L now denotes the length of the string in meters (m). Note that there is no dependence on mass.

A second pendulum is an ideal pendulum for which a single swing, in either direction, takes one second, so it has a period of two seconds. From the previous formula, and using the coincidence that at the surface of the Earth $g \approx \pi^2$ m/s², it turns out that the length of a second pendulum is very close to one meter.

A pendulum can be used to measure the acceleration due to gravity, since g appears in both the above formulas. For an ideal pendulum, measurements of its length and period yield the result for g:

$$g = (2\pi/T)^2 L$$

period The time duration of the minimal repeated segment of an oscillation (i.e., the time duration of a single cycle) is the period of the oscillation. Its SI UNIT is the second (s). As an example, a single cycle of a swinging PENDULUM consists of one swing "to" and another swing "fro," after which the pendulum returns to its initial position. The duration of that motion is the period of the pendulum's oscillation. The period of oscillation is the inverse of the oscillation's FREQUENCY, which is the number of cycles per unit time and whose SI unit is the hertz (Hz):

$$T = 1/f$$

where T denotes the period and f the frequency.

periodic table Named more fully "the periodic table of the elements," the periodic table is a display in which the chemical ELEMENTs are represented and arranged in order of increasing ATOMIC NUMBER in such a way that a pattern of recurring similar properties is emphasized. As an example, the group called alkali metals consists of the elements lithium (atomic number 3), sodium (11), potassium (19), rubidium (37), cesium (55), and francium (87), which possess similar chemical and physical properties. They are all very reactive and appear in nature only in COMPOUND form. They all are soft METALs of low DENSITY and low MELTING point. And they all form chemical compounds by readily giving up a single ELECTRON. For another example, the elements helium (atomic number 2), neon (10), argon (18), krypton (36), xenon (54), and radon (86) compose the group of noble gases, or inert gases. They are indeed all gases under ordinary conditions and are all chemically very inert. Thus, their appearance in nature is in elemental form. Note that in these examples, every element in the group of alkali metals has atomic number—and correspondingly, number of electrons—that is one more than that of an element in the noble gas group. (*See* GAS.)

The periodicity that underlies the periodic table results from the structure of the elements' ATOMs, in which electrons are arranged in shells and subshells. All the noble gases possess full electron shells, which are particularly stable against removing or adding electrons. That explains their chemical inertness. The alkali metals all have a single electron in addition to their full shells, a single valence electron, in the terminology of chemistry. Such an electron is loosely bound and easily lost. So alkali metal atoms readily give up their electron to any atom attracting it, transform into positive IONs, and join negative ions to form MOLECULEs. That is the reason for their reactivity. The understanding of atomic structure and the consequent understanding of the periodic table are among the crowning achievements of QUANTUM MECHANICS.

A copy of the periodic table of the elements appears in Appendix III and can also be found in every introductory chemistry text book, usually inside one of the covers, and also in many introductory physics textbooks. It is often displayed as a large poster in science classrooms and lecture halls.

See also PAULI EXCULSION PRINCIPLE.

permeability Called more fully magnetic permeability, permeability is the constant μ that appears in the BIOT-SAVART LAW for the production of a magnetic FIELD by an electric CURRENT or by a moving electric CHARGE and in all other formulas for magnetic-field production. When the field production takes place in VACUUM, the permeability is that of the vacuum, denoted μ_0, whose value is $4\pi \times 10^{-7}$ tesla·meter per ampere (T·m/A). When a magnetic field is produced in a material, however, the material might respond to the field by developing an induced magnetic dipole moment, or induced magnetization. The net field results from the combined effects of the inducing field and the field of the induced magnetic dipole moment. The latter might enhance the inducing field or might partially cancel it. In any case, the magnitude of the net magnetic field relates to that of the inducing field by a dimensionless factor called the relative permeability of the material and denoted μ_r. This factor then multiplies the μ_0 in the vacuum version of the Biot-Savart law and other formulas, giving the permeability of the material, $\mu = \mu_r\mu_0$. Then the Biot-Savart law and all other formulas for the production of a magnetic field in vacuum become valid for the net field in a material by the replacement of μ_0 with μ. (*See* DIMENSION; ELECTRICITY; INDUCTION; MAGNETISM; MOMENT, MAGNETIC DIPOLE.)

When the induced field tends to partially cancel the inducing field, the effect is called DIAMAGNETISM. This is a weak effect, so $\mu_r < 1$, $\mu_r \approx 1$, and $\mu \approx \mu_0$. There are two enhancing effects. One is PARAMAGNETISM, which is a weak effect as well. So for paramagnetic materials $\mu_r > 1$, $\mu_r \approx 1$, and $\mu \approx \mu_0$. The other enhancing effect is FERROMAGNETISM, which is a strong effect and for which the value of the relative permeability can reach even into the thousands for certain materials and depends on the magnitude of the inducing field.

permittivity Permittivity is the constant ε that appears in COULOMB'S LAW, GAUSS'S LAW, and all formulas relating the electric FIELD to the electric CHARGEs responsible for it. When the field is produced in VACUUM, the permittivity is that of the vacuum, denoted ε_0, whose value is $8.85418781762 \times 10^{-12}$ $C^2/(N{\cdot}m^2)$. When the electric field is produced in a material medium, it might induce electric POLARIZATION, whereby the material's atomic or molecular ELECTRIC DIPOLES become aligned by the electric field. The aligned dipoles themselves produce an electric field, the induced field, that is oppositely directed to the inducing field and partially cancels it. A material that does not normally possess atomic or molecular dipoles might still be polarized by an inducing electric field. That happens through the field's creating dipoles by causing a separation of positive and negative electric charges in each of the material's ATOMs or MOLECULEs. In this case, too, the induced field partially cancels the inducing field. (*See* ELECTRICITY; INDUCTION.)

The magnitude of the net electric field results from the combined effects of the inducing field and the induced field and is smaller than the magnitude of the inducing field by a dimensionless factor called the DIELECTRIC CONSTANT of the medium and denoted κ. The electric FORCEs between pairs of charges are reduced by the same factor inside the medium. This factor then multiplies ε_0 in the vacuum version of Coulomb's law, Gauss's law, and the other formulas, giving the permittivity of the medium, $\varepsilon = \kappa\varepsilon_0$. Then Coulomb's and Gauss's laws and all formulas for the production of an electric field in vacuum become valid for the net field in a material medium by the replacement of ε_0 with ε. (*See* DIMENSION.)

phase The term *phase* is used in physics in two related ways and one very different one. First, a phase is the same as a STATE OF MATTER. Thus we can speak of water, say, in its GAS, LIQUID, or SOLID phase, meaning its vaporous, wet, or hard state, respectively. A PHASE TRANSITION is the transformation from one state to another, a change of state. As examples, BOILING, MELTING, and FREEZING are phase transformations. A phase diagram is a diagram that shows the state of a material for a range of conditions. A common form uses PRESSURE and TEMPERATURE axes and indicates the regions in which the material exists in one state or another and the boundaries between regions along which different states coexist.

Another use of the term *phase* is as a distinct, bounded, component of a mixture. A mixture of oil and water, for instance, maintains two distinct phases, the oil phase and the water phase, such as the fat globules floating in chicken soup. Or, during the processes of freezing and melting, the liquid and solid states (or phases, in the first sense) coexist as distinct phases (in the second sense) in the freezing mixture. Whereas a solution of table salt in water is homogeneous and forms a single phase, a mixture of salt grains and salt water comprises two phases, the solid-salt phase and the saline-solution phase.

The third use of the term *phase* is to denote a stage in the cycle of a repetitive phenomenon. Such a phenomenon consists of repetitions of a minimal unit, its cycle. One property of a cycle is its length, if it is part of a repetitive pattern in SPACE, or its duration, if the repetition is a temporal one (i.e., an OSCILLATION). The length of a spatial cycle is the pattern's repeat distance, sometimes referred to as the WAVELENGTH of the pattern, even when no WAVE is involved. The duration of a temporal cycle is the PERIOD of the oscillation. Phase can be specified in terms of cycle length or duration, such as 2/3 wavelength from the left end of a cycle, or 1/4 period from the cycle's onset. This type of specification is unified by simply referring to fractions of a cycle: 2/3 cycle, or 1/4 cycle. A very common representation of a cycle is in terms of angles. Then a complete rotation—2π radians, or 360 degrees—denotes a full cycle, and phases are specified as fractions of a complete rotation. A phase of a quarter cycle is $\pi/2$ rad = 90°, for example. (*See* TIME.)

A pair of similar repetitive phenomena that have the same repeat distance or period might differ in that one is spatially or temporally displaced with respect to the other. That is commonly specified as a phase difference. A displacement of 1/8 cycle, or $\pi/4$ rad = 45°, for instance, would be indicated as a phase difference of $\pi/4$ rad, or of 45°, or of 1/8 cycle. If there is no phase difference, the phenomena are said to be in phase. Otherwise

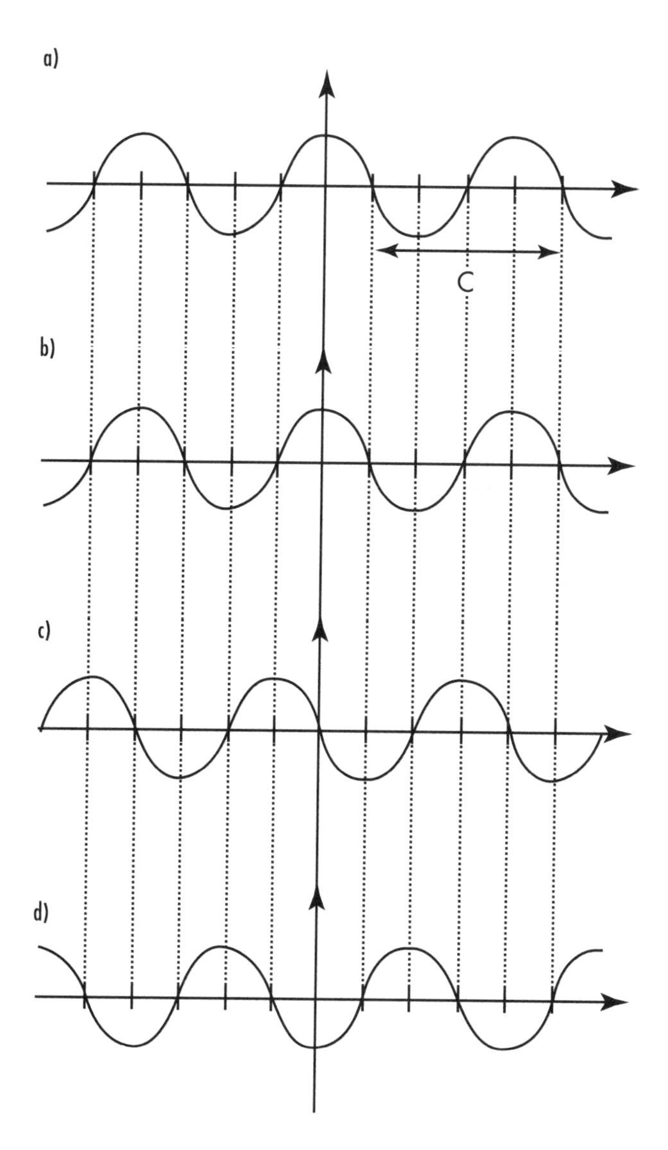

Phase differences between periodic phenomena (in time or in space). The vertical axes denote the values of sinusoidally varying quantities. The horizontal axes all represent either time or position, according to whether periodicity in time or in space is being considered. (a) and (b) are in phase, with a phase difference of 0. The phase difference between (a) and (c) is one quarter cycle, or $\pi/2$ rad (90°). (a) and (d) possess a phase difference of a half cycle, or π rad (180°). They are in antiphase. A single cycle is indicated by C in (a).

they are out of phase (by some fraction of a cycle or by some angle). If their phase difference is a half cycle, they are in antiphase. (Some people use out of phase to mean out of phase by a half cycle, i.e., in antiphase.)

When a repetitive pattern is displaced with respect to itself, we speak of a phase shift. HARMONIC MOTION, for instance, can be represented by:

$$x = A \cos (2\pi ft + \alpha)$$

where x denotes the coordinate describing the motion, A is the amplitude of the oscillation in the same UNITS as x, f is the FREQUENCY in hertz (Hz), and t is the time in seconds (s). The quantity α in radians, with $|\alpha| < 2\pi$, is the phase shift, or phase constant, and determines—for positive α—by how many radians the above process is earlier than, or leads, the process:

$$x = A \cos 2\pi ft$$

(For negative α, the first process is later than, or lags, the second process.) The phase shift expressed in terms of fraction of a cycle is $\beta = \alpha/2\pi$, $|\beta| < 1$, and appears in the above representation in the form:

$$x = A \cos 2\pi(ft + \beta)$$

phase space The abstract space in which a point represents the values of the coordinates of a system for all its DEGREES OF FREEDOM, as well as the values of the corresponding momenta, called conjugate momenta, is known as the system's phase space. The phase space of a system of N point particles is $6N$-dimensional: $3N$ dimensions for the particles' coordinates—three for each particle—and $3N$ additional dimensions for the components of the particles' LINEAR MOMENTA, also three per particle. A rigid body possesses six degrees of freedom, three for its position and three for its orientation. That translates to six coordinates: the x-, y-, z-coordinates of, say, its CENTER OF MASS, and three angles. For each coordinate there is a conjugate momentum, so there are six conjugate momenta: the x-, y-, z-components of linear momentum and the x-, y-, z-components of ANGULAR MOMENTUM. Accordingly, the phase space of a system of N rigid bodies is $12N$-dimensional. (*See* DIMENSION.)

The evolution of the system is described by a curve in phase space, called a trajectory. (This use of the term *trajectory* should not be confused with its meaning as a path in SPACE, such as the trajectory of a missile.) Since a point in phase space represents the amount of information needed to specify a set of INITIAL CONDITIONS for the system, every point uniquely determines a complete trajectory passing through it. Phase space is the mathematical arena for CHAOS theory.

phase speed Also called phase velocity, phase speed is the propagation SPEED of a sinusoidal WAVE. The phase speed, v, relates to the FREQUENCY, f, and WAVELENGTH, λ, of a sinusoidal wave by:

$$v = f\lambda$$

where v is in meters per second (m/s), f is in hertz (Hz), and λ is in meters (m). A dispersive medium is one in which the phase speed depends on the wavelength/frequency. Glass and water are examples of dispersive media. DISPERSION causes the material's INDEX OF REFRACTION to depend on the color of LIGHT, giving rise to prismatic effects in glass and rainbows from water droplets. If the phase speed is the same for all wavelengths/frequencies, the medium is called nondispersive. Electromagnetic waves in VACUUM, for example, all propagate at the same speed. Their speed is the SPEED OF LIGHT, denoted c. (*See* ELECTROMAGNETISM; PRISM.)

See also GROUP SPEED.

phase transition A change of STATE OF MATTER, during which the substance transforms from one of its states, or PHASES, to another, is called a phase transition. Examples of phase transitions are CONDENSATION, BOILING, EVAPORATION, FREEZING, and MELTING. A phase transition for a given substance occurs as the substance reaches a certain TEMPERATURE, the transition temperature, which depends on the PRESSURE. As the temperature of LIQUID water is raised under atmospheric pressure, boiling commences at 100 degrees Celsius (°C), the boiling point of water at atmospheric pressure, as an example. Inversely, under atmospheric pressure, water vapor, the gaseous state of water, starts to condense to the liquid state when the temperature is lowered to 100°C. In an alternative approach, for fixed temperature and changing pressure, the phase transition can occur when a certain pressure, which depends on temperature, is reached. As an example, liquid water can be made to boil at temperatures below its boiling point by sufficiently reducing the pressure on it. (*See* GAS.)

Phase transitions normally involve the absorption or emission of HEAT, the heat of transition of the substance for the specific phase transition. The same amount of heat is absorbed or emitted as the material undergoes the phase transition in one direction or the other. The heat of transition does not cause a change of temperature. When a sample of ice is heated, for instance, its temperature increases to 0°C, upon which it starts melting. Further heating does not raise the temperature but causes more ice to transform to liquid water. When the sample is fully melted, it has absorbed its heat of fusion at 0°C. Further heating raises the temperature of the liquid. When the same amount of liquid water is cooled to 0°C, it starts to freeze. Further cooling (i.e., removal of heat) does not lower the temperature but transforms more liquid to ice. When the sample is fully frozen, it has emitted at 0°C the same amount of heat it absorbed while melting, its heat of fusion. Further cooling lowers the temperature of the ice.

phonon The particle aspect of acoustic, or SOUND, WAVES in MATTER, according to QUANTUM PHYSICS, is manifested in phonons. They are localized entities that possess ENERGY and LINEAR MOMENTUM and exist only within the medium carrying the wave. Phonons are the acoustic analog of PHOTONs, which form the particle aspect of electromagnetic waves. Just as the interaction of matter with electromagnetic radiation is understood in terms of the emission, absorption, and scattering of photons, so the effect of acoustic waves on the ATOMs, ELECTRONs, and HOLEs of matter is comprehended in terms of phonon emission, absorption, and scattering. And in a manner similar to the electromagnetic interaction of matter with matter by photon exchange, the constituents of matter can interact acoustically by exchange of phonons. (*See* ACOUSTICS; ELECTROMAGNETISM; WAVE-PARTICLE DUALITY.)

QUANTUM MECHANICS gives relations between the frequency and wavelength of waves, in this case acoustic waves, on the one hand, and the energy and momentum of the particles associated with the waves, phonons, on the other. For a wave of frequency f in hertz (Hz), the phonons associated with it individually possess energy E in joules (J), such that:

$$E = hf$$

where h is the PLANCK CONSTANT and has the value $6.62606876 \times 10^{-34}$ joule·second (J·s). In addition, the wavelength of a wave, λ, in meters (m) and the magnitude of the momentum of each of its phonons, p, in kilogram·meters per second (kg·m/s) are related by:

$$p = h/\lambda$$

Due to the relation:

$$f\lambda = v$$

where v denotes the speed in meters per second (m/s) of sound of frequency f in the medium, the energy and momentum of a phonon are related by:

$$E = pv$$

phosphorescence The emission of electromagnetic RADIATION by a material as a result of the material's excitation, where the emission persists significantly after the cessation of excitation, is called phosphorescence. The mechanism of phosphorescence is the absorption of ENERGY by the particles constituting the material (ATOMs, MOLECULEs, IONs) and the subsequent release of the energy as electromagnetic radiation emitted in conjunction with the particles' transition from relatively long-lived excited states to lower-energy states. Phosphorescence is a special case of LUMINESCENCE, which is the general phenomenon of emission of electromagnetic radiation by excited materials, whether over an extended period of time or not. If the emission stops immediately with the cessation of excitation, the effect is termed FLUORESCENCE. (*See* ELECTROMAGNETISM.)

photoelectric effect The emission of ELECTRONs from METALs as a result of illumination by LIGHT or other electromagnetic RADIATION is the photoelectric effect. The effect was explained by ALBERT EINSTEIN as a quantum effect and is understood as a process whereby one or more PHOTONs are absorbed in the metal, with practically all of their ENERGY taken up by an electron that subsequently exits the metal. At sufficiently low light INTENSITIES, the kinetic energy of an emitted electron comes from a single photon, and the effect is designated as the single-photon photoelectric effect. (Lasers can produce the multiple-photon photoelectric effect.) In the single-photon case, even for light of a definite FREQUENCY (i.e., monochromatic radiation), each of whose photons possesses the same amount of energy, the emitted electrons carry a range of KINETIC ENERGIES up to some maximal value. (*See* ELECTROMAGNETISM; QUANTUM PHYSICS.)

For all metals, the relation between maximal electron kinetic energy and light frequency has the form:

$$\text{Maximal kinetic energy} = hf - \phi$$

where energy is in joules (J); frequency, f, is in hertz (Hz); and h represents the PLANCK CONSTANT, whose value is $6.62606876 \times 10^{-34}$ joule·second (J·s). The quantity ϕ, in joules (J), is the work function of the metal and differs from metal to metal. This equation is explained by the model of a photon, whose energy is hf, entering the metal and being absorbed by an electron. The photon's energy is practically fully converted to the electron's kinetic energy. If that electron subsequently leaves the metal, it will suffer energy losses along the way to its emergence. Electrons near the surface suffer the least loss, and the minimal amount of energy needed to liberate an electron from the metal is called the work function of the metal, which varies from metal to metal. The electrons that are emitted with the most kinetic energy are those that lost the least of their initial kinetic energy, hf, on the way out. That least is the work function, ϕ. So the above equation is explained.

Because of the work function, every metal possesses a threshold frequency below which incident light produces no single-photon photoelectric effect. A single photon of such radiation simply does not possess sufficient energy to liberate an electron. The threshold frequency, f_0, is given by:

$$f_0 = \phi/h$$

and the frequency dependence of the maximal electron kinetic energy can be expressed as:

$$\text{Maximal kinetic energy} = h(f - f_0)$$

The photoelectric effect stands in stark contrast to what might be expected to occur according to classical electromagnetism. The classical picture has the electromagnetic WAVE cause the electron to oscillate and gradually absorb energy until it possesses sufficient energy to leave the metal. Then it is expected that: (1) the maximal kinetic energy of the emitted electrons will depend on the intensity of the light and should increase with increasing intensity; (2) there will be no simple dependence of the effect on the frequency of the light; (3) at low light intensity, there will be a time delay from the start of the illumination to the onset of electron emission. In reality, however, the situation is very different. (1) The maximal electron kinetic energy is independent of light intensity. Rather, it depends on the frequency of the incident light and only on the frequency (for any particular metal), according to the above equation. What the intensity does affect is the number of electrons emitted per unit time. (2) The dependence of the effect on frequency is indeed a simple one: a linear relation between maximal kinetic energy of the emitted electrons and light frequency. (3) Even at extremely low light intensities, there is no delay between the start of illumination and the onset of electron emission. (*See* CLASSICAL PHYSICS; OSCILLATION.)

Einstein's model explains it all. (1) Greater light intensity means a larger number of photons entering the metal per unit time. That results in a larger number of electrons emitted per unit time. Intensity does not affect the energy of an individual photon, and it is the latter that gives an individual electron its initial kinetic energy, as explained above. (2) The linear relation follows straightforwardly from simple energy considerations. (3) The acquisition of energy by an electron is not a gradual process that requires a time delay but is, rather, an instantaneous process of photon absorption by an electron.

Additional methods of causing electron emission from metals include FIELD EMISSION and THERMIONIC EMISSION.

photon The particle aspect of electromagnetic WAVES, including LIGHT, according to QUANTUM PHYSICS, is manifested by photons. They are localized entities that possess ENERGY and LINEAR MOMENTUM and move at the SPEED OF LIGHT (2.99792458×10^8 meters per second [m/s]). Photons are BOSONs and possess SPIN 1. They are massless particles. The meaning of this is that although photons, by their nature, are never at rest, moving at the speed of light and only at that speed, ALBERT EINSTEIN's special theory of relativity nominally assigns them zero REST MASS. The interaction of MATTER with the electromagnetic FIELD is understood in terms of emission, absorption, and scattering of photons, for which the COMPTON EFFECT and the PHOTOELECTRIC EFFECT can serve as examples. The ELECTROMAGNETIC INTERACTION of matter with matter is comprehended as occurring through exchange of photons. On the other hand, the propagation of electromagnetic RADIATION through SPACE is well described in wave terms. (*See* ELECTROMAGNETISM; MASS; QUANTUM ELECTRODYNAM-

ICS; QUANTUM FIELD THEORY; RELATIVITY, SPECIAL THEORY OF; WAVE-PARTICLE DUALITY.)

The energy, E, of an individual photon is related to the FREQUENCY, f, of the wave by:

$$E = hf$$

where E is in joules (J), f is in hertz (Hz), and h represents the PLANCK CONSTANT, whose value is $6.62606876 \times 10^{-34}$ joule·second (J·s). The magnitude of a photon's momentum, p, relates to the WAVELENGTH, λ, of the wave by:

$$p = h/\lambda$$

Since for electromagnetic waves:

$$f\lambda = c$$

where c denotes the speed of light, the energy and momentum of a photon are related by:

$$E = pc$$

The INTENSITY of electromagnetic radiation is proportional to the photon FLUX, the number of photons passing through a unit area per unit time.

Whether the wave or particle aspect of electromagnetic waves is in play depends on the circumstances. INTERFERENCE and DIFFRACTION effects, for example, are typical wave phenomena. They are free-propagation effects, not involving interaction with matter. Attempts to describe them in terms of photons lead to untenable ideas of photons somehow "splitting" and "interfering with themselves." When electromagnetic radiation interacts with matter, such as in the production of a diffraction pattern on a photographic plate or a charge-coupled device (CCD) chip, the most successful description is in terms of discrete, localized quantum processes involving photons. (*See* YOUNG'S EXPERIMENT.)

See also ELEMENTARY PARTICLE; GAMMA RAY; X RAY.

physicist A physicist is one who has studied physics in depth and in breadth and has made physics his or her profession, where PHYSICS is the human endeavor of attempting to rationally understand nature in its most fundamental aspects. Most often, a physicist is involved in physics research, physics teaching, application of physics, industrial administration, academic administration, consulting, or some combination of those. As a rule, a physicist is either an experimental physicist or a theoretical physicist. The experimentalist probes nature by performing experiments. Every experiment is a question asked of nature. The experimentalist forces nature to give answers. The theoretician tries to find order and pattern in nature's answers and formulate general laws that describe nature's behavior. Proposed laws are confirmed or disproved by additional experiments. Having firm laws of nature in hand, the theoretician attempts to devise explanations, or theories, for the laws. Proposed theories, too, are subject to experimental confirmation or disproof. In brief caricature: experimentalists tinker with designs and apparatus and use equipment; theoreticians tinker with ideas and mathematics and use paper; both use computers. (*See* EXPERIMENTAL PHYSICS; THEORETICAL PHYSICS.)

In this connection, it should be noted that the meaning of the term *theory* in science, and in physics in particular, is different from the common connotation of hypothesis or speculation, as in "It's only a theory." A theory is an explanation. It might indeed be a speculative theory, as yet not well confirmed, or it might be a solid, accepted one. ALBERT EINSTEIN's special theory of relativity, for example, is a very well-confirmed theory and universally accepted. It is not Einstein's special hypothesis of relativity or special speculation. It is a well-founded, unifying explanation of a tremendous number of phenomena. (*See* RELATIVITY, SPECIAL THEORY OF.)

The education of a physicist starts with the most general ideas, concepts, and theories that underlie physics and with the basic mathematical tools needed to handle them. Then the student learns more advanced approaches to the fundamental theories along with the more advanced mathematics required for them. Some ideas are approached again at an even higher level. Along the way, the student gains laboratory experience as well. At some point, the physics student specializes in a particular field of physics and chooses between an experimental path and a theoretical one. The culmination of a physicist's full formal education, leading to a Ph.D. degree, is a supervised research project. Nevertheless, many physicists enter the profession with a bachelor's or master's degree. But even a Ph.D. is not the end of education. Those physicists who are taking part in extending the boundaries of knowledge are gaining new insights into the workings of nature. Other active physicists generally keep abreast of the latest developments in their fields of interest.

Beyond the physics itself, a physics education develops in the student a quantitative approach to dealing with matters, proficiency in applying mathematics to problems, analytic thinking, and, in the case of an experimentalist, laboratory skills as well. In this way, a physics education forms an excellent preparation for much more than solely a career in physics. As an often-presented example, holders of physics degrees have taken jobs in the financial world and have done well there.

physics Physics is the branch of science that deals with the most fundamental aspects of nature. Science—some might prefer the designation "natural science"—

Grist for the Mill: What Physics Studies

Stated briefly, physics studies the most fundamental aspects of nature. But the precise subject of physics research, what physics *really* studies, are the most fundamental of the reproducible and predictable aspects of nature, where the term *nature* refers to the material UNIVERSE with which we can, or can conceivably, interact. Consider these four concepts: materiality, INTERACTION (actual or conceivable), reproducibility, and predictability. (Fundamentality and the universe are discussed in other essays.) The importance of the four to physics and to the other natural sciences (i.e., chemistry and biology) cannot be overemphasized. Each one of these concepts contributes in an essential way to making the natural sciences, and physics among them, as valuable and useful as they are.

Material, as in "material universe," means that physics is concerned with the happenings in the universe that involve only MATTER and ENERGY in their various forms, including FIELDS and WAVES. The supernatural is definitely out. The spiritual aspects of the universe, as important a role as they might play in human affairs, are excluded as well. Physics takes a totally materialist view of the universe and deals only with what is aptly termed the real world. This idea is so ingrained in physicists that it is difficult to even imagine a physicist, as a physicist, attempting to investigate any nonmaterial aspect of the universe.

This characteristic of physics, which is shared by all the natural sciences, means that physics does not deal with deities, spirits, souls, superstitions, or the like, nor with perceptions, emotions, feelings, intuitions, beliefs, or such. Thus, the domain of physics is severely restricted and does not include many very important aspects of human lives. What physics *does* deal with is wholly common to all people. (It is not unreasonable to assume that it is common as well to all intelligent beings, even extraterrestrial ones, although that requires some conjecture.)

The fact that physics is concerned solely with the real world endows physics (and the other natural sciences) with objectivity. The findings of physics (and of the natural sciences) are valid for everybody, irrespective of their world views, their beliefs, their cultures, their feelings, and so on. There is no American physics as opposed to, say, Chinese physics. Neither is there religious physics as opposed to secular physics. The truths of physics are the same for happy people as for depressed ones, for the curious as well as the indifferent.

The notion of *interaction* comprises acting upon and being acted upon. The material universe with which we can interact consists of those components of the material universe that we can affect and that can affect us. What we really need here is to be able to perform observations and measurements and to receive data. That is enabled by actual interaction. In order not to be too limiting, we also allow conceivable interaction. This is when, although we might not be able to interact at present, interaction is not precluded by any principle known to us and is considered attainable through further technological research and development. We might reformulate the meaning of nature as the material universe that we can, or can conceivably, observe and measure.

None of that excludes the possible existence of components of the material universe with which we cannot, even conceivably, interact. Are there any? Physics has nothing to say in that regard. If such components exist, we cannot see them, cannot observe them in any other way, cannot measure their properties, nor are we affected by them in any manner. Neither is any of that even conceivable. So as far as physics is concerned, they may as well not exist. Whether they exist in anybody's imagination, dreams, or belief is of no interest to physics. Thus, any phenomenon a physicist actively investigates forms of necessity an aspect of nature with which she can interact.

For example, BLACK HOLES cannot be seen directly, since no LIGHT or other RADIATION can leave from them. Yet they have very observable effects. They affect the orbits of their binary partners, if they make up part of a binary (i.e., a two-star) system, and they emit characteristic radiation as a result of matter falling into them. The gigantic black holes located at the centers of GALAXIES affect the orbits of all the is the human endeavor of attempting to rationally comprehend, in terms of general principles and laws, the reproducible and predictable aspects of nature. Physics, as a branch of science, concentrates on those aspects of nature that are, or at least appear to be, the most fundamental, in the sense that they underlie all the rest. The notion of fundamentality here is a component of the view that some aspects of nature indeed underlie others, that not all aspects of nature are equal in this regard. That concept is well reflected in the hierarchical organization of science and seems to form a valid picture of nature. As an example, chemistry is the study of matter and its transformations at the atomic, molecular, and larger scales. Chemistry is quite autonomous in

stars of the galaxy. So there is no problem with the existence of black holes.

Reproducibility means that experiments and observations can be repeated by the same and other investigators and produce the same results. That gives data of objective, lasting value about natural phenomena. Reproducibility makes physics a common human endeavor, rather than, say, an incoherent jumble of private efforts. It allows physicists to communicate meaningfully and to progress toward a greater understanding of nature through their joint effort. Thus, reproducibility contributes to making physics (and the other natural sciences) as nearly as possible an objective endeavor of enduring validity. There would seem to be no necessity a priori that nature be reproducible at all, but the very fact that physics is being done, and so successfully, proves that nature indeed possesses reproducible aspects.

This is not to imply that nature is reproducible in all its aspects. But any irreproducible aspects it might possess lie outside the domain of physics. Parapsychological phenomena, for example—extrasensory perception (ESP), telepathy, telekinesis, clairvoyance—if, as some claim, they existed, would form such an irreproducible aspect of nature.

The idea of *predictability* is that, among the natural phenomena investigated, pattern and order can be found, based upon which it is possible to find laws that predict the results of new experiments and observations. That allows making use of nature, although this is not a major goal of physics. More importantly for physics, predictability forms an essential stage in developing an understanding of nature. Just as for reproducibility, no necessity seems to exist for nature to be predictable in any aspect, to exhibit any lawful behavior. The very fact that physics works shows that nature does possess predictable, lawful aspects. It is not excluded that nature possesses unpredictable aspects as well. But if so, such phenomena would lie beyond the domain of physics. Parapsychological phenomena, if in fact there were such effects, would form such an aspect of nature. (*See* LAWS OF NATURE.)

Consider a hypothetical scenario, to see how the concepts of materiality, interaction, reproducibility, and predictability apply. Imagine a physicist discovers that when he shines LIGHT on certain METALS, they emit ELECTRONS. He investigates this phenomenon by trying it for different metals and for different FREQUENCIES and INTENSITIES of light. He might study such as the intensity of electron emission, the ENERGIES of the emitted electrons, the time delay between the start of illumination and the onset of electron emission, and the effect of the identity of the metal on all those. The phenomenon is obviously an aspect of nature (i.e., it is of a *material* character). Moreover, since the physicist is manipulating the phenomenon and performing measurements on it, he can clearly *interact* with it. That much hardly needs mentioning.

Through numerous repetitions of the experiments, by the physicist and by other physicists, it becomes clear that this is a *reproducible* phenomenon. It presents itself as a candidate for serious investigation and is given a name, "the PHOTOELECTRIC EFFECT." Simultaneously with the demonstration of reproducibility, whereby the effect proves its potential for being understood, the data from the experiments are carefully studied. They reveal a pattern of lawful behavior, demonstrating that the effect is *predictable.* New experiments are performed to test the laws, which are invariably confirmed. That forms the first step toward an understanding of the photoelectric effect.

Since this scenario covers materiality, interaction, reproducibility, and predictability, it should end here. But for the sake of completeness, we note that the laws of the photoelectric effect turn out to be remarkably simple. They stand in stark contradiction to the laws that follow from the atomic and electromagnetic theories that are accepted at the time. ALBERT EINSTEIN takes the second and final step toward understanding the photoelectric effect. He realizes that new physics is required and provides a successful theory of the photoelectric effect in terms of PHOTONS, which are particles of light, and their interaction with electrons in the metal. (Note that in the language of science, the word *theory* does not mean speculation or hypothesis, as it does in everyday language, but rather explanation.) Thus Einstein makes a major contribution to the founding of QUANTUM PHYSICS. (*See* ATOM; ELECTROMAGNETISM.)

Achieving Understanding: How Physics Operates

Physics is the branch of science that studies the most fundamental phenomena of nature. In its operation, physics searches for order and pattern among the phenomena; attempts to discover *laws* for those phenomena, based on their order and pattern, which allow the prediction of new phenomena; and tries to explain the laws by means of theories. (Note that in the language of science, the word *theory* does not mean speculation or hypothesis, as it does in everyday language, but rather explanation.) Other branches of science deal with nature in more or less similar ways. What is taught in school as "the scientific method"—involving observation, hypothesis, experiment, and theory—makes up only part of the picture and presents a gross simplification of what actually happens. Here is what actually happens, how physics reaches understanding of fundamental natural phenomena. (*See* LAWS OF NATURE; ORDER.)

Start with a rather simple example of the operation of physics and consider the effect of HEAT on SOLIDS. Over the years, many measurements had been made of the sizes of solid bodies at different TEMPERATUREs. The results of the measurements revealed clear pattern and order: qualitatively, the higher the temperature, the larger the body's size. A quantitative analysis of the data showed that the change in size is proportional both to the temperature increment (as long as the increment is not too large) and to the body's size. The coefficient of proportionality was found to depend on the material the body was made of and to be a characteristic of that material: copper has some value, iron has another, wood yet another, and so on. (*See* THERMAL EXPANSION.)

Consider a body that undergoes a temperature change from T_1 to T_2. Consider a linear dimension of the body, say, the length of a rod, that has the value L_1 at initial temperature T_1 and value L_2 at final temperature T_2. The pattern that was found can be expressed in this way: the relative change of length, $(L_2-L_1)/L$, is proportional to the corresponding temperature change, $T_2 - T_1$. In a formula:

$$(L_2-L_1)/L_1 = \alpha(T_2 - T_1)$$

The proportionality coefficient, α, is the coefficient of linear expansion of the body's substance. The fact that bodies expand when heated is expressed by the fact that the values of α are positive.

Here was a law: the mathematical description of the order that was discovered in the data produced a law. The law not only fitted all the data, but also predicted the results of any number of future measurements. It has since been confirmed, and continues to be confirmed, by very many additional measurements.

So far, we have seen the finding of order and pattern and the discovery of a law. That forms a step toward understanding. But understanding has not yet been achieved. For this, we need an explanation of the law, a theory. A theory of thermal expansion indeed exists and thermal expansion is well understood. The theory is based on the nature of thermal ENERGY in solids as the random VIBRATION of the ATOMs of the substance about their fixed positions in the material: the higher the temperature, the stronger the vibration, the greater the effective size of the atoms, and the larger the interatomic distances at EQUILIBRIUM, causing an increase in size of the whole sample. This example shows the full operation of physics in achieving understanding of thermal expansion. Order and pattern led to a law, which was given an explanation by a theory.

For an additional example of the operation of physics in gaining understanding, consider the experimental setup of a given uniform sphere rolling down a fixed straight, inclined track. The sphere is released from rest, and the distance it rolls, d, and the time it covers that distance, t, are measured. Experiments are run for different distances and times. In this

that it can and does achieve a great deal of understanding without needing to probe into the nature of atoms. So chemistry forms an independent level in the hierarchy of science. Nevertheless, it is understood by all chemists that everything going on in their field of interest ultimately *does* derive from the nature and properties of atoms. Physics studies the nature and properties of atoms, among other phenomena. Thus physics is a more fundamental branch of science than is chemistry. In similar vein, ascending the hierarchy ladder (or descending, depending on which picture one prefers), chemistry underlies biochemistry, which underlies biology, which might even be viewed as underlying psychology. (*See* ATOM; MOLECULE.)

In its study of the atom, physics finds ELECTRONs and NUCLEI and studies them too. Nuclei are discovered to be composed of PROTONs and NEUTRONs, which physics investigates, as it does also the other ELEMENTARY PARTICLEs. Some elementary particles, including the proton and neutron, are found to be composed of QUARKs. Consequently, physics studies quarks. In this manner physics probes nature at smaller way, data, consisting of d-t pairs, are collected. These form the phenomena to be studied. The next step is to search for order and pattern. One way of doing that is by plotting the data in various ways. In most of the plots, the data fall on variously bent curves and nothing of particular interest manifests itself. But in the plot of distance, d, against the *square* of the time interval, t^2, all the data tend to fall on a straight line. Here is order; here is a pattern, and a very simple one at that. The straight-line plot indicates that the relation between d and t is simply that d is proportional to t^2:

$$d = bt^2$$

This mathematical relation between d and t not only correctly describes all the data, but can serve to predict the results of any number of additional runs of the experiment. Now we have a law. Again and again the sphere is rolled, the distance and time measured, and the law tested. Again and again the law passes the test. Finally, the law is considered reliable. Historically, this law was discovered by GALILEO GALILEI. It is explained by ISAAC NEWTON's theory, in the form of his three laws of MOTION and law of GRAVITATION. Newton's theory was not developed for the sole purpose of explaining Galileo's law, but Newton did take this law into account in his work (together with JOHANNES KEPLER's laws of planetary motion). (*See* KEPLER'S LAWS; NEWTON'S LAWS.)

These two examples show the process through which understanding is reached in physics. A set of phenomena is studied for pattern and order. When those are found, a law is formulated to summarize the data and predict new phenomena. The law is confirmed and deemed reliable. Then a theory is found to explain the law. In practice, though, the steps just outlined rarely occur in such clean succession. More often, pattern and order are being searched for and laws are being proposed even as the data are still coming in. At the same time, the tenuous and provisional laws that appear to be emerging are serving as grist for the theorizers' mill. As more data accumulate, the laws are refined or revised, and the theories are modified to accommodate or are replaced. Eventually, a plateau is reached, when a set of laws and theories becomes generally accepted. But only for a while. Further experiments and observations, at more extreme conditions, invariably reveal weaknesses in what has become standard wisdom, and the process repeats itself. (*See* EXPERIMENTAL PHYSICS; THEORETICAL PHYSICS.)

Three fields of physics (but hardly the only ones) in which this process is actively occurring at present are COSMOLOGY, ELEMENTARY PARTICLE physics, and high-temperature SUPERCONDUCTIVITY, as examples. In cosmology, astronomical observations at increasingly greater distances are returning information about the increasingly far past of the UNIVERSE. Those data are continually being digested and processed as described above, with the concomitant emergence of various proposed laws and conditional theories about the evolution of the universe. In the field of elementary particle physics, current experiments are generating data that test the presently standard theory—electroweak theory and quantum chromodynamics—and will serve, through the above process, for the development of better and more general theories. Much work is being invested in research into the phenomenon of high-temperature superconductivity. This type of superconductivity is not explained by the BCS theory, which is the theory that works for low-temperature superconductors. The field is presently in the stage of collecting data and searching for pattern and order. (*See* ELECTROWEAK INTERACTION; STRONG INTERACTION.)

In these three fields, as in other fields of physics, the future holds many exciting surprises in store. They will be revealed as physicists continue to improve our understanding of nature's fundamental phenomena through the process described above: collect data on phenomena, search for order and pattern, find laws to summarize the data and predict new phenomena, and devise theories to explain the laws.

and smaller scales, which translates into higher and higher ENERGIES. On the other hand, physics looks at larger scales by investigating aggregations of particles, atoms, and molecules. In this connection, we return to the relation between physics and chemistry as a further illustration of their relative positions in the science hierarchy. One interest of chemists is the various STATES OF MATTER and transformations from one to another, taking them as given. Physics studies the formation of states of matter and the basic principles underlying their transformation as part of its investigation of the general behavior of conglomerations of atoms and molecules. (*See* PHASE TRANSITION.)

Proceeding toward larger and larger scales, physics studies the formation, properties, and behavior of astronomical bodies, such as planets, moons, and stars. Ever-greater-size groups of such bodies are investigated: solar systems, GALAXIES, clusters of galaxies, and superclusters. That leads to everything, to the UNIVERSE itself, whose study, COSMOLOGY, forms a field of physics. And here physics is revealing an amazing circularity in nature. Since the universe comprises everything, its properties and behavior must tie in with the principles and laws of nature at all scales, and in particular at the smallest and most fundamental scale, that of the elementary particles. Present understanding does indicate an intimate meshing of the largest-scale properties and behavior of the universe with the properties and INTERACTIONS of the elementary particles. That interdependence is understood to have developed during the earliest stages in the evolution of the universe following the BIG BANG. (*See* SOLAR SYSTEM.)

Yet even the universe does not constrain the imagination of physicists. SPACE and TIME themselves are subjects of study, and cosmologists are actively investigating cosmological models in which our universe forms but a part. Examples include: a multiverse, in which other universes accompany ours; extra DIMENSIONS beyond the four dimensions of space and time; cyclic evolution of the universe, with recurring big bangs.

Physics is divided into various and diverse fields of specialization, some of which are included in this work.

See also CLASSICAL PHYSICS; PHYSICIST; PHYSICS AND SOCIETY; QUANTUM PHYSICS.

physics and society The term *physics and society* labels an interdisciplinary field of interest having to do with the mutual effects on each other of PHYSICS and PHYSICISTS, on the one hand, and the larger society in which physicists live and physics is done, on the other. The field is concerned with broad issues such as what physics can and should contribute to social and political matters; the particular rights and obligations of physicists, as physicists, in society; the effects of society and politics on physics and on physicists; and the sociology of the physics community in the context of the larger society. Examples of particular issues that are generally considered as falling into the physics-and-society category are: nuclear power and nuclear weapons; the role of physics and physicists in the formation of national policy; in particular, their role in military affairs and in counterterrorism; space exploration; energy concerns; the environment; and women and minorities in physics.

piezoelectricity The production of ELECTRICITY from PRESSURE and vice versa are termed piezoelectricity. The creation of an electric potential difference, or VOLTAGE, across a CRYSTAL when a mechanical FORCE is applied to a crystal is called the direct piezoelectric effect. The converse piezoelectric effect is when an applied voltage generates a mechanical distortion of a crystal. The effects are the result of the mutual dependence of electric dipole moment and mechanical STRAIN that can occur for crystals of certain classes. (*See* LATTICE; MOMENT, ELECTRIC DIPOLE; POTENTIAL, ELECTRIC.)

Devices based on the direct piezoelectric effect include transducers that convert strain into an electric signal, such as microphones and pressure gauges, as well as certain ignition devices. The inverse piezoelectric effect underlies the operation of some acoustic sources and headphones. Both effects come into play when a vibrating crystal is used as a FREQUENCY reference. (*See* ACOUSTICS; SOUND; VIBRATION.)

See also PYROELECTRICITY.

pion Also called pi meson, the pion is a type of ELEMENTARY PARTICLE. It is a SPIN-0 BOSON belonging to the family of HADRONS. The pion appears in three versions of electric CHARGE: neutral or one fundamental unit of positive or negative charge. Accordingly, it is denoted π^0, π^+, π^-. The latter two are the antiparticles of each other, while π^0 is its own antiparticle. All are unstable. The mass of π^0 is 135 MeV/c^2, and its QUARK composition is a quantum mixture of $u\bar{u}$ and $d\bar{d}$. The π^+ and π^+ are composed of $u\bar{d}$ and of $\bar{u}d$, respectively. Their mass is 140 MeV/c^2. (*See* ANTIMATTER; ELECTRICITY; MESON; QUANTUM PHYSICS.)

While at the most fundamental level the STRONG INTERACTION is mediated by GLUONS, at the nuclear level the action of the strong interaction is described by the exchange of mesons among BARYONS. So the PROTONS and NEUTRONS that compose a NUCLEUS interact via pion exchange. (*See* NUCLEAR PHYSICS.)

pitch Pitch is a mode of human perception of SOUND that is highly correlated with FREQUENCY. Low audible frequencies evoke the sensation of pitches that are commonly described as "low," while the pitches associated with high audible frequencies are correspondingly designated as "high." (The range of audible frequencies for humans is conventionally taken as 20–20,000 hertz (Hz), although there exists individual variability, and the high end of the range decreases noticeably with age.) Pairs of frequencies of which the higher is twice the lower—such a pair defines a frequency interval called an octave—are perceived as pitches that, although different, possess some similarity to each other. That similarity extends to higher-integer-multiple-frequency pairs (i.e., to multiple octaves) as well, but to a somewhat diminished extent. This property of pitch perception allows people with voices in different pitch ranges, such as women and men, or children and adults, to sing the same tune together and feel they are indeed singing as one.

Other modes of sound perception are loudness, which is strongly correlated with INTENSITY, and timbre, which is related to a sound's spectral composition. (*See* SPECTRUM.)

Planck, Max Karl Ernst Ludwig (1858–1947) German *Physicist* A theoretical physicist, Max Planck is famous for his proposal that ENERGY is exchanged in elementary units, called quanta. The PLANCK CONSTANT, the fundamental unit of ACTION that characterizes quantum effects, $h = 6.62606876 \times 10^{-34}$ joule·second (J·s), is named for him, as is the PLANCK SCALE. Planck received his doctorate from the University of Munich, Germany, in 1879. After serving at the Universities of Munich and Kiel, he joined the University of Berlin, Germany, in 1888. In 1930, Planck was appointed president of the Kaiser Wilhelm Institute in Berlin, but he resigned in 1937 in protest against the mistreatment of Jewish scientists by the Nazis. After World War II, Planck was reappointed president of the institute, which was renamed the Max Planck Institute. In his successful attempt to explain the electromagnetic RADIATION from a BLACK BODY, Planck assumed that the electromagnetic FIELD inside a hollow body and the walls of the cavity exchanged energy in discrete bundles. That led to the development of QUANTUM PHYSICS and QUANTUM MECHANICS. Planck was awarded the 1918 Nobel Prize in physics "in recognition of the services he rendered to the advancement of physics by his discovery of energy quanta."

Max Planck proposed that energy is exchanged in discrete units, called quanta. That led to the development of quantum physics and earned him the 1918 Nobel Prize in physics. ***(Photograph by R. Dührkoop, courtesy AIP Emilio Segrè Visual Archives, W. F. Meggers Gallery of Nobel Laureates)***

See also ELECTROMAGNETISM; THEORETICAL PHYSICS.

Planck constant Named for the German physicist MAX KARL ERNST LUDWIG PLANCK and denoted h, the Planck constant is the fundamental physical quantity $6.62606876 \times 10^{-34}$ joule·second (J·s). This quantity, which is very small in SI UNITs, characterizes the quantum character of nature. One of the Planck constant's most important uses is in relating the FREQUENCY, f, of a WAVE to the ENERGY, E, of an individual quantum particle of the wave, according to WAVE-PARTICLE DUALITY:

$$E = hf$$

Similarly, the WAVELENGTH, λ, of a wave and the magnitude of LINEAR MOMENTUM, p, of an individual quantum particle are related by:

$$p = h/\lambda$$

(*See* CONSTANTS, FUNDAMENTAL; QUANTUM PHYSICS.)

The Planck constant also plays a central role in the quantum HEISENBERG UNCERTAINTY PRINCIPLE. There, the quantity $h/(4\pi)$ sets a lower limit to the product of the quantum uncertainties of certain pairs of measurable physical quantities. That is related to the fact that the DIMENSION of the Planck constant is the same as the dimension of [length][momentum], of [TIME][energy], and of [angle][ANGULAR MOMENTUM]. The Planck constant determines the fundamental unit of angular momentum, including SPIN. All angular momenta are integer or half-integer multiples of $h/(2\pi)$.

The Planck constant participates in determining the magnitudes of fundamental units of length, MASS, time, and energy, called, respectively, the Planck length, the Planck mass, the Planck time, and the Planck energy. All these quantities, and others, are included in the PLANCK SCALE.

Planck scale Named for the German physicist MAX KARL ERNST LUDWIG PLANCK, the Planck scale is a set of fundamental units for length, TIME, MASS, ENERGY, TEMPERATURE, etc. that indicate the respective scales at which we expect nature to reveal the quantum character of SPACE, time, and GRAVITATION. On everyday scales, the quantum aspect of nature seems to be played out in the arena of classical SPACE-TIME and gravitation. Yet, physicists expect that even space, time, and gravitation are not immune to nature's fundamentally quantum character. We do not detect their quantum nature, since we are still probing far from the Planck scale. (*See* CLASSICAL PHYSICS; QUANTUM PHYSICS.)

The units of the Planck scale are determined by the following three fundamental physical constants: the SPEED OF LIGHT $c = 2.99792458 \times 10^8$ meters per second (m/s), the PLANCK CONSTANT $h = 6.62606876 \times 10^{-34}$ joule·second (J·s), and the gravitational constant $G = 6.67259 \times 10^{-11}$ N·m²/kg². These constants represent three fundamental aspects of nature, which are, respectively, space-time, nature's quantum character, and gravitation. We obtain the Planck unit of a physical quantity by forming an algebraic combination of the constants that possesses the appropriate DIMENSION. (*See* CONSTANTS, FUNDAMENTAL.)

Planck length. This unit of length is:

$$\text{Plank length} = \sqrt{\frac{Gh}{c^3}} \approx 4.05 \times 10^{-35} \text{ m}$$

For comparison, note than an atomic NUCLEUS is about 10^{-14} m in size. The Planck length is far beyond our present ability to probe and resolve structure. It is generally thought that if we ever do manage to approach the Planck length in resolution, as we probe at decreasing scale we will find that space more and more loses its familiar, classical properties: lengths become blurred and directions become undifferentiated. At the Planck scale, the classical picture of space should disappear altogether. A minority view holds that the Planck scale defines a quantum of space (i.e., a minimal amount of space) and thus a LATTICE structure for space. That would be similar to the appearance of ANGULAR MOMENTUM only in discrete values, which are integer or half-integer multiples of $h/(2\pi)$ (*See* ATOM; RESOLVING POWER.)

Planck time. The Planck time is the time interval during which a PHOTON, which travels at the speed of light, traverses the distance of one Planck length:

$$\text{Plank time} = \sqrt{\frac{Gh}{c^5}} \approx 1.35 \times 10^{-43} \text{ s}$$

This unit of time is exceedingly shorter than the briefest time interval that is presently resolvable. The general consensus is that should we ever be able to resolve time intervals approaching the Planck time, time will be found to more and more lose its classical character: time intervals become undefined and the past-future distinction blurs. At the Planck scale, time should lose its classical character altogether. And at the Planck scale of length and time, space-time should be completely nonclassical, a wildly fluctuating situation often described as space-time foam. Nevertheless, some hold the view that the Planck time is a quantum of time, the minimal significant interval of time. Then all time intervals would be integer multiples of the Planck time.

Planck mass. This unit of mass is:

$$\text{Plank mass} = \sqrt{\frac{hc}{G}} \approx 5.46 \times 10^{-8} \text{ kg}$$

The significance of the Planck mass is not clear. It is much larger than the masses of the ELEMENTARY PARTICLES, while less than, but relatively not very much less than, everyday masses. The energy equivalent of the Planck mass in the Planck energy. (*See* MASS ENERGY.)

Planck energy. Its value is:

$$\text{Plank energy} = \sqrt{\frac{hc^5}{G}} \approx 4.91 \times 10^9 \text{ J} = 3.06 \times 10^{28} \text{ K}$$

Planck temperature. The Planck temperature is the temperature at which the average thermal energy of a particle in matter at that temperature equals the Planck energy. Its value is:

$$\text{Plank temperature} = \sqrt{\frac{hc^5}{Gk^2}} \approx 3.56 \times 10^{32} \text{ K}$$

where k denotes the Boltzmann constant, whose value is $1.3806503 \times 10^{-23}$ joule per kelvin (J/K). (*See* HEAT; KINETIC THEORY.)

plasma A plasma is a STATE OF MATTER similar to a GAS, but in which the ATOMs are partially or fully ionized and the liberated ELECTRONs, together with the ions and NUCLEI, form the constituents. A plasma consists of relatively freely moving particles, as does a gas, but whereas a gas is a collection of electrically neutral particles, a plasma contains negatively charged electrons and positively charged ions and nuclei. As a result, a plasma is strongly affected by electric and magnetic FIELDs and by electromagnetic WAVEs. (*See* CHARGE; ELECTRICITY; ELECTROMAGNETISM; IONIZATION; MAGNETISM.)

Matter becomes a plasma at sufficiently high TEMPERATUREs, when the thermal ENERGY of its constituents is high enough to cause ionization. So plasma exists naturally in the Sun and can be created artificially by heating. Ionization can be brought about in other ways, such as by strong electric fields. An example of that is CORONA DISCHARGE, which involves production of plasma. Another path to plasma is through bombardment by energetic electrons or other particles. (*See* HEAT.)

Poincaré, Jules Henri (1854–1912) French *Mathematician* A distinguished and prolific mathematician, Henri Poincaré is best known in physics for his contributions to the special theory of relativity and to STATISTICAL MECHANICS. He also studied the MECHANICS of the SOLAR SYSTEM, as well as other fields of physics, and contributed to the philosophy of science and mathematics. The POINCARÉ TRANSFORMATION is named for him. Poincaré received his doctorate in mathematics in 1879. In 1881, he was appointed professor of mathematics at the University of Paris, France, and subsequently received additional, concurrent appointments, which he held until his death.

See also RELATIVITY, SPECIAL THEORY OF.

The 19th/20th-century mathematician Jules Henri Poincarè made important contributions to the special theory of relativity and to statistical mechanics. *(Cliché Henri Manuel, courtesy AIP Emilio Segrè Visual Archives, Physics Today Collection)*

Poincaré transformation Named for the French mathematician JULES HENRI POINCARÉ, Poincaré transformations are the most general mathematical relations between the respective SPACE coordinates and TIME variables used by any two observers in relative constant-VELOCITY motion (i.e., in relative MOTION that is at constant SPEED and in a straight line), according to ALBERT EINSTEIN's special theory of relativity. In other words, the special theory of relativity states that the coordinates and time variables referring to any pair of INERTIAL REFERENCE FRAMEs are related by Poincaré transformations. LORENTZ TRANSFORMATIONs are a special case of Poincaré transformations, when the two coordinate origins coincide at some instant, and at that instant both clocks read 0. At sufficiently low relative speeds compared with the SPEED OF LIGHT, Poincaré transformations are well approximated by their corresponding GALILEI TRANSFORMATIONs. (*See* COORDINATE SYSTEM; RELATIVITY, SPECIAL THEORY OF.)

SPACE and TIME are merged to form four-dimensional SPACE-TIME, in which events are represented by their space-time coordinates (x, y, z, t), combining their spatial coordinates (x, y, z) and their time coordinate t. In analogy with the distance between points, an interval, D, is defined for any two events in space-time and is given by:

$$D^2 = (x_2 - x_1)^2 + (y_2 - y_1)^2 + (z_2 - z_1)^2 - c^2(t_2 - t_1)^2$$

where (x_1, y_1, z_1, t_1) and (x_2, y_2, z_2, t_2) denote the space-time coordinates of the two events. The symbol c represents the speed of light, whose value is 2.99792458 × 10^8 meters per second (m/s). Although the coordinates of the same event have different values with respect to different reference frames (i.e., in different coordinate systems) in space-time, the interval between any two events is invariant for reference frames in relative constant-velocity motion. In other words, the interval between a pair of events does not change under Poincaré transformations. If unprimed and primed symbols denote coordinates with respect to two such reference frames, this invariance is expressed as:

$$(x_2' - x_1')^2 + (y_2' - y_1')^2 + (z_2' - z_1')^2 - c^2(t_2' - t_1')^2$$
$$= (x_2 - x_1)^2 + (y_2 - y_1)^2 + (z_2 - z_1)^2 - c^2(t_2 - t_1)^2$$

This invariance assures that the speed of light has the same value with respect to both reference frames.

Poiseuille's law Named for its discoverer, French physician Jean Leonard Marie Poiseuille (1799–1869), this law deals with the steady flow of a viscous FLUID through a cylindrical pipe. Denote the inside radius of the pipe by r and the pipe's length by L. Both are in meters (m). Let the difference in fluid PRESSUREs at the ends of the pipe be Δp in pascals (Pa) and denote the VISCOSITY of the fluid in newton·seconds per square meter (N·s/m²) by η. Then Poiseuille's law states that the volume FLUX, or volume flow rate, of the fluid through the pipe (i.e., the volume of fluid passing through the pipe per unit time) in cubic meters per second (m³/s) is given by:

$$\text{Volume flux} = \frac{\pi r^4}{8\eta} \frac{\Delta p}{L}$$

polarization With regard to WAVEs, polarization is a restriction on the direction of OSCILLATION of a transverse wave. Whereas a transverse wave might oscillate in any direction perpendicular to its direction of propagation, and an unpolarized transverse wave oscillates randomly in all perpendicular directions, a polarized transverse wave has its direction of oscillation limited in any of a number of ways. The most familiar type of polarization is plane polarization, or linear polarization. A plane-polarized, or linearly polarized, wave oscillates only in a single direction. If the wave is propagating horizontally toward the north, for example, it might be plane-polarized vertically, so that it oscillates only up and down. If the wave is a transverse elastic wave in a SOLID, the DISPLACEMENT of the medium is then solely vertical. An electromagnetic wave, such as LIGHT, is transverse by its nature. For a vertically plane-polarized electromagnetic wave, convention has the electric FIELD oscillating vertically. The oscillating magnetic field that accompanies the electric field is perpendicular to both the direction of propagation and the electric field, so it oscillates east and west in the present example. (*See* ELECTRICITY; ELECTROMAGNETISM; MAGNETISM.)

Another type of polarization is circular polarization. In a circularly polarized electromagnetic wave, for instance, at every point along the line of propagation the electric and magnetic field VECTORs rotate in the plane perpendicular to the propagation direction. They rotate at constant angular SPEED and maintain their mutual perpendicularity. A snapshot of the wave at any

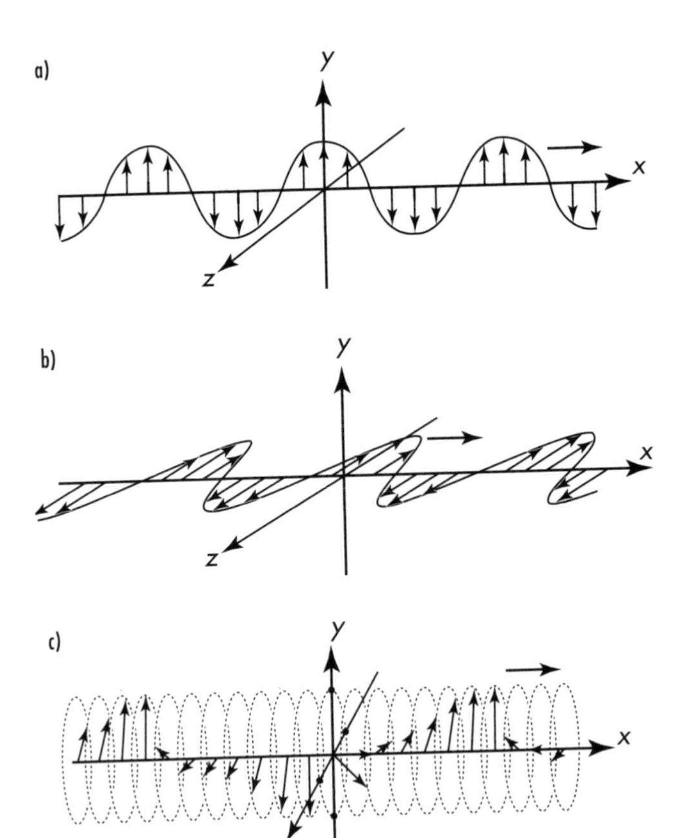

Examples of polarization of a transverse wave propagating in the *x* direction. (a) Vertical plane polarization (in the *y* direction). (b) A wave that is plane-polarized horizontally (in the *z* direction). (c) Circular polarization.

time would show the tip of each field vector describing a helix whose axis lies along the line of propagation, like the thread of a machine screw. Circular polarization can be obtained by having two otherwise identical waves, plane-polarized in perpendicular directions, propagate together at a longitudinal offset of a quarter WAVELENGTH (or equivalently, at a PHASE difference of a quarter cycle = $\pi/2$ rad = 90°). (*See* ROTATION.)

Elliptical polarization is similar to circular polarization, but as the vectors rotate at every point, they simultaneously oscillate in magnitude, so that they are both greatest at the same time in some direction (say, when the electric-field vector is up-down and the magnetic-field vector east-west) and smallest at the same time in the perpendicular direction (east-west and up-down, respectively). In other words, at every point along the line of propagation, the tips of the electric- and magnetic-field vectors periodically traverse ellipses. A snapshot of the wave would show the tip of each field vector describing a somewhat flattened helix. This polarization can be obtained from two waves as for circular polarization, but the two must have different amplitudes.

Unpolarized light can be plane-polarized by REFLECTION or by passing through a polarizing device. Such a device, called a polarizer, is either a special crystal or is made from many aligned microscopic crystals. It allows to pass only the component of the light that is in one direction of polarization, while absorbing the perpendicular component. A circular polarizer lets both components through equally, but retards one by a quarter wavelength with respect to the other. An elliptical polarizer operates similarly, but also attenuates one component more than the other. (*See* ABSORPTION; BREWSTER'S LAW.)

A different use of the term *polarization* is for the creation or alignment of ELECTRIC DIPOLEs or MAGNETIC DIPOLEs. An applied electric field polarizes a sample of material when it aligns the existing electric dipoles there or creates dipoles where there were none and aligns them. This polarization brings about, or induces, an internal electric field that opposes and partially cancels the applied field inside the material. In a similar way, an applied magnetic field can polarize a sample of material by aligning and creating magnetic dipoles in it, with a range of possible results, including a reinforcement of the applied field (in which case the polarization is called magnetization). (*See* DIAMAGNETISM; DIELECTRIC CONSTANT; FERROMAGNETISM; INDUCTION; PARAMAGNETISM; PERMEABILITY; PERMITTIVITY.)

A third use of the term *polarization* is for the aligning of the SPINs of particles in the beams of particle accelerators. If all the ELECTRONs in an electron beam, for instance, have their spins pointing in the same direction, the situation is referred to as a polarized electron beam, or a beam of polarized electrons. (*See* ACCELERATOR, PARTICLE.)

See also MALUS'S LAW.

polycrystalline solid A polycrystalline solid is a SOLID material that is composed of a densely packed conglomeration of many small, randomly oriented CRYSTALs. In a polycrystalline solid, any direction-dependent properties that a single crystal of the substance might possess are averaged out in the polycrystalline form. Many minerals are found in nature in polycrystalline form.

positron The positron is a kind of ELEMENTARY PARTICLE that is the antiparticle of the ELECTRON. The positron is similar to the electron in many ways: they possess the same MASS of 9.10938×10^{-31} kilogram (kg) = 0.5110 MeV/c^2 and the same SPIN of a half unit, and they both appear to be structureless. On the other hand, the positron carries a positive electric CHARGE of $1.602176462 \times 10^{-19}$ coulomb (C), while the electron's charge is negative, but of the same magnitude. Positrons are not normally and stably present in nature, but are created, both naturally and artificially, through the process of PAIR PRODUCTION. A positron and an electron (like any particle-antiparticle pair), upon colliding with each other, undergo mutual annihilation, whereby they both dematerialize and convert to a pair of GAMMA RAYs (high-ENERGY PHOTONs). Since there are many electrons in ordinary MATTER, a positron, once created, does not survive for very long before its demise. Nevertheless, a short-lived configuration of electron and positron, called positronium, can come about before mutual annihilation. It is similar to a hydrogen ATOM, but with the positron replacing the PROTON. A positron and an antiproton can form an antihydrogen atom. (The creation of antihydrogen was recently announced.) Such a structure is stable in itself, although it is eventually destroyed through mutual annihilation with the electrons and protons in ordinary matter. (*See* ANTIMATTER; ELECTRICITY.)

potential, electric The electric potential, or simply potential, at any location is, by definition, the electric POTENTIAL ENERGY of a positive unit CHARGE at that location. The electric potential energy of a charge at any location equals the WORK required to move the charge from infinity to that location against electric FORCES. The electric potential is thus a scalar FIELD. The potential difference, or VOLTAGE, between two locations is the algebraic difference of the values of the potential at the locations. The SI UNIT of electric potential and of potential difference is the volt (V), equivalent to joule per coulomb (J/C). (*See* ELECTRICITY; SCALAR.)

The effect that the electric potential has on a charge is to determine its potential energy. In that capacity, a charge is referred to as a test charge. The potential energy, E_P, in joules (J) of a test charge, q, in coulombs (C) at a location where the potential has the value V is:

$$E_p = qV$$

Electric potential is produced by electric charges. In their capacity as sources of potential, charges are called source charges. The contribution, V, that a source charge, q, makes to the potential at any point is:

$$V = \frac{1}{4\pi\kappa\varepsilon_0}\frac{q}{r}$$

where r denotes the distance in meters (m) from the source charge to the point; ε_0 is the PERMITTIVITY of the vacuum, whose value is $8.85418782 \times 10^{-12}$ $C^2/(N{\cdot}m^2)$; and κ is the DIELECTRIC CONSTANT of the medium in which this is taking place ($\kappa = 1$ in the VACUUM). The total potential at a location is the algebraic sum of the contributions from all source charges. A test charge cannot also serve as a source charge at the same time, however (i.e., a charge cannot affect itself).

A useful device for describing the spatial configuration of the electric potential is the EQUIPOTENTIAL SURFACE. This is a surface in space such that, at all points on it, the electric potential has the same value. As an example, the equipotential surfaces of a point source charge are spherical shells centered on the charge. For any point in space, the electric field line passing through the point is perpendicular to the equipotential surface containing the point. A quantitative relation between the electric field and equipotential surfaces is that for any pair of close equipotential surfaces: (1) the direction of the electric field at any point between them is perpendicular to them and pointing from the higher-potential surface to the lower, and (2) the magnitude of the field, E, in newtons per coulomb (N/C) (equivalent to volts per meter [V/m]) is given by:

$$E = |\Delta V/\Delta s|$$

where ΔV and Δs are, respectively, the potential difference between the equipotential surfaces and the (small) perpendicular distance between them in meters (m).

In this connection, the electric potential at any location can be expressed in terms of the electric field, **E**, as an integral:

$$V(\text{location}) = -\int_{\text{infinity}}^{\text{location}} \mathbf{E} \cdot d\mathbf{s}$$

where $d\mathbf{s}$ denotes a directed infinitesimal element of path length in meters (m), $\mathbf{E}{\cdot}d\mathbf{s}$ is the scalar product of the VECTORS **E** and $d\mathbf{s}$, and the integration is performed along any path from infinity to the location where the potential is being evaluated.

potential energy ENERGY of any form that is not kinetic is termed potential energy. Equivalently, potential energy is energy of a system that is due to any property of the system except MOTION. It is called potential because it possesses the potential to produce motion (through the system's performing WORK), thus converting to KINETIC ENERGY. A form of potential energy can be associated with any CONSERVATIVE FORCE. There are many and diverse forms of potential energy.

power The rate of performance of WORK (i.e., work per unit time) as well as the rate of transfer of ENERGY (which is energy per unit time) is called power. It is a SCALAR quantity. The SI UNIT of power is the watt (W), equivalent to a joule per second (J/s). As an example, if a FORCE **F** in newtons (N) is acting on a body whose instantaneous VELOCITY is **v** in meters per second (m/s), the instantaneous power of the work the force is performing on the body is given by the scalar product:

$$\text{Power} = \mathbf{F}{\cdot}\mathbf{v}$$

As another example, if a BATTERY, whose terminal VOLTAGE is V in volts (V), is delivering electric CURRENT i in amperes (A) to a circuit, the power being delivered to the circuit is:

$$\text{Power} = Vi$$

See also ELECTRICITY.

precession The term *precession* refers to the motion of a GYROSCOPE, whereby its axis swings, or precesses, around a fixed direction. Precession is the result of a TORQUE acting on the rotating gyroscope. When a gyroscope is rotating at a sufficiently high angular SPEED ω and has its off-vertical shaft supported at some point that is not the gyroscope's CENTER OF MASS, the gyroscope's WEIGHT produces a torque that causes the axis to precess at angular speed Ω, where:

$$\Omega = \frac{mgr}{I\omega}$$

Here Ω and ω are both in radians per second (rad/s); m denotes the gyroscope's MASS in kilograms (kg); g is the ACCELERATION due to gravity, whose nominal value at the surface of the Earth is 9.8 meters per second per second (m/s^2); r is the distance in meters (m) from the gyroscope's center of mass to the point of support; and I represents the gyroscope's MOMENT OF INERTIA about its axis in kilogram·meter^2 (kg·m^2). (*See* GRAVITATION; ROTATION.)

Any spinning object can undergo precession. The Sun and Moon exert a net torque on the Earth due to its nonspherical shape. That causes the rotation axis of the Earth to precess, so that it takes about 26,000 years to complete a single sweep. As a result, the star Polaris—the pole star, located directly over the Earth's north pole—was not the pole star referred to in the recorded history of a few thousand years ago, nor will it lie along Earth's axis anymore in the not-too-distant future.

pressure The quantity termed pressure has to do with FORCEs distributed and acting over surfaces and is the force acting perpendicularly on a surface per unit AREA of surface. Pressure is a SCALAR quantity, whose SI UNIT is the pascal (Pa), equivalent to newton per square meter (N/m^2).

Pressure is relevant to FLUIDs (i.e., to LIQUIDs and GASes), which exert pressure on the walls of their container as well as on any object immersed in them. If the fluid pressure at a point on a surface that is in contact with the fluid is denoted p, then the magnitude of the force acting on an infinitesimal area of surface, dA, due to the pressure at that point is given by:

$$dF = p\,dA$$

Here dF denotes the infinitesimal magnitude of force in newtons (N), and dA is in square meters (m^2). The

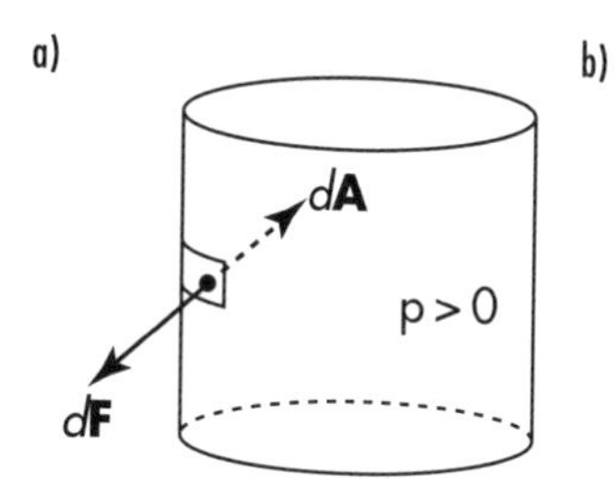

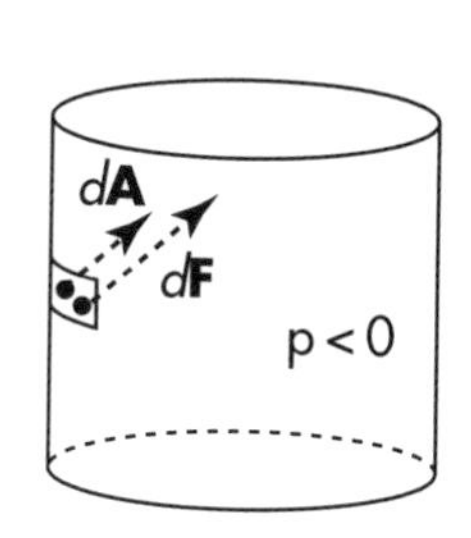

The pressure in a fluid causes a force on a surface with which the fluid is in contact. The force is perpendicular to the surface and points away from the fluid for positive pressure and into the fluid for negative pressure. The infinitesimal magnitude of the force on an infinitesimal element of area dA is $dF = |p|\,dA$, where p denotes the pressure. The vector relation among pressure, surface area, and force is $d\mathbf{F} = -p\,d\mathbf{A}$, where $d\mathbf{F}$ represents the infinitesimal force vector and $d\mathbf{A}$ is the vector of magnitude dA that is perpendicular to the area element and pointing toward the fluid. The situation is depicted for fluid in a container, showing the direction of force on an element of inner wall surface for positive fluid pressure (a) and for negative pressure (b).

direction of the force due to the pressure is perpendicular to the surface and pointing outward from the fluid toward the surface. We are assuming positive pressure, which is the usual case. Negative pressure has an inward-pulling effect, with the surface being sucked toward the fluid. To take negative pressure into account, the above relation might better be written:

$$dF = |p|\,dA$$

The VECTOR form of this equation, correct for positive and negative pressures, is:

$$d\mathbf{F} = -p\,d\mathbf{A}$$

where $d\mathbf{A}$ is a vector whose magnitude is dA and whose direction is perpendicular to the surface at the point in question and pointing from the surface inward toward the fluid. The vector $d\mathbf{F}$ denotes the infinitesimal force resulting from the action of the pressure.

The total force on a finite area of surface is obtained by integrating this relation over the area upon which the pressure acts. For the simple situation in which the pressure is uniform and the surface is flat, the magnitude of the force, F, on the surface area is:

$$F = |p|A$$

with A denoting the area.

A perpendicular pressure force is the only force a fluid at rest exerts on a surface it is in contact with. A moving fluid exerts also a tangential force on a surface (i.e., a force parallel to the surface) due to VISCOSITY.

See also BERNOULLI'S EQUATION; HYDRODYNAMICS; HYDROSTATICS; PASCAL'S PRINCIPLE.

prism A prism is an optical device made of a block of uniform transparent material with polished planar faces. Prisms are made in different sizes and of diverse shapes, according to their purpose. One use of a prism is to change the direction of LIGHT rays, with or without inverting the image that is being transmitted. To this end, the effect of total internal reflection is exploited, whereby the light entering the prism is reflected once or more within the volume of the device before exiting. Another use of a prism is to separate light into its component colors (or, equivalently, FREQUENCIES, or WAVELENGTHS), into a SPECTRUM. This makes use of the effect of DISPERSION, whereby the SPEED OF LIGHT in the material from which the prism is formed depends on the frequency of the light. As a result, the extent of bending of light rays passing through the prism—their deviation, which results from their REFRACTION at the prism's surfaces—also depends on the light's frequency. (*See* OPTICS; REFLECTION; REFRACTION.)

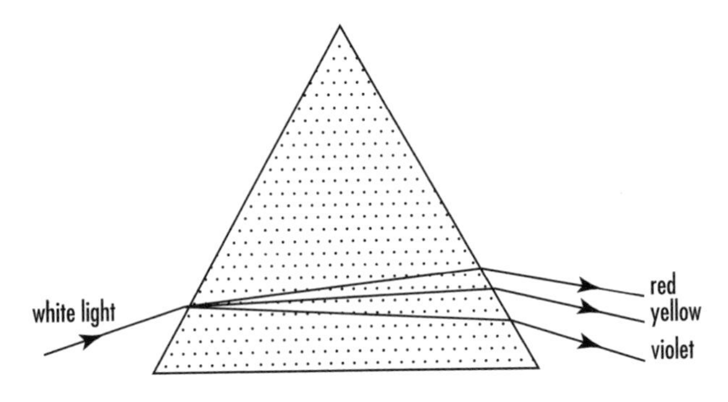

Use of a triangular prism to produce a spectrum by means of dispersion. When white light enters the prism, its color components are refracted at different angles at each face, causing the ray to separate into a continuous spread of colors. Some of the color components are indicated.

proton The proton is a type of ELEMENTARY PARTICLE that belongs to the family of BARYONs and to the wider family of HADRONs. The conventional symbol for the proton is p. Protons and NEUTRONs form the constituents of atomic NUCLEI and together are referred to as NUCLEONs. The proton carries one positive fundamental unit of electric CHARGE, so its charge is $1.602176462 \times 10^{-19}$ coulomb (C). It is a FERMION with a half unit of SPIN. The proton's MASS is 1.673×10^{-27} kilogram (kg) = 938.3 MeV/c^2. The proton is not strictly elementary, as it is composed of three QUARKS: two up quarks and one down quark (uud). As far as is presently known, the proton is a stable particle. Although no proton has yet been observed to decay, there are versions of GRAND UNIFIED THEORIES (GUTs) that do predict proton decay. Protons are affected by all the fundamental INTERACTIONs. (*See* ATOM; ELECTRICITY.)

pulsar A pulsar is any of a class of NEUTRON STARS from which radio pulses reach Earth at very regular intervals. The mechanism of pulse generation is understood as follows. The neutron star is rotating about its axis at a rate of from once every few seconds to as often as hundreds of times a second. It is generating, in a manner not yet clear, copious electromagnetic RADIATION that is channeled by the pulsar's intense magnetic FIELD into two oppositely directed beams leaving the star at its magnetic poles. The pulsar's magnetic axis does not coincide with its rotation axis, so the beams are sweeping the sky like light from a lighthouse as the star rotates. For some of these stars, the Earth happens to lie in just the right direction so that it is "illuminated" by the pulsar beam once during every rotation of the pulsar. Although the pulsars' angular SPEEDS, and thus their pulse FREQUENCIES, are very steady, they are gradually decreasing. The radiation carries away ANGULAR MOMENTUM, so the stars' rotation is slowing down little by little. (*See* ELECTROMAGNETISM; MAGNETISM; ROTATION.)

pyroelectricity The production of an electric potential difference, or VOLTAGE, across a CRYSTAL when the crystal is heated is termed pyroelectricity. The HEAT causes a rearrangement of the permanent ELECTRIC DIPOLES in the crystal, which creates an electric dipole moment for the whole crystal. That brings about a positive electrically charged surface and an oppositely located negative surface, between which a voltage exists. Over time, IONs from the air collect on the charged surfaces and cancel their surface CHARGE. Then another applied TEMPERATURE change can again rearrange the electric dipoles and produce fresh surface charges. (*See* ELECTRICITY; MOMENT, ELECTRIC DIPOLE; POTENTIAL, ELECTRIC.)

See also PIEZOELECTRICITY.

QED See quantum electrodynamics.

quantization Quantization refers to the process of achieving understanding of a physical system in terms of QUANTUM PHYSICS, rather than of CLASSICAL PHYSICS, and according to the rules of QUANTUM MECHANICS. One aspect of quantization is that some physical quantities possess a discrete set of possible values rather than a continuous range of values. In classical physics, a HARMONIC OSCILLATOR, for example,

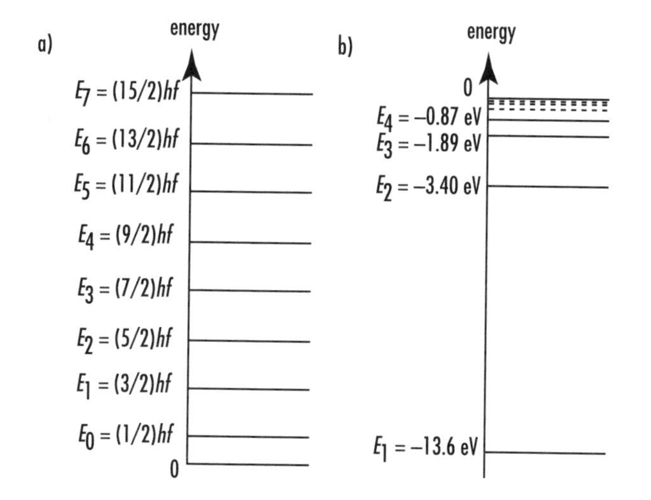

Energy quantization. (a) The lowest energy levels of a harmonic oscillator of frequency *f*. The levels are evenly spaced and are given by $E_n = (n + {}^1/_2)hf$, for $n = 0, 1, \ldots$, where *h* denotes the Planck constant. (b) The lowest energy levels of the hydrogen atom, whose energy levels in electron volts are given by $E_n = -13.6/n^2$ eV, for $n = 1, 2, \ldots$.

may have any value of ENERGY from zero on up. The quantized harmonic oscillator, on the other hand, may only possess energy values given by the formula:

$$E_n = (n + {}^1/_2)hf \text{ for } n = 0, 1, \ldots$$

where E_n denotes the nth energy level in joules (J), f represents the classical FREQUENCY of the oscillator in hertz (Hz), and h is the PLANCK CONSTANT, whose value is $6.62606876 \times 10^{-34}$ joule·second (J·s). (Note that the value zero is not allowed for the energy.) The number n is referred to as a QUANTUM NUMBER, the energy quantum number in this case.

An additional example of quantization is the system comprising a PROTON and an ELECTRON bound by the force of their electric attraction (i.e., the hydrogen ATOM). Its energy levels in electron volts are:

$$E_n = -13.6/n^2 \text{ eV for } n = 1, 2, \ldots$$

(*See* COULOMB'S LAW; ELECTRICITY.)

That leads to the common expression that energy is quantized for the harmonic oscillator and the hydrogen atom. Similarly, one says that any physical quantity is quantized when quantum mechanics allows it only a discrete set of possible values, whereas in classical physics it is a continuous variable. ANGULAR MOMENTUM is another example of such a quantity. In classical physics, angular momentum can have any value. But in the quantum domain it is quantized, as its component in any direction is restricted to integer multiples of $h/(2\pi)$ (i.e., to $mh/(2\pi)$, where $m = 0, \pm1, \pm2, \ldots$). The set of allowed values for a particular angular momentum depends on the magnitude of the angular momentum, which is also quantized and can have only values $\sqrt{l(l+1)}\,h/(2\pi)$, where $l = 0, 1, 2, \ldots$ is the quantum number for angular momentum. For a particular value of l, the allowed values of the component in any direction are $lh/(2\pi)$, $(l-1)h/(2\pi)$, $\ldots$, 0, $\ldots$, $-(l-1)\,h/(2\pi)$, $-lh/(2\pi)$.

The angular momentum that is associated with the SPIN of ELEMENTARY PARTICLES is similarly, but not identically, quantized. Its component in any direction can only possess a value that is a half-integer multiple of $h/(2\pi)$, or $(n/2)\,h/(2\pi)$, where $n = 0, \pm1, \pm2, \ldots$ The exact set of values for a particular type of particle is determined by the spin of the particle. For a particle of spin s (i.e., whose spin quantum number is $s = 0, 1/2, 1, 3/2, \ldots$), the allowed values of the component of its spin angular momentum in any direction are $sh/(2\pi)$, $(s-1)\,h/(2\pi)$, $\ldots$, $-(s-1)\,h/(2\pi)$, $-sh/(2\pi)$. The allowed values of the spin component of an electron, for instance, which is particle with a half unit of spin, are $h/(4\pi)$ and $-h/(4\pi)$.

Another aspect of quantization is that some physical quantities do not possess sharp values, but rather have probabilities for each of their various allowed values to be detected when a measurement is performed on the system. In certain situations, the quantized harmonic oscillator, for example, might possess a definite value for its energy, one of its E_n. In other situations, however, it might not have a sharp value of energy, having instead only a set of probabilities for each of the values E_n to be found when the system's energy is measured and a single value is determined. In such a case, it is not that the system possesses some value of the quantity that we do not know and discover only by measuring it. Rather, the system does not *have* a value for the quantity, and it is the act of measurement that endows it with a value. What the system does possess in such a case is a set of probabilities for the various allowed values to turn up in a measurement.

To continue with the same example, even when the quantized harmonic oscillator does have a definite value for its energy, its position is not sharp. Its position is not restricted and may have any value, but it does not have a definite value until a position measurement "forces" it to take one, according to the system's probabilities for position.

A further aspect of quantization is that certain pairs of physical quantities are mutually exclusive, according to the HEISENBERG UNCERTAINTY PRINCIPLE. This means that the sharpness of one is at the expense of the sharpness of the other and vice versa. The principle is expressed in terms of the uncertainty of a quantity, which is inversely related to its sharpness: the less the uncertainty, the greater the sharpness, and the greater the uncertainty, the less sharp the quantity. One such pair is position and LINEAR MOMENTUM. Let Δx denote the uncertainty in the x-coordinate of position and Δp_x the uncertainty in the x-component of momentum. The uncertainty relation states that the product of these two uncertainties cannot be less than a certain amount, specifically:

$$\Delta x \Delta p_x \geq \frac{h}{4\pi}$$

Thus, given the uncertainty of one of such a pair, say Δx in this example, the least possible uncertainty of the other is determined:

$$\Delta p_x \geq \frac{h}{4\pi \Delta x}$$

When one quantity of such a pair is sharp (i.e., its uncertainty is zero), the uncertainty of the other is infinite. That means its range of values is unlimited; it is as unsharp as possible. So if the position of a particle is precisely known, the particle's momentum is maximally uncertain. And conversely, if the momentum is sharp, the particle has no location and can be anywhere.

quantum electrodynamics (QED) The application of QUANTUM MECHANICS to the electromagnetic FIELD, its INTERACTION with MATTER, and the ELECTROMAGNETIC INTERACTION of matter with matter is called quantum electrodynamics (QED). In very rough outline, electromagnetic WAVES are viewed as flows of PHOTONS; the interaction of the electromagnetic field with matter is understood as the emission, absorption, and scattering of photons by matter; and the electromagnetic interaction of matter with matter is comprehended in terms of exchange of photons. In a particular approach to performing calculations in QED, called perturbative QED, FEYNMANN DIAGRAMS are used. With their aid, physicists have obtained highly accurate results. Quantum electrodynamics is a very successful theory and is, in fact, generally considered to be the best-confirmed theory in physics. (*See* ELECTROMAGNETISM; QUANTUM FIELD THEORY.)

quantum field theory The application of QUANTUM MECHANICS to the understanding of FIELDs is termed quantum field theory. The underlying idea of quantum field theory is that all INTERACTIONS are described by fields that possess a particle aspect and that all MATTER particles are described by fields, such that the principles of QUANTUM PHYSICS and the rules of quantum mechanics are obeyed. (*See* ELEMENTARY PARTICLE; FIELD THEORY; WAVE-PARTICLE DUALITY.)

QUANTUM ELECTRODYNAMICS (QED) is the very successful QUANTIZATION of the electromagnetic field, its interaction with matter, and the ELECTROMAGNETIC INTERACTION of matter with matter. According to QED, the particle aspect of the electromagnetic field is the PHOTON; electromagnetic WAVES are viewed as flows of photons; the interaction of the electromagnetic field with matter is understood as the emission, absorption, and scattering of photons by matter; and the electromagnetic interaction of matter with matter is comprehended in terms of exchange of photons. The electromagnetic field also can be explained (approximately) by a classical theory. (*See* CLASSICAL PHYSICS; ELECTROMAGNETISM.)

The STRONG INTERACTION and the WEAK INTERACTION cannot be explained by classical theories, and they are most successfully dealt with by quantum field theory. The interaction field of the strong interaction is the GLUON field, of which gluons form the particle aspect. The QUARKs are the elementary particles most directly affected by the gluon field, and they are described by their corresponding fields. The quantum field theory of the strong interaction is called quantum chromodynamics (QCD). The particle aspect of the weak-interaction field comprises the intermediate vector BOSONs, which are named and denoted W^+, W^-, and Z^0.

While the gravitational field can be explained by a classical theory, which is the general theory of relativity, its quantization is problematic and has not yet been successfully carried out. This appears to be due to the fact that, while the other fields seem merely to play their various roles in the arena of SPACE-TIME, so to speak, the gravitational field is clearly intimately linked to space-time, so that its quantization would seem to involve the quantization of space-time itself. Although it has not yet been experimentally detected, the graviton is the putative particle aspect of the gravitational field. (*See* GRAVITATION; PLANCK SCALE; RELATIVITY, GENERAL THEORY OF.)

See also GAUGE THEORY.

quantum gravity *Quantum gravity* is the name commonly applied to the quantum aspect of GRAVITATION, in general, and to the QUANTIZATION of the gravitational FIELD, in particular. The gravitational field has not yet been successfully quantized, however, although various possibilities are under investigation. The problem is that, while the fields that have been quantized appear merely to play their roles in the arena of SPACE-TIME, so to speak, the gravitational field is intimately linked to space-time, as ALBERT EINSTEIN's general theory of relativity tells us. So quantum gravity would seem to involve the quantization of space-time itself. Some inkling of what quantum gravity might imply is offered by the PLANCK SCALE. (*See* QUANTUM FIELD THEORY; QUANTUM PHYSICS; RELATIVITY, GENERAL THEORY OF.)

quantum mechanics The realization of the principles of QUANTUM PHYSICS as they apply to actual physical systems is known as quantum mechanics. Here, we briefly present some of the ideas, concepts, and terminology that are involved in quantum mechanics.

In quantum mechanics, a physical system is described by a wave function, Ψ, which specifies the state of the system at any time. The wave function is a complex function (i.e., its values are complex numbers) that depends on the system's generalized coordinates and on time, t. Generalized coordinates are the set of physical quantities needed to specify a state. In the case of a system of particles, for instance, the generalized coordinates consist of all the spatial coordinates of the individual particles. For the sake of simplicity, let us assume the system comprises a single particle, whose position is specified by the position VECTOR $\mathbf{r}$, which might be expressed in terms of its Cartesian coordinates (x, y, z). Then the wave function is written as $\Psi(\mathbf{r}, t)$ or $\Psi(x, y, z, t)$. In a one-dimensional situation the wave function is simply $\Psi(x, t)$. (*See* COORDINATE SYSTEM; DIMENSION.)

The wave function contains all the information about the system that there is. One most important property of the wave function is that it gives probabilities for the system. In the one-dimensional case, the probability that the particle is found in the interval between x and $x + dx$ at time t, denoted by $P(x, t)\, dx$, is given by the square of the absolute value of the wave function, $|\Psi(x, t)|^2 = \Psi(x, t)^*\Psi(x, t)$, where $\Psi(x, t)^*$ denotes the complex conjugate of $\Psi(x, t)$. The relation is:

$$P(x, t)\, dx = |\Psi(x,t)|^2\, dx$$

From this, it follows that the probability for the particle to be found in the finite interval $a \leq x \leq b$ at time t is:

$$\text{Probability} = \int_a^b |\Psi(x,t)|^2 \, dx$$

For the wave function to correctly serve in this capacity, it must be normalized in order to ensure that the probability for the particle to be *somewhere* is unity. The normalization condition for the wave function is then:

$$\int_{-\infty}^{+\infty} |\Psi(x,t)|^2 \, dx = 1$$

This condition is appropriately generalized in more general cases.

Physical Quantities

Physical quantities are represented by operators, which are mathematical expressions that "operate" on the wave function. One use of operators is as follows. If O_Q denotes the operator representing the physical quantity Q, then the average of the values of Q that are found by repeated measurements of Q when the system is in the state specified by $\Psi(x, t)$, called the expectation value of Q and denoted by $<Q>$, is given by:

$$<Q> = \int_{-\infty}^{+\infty} \Psi(x,t)^* \, O_Q \Psi(x,t) dx$$

in the one-dimensional case. The operator corresponding to position, x, for example, is simply multiplication by x. Thus, the expectation value of position is given by:

$$<x> = \int_{-\infty}^{+\infty} \Psi(x,t)^* \, x \Psi(x,t) dx$$

The LINEAR-MOMENTUM operator is given by $[h/(2\pi i)](\partial/\partial x)$, where h denotes the PLANCK CONSTANT, whose value is $6.62606876 \times 10^{-34}$ joule second (J·s), so the expectation value of the particle's momentum, p, is:

$$<p> = \int_{-\infty}^{+\infty} \Psi^* \, \frac{h}{2\pi i} \frac{\partial \Psi}{\partial x}$$

For another example, the operator corresponding to the particle's total ENERGY, E, is $[ih/2\pi](\partial/\partial t)$ giving for the energy expectation value:

$$<E> = \int_{-\infty}^{+\infty} \Psi^* \, \frac{ih}{2\pi} \frac{\partial \Psi}{\partial t} \, dx$$

In general, a physical quantity does not possess a sharp value for an arbitrary state of a system. As mentioned, the expectation value then gives the average of the various values that are found for the quantity when it is measured for the same state. It might happen, however, that the action of an operator O_Q—representing the physical quantity Q—on the wave function, $\Psi(x,t)$, specifying a state results in the same wave function multiplied by a number. That number is just the value of the quantity Q, q, for that state, for which the quantity then possesses a sharp value. In a formula:

$$O_Q \, \Psi(x,t) = q \, \Psi(x,t)$$

In such a case, the wave function is called an eigenfunction of the operator O_Q with eigenvalue q, and the state specified by the wave function is called an eigenstate of the quantity Q, also with eigenvalue q. Then the expectation value of Q equals its sharp value q:

$$\begin{aligned} <Q> &= \int_{-\infty}^{+\infty} \Psi(x,t)^* \, O_Q \, \Psi(x,t) dx \\ &= \int_{-\infty}^{+\infty} \Psi(x,t)^* \, q \Psi(x,t) dx \\ &= q \int_{-\infty}^{+\infty} \Psi(x,t)^* \, \Psi \, (x,t) \, dx = q \end{aligned}$$

The last equality follows from the normalization condition for the wave function.

Uncertainty

The uncertainty of a physical quantity Q is denoted ΔQ and is defined by:

$$\Delta Q = \sqrt{< Q^2 > - < Q >^2}$$

It expresses the spread of the values found by repeated measurements of Q about the average of those values, for whatever state of the system is being considered. If Q has a sharp value, its uncertainty is zero, $\Delta Q = 0$.

There exist pairs of physical quantities that obey the HEISENBERG UNCERTAINTY PRINCIPLE. Every generalized coordinate and its generalized momentum form such a pair. If we denote a generalized coordinate by X, its generalized momentum is the physical quantity represented by the operator $[(h/(2\pi i)](\partial/\partial X)$. The x-coordinate of a particle's position, x, and the x-component of the particle's linear moment, p_x, represented by the operator $[(h/(2\pi i)] \, (\partial/\partial x)$, from such a pair, for example. Their uncertainty relation takes the form:

$$\Delta x \Delta p_x \geq \frac{h}{4\pi}$$

where x is in meters (m), and p_x is in kilogram·meters per second (kg·m/s). This tells us that the minimal uncertainty of one of the pair, say x, is inversely proportional to the uncertainty of the other, p_x:

$$\Delta x \geq \frac{h}{4\pi \Delta p_x}$$

In other words, the more certain the one, the less certain is the other. If one of the pair is sharp, with zero uncertainty, then the other is maximally uncertain. In this example, if a particle has a sharp value of momentum, then its location is completely ambiguous, and it might be found anywhere.

Schrödinger Equation

The wave function is a function of time. Its temporal evolution is described by the SCHRÖDINGER EQUATION. For a single particle whose POTENTIAL ENERGY is a function of position, $U(\mathbf{r})$, the Schrödinger equation for the wave function, $\Psi(\mathbf{r}, t)$, takes the form:

$$\frac{ih}{2\pi}\frac{\partial \Psi}{\partial t} = -\frac{h^2}{8\pi^2 m}\nabla^2\Psi + U(\rho)\Psi$$

(This is actually an operator equation expressing the fact that the total energy equals the sum of the KINETIC ENERGY and the potential energy for the state specified by $\Psi(\mathbf{r}, t)$.) Expressed in terms of coordinates, the equation takes the form:

$$\frac{ih}{2\pi}\frac{\partial \Psi}{\partial t} = -\frac{h^2}{8\pi^2 m}\left(\frac{\partial^2}{\partial x^2} + \frac{\partial^2}{\partial y^2} + \frac{\partial^2}{\partial z^2}\right)\Psi + U(x,y,z)\Psi$$

In the one-dimensional case, the Schrödinger equation for the wave function, $\Psi(x,t)$, is:

$$\frac{ih}{2\pi}\frac{\partial \Psi}{\partial t} = -\frac{h^2}{8\pi^2 m}\frac{\partial^2 \Psi}{\partial x^2} + U(x)\Psi$$

quantum number In QUANTUM MECHANICS, a quantum number is a PARAMETER, most often an integer number, that specifies and enumerates the allowed values of a physical quantity. As an example, the allowed values of ENERGY for the quantized HARMONIC OSCILLATOR are given by the formula:

$$E_n = (n + {}^1\!/_2)hf \text{ for } n = 0, 1, 2, \ldots$$

where E_n denotes the nth energy level in joules (J), f represents the classical FREQUENCY of the oscillator in hertz (Hz), and h is the PLANCK CONSTANT, whose value is $6.62606876 \times 10^{-34}$ joule·second (J·s). The number n is the quantum number here. (*See* CLASSICAL PHYSICS; QUANTIZATION.)

For an additional example, the energy levels of the hydrogen ATOM in electron volts are:

$$E_n = -13.6/n^2 \text{ eV for } n = 1, 2, \ldots$$

where n is the quantum number, called the principle quantum number in this case. The hydrogen atom possess an orbital ANGULAR MOMENTUM quantum number, l, whose allowed values are 0, 1, . . ., $n - 1$. The magnitude of angular momentum that is specified by this quantum number is $\sqrt{l(l+1)}\, h/(2\pi)$. A third quantum number for the hydrogen atom is the so-called magnetic quantum number, m_l, which specifies the value of the component of the angular momentum in any given direction. Its range of values is l, $l - 1$, . . ., 0, . . ., $-(l - 1)$, $-l$. The value of angular momentum component specified by m_l is $m_l h/(2\pi)$. The fourth and last hydrogen atom quantum number is the SPIN quantum number, m_s, which specifies the value of the component of ELECTRON spin in any given direction. Its values are $\pm{}^1\!/_2$, and the value it specifies is $m_s h/(2\pi)$. So the hydrogen atom possesses the four quantum numbers n, l, m_l, and m_s.

quantum physics Quantum physics is the most general, the most widely applicable physics that we have. Quantum physics deals with nature on a broad range of scales, both the larger scales that form the domain of CLASSICAL PHYSICS and the smaller scales—the atomic and molecular scales and the scale of ELEMENTARY PARTICLES—where classical physics does not apply. Its possible applicability on the PLANCK SCALE is not clear at present. (*See* ATOM; MOLECULE.)

Quantum physics is characterized by the PLANCK CONSTANT, $h = 6.62606876 \times 10^{-34}$ joule·second (J·s), which is nature's elementary unit of ACTION. Some further characteristics of quantum physics, which apparently reflect fundamental properties of nature, are these:

Uncertainty. Physical quantities do not in general possess sharp values. They generally do not even have *any* value until they are measured. It is not that they have definite values that we do not know but become revealed to us by measurement. Rather, they actually have no value at all until a measurement "forces" them to exhibit one and thus, so to speak, endows them with a value. When repeated measurements of a physical quantity are performed on the same state of a system, the results are generally different, giving an uncertainty in the value of the quantity. In particular situations, however, the results can be the same, whereby the quantity has a sharp value. Certain pairs of physical quantities obey the HEISENBERG

UNCERTAINTY PRINCIPLE, according to which the minimal uncertainty of one is inversely proportional to the uncertainty of the other. Thus, if one such quantity possesses a sharp value, the other has maximal uncertainty and can show any allowed value upon measurement.

Indeterminism. Given an initial state of an isolated system, what evolves from that state at future times is only partially determined by the initial state. The system is described by a wave function, which evolves deterministically until the system spontaneously and suddenly undergoes a transition, such as radioactive decay, or until a measurement is performed on it. The occurrence of transitions and the values of physical quantities are not in general uniquely determined. Rather, the wave function gives the probabilities of transitions and the probabilities that the system's physical quantities, when measured, will have any of their allowed values. It is the probability, then, that is determined, while indeterminism reigns with regard to the actual occurrence of transitions and to the values of quantities. From this state of affairs, it follows that quantum predictability exists only with respect to probabilities. The time of occurrence of a spontaneous transition or the actual value measured for some physical variable are, in general, undetermined and therefore unpredictable. (*See* DECAY PROCESS; INITIAL CONDITIONS; RADIOACTIVITY.)

Discontinuity and discreteness. An isolated physical system, when left to its own devices, evolves continuously in time for some while. That continuity ends when the system undergoes spontaneous transition, sometimes called a "quantum jump," or when a measurement is performed on it. An additional type of quantum discontinuity is for certain physical quantities whose range of possible values, according to classical physics, is continuous. According to quantum physics, those quantities possess a range of allowed values that can include, or possibly even consist wholly of, a set of discrete values.

As an example, the allowed ENERGY values of the bound ELECTRON and PROTON that constitute a hydrogen atom form a discrete set and are given in electron volts by:

$$E_n = -13.6/n^2 \text{ eV for } n = 1, 2, \ldots$$

Here n is the QUANTUM NUMBER for energy. The unbound electron and proton (i.e., a positive hydrogen ION and an electron wandering off somewhere) can possess any positive value of energy. So the allowed energy values of the electron-proton system consist of both a discrete set of negative values and a continuous range of positive values.

Nonlocality. The situations at different locations can be entangled, so that what happens at one affects what happens at the other in a manner that is not explainable in terms of an influence propagating from one to the other. This is expressed by the statement that a quantum state is nonlocal. Consider this example. Two particles might be emitted by a common source such that their SPINs are in opposite directions. When they are very far apart, the spin direction of one is measured. Then, within a time interval too short for a signal to pass from one location to the other, the spin direction of the other is measured. It is invariably found that, indeed, the spin directions are opposite. What distinguishes this effect as a nonlocal quantum effect is based on the fact that a particle does not possess a spin direction until it is measured. So the first measurement affects not only the particle whose spin direction is first measured, but affects the whole quantum state that involves both particles. The second-measured particle instantaneously "knows" the spin direction it must have in order for it to be opposite that of the first-measured one. (*See* EPR.)

Wave-particle duality. A WAVE is a propagating disturbance, possibly characterized by such as FREQUENCY and WAVELENGTH. It possesses spatial extent, so it is not a localized entity. A particle, on the other hand, is localized. It is characterized by MASS, VELOCITY, energy, etc. Every wave possesses a particle aspect and every particle has a wave aspect. The PHOTON and the PHONON, for example, form the particle aspects of electromagnetic and acoustic waves, respectively. On the other hand, the electron and the proton, normally viewed as particles, are each related to a corresponding wave. Which aspect is manifested depends on the phenomenon being observed. When a wave is exchanging energy with some system, for instance, it does so in distinct units, so its particle aspect is apparent. When a beam of particles is split and rejoined, INTERFERENCE takes place, and it is the wave aspect that is revealing itself. (*See* ACOUSTICS; ELECTROMAGNETISM; WAVE-PARTICLE DUALITY.)

Particle indistinguishability. Identical particles are fundamentally indistinguishable. That leads to the PAULI

EXCLUSION PRINCIPLE, stating that no more than a single FERMION can occupy the same state, and to BOSE-EINSTEIN STATISTICS and FERMI-DIRAC STATISTICS for many-particle systems of BOSONs and fermions, respectively.

QUANTUM MECHANICS is the realization of the principles of quantum physics as they apply to actual physical systems. In quantum mechanics, systems are described by wave functions, which evolve according to the SCHRÖDINGER EQUATION. For FIELDs in particular, QUANTUM FIELD THEORY is the application of quantum mechanics to them.

Although quantum physics gives the best description of nature we currently possess, classical physics can well approximate quantum physics, while being considerably simpler than it, in the appropriate domain. That domain, called the classical domain, is characterized by lengths, durations, and masses that are not too small, that are in general larger, say, than the atomic and molecular scales.

See also QUANTIZATION.

quark The building blocks of the HADRONs, according to present understanding, are the quarks. They are particles with a half unit of SPIN and thus are FERMIONs. One of the quarks' most intriguing properties is that under ordinary conditions, as well as under a wide range of extraordinary conditions, quarks do not exist as free, individual particles. No single quark has ever been discovered. Indirect evidence for their existence abounds, however, and the quark model gives a good understanding of the properties of the hadrons and of the STRONG INTERACTION. (*See* ELEMENTARY PARTICLE.)

According to quantum chromodynamics (QCD), which is the QUANTUM FIELD THEORY of the strong interaction, the quarks possess a kind of CHARGE called color, which takes the values red, blue, and green. (This does not refer to real color; it is the fruit of physicists' whimsy.) The FORCE particles of the strong interaction, called GLUONs, couple to this charge. (The strong interaction among quarks is also called the color force.) Various combinations of bound quarks and antiquarks form the hadrons (i.e., the MESONs and the BARYONs). (*See* ANTIMATTER.)

There are six types of quark. Each comes in all three color charges. Their MASSes range from less than a percent of to over a hundred times the mass of the PROTON. They all possess electric charge whose magnitude is either one- or two-thirds of the fundamental charge (i.e., of the magnitude of the ELECTRON's charge, $e = 1.602176462 \times 10^{-19}$ coulomb [C]). There is an antiparticle type for each quark type, possessing the same mass and spin, the opposite sign of electric charge, and the opposite color charge. (The color charge opposites are named antired, antiblue, and antigreen.) The different types of quark are referred to as flavors. The quark flavors are divided into three generations, in a manner similar to the generations of the LEPTONs. That parallelism seems to indicate a deep connection between the quarks and the leptons. (*See* ELECTRICITY; GRAND UNIFIED THEORY [GUT].)

Here is a list of the quark flavors, with symbol and electric charge indicated. In the first generation there are: up (u, $+2e/3$) and down (d, $-e/3$). In the second generation there are: strange (s, $-e/3$) and charmed (c, $+2e/3$). In the third generation there are: bottom (b, $-e/3$) and top (t, $+2e/3$). The corresponding antiquark types are these. In the first generation there are: antiup ($\bar{u}$, $-2e/3$) and antidown ($\bar{d}$, $+e/3$). In the second generation there are: antistrange ($\bar{s}$, $+e/3$) and anticharmed ($\bar{c}$, $-2e/3$). In the third generation there are: antibottom ($\bar{b}$, $+e/3$) and antitop ($\bar{t}$, $-2e/3$).

The mesons are each composed of a quark and an antiquark. The PIONs, for instance, have these quark compositions: π^+ ($u\bar{d}$), π^- ($\bar{u}d$), π^0 (a quantum mixture of $u\bar{u}$ and $d\bar{d}$). The π^+ and π^- are the antiparticles of each other, while the π^0 is its own antiparticle. The baryons consist of three quarks each. Taking the NUCLEONs as an example, the composition of the proton (p) is (uud) and that of the NEUTRON (n) is (udd). This holds, correspondingly, for the antiparticles $\bar{p}$ ($\bar{u}\bar{u}\bar{d}$) and $\bar{n}$ ($\bar{u}\bar{d}\bar{d}$). (*See* QUANTUM PHYSICS.)

quasar The term *quasar* is a shortened version of "quasi-stellar radio source" and was coined when a number of very powerful astronomical radio sources, apparently very compact and very distant, were discovered. Those objects have since joined others as members of a class of compact, very distant, and very powerful astronomical sources of ENERGY in the whole range of the electromagnetic SPECTRUM, called quasars. The brightest of them are more that 1,000 times brighter than our own GALAXY, the Milky Way. It is generally accepted that quasars are active cores, or nuclei, of galaxies, consisting of gigantic BLACK HOLEs and their accretion disks of infalling material. As the material falls into the black hole, it heats up to very high TEMPERATUREs and radiates copiously in ultraviolet rays and X RAYs. This RADIATION then excites matter in surrounding space, which in turn radiates at longer WAVELENGTHs. (*See* ELECTROMAGNETISM.)

radiation The transfer of ENERGY by WAVEs or by particles is termed radiation. Radiation by waves might take the form of electromagnetic waves. That includes visible LIGHT and is one of the modes of HEAT transfer (together with CONDUCTION and convection). Another type of wave radiation is acoustic, or sonic, radiation, which is energy transfer by SOUND waves. Similarly, in gravitational radiation, energy is transferred by means of gravitational waves. (Their existence has not yet been directly confirmed, however.) Whereas electromagnetic and gravitational radiation can occur in a VACUUM, acoustic radiation requires a material medium. (*See* ACOUSTICS; ELECTROMAGNETISM; GRAVITATION.)

RADIOACTIVITY is a natural source of particle radiation, most commonly of ALPHA PARTICLES, BETA PARTICLES, or GAMMA RAYs. The latter are high-energy PHOTONs, which manifest the particle aspect of electromagnetic radiation. Due to WAVE-PARTICLE DUALITY, the distinction between wave radiation and particle radiation can be somewhat blurred.

The term *radiation* also refers to the waves and particles involved in the transfer of energy.

See also FLUX; INTENSITY.

radioactivity The term *radioactivity* refers to the DECAY of an unstable NUCLEUS, the parent nucleus, to a daughter nucleus with concomitant RADIATION in the form of the emission of one or more particles. (*See* DECAY PROCESS.)

Among the various types of radioactivity, five are common enough to have been given special names:

Alpha decay. The nucleus emits a helium nucleus, called an ALPHA PARTICLE in this context, and transforms into a nucleus with ATOMIC NUMBER lower by two and ATOMIC MASS lower by four than those of the parent nucleus.

Beta decay. The nucleus emits an ELECTRON, also called a BETA PARTICLE, and an electron-type antineutrino. The daughter nucleus has the same atomic mass as does the parent and an atomic number greater by one. (*See* ANTIMATTER; NEUTRINO.)

Inverse beta decay. The nucleus emits a POSITRON, the antiparticle of the electron, and an electron-type neutrino. The daughter nucleus has the same atomic mass as does the parent and an atomic number smaller by one.

Electron capture. The nucleus absorbs one of the atomic electrons and emits an electron-type neutrino. Similarly to inverse beta decay, the daughter nucleus has the same atomic mass as does the parent and an atomic number smaller by one. The basic process is one in which the "captured" electron combines with a proton to produce a neutron, which remains in the nucleus, and an electron-type neutrino, which leaves the nucleus.

Gamma decay. The decaying nucleus is itself the result of some nuclear process and is produced in an excited state (i.e., in a state that possesses higher ENERGY than does the GROUND STATE of that nucleus). The nucleus emits a photon, referred to as a GAMMA RAY, thereby releasing energy and transforming to a lower-energy state, possibly to its ground state. There is no change of atomic number or atomic mass. (*See* NUCLEAR PHYSICS.)

Many ISOTOPEs, whether naturally occurring or artificially produced through nuclear reactions, are radioactive. All nuclei of atomic number greater than 83 are unstable and thus radioactive. The half-lives of the various radioactive isotopes range widely, from the exceptionally short, about 10^{-16} second, to as long as around 10^{21} years, which is much longer than the currently generally accepted value for the age of the UNIVERSE (approximately 15 billion years, give or take a few billion).

Rayleigh, Lord (**John William Strutt**) (1842–1919) British *Physicist* A versatile physicist, John Strutt is known today mostly for RAYLEIGH SCATTERING and for the Rayleigh criterion. The former has to do with the scattering of LIGHT and explains, among other things, why the sky is blue. The Rayleigh criterion forms part of the definition of the RESOLVING POWER of an optical system. Strutt had the good fortune to be financially independent, and he carried out his research mostly at his home and in his personal laboratory. He was especially interested in WAVE phenomena, both in OPTICS and in ACOUSTICS. Strutt also contributed to THERMODYNAMICS, ELECTROMAGNETISM, including the BLACK BODY, and MECHANICS. In his investigations of GASes, he discovered the noble gas argon. Strutt witnessed the birth of QUANTUM PHYSICS and relativity, but was not accepting of the new ideas. In 1904, he was awarded the Nobel Prize in physics "for his investigations of the densities of the most important gases and for his discovery of argon in connection with these studies."

See also RELATIVITY, SPECIAL THEORY OF.

A versatile physicist, Lord Rayleigh (John William Strutt) investigated many areas of physics, especially wave phenomena, and explained the blue color of the sky. His work with gases led to his discovery of the element argon and to the 1904 Nobel Prize in physics. *(AIP Emilio Segrè Visual Archives, Physics Today Collection)*

Rayleigh scattering Named for the British physicist John William Strutt (LORD RAYLEIGH), Rayleigh scattering is the scattering of LIGHT by a low-density GAS, or similar scattering of any RADIATION by particles smaller than the WAVELENGTH of the radiation. Due to the low DENSITY of scattering centers (gas MOLECULES or small particles), INTERFERENCE effects from the radiation scattered by different centers are negligible. As a beam of radiation traverses the medium, the centers scatter radiation uniformly in all directions. That results in attenuation of the beam. The scattered radiation possesses the same FREQUENCY (and, correspondingly, the same wavelength) as the original beam. (Such scattering is called *elastic scattering.*) The INTENSITY of the scattered radiation is proportional to the fourth power of the radiation's frequency (or equivalently, is inversely proportional to the fourth power of the wavelength).

As light from the Sun passes through the Earth's atmosphere, Rayleigh scattering results in preferential scattering of the higher frequencies (shorter wavelengths, bluer colors). That leaves transmitted light that is redder than the original light (with more of the lower-frequency, longer-wavelength light remaining) and creates a blue sky (with a greater proportion of higher frequencies, or shorter wavelengths). The effect is especially pronounced at sunrise and sunset, when the light from the Sun traverses the greatest amount of air. Without such scattering, the sky would appear black, as it does at night from Earth and at all times from the Moon, which has no atmosphere.

reactance The IMPEDANCE in an AC (alternating CURRENT) circuit of a CAPACITOR or an INDUCTOR or any combination of capacitors and inductors is known as reactance. The reactance of a single capacitor, X_C, called capacitive reactance, is given by:

$$X_C = \frac{1}{2\pi fC}$$

where X_C is in ohms (Ω), f is the AC FREQUENCY in hertz (Hz), and C is the CAPACITANCE in farads (F). The instantaneous VOLTAGE across a capacitor lags behind the instantaneous CURRENT through it by a quarter cycle ($\pi/2$ radians (rad), or 90°). (*See* ELECTRICITY; PHASE.)

The reactance of a single inductor, X_L, is referred to as inductive reactance, and is given by:

$$X_L = 2\pi fL$$

where X_L is in ohms (Ω), and L is the INDUCTANCE in henrys (H). The instantaneous voltage across an inductor leads the instantaneous current through it by a quarter cycle ($\pi/2$ rad, or 90°).

The total impedance of combinations of electric components of various kinds must take into account the AC phase differences that occur. One method of doing that is based on complex numbers. Another method, the phasor method, represents AC quantities by two-dimensional VECTORs, with the time-varying quantities—the instantaneous voltages and currents—appearing as rotating vectors. In the simple case of capacitors and inductors connected in series, the total reactance, X, in ohms (Ω) is:

$$X = \Sigma X_L - \Sigma X_C$$

where the individual inductive reactances and capacitive reactances are summed separately. When, for example, such a combination of components is connected in series with a resistor or a combination of RESISTORs, the total impedance, Z, in ohms (Ω) is given by:

$$Z = \sqrt{R^2 + X^2}$$

where R denotes the RESISTANCE in ohms (Ω) of the resistor or the combination of resistors. (*See* DIMENSION.)

reaction force The term *reaction force* refers to a FORCE that comes into being as a result of an action force, according to ISAAC NEWTON's third law of MOTION, also called the law of action and reaction. The law states that when a body exerts a force on another body, the second body exerts on the first a force of equal magnitude and of opposite direction. Forces come in pairs, where one is the action force and the other the reaction force. Note that the forces of an action-reaction pair act on different bodies. Either of such a pair can be considered the action force, although one usually thinks of the force we have more control over as the action, with the other serving as its reaction. If you push on a wall, for instance, the wall pushes back on you with equal magnitude and in opposite direction. We usually think of your push as the action and the wall's push as its reaction. But the reaction force is no less real than the action force. If you are standing on a skateboard and push on a wall, you will be set into motion away from the wall by the wall's force on you, according to Newton's second law of motion. An important pair of action and reaction forces are CENTRIPETAL FORCE and CENTRIFUGAL FORCE. (*See* NEWTON'S LAWS.)

reactor More fully called a nuclear reactor, a reactor is a device for producing and maintaining controlled self-sustaining nuclear reactions. There are two major types of reactor, depending on the type of reaction involved: FISSION reactors and fusion reactors. In the former, the reaction is one in which heavy NUCLEI split into fragments with a release of ENERGY. Typically, a nucleus of the ISOTOPE uranium-235, ^{235}U, absorbs a slow NEUTRON, n, and transforms into a nucleus of uranium-236 in an excited state, denoted $^{236}U^*$. This quickly splits into two fragments, usually, and a number of neutrons:

$$n + {}^{235}U \rightarrow {}^{236}U^* \rightarrow 2 \text{ fragments} + n\text{'s}$$

An example of such a scenario is:

$$n + {}^{235}U \rightarrow {}^{236}U^* \rightarrow {}^{141}Ba + {}^{92}Kr + 3n$$

On the average, 2.5 neutrons are emitted for every ^{235}U nucleus that undergoes fission in this way. The emitted neutrons might then be absorbed by additional ^{235}U

nuclei and cause them to undergo fission, and so on. Thus a chain reaction might take place. When one emitted neutron per fission event causes additional fission, the chain reaction is self-sustaining and can be controlled. The reactor is then said to be critical. When more than a single emitted neutron bring about further fission, the chain reaction intensifies exponentially and develops into an explosion. The reactor is then supercritical. On the other hand, when less than one neutron per fission causes another fission, the reactor is subcritical, and the chain reaction dies out. (*See* FUSION, NUCLEAR; NUCLEAR PHYSICS.)

The elements containing the uranium fuel form the essential component of the reactor's core. In order to reduce the energy of the emitted neutrons and make their capture by ^{235}U nuclei more efficient, suitable moderating material, such as water or carbon, is placed among the fuel elements in the core. Control of the chain reaction is achieved with the help of control rods that can be inserted into and withdrawn from the core. These are made of efficient neutron-absorbing material, such as cadmium. There are arrangements for cooling the core, for utilizing the energy produced, for shielding the core to protect the environment from RADIATION, and for containing the core in case of a leak. For safety's sake, there are also backup mechanisms to stop the chain reaction in case of loss of control and to cool the core should it overheat. Fission reactors serve for the generation of electricity, for the production of isotopes, and for research purposes.

A fusion reactor is a device in which nuclear fusion occurs. Sustained fusion reactions have not yet been achieved, and the field is in a state of development. It is hoped that controlled sustained nuclear fusion will eventually serve as an energy source, since the required raw materials are abundant.

red shift A decrease in the FREQUENCY of an electromagnetic WAVE, including LIGHT, from the wave's emission to its observation, is called a red shift. The term is derived from the fact that when the frequency of light is decreased, the light's color becomes redder. There are three causes of red shift: relative MOTION of the source and the observer, GRAVITATION, and the expansion of the UNIVERSE. (*See* ELECTROMAGNETISM.)

Relative motion. Red shift due to relative motion of the source and observer is the DOPPLER EFFECT for electromagnetic waves, with the source and observer moving away from each other. When they are approaching each other, the observed frequency is greater than that of the source, an effect called a blue shift. When the source is moving perpendicularly to the line connecting the observer and the source, so their distance is not changing, there is still a red shift, called the relativistic Doppler effect, or relativistic red shift, due to TIME DILATION. (*See* RELATIVITY, SPECIAL THEORY OF.)

Gravitation. This effect, called the gravitational red shift, is described by ALBERT EINSTEIN's general theory of relativity. When light travels from lower to higher gravitational potential, such as from the surface of a star to the surface of Earth, it is red shifted. Conversely, light is blue shifted when traveling from higher to lower gravitational potential. This can also be understood in terms of the change of a PHOTON's ENERGY as it moves through the gravitational FIELD and the effect of that on the frequency of the wave, according to the relation:

$$E = hf$$

where E denotes the photon's energy in joules (J), f is the frequency of the electromagnetic wave in hertz (Hz), and h is the PLANCK CONSTANT, whose value is $6.62606876 \times 10^{-34}$ joule·second (J·s). As the energy of the photon decreases or increases, the wave's frequency proportionally decreases (red shift) or increases (blue shift), respectively. (*See* RELATIVITY, GENERAL THEORY OF.)

Expansion of the universe. Called the cosmological red shift, this red shift is caused by the stretching of SPACE, and the concomitant lengthening of WAVELENGTHs of all electromagnetic radiation reaching Earth from distant galaxies as the universe expands, according to relativistic COSMOLOGY. Wavelength and frequency are inversely proportional to each other:

$$f = c/\lambda$$

where λ denotes the wavelength in meters (m), and c is the SPEED OF LIGHT, with value 2.99792458×10^8 meters per second (m/s). So as the wavelength is increased, the frequency is reduced by the same proportion, and the radiation is red shifted.

See also HUBBLE EFFECT.

reference frame A reference frame is a COORDINATE SYSTEM that serves as a reference—as a standard—for position, rest, MOTION, VELOCITY, SPEED, ACCELERATION, INERTIA, etc. When we say that a body is at rest, for instance, with respect to what do we mean it is at rest? If it is at rest with respect to A while B is running by, then it will not be at rest with respect to B; B will see it moving. Or, if a body is accelerating with respect to B, A might be moving in such a way as to see the body moving at constant speed. A reference frame might be attached to, or related to, one or more bodies, such as to A and to B in the examples. We commonly use a coordinate system attached to the Earth—involving latitude, longitude, and altitude—as a reference frame. In COSMOLOGY, a widely used reference frame is the comoving coordinate system, which is a coordinate system with respect to which the GALAXIES are, on the average, at rest.

EINSTEIN's special and general theories of relativity are formulated in terms of reference frames and how the laws of physics, as observed in different reference frames, relate to each other. (*See* RELATIVITY, GENERAL THEORY OF; RELATIVITY, SPECIAL THEORY OF.)

See also REFERENCE FRAME, INERTIAL.

reference frame, inertial An inertial reference frame is a REFERENCE FRAME with respect to which ISAAC NEWTON's first law of MOTION, his law of INERTIA, is valid. So in an inertial reference frame, a body upon which the net FORCE acting is zero is either at rest or moving at constant SPEED in a straight line (i.e., undergoing inertial motion, which is motion with no ACCELERATION). Conversely, in an inertial reference frame, a body in inertial motion has no net force acting on it. So in an inertial reference frame, Newton's first law can serve as a definition of force as that which causes noninertial motion, which is accelerated motion. (*See* NEWTON'S LAWS.)

Alternatively, Newton's first law can serve as a definition of inertial reference frames. One assumes that the presence of forces acting on a body is readily identifiable independently of their effect on the body, such as by contact with another body or by the presence of nearby bodies affecting it gravitationally or electromagnetically. When forces are known to be absent and a body's motion is unaccelerated, the reference frame in use is an inertial one. But if in the absence of forces a body's motion is accelerated, it is being observed in a noninertial reference frame. Such accelerations are sometimes "explained" by fictional forces, such as centrifugal force and Coriolis force. (*See* CENTRIPETAL FORCE; CORIOLIS ACCELERATION; ELECTROMAGNETISM; GRAVITATION.)

reflection When RADIATION, whether of WAVES or of particles, impinges upon the interface surface between two different media, some of the radiation might propagate back from the surface into the medium from which it originates. That phenomenon is reflection. (As for the radiation that is transmitted through the interface into the second medium, see REFRACTION.) Even when the radiation forms a directed beam, if it is incident upon a rough surface, the reflection will occur in all directions. This effect is termed diffuse reflection. When a directed beam falls on a smooth surface, the reflected beam possesses a definite direction, an effect called specular reflection.

The law of reflection describes the reflection of a narrow beam of radiation, a ray, from a smooth plane surface. Consider the perpendicular to the reflecting plane at the ray's point of incidence on the plane. This perpendicular is called the normal. The smaller angle

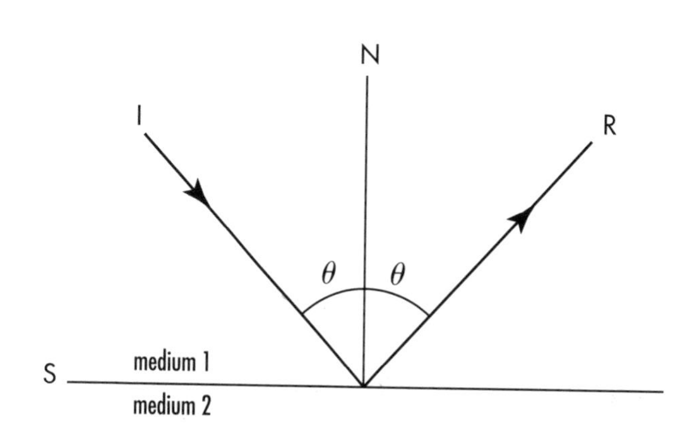

Specular reflection of a ray from the interface surface S between two media. The incident ray I, the normal N (perpendicular to the reflecting surface at the point of incidence), and the reflected ray R all lie in the plane of incidence, which is the plane of the figure. The angle of reflection equals the angle of incidence, both denoted by θ.

between the incident ray and the normal is the angle of incidence. The incident ray and the normal together define a plane, the plane of incidence. This plane is perpendicular to the reflecting plane and contains both the incident ray and the normal. The reflected ray is emitted from the reflecting plane at the point of incidence and lies in the plane of incidence. It is situated on the other side of the normal from the incident ray, and its smaller angle with the normal, the angle of reflection, equals the angle of incidence. In brief, the law of reflection states that (1) the incident ray, the reflected ray, and the normal to the reflecting plane all lie in the same plane, with the two rays on opposite sides of the normal, and (2) the angle of reflection equals the angle of incidence. For reflection from a smooth curved surface, the law of reflection holds with respect to the plane that is tangent to the surface at the point of incidence.

A MIRROR is a device whose purpose is to reflect radiation, in particular waves, in a controlled manner.

refraction The change in the propagation direction of WAVES upon passing through the interface surface between two media in which the wave possesses different SPEEDs of propagation is called refraction. Consider a narrow beam, a ray, that impinges on a smooth plane interface surface and penetrates into the second medium, and consider the perpendicular to the plane at the point of incidence on the plane. This perpendicular is called the normal. The smaller angle between the incident ray and the normal is the angle of incidence. The incident ray and the normal together define a plane, the plane of incidence. This plane is perpendicular to the interface plane and contains both the incident ray and the normal. The refracted ray is transmitted into the second medium from the point of incidence on the interface plane and lies in the plane of incidence. It is situated on the same side of the normal as the geometric extension of the incident ray into the second medium. The smaller angle that the refracted ray makes with the normal is the angle of refraction.

SNELL'S LAW gives the relation between the angle of refraction and the angle of incidence as follows. Denote the angles of incidence and refraction by θ_i and θ_r, respectively, and the INDEXes OF REFRACTION of the corresponding media by n_i and n_r. Then:

$$n_i \sin \theta_i = n_r \sin \theta_r$$

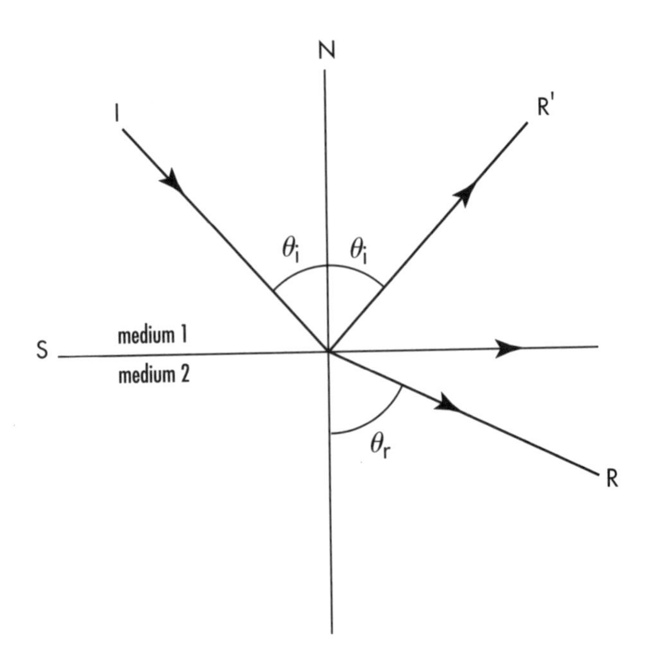

Refraction of a ray at the interface surface S between two media. The incident ray I, the normal N (perpendicular to the interface surface at the point of incidence), and the refracted ray R all lie in the plane of incidence, which is the plane of the figure. The angles of incidence and refraction are denoted θ_i and θ_r, respectively. They are related through Snell's law, $n_i \sin \theta_i = n_r \sin \theta_r$, where n_i and n_r denote the indexes of refraction of medium 1 and medium 2, respectively. The reflected ray R′ is shown as well. (The angle of reflection equals the angle of incidence.)

The law of refraction is then, in brief, that (1) the incident ray, the refracted ray, and the normal to the interface plane all lie in the same plane, with the refracted ray on the same side of the normal as the geometric extension of the incident ray into the second medium, and (2) the angles of incidence and refraction obey Snell's law. Note that a ray impinging on an interface surface perpendicularly ($\theta_i = 0$) is always transmitted with no deviation ($\theta_r = 0$), and as the angle of incidence increases, so does the angle of refraction. Further, upon passing from a medium of lower index of refraction into one of higher index, a ray is bent toward the normal ($\theta_r < \theta_i$), while in passing from higher to lower index of refraction it is bent away from the normal ($\theta_r > \theta_i$).

In general, a ray that is incident on a smooth interface surface undergoes both REFLECTION from the surface and transmission into the second medium. But when a wave passes from a medium of higher index of refraction to one of lower index, and is thus bent away from the normal, there exists a range of angles of incidence for which there is no transmission at all. This effect is called total internal reflection. It comes about as follows. According to Snell's law:

$$\sin \theta_r = (n_i/n_r) \sin \theta_i$$

At the incidence angle $\theta_i = \theta_c = \sin^{-1} (n_r/n_i)$, called the critical angle, $\sin \theta_r$ equals 1 and the angle of refraction equals 90°. Then the refracted ray is bent as far from the normal as possible: it is parallel to and just grazing the interface surface. For angles of incidence greater than the critical angle, Snell's law cannot be fulfilled (it then requires $\sin \theta_r > 1$), and no transmission occurs.

Devices that are designed to control and utilize refraction include LENSes and PRISMS.

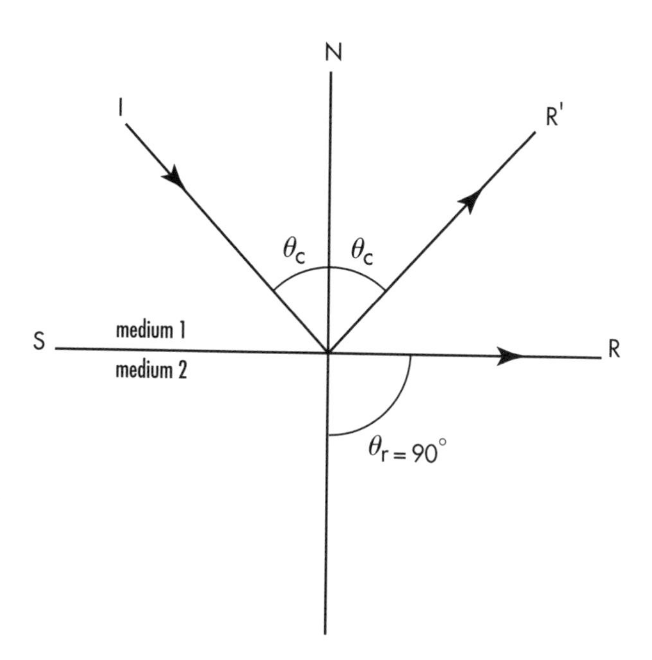

Critical-angle incidence of a ray at the interface surface S between two media. The incident ray I, the normal N (perpendicular to the interface surface at the point of incidence), the reflected ray R′, and the refracted ray R all lie in the plane of incidence, which is the plane of the figure. The critical angle θ_c is that angle of incidence for which the angle of refraction θ_r equals 90°, which is the largest value it can have. According to Snell's law, $\theta_c = \arcsin (n_r/n_i)$, where n_i and n_r are the indexes of refraction of medium 1 and medium 2, respectively, and $n_i > n_r$. At angles of incidence greater than the critical angle, there is no refracted ray and the incident ray is totally reflected, an effect called total internal reflection. The angle of reflection always equals the angle of incidence, both indicated by θ_c in the figure.

relativity, general theory of ALBERT EINSTEIN's general theory of relativity is a theory of GRAVITATION. A generalization of the special theory of relativity, this theory, too, is formulated in terms of four-dimensional SPACE-TIME. Space-time is not flat as it is in the special theory, however, but rather curves as an effect of and in the vicinity of MASSes and ENERGIES. According to the theory, gravitation is not a FORCE, like, say, the magnetic force, but rather is an effect of the curved geometry of space-time. That comes about in this way. Force-free bodies obey a generalization of ISAAC NEWTON's first law of MOTION. Their free motion, or inertial motion, however, is not in a straight line, but follows a path that is the closest to a straight line in curved space-time. This is referred to as geodesic motion. The result is motion that is neither in a straight line nor at constant speed, in general, and manifests the effect of gravitation on the bodies. (*See* DIMENSION; INERTIA; MAGNETISM; MOTION; NEWTON'S LAWS; RELATIVITY, SPECIAL THEORY OF.)

Here is a description of an often-presented model, to give some idea of what is going on. Set up a horizontal stretched sheet of rubber. It represents flat space-time. Place a heavy marble on the sheet. The marble represents a massive body. Due to the marble's weight, the rubber sheet is stretched downward at the location of the marble and assumes a somewhat conical shape, with the marble at the apex. This shape represents curved space-time in the presence of and in the vicinity of mass. Now, toss light marbles or ball bearings onto the rubber sheet so they roll at various speeds and in various directions. Some might orbit the marble. Low-speed ones will spiral into the marble. High-speed ones will escape capture but still have their trajectory affected. This behavior represents the effect of space-time curvature on bodies as an explanation of gravitation.

The EQUIVALENCE PRINCIPLE forms an essential ingredient of the general theory of relativity. In addition, Einstein was strongly influenced by MACH'S PRINCIPLE. One of the basic postulates of the general theory, which is a straightforward generalization of a basic postulate of the special theory, is that the laws of physics are the same in all REFERENCE FRAMEs, *no matter what their relative motion*. (In the special theory, the reference frames are restricted to relative motion at constant VELOCITY.) Furthermore, the general theory is designed to produce Newton's law of gravitation, under appropriate conditions. (*See* INVARIANCE PRINCIPLE.)

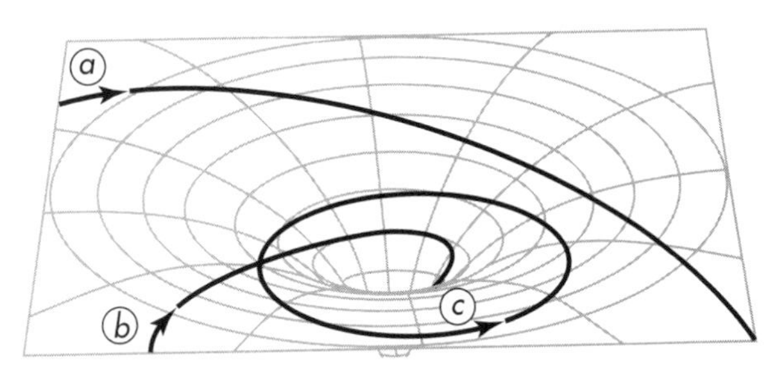

A rubber-sheet model of gravitation according to the general theory of relativity. A heavy object, such as a marble, placed on a stretched horizontal rubber sheet, deforms it as in the figure. The motion of freely rolling ball bearings, or other light spherical objects, across the sheet is no longer the straight line it would have been in the absence of the distortion. Three sample trajectories are shown: (a) the rolling object approaches the marble and leaves it, (b) the rolling object falls onto the marble, and (c) the rolling object orbits the marble.

The general theory of relativity has passed all the tests it has been put to. It explains the deviation of the orbit of the planet Mercury from the prediction of Newtonian MECHANICS. It also correctly gives the amount of deflection from a straight-line trajectory of starlight as it passes near the Sun. Its prediction of the gravitational RED SHIFT has been confirmed. The theory also predicts the existence of GRAVITATIONAL WAVES. There is indirect evidence for such waves, although they have not yet been detected by the special detectors designed for that purpose. The general theory of relativity predicts BLACK HOLES. Astronomical observations indicate their existence as a common phenomenon. The theory predicts gravitational lensing, and the effect is observed. The general theory of relativity is incorporated into the standard theoretical framework for the study of COSMOLOGY, called relativistic cosmology. And as for practical applications, the theory is routinely invoked in connection with the navigation of spacecraft and in the design and operation of the global positioning system (GPS). *See* GRAVITATIONAL LENS; LIGHT.

See also MASS ENERGY.

relativity, special theory of ALBERT EINSTEIN's special theory of relativity deals with the fundamental concepts of SPACE and TIME and how distances and time intervals are measured and relate to each other. It is a theory about theories, a supertheory, in that it lays down conditions that other physics theories must fulfill. It has two basic postulates: (1) The laws of physics are the same in all REFERENCE FRAMEs that move at constant VELOCITY with respect to each other. In other words, every pair of observers who are in relative MOTION in a straight line and at constant SPEED discover the same laws of physics. (2) One such law is that LIGHT (and all other electromagnetic RADIATION as well) moves at the speed of $c = 2.99792458 \times 10^8$ meters per second (m/s), called the SPEED OF LIGHT. (The null result of the MICHELSON-MORLEY EXPERIMENT served Einstein as evidence for the validity of this postulate.) (*See* ELECTROMAGNETISM; INVARIANCE PRINCIPLE.)

It is useful to introduce the concepts of relative and absolute. In the present context, the term *relative* means observer dependent. If different observers can validly claim different values for the same physical quantity, then that quantity is relative. On the other hand, an absolute quantity is one that has the same value for all observers. The special theory of relativity, in its second postulate, states that the speed of light is an absolute quantity. Note that the terms *relative* and *absolute* in the special theory of relativity mean no more and no less than observer dependent and observer independent, respectively. There is no claim whatsoever that "everything is relative." Note that the basic postulates of the theory concern absoluteness rather than relativity.

The first postulate implies that there exists no absolute, observer-independent, reference frame for constant-velocity motion. In other words, a state of absolute rest is a meaningless concept. This idea was not new from Einstein and is included in ISAAC NEWTON's laws of motion, which predate Einstein by some 200 years. The second postulate was revolutionary. It runs counter to everyday experience with the composition of velocities. As an example, assume an observer has a beam of light going past her at speed c and has another observer passing her at speed $0.9c$ in the direction of the light beam. According to Newtonian MECHANICS and everyday experience with bodies in relative motion, the second observer should observe the light beam passing him at speed $c - 0.9c = 0.1c$. But the second postulate states that the second observer, too, will observe speed c for the light beam. In other words, the velocity of light is different from that of material bodies and holds a unique status in the special theory of relativity. In Newtonian mechanics it is relative, as are all velocities, while in the special theory of relativity it is absolute. (*See* NEWTON'S LAWS.)

The differences between relativistic physics and Newtonian physics become manifest at speeds that are a considerable fraction of the speed of light and become extreme as speeds that approach the speed of light. For speeds that are sufficiently small compared with the speed of light, relativistic physics becomes indistinguishable from Newtonian physics. This limit is called the nonrelativistic limit.

The special theory of relativity views matters in terms of events in SPACE-TIME. An event is an occurrence at some location at some instant. Space-time, also referred to as Minkowski space (after the German mathematician and physicist Hermann Minkowski [1864–1909]) in the context of the special theory of relativity, is a four-dimensional abstract space consisting of three spatial dimensions and one temporal one. An event is specified by its location—represented by three coordinates with respect to three spatial axes, say x, y, and z—and by its instant of occurrence, represented by a single coordinate with respect to a single temporal axis, t. So an event is represented by a point in space-time, specified by its four coordinates (x, y, z, t). A reference frame is defined by a specific coordinate system. When reference frames are in relative motion, their respective coordinate systems are moving with respect to each other. The mathematical relations between an event's coordinates with respect to any coordinate system and the same event's coordinates with respect to a coordinate system moving at constant velocity relative to the first one are given, in the special theory of relativity, by POINCARÉ TRANSFORMATIONs and LORENTZ TRANSFORMATIONs. (*See* COORDINATE SYSTEM; DIMENSION.)

These transformations mix spatial and temporal coordinates. As a result, a pair of events that is observed as simultaneous (i.e., occurring at the same time) by one observer might not appear so to another observer. Other observers, depending on their motion, might measure one event as taking place before the other or the other before the one. In other words, simultaneity is relative. In Newtonian mechanics, on the other hand, simultaneity is absolute. (*See* SIMULTANEITY.)

Relative Quantities

Another result is that length is relative; the same object is measured by different observers to have different values for its length. Let an observer for whom the object is at rest measure length L_0, the object's rest length. Then an observer moving at speed v in the direction of the length (i.e., an observer for whom the object is moving longitudinally at speed v), measures a shorter length, L, where:

$$L = L_0 \sqrt{1 - v^2/c^2}$$

Here L and L_0 are expressed in the same UNIT of length, and v is in meters per second (m/s). Note that in the limit of v approaching c, L contracts down to zero. At the other extreme, in the nonrelativistic limit (v small compared with c), L is indistinguishable from L_0. The length-contraction effect is mutual, as an object at rest for the second observer is observed by the first to be contracted. Length is absolute in Newtonian mechanics. (*See* LENGTH CONTRACTION.)

Time intervals are relative as well. The same clock runs at different rates for different observers. Assume that a clock (or heart, etc.) is ticking off time intervals of duration T_0, as measured by an observer for whom the clock is at rest. Then an observer moving at speed v relative to the first one (i.e., an observer for whom the clock is moving at speed v), measures longer time intervals, T, for the same ticks (or heartbeats), where:

$$T = \frac{T_0}{\sqrt{1 - v^2/c^2}}$$

with the two time intervals expressed in the same unit. In other words, a moving clock is measured as a slower clock. This effect is called TIME DILATION. Similarly, a moving organism is observed to live longer than if it were at rest, and unstable ELEMENTARY PARTICLEs have longer lifetimes (or half-lives) when they DECAY in flight than when they are at rest. Note that as v approaches c, T tends to infinity, and the clock is observed to have almost stopped, the organism is in quasi-suspended animation, and the unstable particles almost never decay. In the nonrelativistic limit (v/c approaches 0) time dilation is negligible. In Newtonian mechanics time intervals are absolute. Time dilation is a mutual effect. (*See* TWIN PARADOX.)

Even MASS is not immune to relativistic effects, as the mass of the same body is measured by different observers as having different values. Let m_0 denote the mass of a body for an observer for whom the body is at rest. This is called the REST MASS of the body. An observer moving at speed v with respect to the first observer, and who therefore observes the body moving at speed v, measures a greater mass, m, where:

$$m = \frac{m_0}{\sqrt{1 - v^2/c^2}}$$

with m and m_0 expressed in the same unit. Note that the mass of a moving body increases with its speed and approaches infinity as the speed approaches the speed of light. This means that a body's INERTIA increases with its speed and approaches infinity as v approaches c. In the nonrelativistic limit, the rest mass adequately represents the body's mass. In Newtonian mechanics, mass is absolute.

Relativistic LINEAR MOMENTUM has the form:

$$\mathbf{p} = m\mathbf{v} = \frac{m_0 \mathbf{v}}{\sqrt{1 - v^2/c^2}}$$

where **p** and **v** denote the momentum and velocity VECTORs, respectively, and v is the magnitude of **v**. The SI unit of momentum is kilogram·meter per second (kg·m/s), and the masses are in kilograms (kg). As v approaches c, the momentum approaches infinity, while in the nonrelativistic limit we obtain the Newtonian expression for momentum.

Mass and Energy

The special theory of relativity assigns an ENERGY, E, to a body:

$$E = mc^2 = \frac{m_0 c^2}{\sqrt{1 - v^2/c^2}}$$

where E is in joules (J). Note, very importantly, that even a body at rest possesses energy, its rest energy:

$$E_0 = m_0 c^2$$

So every mass is assigned an equivalent energy, its MASS ENERGY. This energy can be tapped and made use of through the nuclear processes of FISSION and fusion. Conversely, every energy, E, possesses a mass equivalent, m, whose value is given by:

$$m = E/c^2$$

So energy behaves as mass: it possesses inertia and is affected by the gravitational FIELD. (*See* FUSION, NUCLEAR; GRAVITATION; NUCLEAR PHYSICS; NUCLEUS.)

A body's KINETIC ENERGY is its total energy less its rest energy, or:

$$\text{Kinetic energy} = E - E_0 = m_0 c^2 \left(\frac{1}{\sqrt{1 - v^2/c^2}} - 1 \right)$$

It can be shown that in the nonrelativistic limit, this expression becomes the Newtonian expression for kinetic energy, $m_0 v^2/2$.

A notable relativistic effect is that the speed of light is a limiting speed, in the sense that bodies cannot be accelerated to the speed of light and cannot move at that speed. The impossibility of accelerating a body to the speed of light can be understood in terms of the body's inertia approaching infinity as its speed approaches c, as seen above, so the accelerating effect of a FORCE acting on the body tends to zero. It turns out that the speed of light is a limiting speed for the transfer of information and for the propagation of causal effects, as well. Only light, and presumably also gravitational WAVEs, propagate at the speed of light. Accordingly, PHOTONs (the particles of light) and gravitons (the putative particles of gravitational RADIATION) move at that speed also and only at that speed. (*See* ACCELERATION.)

Relativistic Mechanics

The previous discussion ties into the way velocities combine in the special theory of relativity. Let there be two reference frames, R and R′, such that R′ is moving at speed v relative to R. Assume a body is moving in the same direction as R′ and with speed u as observed in frame R, and let u be greater than v, for simplicity of discussion. Denote by u' the body's speed as observed in frame R′. In Newtonian physics, these relative speeds combine by simple addition as:

$$u = v + u'$$

which conforms with our everyday experience. But if this relation were true for high speeds, we could have speeds that exceed the speed of light. If, for instance, reference frame R′ is moving at speed $0.6c$ relative to R and the body is moving at $0.5c$ relative to R′, then the body would be observed moving at $1.1c$ relative to R, faster than the speed of light. In the special theory of relativity, the formula for the addition of speeds is, instead:

$$u = \frac{v + u'}{1 + vu'/c^2}$$

In the nonrelativistic limit, this formula does indeed reduce to the Newtonian one, but for high speeds its results are quite different. In the above example, the body is observed moving only at speed $0.85c$

relative to frame R. In fact, no matter how large the speeds v and u' are, as long as they are less than c, the result of their relativistic addition, u, is always less than c. If a light beam is being observed rather than a moving body, so that $u' = c$, then no matter what the relative speed of the reference frames, v, is, the formula gives $u = c$. This result conforms with the second postulate of the special theory of relativity, that all observers measure speed c for light.

Newton's second law of motion is commonly expressed as:

$$\mathbf{F} = m_0\mathbf{a} = m_0\, d\mathbf{v}/dt$$

where **F** denotes the force vector in newtons (N), **a** is the acceleration vector in meters per second per second (m/s^2), and t is the time in seconds (s). Actually, Newton's own formulation was in terms of momentum:

$$\mathbf{F} = d(m_0\mathbf{v})/dt$$

which is equivalent to the former when the mass does not change in time. The latter is the valid form of the law even for cases when a body accrues or loses mass over time (such as a rocket). We use m_0 here rather than the usual notation, m, since in Newtonian physics, the mass does not depend on the body's speed and is therefore the rest mass.

Newton's second law does not conform with the special theory of relativity. An indication of this is the result that, according to the formula, a constant force acting for sufficient time will accelerate a body to any speed whatsoever, even exceeding the speed of light. The relativistic form of Newton's second law is based on its momentum formulation, but expressed in terms of relativistic momentum:

$$\mathbf{F} = d\mathbf{p}/dt = d(m\mathbf{v})/dt = m_0 \frac{d}{dt} \frac{\mathbf{v}}{\sqrt{1 - v^2/c^2}}$$

Although Newtonian mechanics requires modification to conform with the special theory of relativity, as we just saw, the laws of electromagnetism, as expressed by MAXWELL'S EQUATIONS, require no modification at all. These laws were formulated well before Einstein's work, yet they turned out to be perfectly relativistic as they are. They fulfill the second postulate of the theory by predicting that all observers, whatever their velocities relative to each other, will indeed observe the same speed, c, for light. They even predict the correct value of c.

Space-Time Structure

Space-time turns out to have different properties in different directions, according to the special theory of relativity. That follows from the fundamental difference between space and time, although both are combined in space-time. Let us denote any given space-time point by P. All those points that can be reached by a light ray emitted from P together with all the points that can be reached by a particle leaving P and moving more slowly than c constitute the *future lightcone* of P. All observers, whatever their state of motion, agree that all events in the future lightcone of P are in P's future, i.e., that an event at P occurs before any event in its future lightcone. In similar vein, all those points from which a light ray can reach P and all points from which a particle moving more slowly than c can reach P make up the *past lightcone* of P. All observers agree that all events in the past lightcone of P are in P's past, i.e., that an event at P occurs *after* any event in P's past lightcone. Thus a pair of events that lie in each other's lightcones possess an observer-independent—absolute—temporal order: one is absolutely in the other's future, while the latter is absolutely in the past of the former.

When two events, or points in space-time, are related in such a way that a particle moving at less than c can connect them, they are said to possess a timelike separation, or timelike interval. As mentioned, such events lie within each other's lightcones and have an absolute temporal ordering. As such, they are said to be causally related, in that the earlier event can affect the later one. For such a pair of events, a reference frame can always be found in which both events occur at the same spatial location (at different times). When two events, or space-time points, can only be connected by a light ray, or photon, moving at speed c, they have a lightlike separation, or lightlike interval. They are then said to lie on (rather than within) the light cones of each other. These events, too, possess an absolute temporal ordering and are causally related in that the earlier can affect the later.

On the other hand, when a pair of events, or points in space-time, have a spatial separation that is so large compared with their time difference that any influence between them would have to exceed the speed of light, they are referred to as having a spacelike separation, or spacelike interval. They lie outside each other's lightcones and are not causally related: neither can affect the other. Moreover, their temporal order is relative. With respect to different reference frames moving at different velocities, one event might be observed to precede the other or vice versa, or both events might be measured as occurring simultaneously. A reference frame can always be found in which such a pair of events occurs at the same time (at different spatial locations). This relates to an earlier discussion on the relativity of simultaneity.

Einstein's special theory of relativity is a very well-supported theory. It forms an essential part of the design and operation of such as particle accelerators, the Global Positioning System (GPS), and nuclear devices. In the nonrelativistic limit, it reduces to Newtonian physics, which is adequate for most everyday applications. The special theory of relativity was generalized by Einstein to a theory of gravitation, known as his general theory of relativity. (*See* ACCELERATOR, PARTICLE; RELATIVITY, GENERAL THEORY OF.)

resistance Electric resistance is the hindrance that MATTER sets to the passage of an electric CURRENT through it. That might be due to the unavailability of current carriers (which in SOLIDs are ELECTRONs and HOLEs) or to impediments to the motion of current carriers (such as, again in solids, impurities and LATTICE defects and VIBRATIONs). As a physical quantity, the resistance of a sample of material, R, in ohms (Ω) is the ratio of the VOLTAGE across the sample, V, in volts (V) to the current through the sample, i, in amperes (A):

$$R = V/i$$

(*See* ELECTRICITY.)

The resistance of a sample depends both on the nature of the material and on the sample's shape. In the case of a homogeneous material with a shape that has a definite length and cross-section, the resistance is given by:

$$R = L\rho/A$$

where L denotes the length in meters (m), A is the cross-section AREA in square meters (m^2), and ρ is the RESISTIVITY of the material in ohm·meters (Ω·m), which is a characteristic of the material.

When a current passes through matter, electric ENERGY is converted to HEAT. The rate of heat generation, or POWER, P, in watts (W) is given by:

$$P = Vi$$

where i denotes the current through the sample, and V is the voltage across the sample. This can be expressed equivalently in terms of the resistance of the sample as:

$$P = i^2R = V^2/R$$

This result is known as Joule's law, and the heat thus produced as Joule heat. (*See* JOULE, JAMES PRESCOTT.)

Resistance is the inverse of CONDUCTANCE. Resistance is TEMPERATURE dependent through the temperature dependence of resistivity. A RESISTOR is a device that is designed to possess a definite resistance and be able to function properly under a definite rate of heat production.

See also IMPEDANCE; OHM'S LAW; SUPERCONDUCTIVITY.

resistivity Resistivity is a measure of the characteristic opposition of a material to the flow of electric CURRENT. Resistivity is defined as the electric RESISTANCE of a 1-meter cube of the material when a VOLTAGE is maintained between two opposite faces of the cube. Its SI UNIT is the ohm-meter (Ω·m). So a resistivity of 1 Ω·m for a material means that when a voltage of 1 volt (V) is applied to a cube of the material as described, a current of 1 ampere (A) flows between the opposite faces. (*See* ELECTRICITY.)

Resistivity is TEMPERATURE dependent in general. For almost all SOLIDs, the resistivity increases with increasing temperature. For small temperature changes, the change in resistivity is proportional to the change in temperature. At sufficiently low temperatures, some materials enter a superconducting state, in which their resistivity is precisely zero. (*See* SUPERCONDUCTIVITY.)

resistor An electric device that is designed to possess a definite RESISTANCE and be able to function properly under a definite rate of HEAT production is called a resistor. A resistor is characterized by the value of its resistance in ohms (Ω) and by the value of its maximal allowed rate of heat production in watts (W). The relation between the VOLTAGE on a resistor and the current through it is given by OHM'S LAW:

$$V = iR$$

where V denotes the voltage in volts (V), i the current in amperes (A), and R the resistance of the resistor. (*See* ELECTRICITY.)

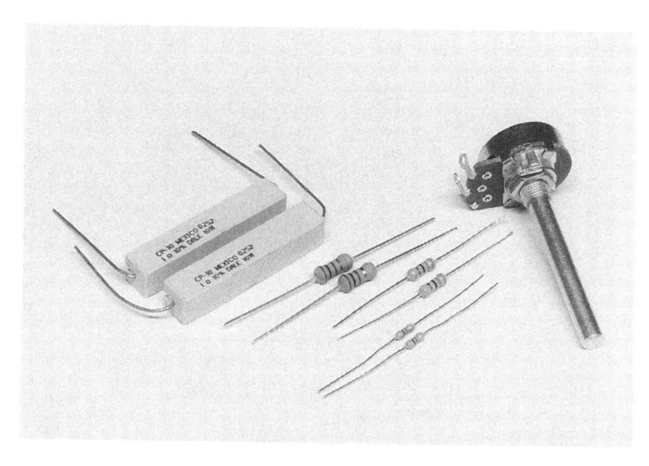

A collection of various resistors. The resistance, the tolerance (i.e., the precision of the resistance specification), and the maximal allowed rate of heat production are indicated on each. On small resistors, those quantities are shown by means of colored bands, according to a standard code. The device on the far right is an adjustable resistor of a type called "potentiometer." It is used, for instance, as the volume control of an amplifier. To set the scale of the figure, the length of the potentiometer's shaft is 4.4 centimeters (almost 1.8 inch). ***(Photo by Frost-Rosen)***

When a current flows through a resistor, electric ENERGY is converted to heat. The rate of heat generation, or POWER, P, in watts (W) is given by:

$$P = Vi$$

Through Ohm's law this can be expressed equivalently in terms of the resistance as:

$$P = i^2R = V^2/R$$

This result is known as Joule's law, and the heat thus produced as Joule heat. (*See* JOULE, JAMES PRESCOTT.)

When resistors are connected in series (i.e., end to end in linear sequence, so that the same current flows through them all), the resistance of the combination—their equivalent resistance—equals the sum of their individual resistances. So if we denote the resistances of the resistors by R_1, R_2, . . . and the equivalent resistance of the series combination by R, then:

$$R = R_1 + R_2 + \ldots$$

A parallel connection of resistors is when one end of each is connected to one end of all the others and the other end of each one is connected to the other end of all the others. In that way, the same voltage acts on all of them. The equivalent resistance, R, of such a combination is given by:

$$1/R = 1/R_1 + 1/R_2 + \ldots$$

See also IMPEDANCE.

resolving power The ability of an optical imaging system to exhibit fine detail that exists in the object being imaged is called the resolving power of the system. Resolving power is affected by the quality of the components, the LENSes and MIRRORs, that compose the system. Even when the components are as nearly perfect as possible, there remains an inherent limit to resolving power due to DIFFRACTION effects. The underlying factor is that the image of a point source does not appear as a point image, but rather as a blob surrounded by concentric rings. So the images of nearby points on an object might blend together and not appear separated if the points are too close to each other. (*See* OPTICS.)

The Rayleigh criterion for minimal image resolution of two point light sources is that the center of the image of one falls on the first diffraction ring of the second. What this turns out to mean is that for two point sources to be at least marginally resolvable by an imaging system, the angle between them as viewed by the system must not be less than a certain minimal angle, called the angular limit of resolution, and given in radians (rad) by:

$$\text{Angular limit of resolution} = 1.22\lambda/D$$

Here λ denotes the WAVELENGTH in meters (m) of the radiation used for imaging, and D is the diameter of the imaging system's aperture (or, opening), also in meters (m). The resolving power of the system, in the

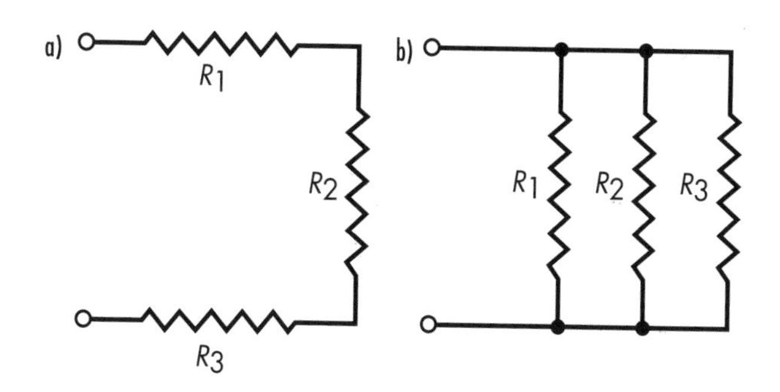

Resistors in combination. (a) Three resistors, R_1, R_2, and R_3, connected in series. The equivalent resistance of the combination equals $R_1 + R_2 + R_3$. (b) Three resistors in a parallel connection. Their equivalent resistance equals $\frac{1}{1/R_1 + 1/R_2 + 1/R_3}$.

diffraction limit, is just the inverse of the angular limit of resolution:

$$\text{Resolving power} = D/(1.22\lambda)$$

(*See* RAYLEIGH, LORD.)

It follows from this expression that resolving power can be increased both by increasing the aperture and by decreasing the wavelength. As for the former, satellite-borne spy cameras, which, we are told, can take pictures of the license plate numbers of motor vehicles, are constructed with enormously wide lenses. And birds of prey that search for their food from up high have especially wide pupils. Wavelength dependence is utilized through the use of ELECTRON microscopes, for example, which can image molecular structures that cannot be resolved by LIGHT microscopes. The wavelengths of electron beams, according to WAVE-PARTICLE DUALITY, can be considerably smaller than those of visible light. (*See* ELECTRON MICROSCOPY; MOLECULE.)

See also ABERRATION.

resonance Resonance is the effect that an oscillating system responds strongly when it is acted on, or driven, by a continuing excitation at any one of its natural FREQUENCIES, also called resonant frequencies. A common example of resonance is pushing a child in a swing. When the pusher pushes at the frequency of the swing's oscillation, the swing absorbs ENERGY from the pushes and develops a large amplitude. Any deviation from that frequency causes the response to be considerably depressed. Then only some of the pushes add energy to the swing, while others reduce its energy. For another example, the air in a shower stall can undergo PRESSURE oscillations with an infinite number of natural frequencies. When a shower taker's singing hits any of those frequencies, a noticeable amplification is heard. (*See* FREQUENCY, NATURAL; OSCILLATION.)

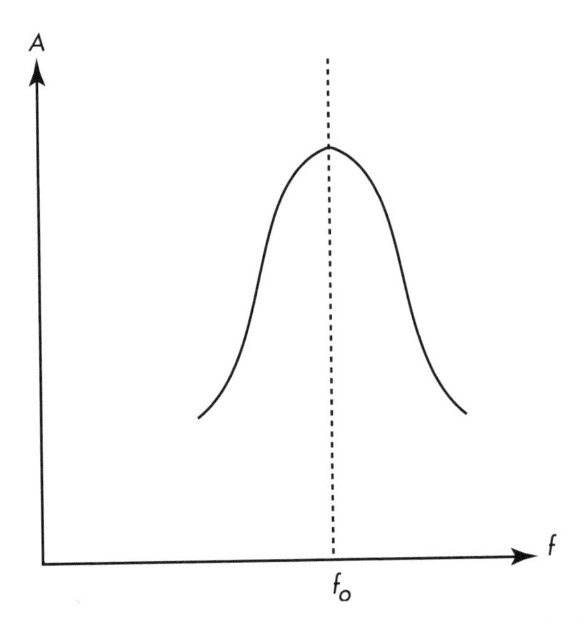

A typical resonance response curve, showing the amplitude of a system's response, *A*, as a function of the excitation frequency, *f*, near one of the system's natural frequencies, f_0. The actual shape of the curve (height and width) depends on various parameters of the system and of the excitation. (In certain cases, the frequency of maximal response can be somewhat less than f_0.)

Resonance can be beneficial as well as objectionable. As for its benefits, it forms an essential component of the operation of acoustic systems, for instance, such as musical instruments and animal voice production. Also, resonance is crucial in many electronic circuits, such as those of amplifiers and tuners. On the other hand, resonance must be taken into account in the design of structures, such as bridges, buildings, ships, and aircraft, in order to prevent undesirable, and even potentially damaging, large-amplitude oscillation and VIBRATION. DAMPING serves as one means of dealing with those problems. As an additional example of objectionable resonance, in loudspeakers it can cause certain pitches to be unduly emphasized and must be compensated for. (*See* ACOUSTICS.)

rest mass The MASS of a body as measured in a REFERENCE FRAME in which the body is at rest is termed the body's rest mass. When the mass of a body is referred to, it is usually the body's rest mass that is intended. According to ALBERT EINSTEIN's special theory of relativity, every mass possesses an ENERGY equivalent, its MASS ENERGY, given by:

$$E = mc^2$$

where m denotes the mass in kilograms (kg), E represents its mass energy in joules (J), and c is the SPEED OF LIGHT, whose value is 2.99792458×10^8 meters per second (m/s). Thus every body possess also a rest energy, given by this relation with m taken to be the body's rest mass, m_0:

$$E = m_0c^2$$

(*See* RELATIVITY, SPECIAL THEORY OF.)

Since the total energy of a body depends on its SPEED and can depend on other factors as well, the mass-energy relation can be used in reverse to attribute to a body a mass that is the mass equivalent of its energy. According to the special theory of relativity, the mass equivalent, m, of a body's total energy, due only to its rest mass and its motion, is given by the expression:

$$m = \frac{m_0}{\sqrt{1 - v^2/c^2}}$$

Here v denotes the body's SPEED in meters per second (m/s). For speeds that are sufficiently small compared with the speed of light, this relation takes the form:

$$m = m_0 + (1/2)\, m_0 v^2/c^2 + \text{negligible terms}$$

which shows the mass equivalent of the total energy as the sum of the rest mass and the mass equivalent of the body's kinetic energy (in its low-speed form). (*See* KINETIC ENERGY; POTENTIAL ENERGY.)

reversible process A process during which a system is in EQUILIBRIUM at every stage is known as a reversible process. This concept is an idealization, although a very useful one, especially in THERMODYNAMICS. If a system were indeed in equilibrium, it would not change and there would be no process. In the idealization, it is assumed that the conditions of the system are changed infinitesimally from those of equilibrium in order to nudge the system into a further infinitesimal change, while the system remains infinitesimally close to equilibrium. Such a process would take an infinite amount of time to complete. By reversing the infinitesimal changes imposed on the system, this process can be made to proceed exactly in reverse, so that the system reverts to its initial state.

A necessary condition for a process to be reversible is that it involve no dissipative FORCES, such as FRICTION and VISCOSITY, or other dissipative effects. Dissipation depletes the orderly ENERGY of a system and generates HEAT in its stead. That precludes reversibility, since heat cannot be fully reconverted to orderly forms of energy, according to the second law of thermodynamics. (*See* DISSIPATION.)

As an example of an idealized approximation to a reversible process, imagine a vertical cylinder with thermally conducting walls and with a frictionless piston in it that presses down on a gas contained within the cylinder. (The frictionless piston is already an idealization.) Start with the WEIGHT of the piston supported by the PRESSURE of the gas and the TEMPERATURE of the gas the same as that of the surroundings. Here is a description of a nearly reversible ISOTHERMAL PROCESS of compression (i.e., a process in which the gas is compressed at constant temperature). Very slowly, say one grain per hour, dribble extremely fine sand onto the top of the piston. As the weight of the piston-plus-sand increases, the piston gradually descends, and the pressure of the gas increases to support the increased weight. The gas's temperature remains very nearly constant, since the process is so slow that there is sufficient time for the gas's temperature to equalize almost perfectly with that of the surroundings. The process ends when a given amount of sand has been placed on the piston. At any time during the process, the state of the system deviates from equilibrium at most in proportion to the ratio of the weight of a single grain of sand to the weight of the piston-plus-sand (i.e., by very little indeed). "Very little indeed" here is an approximation to infinitesimal, so the process is approximately reversible. (*See* CONDUCTION.)

To run the process in reverse, very slowly, at the rate of one grain per hour, remove the sand from the top of the piston. The gas will then isothermally decompress and, when all the sand is gone from the piston, will very closely regain its initial VOLUME, pressure, and temperature, and the piston will very closely return to its initial position.

An irreversible process is a process that is not reversible. In fact, all processes in macroscopic systems are irreversible, since dissipation cannot be totally prevented and processes are completed within finite times. Processes involving small numbers of ELEMENTARY PARTICLES can, at least in principle, be reversible, if sufficient control of their states is achieved. This is reversibility in the sense of being able to run in reverse. Due to the small number of constituents of such systems, the concept of equilibrium is irrelevant to them and thermodynamics does not apply.

Reynolds number Named for the British engineer Osborne Reynolds (1842–1912), the Reynolds number is a PARAMETER that characterizes the flow of an incompressible, viscous FLUID. It is a dimensionless quantity, defined as $\rho v l/\eta$, where ρ denotes the fluid's DENSITY in kilograms per cubic meter (kg/m^3), v is the flow SPEED in meters per second (m/s), l is a length in meters (m) that is characteristic of the geometry of the system, and

η represents the fluid's VISCOSITY in newton·seconds per square meter (N·s/m^2). If a fluid of density ρ and viscosity η is flowing at speed v through a long pipe, for example, then l could be taken as the diameter of the pipe. Or if an aircraft is moving through air or a boat through water, l denotes a characteristic size of the moving object. (Such situations are equivalent to the fluid moving past the object.)
(*See* DIMENSION.)

The Reynolds number has two important uses. One use is for scaling. If it is desired to model a flow situation, such as in a wind tunnel, the model flow will correctly represent the real-world flow only if the Reynolds numbers of the two are equal. So if the wind-tunnel model of an aircraft is a hundredth the size of the original, for instance, then, since the same fluid is being used (air), the wind tunnel would have to be run at 100 times the speed of the real aircraft in order to obtain a truly representative flow pattern.

Another use of the Reynolds number is to distinguish regimes of steady flow, also called laminar flow, and of turbulent flow. In the former, the flow VELOCITY and PRESSURE throughout the fluid are constant in time or vary only slowly. That occurs when the Reynolds number is less than about 2000. For values of Reynolds number over around 3000, the flow is turbulent, with velocity and pressure varying wildly in time. (*See* TURBULENCE.)

rotation Change of spatial orientation taking place over TIME is called rotation. It is one type of MOTION that a body can undergo, the other being translation, which is change of position. It is convenient and common to consider separately the motion of a body's CENTER OF MASS and the body's rotation about its center of mass. If the center of mass moves, then the body is undergoing translational motion. Otherwise it is not. The most general motion of a body involves simultaneous motion of its center of mass and rotation about the center of mass. As an example, consider the motion of the Moon. It is revolving around the Earth while keeping its same side facing the Earth. So as its center of mass performs nearly circular translational motion around the Earth with a PERIOD of about a month, the Moon is simultaneously rotating about its center of mass at exactly the same rate, as it happens. Similarly, the Earth is traveling through space in elliptical translational motion about the Sun, with a period of a year, while also rotating about its axis, with a period of about a day (exactly one sidereal day, to be precise). In the simplest kind of rotation, a body rotates about a line of fixed direction passing through its center of mass, a fixed axis of rotation. Then all particles of the body undergo circular motion about that line. That is what the Moon and the Earth are doing, to a reasonable approximation. (*See* CIRCULAR MOTION; SPACE.)

See also ANGULAR MOMENTUM; DYNAMICS; KINEMATICS; MOMENT OF INERTIA; SPEED; TORQUE; VELOCITY.

Rutherford, Ernest (1871–1937) New Zealand/British *Physicist* An experimental physicist, Ernest Rutherford is famous for his discovery of the NUCLEUS of ATOMS and for his studies of natural and induced RADIOACTIVITY. He earned academic degrees in New Zealand and joined the Cavendish Laboratory at Cam-

Ernest Rutherford was an experimentalist who studied natural and induced radioactivity and discovered that atoms possess a nucleus. He was awarded the 1908 Nobel Prize in chemistry. *(AIP Emilio Segrè Visual Archives)*

bridge University, England, for the period 1895–98. During 1898–1907 Rutherford served as professor of physics at McGill University in Montreal, Canada, and then returned to England to the University of Manchester. Throughout World War I, he worked on problems of submarine detection for the British Admiralty. In 1919, Rutherford returned to the Cavendish Laboratory as its director and remained there for the rest of his life. It was at Manchester that he discovered, by studying the scattering of ALPHA PARTICLES from atoms, that atoms possess a nucleus, in which most of their MASS is concentrated. That discovery led to the BOHR THEORY of the hydrogen atom. Rutherford also found that induced FISSION of nuclei changes the chemical identity of the ELEMENT. In 1908, he was awarded the Nobel Prize in chemistry "for his investigations into the disintegration of the elements, and the chemistry of radioactive substances." (*See* BOHR, NIELS HENRIK DAVID; EXPERIMENTAL PHYSICS.)

S

scalar A physical quantity that is not characterized by direction is designated a scalar. Compare with a VECTOR, which is a directional physical quantity. TEMPERATURE, for instance, is a scalar. So are WORK, ENERGY, SPEED, and the PLANCK CONSTANT, as additional examples.

Some scalars are derived from vectors. The scalar product of two vectors is a scalar. Let (u_x, u_y, u_z) and (v_x, v_y, v_z) denote the x-, y-, and z-components of vectors **u** and **v**, respectively. Then their scalar product, **u**·**v**, equals:

$$\mathbf{u}\cdot\mathbf{v} = u_x v_x + u_y v_y + u_z v_z$$

Work is an example of such a scalar. A vector's magnitude, $|\mathbf{v}|$, is a scalar. It equals the square root of the vector's scalar product with itself:

$$|\mathbf{v}| = \sqrt{\mathbf{v}\cdot\mathbf{v}} = \sqrt{v_x^2 + v_y^2 + v_z^2}$$

As an example of this, speed is the magnitude of the VELOCITY vector. The divergence of a vector FIELD is another kind of scalar that is derived from a single vector. For the vector field $\mathbf{v}(x, y, z)$, the divergence, denoted div **v**, or $\nabla\cdot\mathbf{v}$, is given by:

$$\text{div } \mathbf{v}(x, y, z) = \partial v_x(x, y, z)/\partial x + \partial v_y(x, y, z)/\partial y + \partial v_z(x, y, z)/\partial z$$

Conversely, every scalar field, $s(x, y, z)$, defines a vector, the gradient of the scalar field. This vector is denoted grad s, or ∇s, and its x-, y-, and z-components are given by:

$$(\text{grad } s)_{x, y, z} = (\partial s(x, y, z)/\partial x, \partial s(x, y, z)/\partial y, \partial s(x, y, z)/\partial z)$$

Schrödinger, Erwin (1887–1961) Austrian *Physicist* One of the founders of QUANTUM PHYSICS, Erwin Schrödinger is known for the equation that bears his name, the SCHRÖDINGER EQUATION. In 1910, he received his doctorate from the University of Vienna, Austria, and took a position there. Schrödinger served as an artillery officer during World War I, after which he

Erwin Schrödinger was a theoretician and one of the founders of quantum physics. He is famous for his equation for the quantum wave function, the Schrödinger equation. He shared the 1933 Nobel Prize in physics. ***(AIP Emilio Segrè Visual Archives)***

resumed his position at Vienna. Schrödinger spent the years 1921–27 very fruitfully at the University of Zurich, Switzerland. There he investigated, among other issues of THEORETICAL PHYSICS, THERMODYNAMICS and atomic SPECTRA. Having become dissatisfied with the BOHR THEORY of the hydrogen ATOM, Schrödinger devised his equation for the atom's wave function, known as the Schrödinger equation. He moved to the University of Berlin, Germany, in 1927, but left Germany upon Hitler's rise to power in 1933. Then Schrödinger held various positions in Europe and the United States until setting at the Institute for Advanced Studies in Dublin, Ireland, in 1939. He stayed there until 1956, when he returned to the University of Vienna. Schrödinger shared with PAUL ADRIEN MAURICE DIRAC the 1933 Nobel Prize in physics "for the discovery of new productive forms of atomic theory."

See also QUANTUM MECHANICS.

Schrödinger equation Named for its discoverer, the Austrian physicist ERWIN SCHRÖDINGER, the Schrödinger equation is a fundamental equation of QUANTUM MECHANICS. This equation determines the evolution in TIME of the wave function of a system, given the value of the wave function at any time.

In quantum mechanics, a physical system is described by a wave function, Ψ, which specifies the state of the system at any time and contains all the information about the system that there is. The wave function is a complex function (i.e., its values are complex numbers) that depends on the system's generalized coordinates and on time, t. Generalized coordinates are the set of physical quantities needed to specify a state. In the case of a system of particles, for instance, the generalized coordinates consist of all the spatial coordinates of the individual particles. For more detail concerning the wave function, see QUANTUM MECHANICS.

The Schrödinger equation is a differential equation for the wave function, involving the generalized coordinates and time. When the value of the wave function is known for a given time, solving the Schrödinger equation gives the values of the wave function for all other times, as long as the system evolves independently during those times.

For the sake of simplicity of discussion, let us assume the system comprises a single particle, whose position is specified by the position VECTOR $\mathbf{r}$, which might be expressed in terms of its Cartesian coordinates (x, y, z). Then the wave function is written as $\Psi(\mathbf{r}, t)$ or $\Psi(x, y, z, t)$. For a single particle whose POTENTIAL ENERGY is a function of position, $U(\mathbf{r})$, the Schrödinger equation for the wave function, $\Psi(\mathbf{r}, t)$, takes the form:

$$\frac{ih}{2\pi}\frac{\partial\Psi}{\partial t} = -\frac{h^2}{8\pi^2 m}\nabla^2\Psi + U(\mathbf{r})\Psi$$

where ∇^2 denotes the Laplacian operator, which is:

$$\nabla^2 = \partial^2/\partial x^2 + \partial^2/\partial y^2 + \partial^2/\partial z^2$$

and h represents the PLANCK CONSTANT, whose value is $6.62606876 \times 10^{-34}$ joule·second (J·s). (*See* COORDINATE SYSTEM.)

One method of solving this equation is to look for stationary states, which are states whose wave function takes the form:

$$\Psi(\mathbf{r}, t) = \psi(\mathbf{r})\, e^{-2\pi i f t}$$

Then the function $\psi(\mathbf{r})$ obeys the time-independent Schrödinger equation:

$$-\frac{h^2}{8\pi^2 m}\nabla^2\psi(\mathbf{r}) + U(\mathbf{r})\psi(\mathbf{r}) = E\psi(\mathbf{r})$$

Stationary states are characterized by a definite value for the particle's ENERGY, which is the constant E appearing in the equation, where:

$$E = hf$$

In a one-dimensional situation the wave function is simply $\Psi(x, t)$. The Schrödinger equation is then:

$$\frac{ih}{2\pi}\frac{\partial\Psi}{\partial t} = -\frac{h^2}{8\pi^2 m}\frac{\partial^2\Psi}{\partial x^2} + U(x)\Psi$$

The wave function for stationary states takes the form:

$$\Psi(x, t) = \psi(x)\, e^{-2\pi i f t}$$

where $\psi(x)$ obeys the time-independent Schrödinger equation:

$$-\frac{h^2}{8\pi^2 m}\frac{d^2\psi(x)}{dx^2} + U(x)\psi(x) = E\psi(x)$$

See DIMENSION.

See also QUANTUM PHYSICS.

selection rule A rule that governs the possibility of a physical process is called a selection rule. Such a rule is formulated in terms of the values of certain physical quantities before and after the process and often involves quantities that take discrete values. As a typical example, when a hydrogen ATOM undergoes a transition from one state to another by emission or absorption of a single PHOTON, called a single-photon optical transition, its orbital angular momentum quantum number, conventionally denoted by l, must change by 1. In symbols:

$$\Delta l = \pm 1$$

This selection rule follows from the conservation of angular momentum together with the fact that the photon carries one elementary unit of ANGULAR MOMENTUM due to its SPIN. An allowed transition is one that obeys all applicable selection rules, while a transition that violates a selection rule is called a forbidden transition. Forbidden transitions are not always absolutely forbidden, but then the probability of their occurrence is strongly suppressed. Selection rules are used in the interpretation of atomic and molecular spectra. (*See* CONSERVATION LAW; MOLECULE; QUANTUM NUMBER; SPECTRUM.)

self-induction Self-induction is the INDUCTION of an ELECTROMOTIVE FORCE (emf) in an electric circuit component, usually a coil, by a time-varying electric CURRENT flowing through the component. Self-induction comes about due to AMPERE'S LAW and the BIOT-SAVART LAW. The former determines that the induced emf is proportional to the time rate of change of the magnetic FLUX through the component, while from the latter it follows that the magnetic flux is proportional to the current. (*See* ELECTRICITY; ELECTROMAGNETISM; INDUCTANCE; MAGNETISM.)

See also MUTUAL INDUCTION.

semiconductor A crystalline SOLID that has an ENERGY gap of about 1 eV separating its VALENCE BAND (lower energy) from its CONDUCTION BAND (higher energy) is termed a semiconductor. At ordinary TEMPERATURES the conduction band is sufficiently populated, due to thermal excitation of ELECTRONs from the valence band, for electric CONDUCTION to take place with a small applied VOLTAGE. The higher the temperature, the more electrons are excited into the conduction band, and thus the higher the material's CONDUCTIVITY. When an electron is excited into the conduction band, it leaves behind a vacancy in the valence band, called a HOLE. Holes are mobile and behave as positively charged current carriers. In a pure semiconductor, there are as many holes as electrons contributing to the conductivity. Such a semiconductor is called an intrinsic semiconductor. (*See* CRYSTAL; ELECTRICITY; HEAT.)

For the fabrication of semiconductor devices, it is necessary to have semiconducting materials of various types of conductivity (i.e., those with more holes than conducting electrons and those with more of the latter than the former). Such imbalance is achieved by the controlled addition of impurities to the crystal, a process referred to as doping, and the result is termed an extrinsic semiconductor. Semiconductors are made from crystals of silicon or germanium, both of whose ATOMs possess four valence electrons each. When trivalent atoms, such as of indium or aluminum, are added to the crystal to replace some of the tetravalent ones, holes are thus introduced into the material, giving a higher number of positive charge carriers than negative. Such a semiconductor is a p-type semiconductor. On the other hand, pentavalent doping atoms, such as of arsenic, might be used to achieve a predominance of negative current carriers over positive. Those are n-type semiconductors.

The fruits of semiconductor technology include components and devices such as diodes, LIGHT-EMITTING DIODES (LEDs), light-absorbing diodes (which might serve as solar cells), TRANSISTORS, and semiconductor LASERS.

shear The situation in which a pair of equal-magnitude, oppositely directed (i.e., antiparallel) FORCEs act on a material in such a way that their lines of action

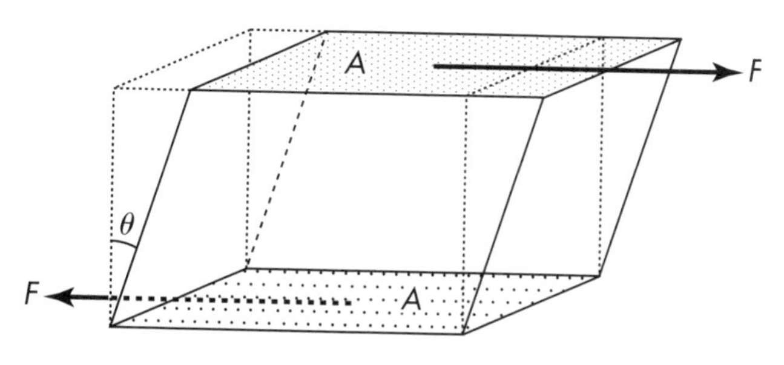

Shear is a deformation of material caused by a pair of equal-magnitude, oppositely directed forces, whose lines of action are displaced from each other. Each force, of magnitude *F*, acts parallel to an area of magnitude *A*. The shear stress is *F/A*, while the shear strain equals the angle of deformation, θ (expressed in radians).

are displaced from each other is called shear. The result of shear can be an angular deformation of the material, possibly including twisting, or even separation of the material. We produce shear when we tear paper or use a paper punch, for instance. Scissors (also called *shears*) and electric shavers operate by means of shear. As examples of noncatastrophic deformation, the joining together of the bare ends of a pair of electric wires by twisting them is a shear effect, as is the realignment of the hook of a wire coat hanger by twisting it.

The shear of the force pair is described by a physical quantity called shear STRESS, whose value is the ratio of the magnitude of one of the forces to the area of the surface parallel to the forces upon which the forces are acting. Its SI UNIT is the pascal (Pa), equivalent to newton per square meter (N/m^2). Note that although shear stress is force per unit area, it is not PRESSURE. The latter is force per unit area perpendicular to the force, while in the present case the force is parallel to the area upon which the force pair is acting. The result of noncatastrophic shear, the deformation of the material, is described by shear STRAIN, which equals the angle in radians (rad) through which the material is deformed. For elastic deformation, the material's resistance to shear is measured by its shear modulus, or modulus of rigidity, whose value is the ratio of shear stress on the material to the resulting shear strain. Its SI unit is the pascal (Pa). (*See* ELASTICITY.)

shock wave A shock wave is a WAVE whose effect at every point it reaches is a sudden onset of high-INTENSITY disturbance. In other words, a shock wave involves a high-intensity WAVEFRONT that is preceded by calm. One way shock waves are generated is by means of a wave source that is moving faster than the propagation SPEED of the wave. An example of that is a supersonic aircraft (i.e., a plane flying faster than the speed of SOUND in air). Since it outpaces the sound waves that it emits, the plane continuously radiates a shock wave (or possibly more than one), whose wavefront forms a conical surface extending behind the plane with the plane at its apex. That is the source of the "supersonic booms" associated with the passage of supersonic aircraft (and not, as is so often stated, the plane's "breaking the sound barrier"). After the plane passes by, the wavefront of its shock wave, which is "sweeping" the earth behind the plane, reaches us and we hear the "boom." (Actually, sometimes two "booms," one from the plane's nose and one from its rear.) A similar effect is the bow wave produced by a boat moving faster than the speed of surface waves on water. In both cases the angle, α, between the wavefront of the shock wave and the direction of the source's motion is given by:

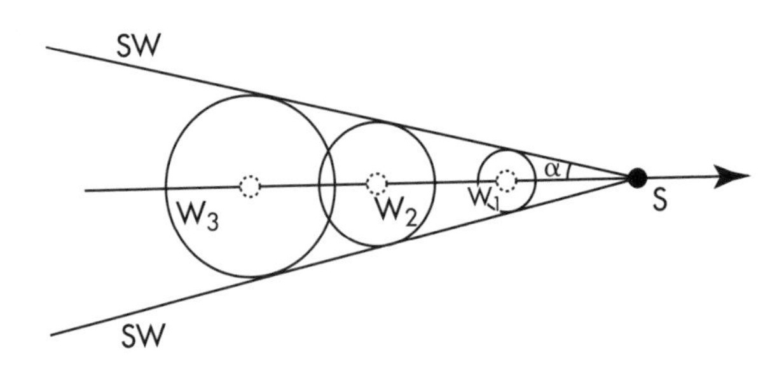

One origin of a shock wave is a wave source moving faster than the propagation speed of the waves. Such a source S, moving to the right, is shown, as are the current positions of the wavefronts W_1, W_2, W_3 that were emitted by the source when it was in three earlier locations. The shock wave SW is tangent to the wavefronts. The angle α is given by $\sin \alpha = u/v$, where u denotes the speed of wave propagation and v is the speed of the wave source.

$$\sin \alpha = u/v$$

where u denotes the speed of wave propagation in the medium and v is the speed of the wave source. Both speeds are expressed in the same UNITS.

See also ČERENKOV RADIATION; MACH NUMBER.

SI *See* INTERNATIONAL SYSTEM OF UNITS.

simultaneity The occurrence of two or more events at the same TIME is termed simultaneity. It is of particular concern in ALBERT EINSTEIN's special theory of relativity, since the theory shows that simultaneity is observer dependent and thus relative (according to the meaning of the term *relative* in the special theory of relativity). Two events might possess the relation that a LIGHT signal cannot connect them, that if such a signal were to be emitted at one event, it would reach the location of the second event only after that event occurred. Such events are said to have a spacelike separation. For such pairs of events, and only for such pairs, some observers might measure their times of occurrence and find them to be simultaneous. Other observers, in MOTION with respect to those, find one of the events occurring before the second. Depending on the direction of an observer's motion, either event might be the earlier. So for events that possess spacelike separation, their temporal order, including their simultaneity, are observer dependent, and thus relative. (*See* RELATIVITY, SPECIAL THEORY OF.)

The following imaginary scenario shows how the relativity of simultaneity comes about. It is based on one of the fundamental postulates of the special theory of relativity, that the SPEED OF LIGHT has the same value, $c = 2.99792458 \times 10^8$ meters per second (m/s), for all pairs of observers that are moving with respect to each other at constant VELOCITY (i.e., at constant speed in a straight line). Let a train car be moving past a station platform at constant speed, lower than c. An observer sits in the center of the car. On the platform another observer has arranged electric contacts and circuits so that when the ends of the car simultaneously pass and close two sets of contacts, two simultaneous light flashes are produced, one at each end of the car. When that happens, the platform observer sees light pulses from the flashes rushing toward each other and meeting in the middle, where the car observer was sitting at the instant of contact. But in the meantime the car observer moves some distance. As a result, the light pulse from the front of the car reaches and passes her before meeting its partner in the middle, and the one from the rear catches up with her only after meeting and passing its partner from the front. So the platform observer sees the car observer detecting the light pulse from the front of the car before detecting the one from the rear and understands how that comes about.

Let us now put ourselves in the position of the car observer. She sees the light pulse from the front of the car first, followed by the pulse from the rear. She knows that both pulses traveled the same distance (half the length of the car). She also knows that both light pulses traveled at the same speed, c. This latter fact is crucial. It is where the special theory of relativity enters the scenario. The car observer thus knows that both pulses were in transit for the same time interval (same distance divided by same speed). Since she detected the pulse

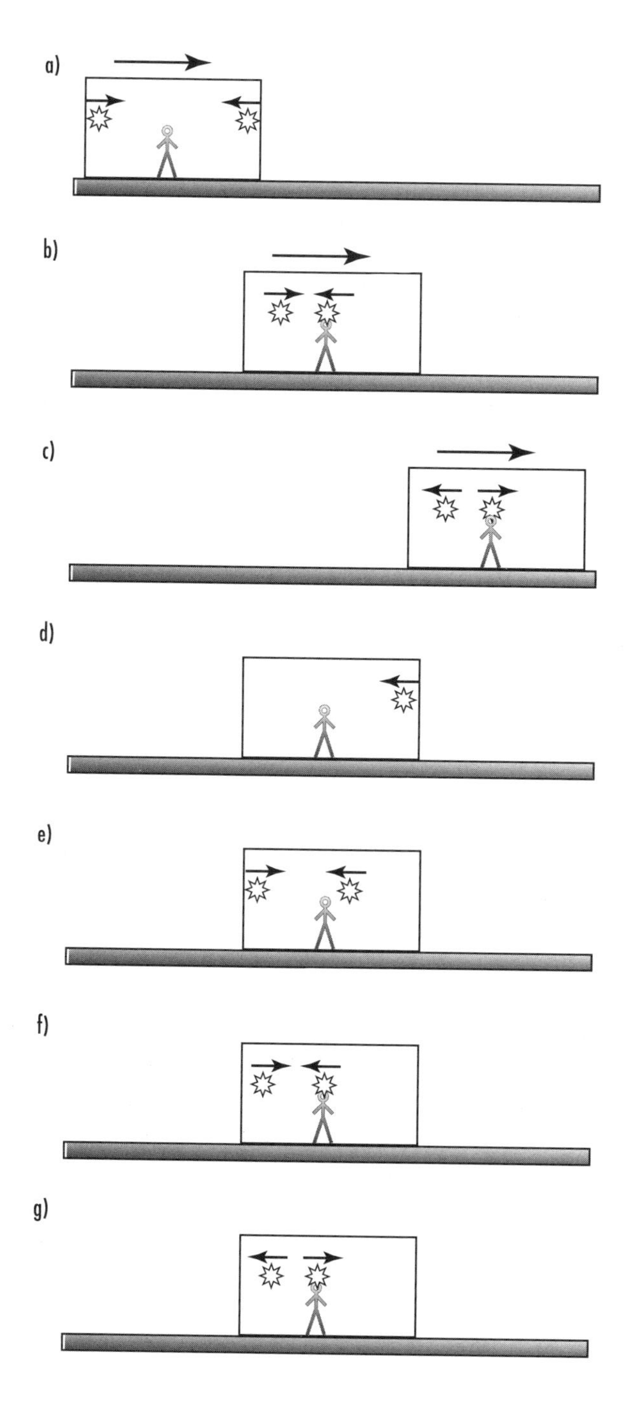

Simultaneity is relative (i.e., observer dependent), according to the special theory of relativity. A train car, moving to the right, with an observer at its center passes a station platform. As observed from the platform: (a) Two light pulses leave the ends of the car simultaneously. (b) The pulse from the front of the car reaches the car observer first, since she is moving toward it. (c) After the pulses pass each other, the one from the rear of the car overtakes the car observer. As the car observer sees it: (d) A light pulse leaves the front of the car. (e) Later a light pulse leaves the rear of the car. (f) The pulse from the front reaches the observer. (g) Then the pulse from the rear reaches the observer, after the pulses have passed each other. The light pulses' leaving the ends of the car occurs simultaneously for a platform observer, while for the car observer the pulse from the front leaves before the pulse from the rear does. Both observers agree that the pulse from the front reaches the car observer before the one from the rear, but this fact is interpreted differently by the observers.

from the front before she detected the one from the rear, the pulse from the front was produced before the pulse from the rear. Here is the relativity of simultaneity. According to the platform observer both flashes are simultaneous, while the car observer has the one in front occurring before the one in the rear. As a general rule, if one observer detects two events as occurring simultaneously, then a second observer, in motion along the line connecting the locations of the events, detects the event at the foremost location as occurring before the event at the rearmost location.

Snell's law Named for its discoverer, the Dutch physicist and mathematician Willebrod Snell (1580–1626), Snell's law concerns the REFRACTION of LIGHT, and actually the refraction of any WAVE, upon its passage through an interface between two media in which the wave possesses different SPEEDs of propagation. Consider a narrow beam, a ray, that impinges on a smooth plane interface surface and penetrates into the second medium, and consider the perpendicular to the plane at the point of incidence on the plane. This perpendicular is called the normal. The smaller angle between the incident ray and the normal is the angle of incidence. The incident ray and the normal together define a plane, the plane of incidence. This plane is perpendicular to the interface plane and contains both the incident ray and the normal. The refracted ray is transmitted into the second medium from the point of incidence on the interface plane and lies in the plane of incidence. It is situated on the same side of the normal as the geometric extension of the incident ray into the second medium. The smaller angle that the refracted ray makes with the normal is the angle of refraction. Snell's law gives the relation between the angle of refraction and the angle of incidence as follows. Denote the angles of incidence and refraction by θ_i and θ_r, respectively, and the propagation speeds of the wave in the corresponding media by v_i and v_r. Then:

$$\frac{\sin\theta_i}{v_i} = \frac{\sin\theta_r}{v_r}$$

For light, this can be expressed equivalently in terms of the INDEXes OF REFRACTION of the media, n_i and n_r, where $n = c/v$, with c the SPEED OF LIGHT, whose value is 2.99792458×10^8 meters per second (m/s). Snell's law then takes the form:

$$n_i \sin\theta_i = n_r \sin\theta_r$$

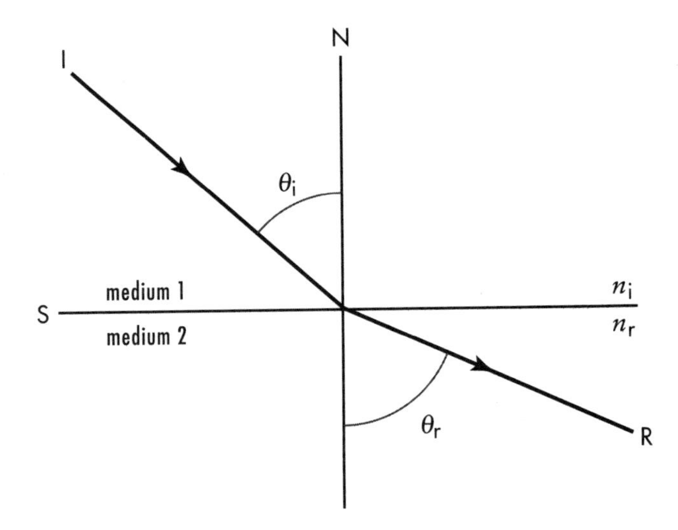

The ingredients of Snell's law of refraction of a ray at the interface surface S between two media. The incident ray I, the normal N (perpendicular to the reflecting surface at the point of incidence), and the refracted ray R are shown. The angles of incidence and refraction are denoted θ_i and θ_r, respectively. The index of refraction of medium 1, in which the incident ray propagates, is n_i, and that of the medium 2, carrying the refracted ray, is n_r. Snell's law states that $n_i \sin\theta_i = n_r \sin\theta_r$.

Note that a ray impinging on an interface surface perpendicularly ($\theta_i = 0$) is always transmitted with no deviation ($\theta_r = 0$), and as the angle of incidence increases, so does the angle of refraction. Further, upon passing from a medium of higher propagation speed into one of lower speed (from lower index of refraction to higher), a ray is bent toward the normal ($\theta_r < \theta_i$), while in passing from lower to higher speed (higher to lower index of refraction) it is bent away from the normal ($\theta_r > \theta_i$). In this connection, see the REFRACTION entry for a discussion of total internal reflection.

solar system The astronomical system consisting of the Sun, the Earth, and all the other objects that are gravitationally bound to the Sun is called the solar system. Here is a brief and rough census of the solar system. There are nine planets revolving around the Sun. From the inner to the outer, they are Mercury, Venus, Earth, Mars, Jupiter, Saturn, Uranus, Neptune, and Pluto. The planets' orbits are mostly nearly circular and lie more or less in a common plane, with Pluto's forming a notable exception on both counts. The radius of Earth's orbit is about 1.5×10^8 kilometers (km). The radii of the other planets' orbits range from about 0.4 to nearly 40 times the radius of Earth's orbit, while their PERIODs of revolution—their "years"—span around three Earth months (for Mercury) to about 250 Earth years (for Pluto). The planets' orbits and motions obey KEPLER'S LAWS. Many of the planets possess one or more moons, which orbit the planets. They are, with the number of moons (as presently known): Earth, 1 (the Moon); Mars, 2; Jupiter, 16; Saturn, 18; Uranus, 17; Neptune, 8; and Pluto, 1. All the moons of a single planet obey Kepler's laws among themselves. Four of the planets—Jupiter, Saturn, Uranus, Neptune—are known to also have rings, with the rings of Saturn being the most visible. (*See* GRAVITATION.)

The solar system also contains thousands of asteroids, which are rocky bodies that are smaller than planets. They are mostly concentrated in the asteroid belt, where they revolve around the Sun between the orbits of Mars and Jupiter. Although still officially categorized as a planet, it seems that Pluto has more in common with the asteroids than with the planets. Even more numerous than the asteroids are the comets and the meteoroids. The comets are of similar MASS to the asteroids, but are composed mostly of dust and ice. Their orbits are of various elongation. They appear to be concentrated in the Kuiper belt beyond the orbit of Neptune and in the Oort cloud far beyond Pluto's orbit. Even smaller are the meteoroids, which appear to originate in the asteroid belt and in the breakup of comets.

Other stars than the Sun are known to possess planetary systems and presumably also other gravitationally bound objects.

solar wind The flow of particles outward from the Sun is known as the solar wind. This flow consists mostly of PROTONs and ELECTRONs, along with various kinds of IONs and ELEMENTARY PARTICLEs. Although the solar wind causes a continual loss of MASS from the Sun, the effect is small in relation to the Sun's total mass. The solar wind is the reason that comets' tails point away from the Sun rather than trailing behind the comets. The solar wind sweeps along with it the gaseous matter that the comet ejects as a result of heating by the Sun. A terrestrial effect of the solar wind is the AURORA. (*See* GAS.)

solid The term *solid* refers to one of the ordinary STATES OF MATTER. A solid is characterized by having fixed VOLUME (i.e., it has low compressibility), to a good approximation, and a fixed shape, in that it does not flow like a LIQUID or a GAS. The particles that constitute a solid—the ATOMs, IONs, or MOLECULEs of which it is composed—occupy fixed positions within the structure of the material. A homogeneous solid might be crystalline, when the constituent particles are arranged in a regular spatial array. It might be polycrystalline, composed of many small crystals. Or it might be amorphous, in which case it possesses no structural regularity at all. (*See* CRYSTAL; POLYCRYSTALLINE SOLID.)

solid-state physics The subfield of CONDENSED-MATTER PHYSICS that deals with matter in the solid state is called solid-state physics. It is concerned with the microscopic properties of solids, such as their interatomic and intermolecular FORCEs, and the bulk properties that are determined by the microscopic properties. (The study of how microscopic properties determine bulk properties is the domain of STATISTICAL MECHANICS.) Subjects of investigation for solid-state physics include HEAT conduction, CONDUCTION of ELECTRICITY, SUPERCONDUCTIVITY, crystalline structure, and SEMICONDUCTORs. (*See* ATOM; CRYSTAL; MOLECULE.)

sound A mechanical WAVE propagating through a material medium is termed sound. ACOUSTICS is the field of physics that studies sound. The nature of a sound wave depends on the type of medium. SOLIDs can transmit longitudinal waves, transverse waves, as well as other versions, such as torsional waves. GASes and LIQUIDs, which do not support SHEAR deformation, carry only PRESSURE waves, which can also be described as longitudinal waves.

Speed of Sound

The propagation SPEED, v, of a sound wave is given by an expression of the form:

$$v = \sqrt{\frac{\text{modulus of elasticity}}{\text{density}}}$$

where the modulus of ELASTICITY represents the stiffness of the medium, the extent to which it resists deformation, and the DENSITY represents the medium's INERTIA.

The appropriate modulus of elasticity is used for each type of wave. In the case of a sound wave in a GAS, such as air, for example, the appropriate modulus of elasticity is the bulk modulus, B, which takes the form:

$$B = \gamma p$$

Here p denotes the pressure of the gas in pascals (Pa) and γ represents the ratio of the gas's specific HEAT CAPACITY at constant pressure to that at constant VOLUME. So the speed of sound in a gas is given by:

$$v = \sqrt{\frac{\gamma p}{\rho}}$$

where ρ is the density of the gas in kilograms per cubic meter (kg/m^3) and v is in meters per second (m/s). For an ideal gas, the expression becomes:

$$v = \sqrt{\frac{\gamma RT}{M}}$$

where R denotes the gas constant, whose value is 8.314472 joules per mole per kelvin (J/[mol·K]); T is the absolute TEMPERATURE in kelvins (K); and M is the MASS of a single MOLE of the gas in kilograms (kg). The latter is the number of kilograms that equals one thousandth of the MOLECULAR WEIGHT (or effective molecular weight for a mixture such as air) of the gas. Note that for a given gas, the speed of sound increases with increasing temperature. For the same temperature and value of γ, the speed of sound increases as the molecular weight of the gas decreases. The speed of sound in dry air at comfortable temperatures is around 340 meters per second. (*See* GAS, IDEAL.)

Audible Sound

For sound to be audible to humans, the FREQUENCY of the sound wave impinging on the eardrum must lie within a certain range, nominally taken as 20 hertz (Hz) to 20 kilohertz (kHz). The frequency of sound correlates strongly with the sound's perceived pitch. The human female voice typically involves higher frequencies than does the voice of an adult male and is generally perceived as having a "higher" pitch than that of the adult male. The perception of loudness is strongly correlated with the physical quantity of INTENSITY and is reasonably well described by a logarithmic function of intensity, called intensity level and most often expressed in decibels (dB). The intensity level is calibrated to zero for an intensity of 10^{-12} watt per square meter (W/m^2), called the threshold of hearing. The perception of tone quality, or timbre, is determined by the spectral composition of the sound (i.e., by the relative POWER of each of the various frequency components that make up the sound wave). That has to do with the difference between, say, the tone of a trumpet and that of an oboe. A pure tone is one consisting of only a single frequency. The study of sound perception is called psychoacoustics. The observed frequency, and thus pitch, of a sound is affected by the MOTION of the source and of the observer relative to the propagation medium through the DOPPLER EFFECT. (*See* SPECTRUM.)

space The concept of space is a difficult one, with deep philosophical ramifications. For the purpose of PHYSICS, however, much of that difficulty must be ignored. In order to make progress in physics, PHYSICISTS cannot, in general, allow themselves to become mired in philosophical detail. So they make do with whatever general understanding is sufficient for carrying out the goals of physics.

In physics, space can be described, even defined, as the DIMENSION of *being,* where the word *dimension* is used in the sense of the possibility of assigning a measure. Objects and events possess the property of location, in that it is meaningful to ask the question "Where?" about them. This question can be answered by giving three numbers, the three coordinates required for a full specification of location, say the x-, y-, and z-coordinates of the object or event. In this way, a measure—the measure of location—can be assigned to existing objects and occurring events. That possibility is what space is about. Another way of looking at this—one that is nearly or wholly equivalent—is that space is the web of positional relations among objects and events. Since three numbers are required to specify a location, space actually involves three measures and is thus said to be three-dimensional. (*See* COORDINATE system.)

Technicalities aside, space is generally viewed by physicists as the arena in which objects exist and events occur, the background for what is and what happens. It seems as if space possesses some kind of absolute reality that is independent of the objects and events in it, that it is, in some sense, an entity in itself. If so, can it be detected? Is there some REFERENCE FRAME that has a privileged position with respect to absolute space? There certainly are privileged reference frames; there are, for instance, inertial reference frames and noninertial ones. As an example, a nonrotating bucket of water and a rotating one are easily differentiated: the water in the former has a flat surface, while that of the latter is concave. The nonrotating bucket defines an inertial reference frame, while a frame attached to the rotating bucket is noninertial. Does this indicate that rotation is taking place with respect to (1) absolute space? or (2) the other matter in the UNIVERSE? or (3) what? These are open questions, although modern thought generally tends to option (2). (*See* MACH'S PRINCIPLE; REFERENCE FRAME, INERTIAL.)

ALBERT EINSTEIN's special and general theories of relativity reveal that space and TIME are not independent of each other and that fundamentally they are best considered together as four-dimensional SPACE-TIME. (*See* RELATIVITY, GENERAL THEORY OF; RELATIVITY, SPECIAL THEORY OF.)

space-time The four-dimensional merger of SPACE and TIME is called space-time. This is a merger, but not a total blending, since the temporal DIMENSION of space-time remains distinct from its three spatial dimensions. Space-time is the arena for events, each of which possesses a location, specified by three coordinates, and an instant, or time, of occurrence, whose specification requires a single coordinate, thus together composing four coordinates. These coordinates can be taken as (x, y, z, t), which combine an event's spatial coordinates, (x, y, z), and its time coordinate, t. In the context of ALBERT EINSTEIN's special theory of relativity, space-time is often referred to as Minkowski space. (*See* COORDINATE SYSTEM; RELATIVITY, SPECIAL THEORY OF.)

The coordinates of the same event can have different values in different coordinate systems (i.e., with respect to different REFERENCE FRAMES) in space-time. (Quantities, such as coordinates, whose values depend on the reference frame are described as relative quantities.) The special theory of relativity is concerned with reference frames in relative motion at constant VELOCITY. The mathematical relations between the coordinates of an event with respect to a pair of such reference frames are given by POINCARÉ TRANSFORMATIONS or LORENTZ TRANSFORMATIONS. Einstein's general theory of relativity allows arbitrary reference frames, and the corresponding transformations are accordingly much more general. In the following, we confine our considerations to those of the special theory. (*See* RELATIVITY, GENERAL THEORY OF.)

In analogy with the distance between points in space, every pair of events in space-time has an interval, D, which is given by:

$$D^2 = (x_2 - x_1)^2 + (y_2 - y_1)^2 + (z_2 - z_1)^2 - c^2(t_2 - t_1)^2$$

where (x_1, y_1, z_1, t_1) and (x_2, y_2, z_2, t_2) are the space-time coordinates of the two events. The symbol c represents the SPEED OF LIGHT, whose value is 2.99792458×10^8 meters per second (m/s). Although the coordinates of the same event have different values with respect to different reference frames (i.e., in different coordinate systems) in space-time, the interval between any two events is invariant for reference frames in relative constant-velocity motion. (Quantities, such as intervals, whose values are independent of the reference frame in which they are evaluated, are described as absolute quantities.) In other words, the interval between a pair of events does not change under Poincaré or Lorentz transformations. If unprimed and primed symbols denote coordinates with respect to two such reference frames, this invariance is expressed as:

$$(x_2' - x_1')^2 + (y_2' - y_1')^2 + (z_2' - z_1')^2 - c^2(t_2' - t_1')^2$$
$$= (x_2 - x_1)^2 + (y_2 - y_1)^2 + (z_2 - z_1)^2 - c^2(t_2 - t_1)^2$$

This invariance ensures that the speed of light has the same value with respect to both reference frames.

Structure of Space-Time

Two events, for which $D^2 < 0$, are said to possess a timelike interval or timelike separation. This relation is absolute (i.e., independent of reference frame). For two such events, there are reference frames with respect to which both events occur at the same location and only their times of occurrence differ. Such events are said to lie within each other's light cone. Of two such events, one occurs after the other. This past-future distinction, too, is absolute. The later-occurring event is said to lie within the future light cone of the earlier one, and the earlier event is within the past light cone of the later. Such a pair of events possesses the property that a particle emitted from the earlier event and moving slower than c can reach the later event. Equivalently, a PHOTON, which travels at speed c, that is emitted from the earlier event reaches the location of the later event before it occurs.

Another possibility for two events is that $D^2 = 0$. They are then said to have a lightlike interval, or lightlike separation. Such a pair of events are also said to lie on (rather than within) each other's light cone. Here, too, the past-future distinction is absolute. The later event lies on the future light cone of the earlier event, and the earlier lies on on the past light cone of the later. A pair of events possessing lightlike separation can be connected by a photon emitted from the earlier and arriving at the later.

The last case is that for two events $D^2 > 0$. Such a pair of events have a spacelike interval, or spacelike separation. They are said to lie outside each other's light cone. Their time ordering is relative (i.e., depends on the reference frame). With respect to certain reference frames, one event occurs before the other. In other frames, it is the other that takes place before the one. And there are reference frames with respect to which both events occur at the same time (i.e., they are simultaneous). There are no reference frames in which the events take place at the same location. No material particle or photon emitted from one event can reach the location of the other event until after it occurs. So no signal or information can pass between events possessing spacelike separation, and they cannot affect each other. (*See* SIMULTANEITY.)

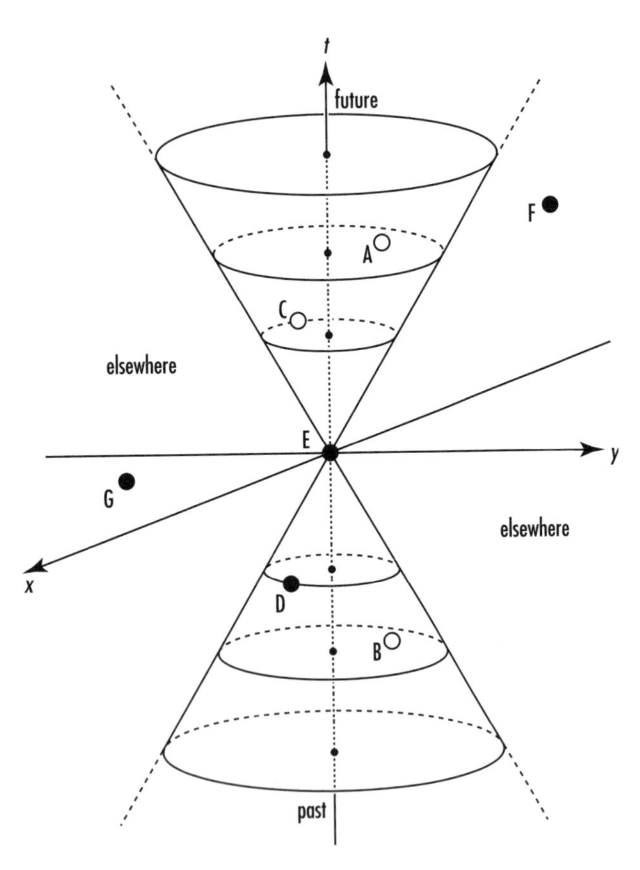

The light cone of event E in space-time. Only two spatial dimensions, *x* (perpendicular to the page) and *y* (horizontal), are shown. The time axis, *t*, is vertical in the figure and forms the axis of the light cone. Events A and B lie within E's light cone, events C and D lie on the light cone, while events F and G lie outside the light cone (in the region labeled "elsewhere"). A and C are in E's absolute future; B and D are in its absolute past. F and G have no absolute temporal relation with E. E's separation from A and B is timelike, from C and D is lightlike, and from F and G is spacelike.

Space-time is not isotropic; its properties are different in different directions. It possesses a light-cone structure. From any event, *E*, taken as origin, there are directions within its future light cone, directions on its future light cone, directions outside its light cone, directions on its past light cone, and directions that lie within its past light cone. E's light cone is independent of reference frame. The different categories of direction possess different properties, as we saw in the previous discussion. An event lying outside E's light cone occurs elsewhere from E and has no reference-frame-independent temporal relation to E. Such events do not affect E, nor are they affected by E. On the other hand, an event lying within or on E's light cone can be said to occur "elsewhen" to E. That event occurs either in E's past or in E's future, and the difference is independent of reference frame. E can affect events in its future, and events in E's past can affect E.

World Line

The events that constitute the life of a massive particle form a continuous curve in space-time, called the particle's world line. Such a world line lies wholly within the light cone of any event on it. Massive particles that are created and then annihilated have their world lines starting at some event and ending at a later event, lying within the future light cone of the earlier event. The world line of a photon lies on the light cone of any of the events forming it. Processes involving a number of particles are represented by bundles of world lines. A body's existence appears as a dense skein of world lines, those of all its constituent particles. Collisions are represented by world lines converging toward each other, possibly actually intersecting each other, and diverging into the future.

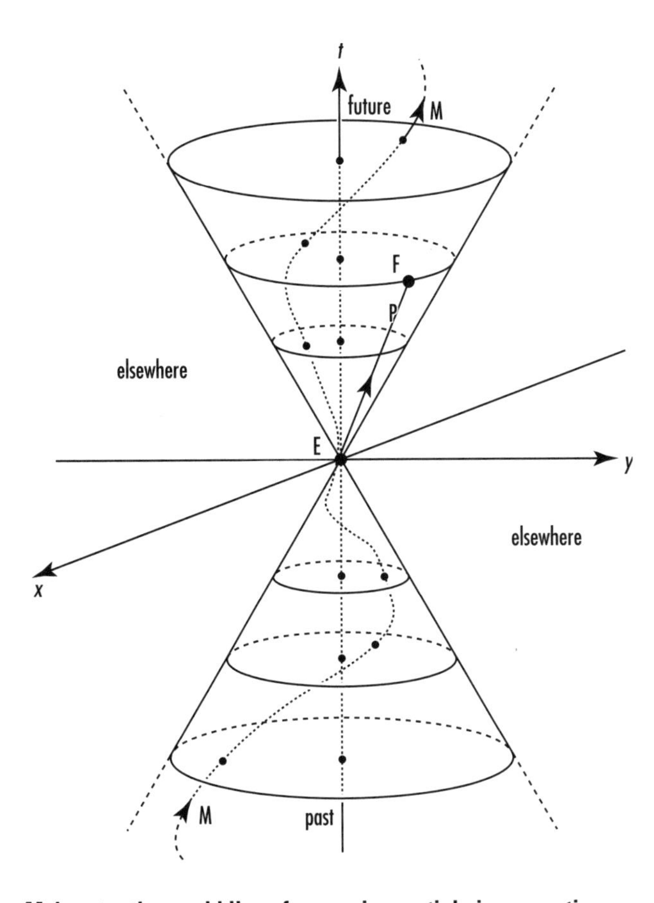

M denotes the world line of a massive particle in space-time. The world line lies wholly within the light cone of any event, E, in the life of the particle. Only two spatial dimensions, *x* (perpendicular to the page) and *y* (horizontal), are shown. The time axis, *t*, is vertical in the figure and forms the axis of E's light cone. The world line of a photon that is emitted at E and absorbed at event F lies on E's light cone and is labeled P.

specific gravity The term *specific gravity* refers to the ratio of the DENSITY of any material to the density of water. The ratio is convenient for three reasons: (1) It is dimensionless, so its value does not depend on the units used. (2) It gives immediate indication about whether a body made of the material will sink in water (when its specific gravity is greater than 1) or float (specific gravity less than 1). (3) It is quite easy to measure, at least for materials that are denser than water. Its measurement involves ARCHIMEDES' PRINCIPLE, according to which the magnitude of the buoyant FORCE acting on a totally submerged body equals the magnitude of the WEIGHT of an amount of water that has the same VOLUME as the body. Although weight is a VECTOR, for brevity we will use "weight" to mean the magnitude of a weight in the following. The procedure is to weigh an object made of the material in question, once in air and once when the object is completely submerged in water. The difference of the two weights equals the magnitude of the buoyant force, which is the weight of the same volume of water. The ratio of the weight in air to the difference of weights equals the ratio of the weight of the object to the weight of the same volume of water. This, in turn, equals the ratio of the density of the object's material to the density of water, or the specific gravity of the material. (*See* BUOYANCY; DIMENSION.)

Let us show that in formulas. Denote the object's weight in air by W, its submerged weight by W_s, and the weight of an equal volume of water by W_w. According to Archimedes' principle:

$$W - W_s = W_w$$

We then have the equality:

$$\frac{W}{W - W_s} = \frac{W}{W_w}$$

Now, weight equals the product of density, volume (which together give the MASS), and g, the ACCELERATION due to gravity, whose nominal value is 9.8 meters per second per second (m/s^2). So if we denote the density of the material by ρ, that of water by ρ_w, and the volume (of the object and of the water) by V, then we have:

$$\frac{W}{W_w} = \frac{\rho Vg}{\rho_w Vg}$$

Canceling Vg in the ratio on the right-hand side leaves the ratio ρ/ρ_w, which is just the specific gravity of the material of which the object is made:

$$\frac{\rho Vg}{\rho_w Vg} = \frac{\rho}{\rho_w} = \text{specific gravity}$$

From the last three equations, we have:

$$\text{Specific gravity} = \frac{W}{W - W_s}$$

as claimed above. (*See* GRAVITATION.)

specific heat *See* HEAT CAPACITY.

spectroscopy Most commonly, spectroscopy is the study of electromagnetic SPECTRA, i.e., the analysis of the electromagnetic RADIATION reaching a detector with regard to the various FREQUENCIES (equivalently, WAVELENGTHs) that compose it and the relative intensities of the radiation at those frequencies. Devices used for such analysis are called spectrometers or spectroscopes. (*See* ELECTROMAGNETISM.)

Spectroscopy can reveal a number of things:

- *The nature of the source.* Dense sources produce continuous spectra that depend on TEMPERATURE. ATOMs, IONs, and MOLECULEs emit characteristic discrete spectra, which, like fingerprints, allow identification of their emitters and also indicate temperature. For radiation from stars, a shift of spectrum frequencies to lower values than normal can indicate the gravitational situation on the surface of the source through the gravitational RED SHIFT. (*See* BLACK BODY; GRAVITATION.)
- *Motion of the source.* The DOPPLER EFFECT on the spectrum frequencies of the source indicates radial MOTION, and the magnitude of the effect gives the radial SPEED. When a red shift is detected, the source is moving away from the spectrometer, while a blue shift shows motion toward the spectrometer. A "smearing" of spectrum frequencies, which results from a continuous range of simultaneous red and blue shifts, gives information on the random motion of many similar sources in a collection of such sources, such as the atoms of a GAS or the stars composing a GALAXY.
- *Transmission conditions.* As electromagnetic radiation propagates from the source to the spectrometer, it is affected by any matter it passes through. That allows observers to draw conclusions about the nature of such matter. The matter might selectively absorb certain frequencies, for example, thus producing an absorption spectrum, whose character can identify the composition of the material. Radiation from other galaxies, during its travel, has its wavelength stretched by the expansion of the UNIVERSE, which produces a cosmological red shift. That is related to the HUBBLE EFFECT. (*See* ABSORPTION; COSMOLOGY.)

For visible-LIGHT spectroscopy, the spectrometer breaks up the incoming light into its component frequencies—corresponding to colors—by means of either a PRISM or a diffraction grating. The former makes use of DISPERSION in glass to bend the light through different angles for different colors. A diffraction grating uses INTERFERENCE to achieve the same effect, where the light is reflected from it or transmitted through it. For electromagnetic radiation other than light, a diffraction grating can be used in reflection mode. (*See* DIFFRACTION; REFLECTION.)

Another use of the term *spectroscopy* is for MASS SPECTROSCOPY. This is the analysis of materials for their chemical composition by measuring—in a mass spectrometer—the masses of their constituent atoms and molecules. A sample of the material to be analyzed is vaporized and ionized. The ions are accelerated in an electric FIELD and then introduced into a magnetic field, which causes their trajectories to bend by an amount that depends on their mass. That brings about a separation of the material's components by their mass, which allows a quantitative analysis of the material's composition. (*See* ACCELERATION; ELECTRICITY; IONIZATION; MAGNETISM; MASS.)

spectrum Most often, a spectrum is the range or set of FREQUENCIES that compose a WAVE, such as an electromagnetic wave or a SOUND wave, or any other phenomenon that is described in terms of frequency. A power spectrum displays the relative INTENSITIES of the various frequencies. (*See* ELECTROMAGNETISM; POWER.)

As examples, a solid body emits electromagnetic RADIATION in a continuous range of frequencies, or in a continuous spectrum, whose character depends on TEMPERATURE. ATOMs that are excited to ENERGY states that are higher than their GROUND STATE undergo DECAY and emit electromagnetic radiation in a discrete set of frequencies, or in a discrete spectrum, which is, like fingerprints, characteristic of each chemical ELEMENT and allows its identification. All those are examples of emission spectra. When radiation passes through material and the material absorbs some of it, causing ranges or sets of frequencies to be lacking from the wave, the result is an absorption spectrum. (*See* ABSORPTION; BLACK BODY.)

An additional use of the term *spectrum* is for the range of values of some property. As an example, the values of the MASSes of the various kinds of atoms and MOLECULEs that make up some material form the mass spectrum of the material.

See also MASS SPECTROSCOPY; SPECTROSCOPY.

speed The magnitude of VELOCITY is called speed, or instantaneous speed. It is a SCALAR quantity, and its SI UNIT is meter per second (m/s). If we denote by $s(t)$ the length of the path that a body follows as a function of TIME, t, then the body's speed, $v(t)$, as a function of time is the time derivative of $s(t)$:

$$v(t) = ds(t)/dt$$

Here s is in meters (m) and t is in seconds (s). Average speed, v_{av}, is the ratio of the distance traveled along the path of the motion, Δs, during a finite time interval, Δt, to the time interval:

$$v_{av} = \Delta s/\Delta t$$

Relative speed is the speed of a body as measured with respect to a moving REFERENCE FRAME, such as one attached to another moving body.

For rotational MOTION there is angular speed. In the case of rotation about an axis of fixed direction, a body's orientation is specified by a single angle in radians (rad) that is, in general, a function of time, $\theta(t)$. The time derivative of this function gives the angular speed, $\omega(t)$, in radians per second (rad/s):

$$\omega(t) = d\theta(t)/dt$$

Average angular speed, ω_{av}, for the time interval Δt is the angular DISPLACEMENT of the body, $\Delta\theta$, during that interval divided by the time interval:

$$\omega_{av} = \Delta\theta/\Delta t$$

See also KINEMATICS; ROTATION.

speed of light The propagation SPEED of electromagnetic WAVEs in VACUUM is referred to as the speed of light. It is conventionally denoted by c and has the same value, 2.99792458×10^8 meters per second (m/s), for all kinds of electromagnetic RADIATION. This fundamental constant is understood to possess such an especially fundamental character that its value is no longer measured but, rather, defined. Since the SI UNIT of TIME, the second, is defined as well, measurements of c are actually measurements of the length of the meter. (*See* BASE QUANTITY; CONSTANTS, FUNDAMENTAL; ELECTROMAGNETISM.)

MAXWELL'S EQUATIONS relate the speed of light to two constants of ELECTRICITY and MAGNETISM: the PERMITTIVITY of the vacuum, ε_0, whose value is $8.85418782 \times 10^{-12}$ $C^2/(N \cdot m^2)$, and the PERMEABILITY of the vacuum, μ_0, with value $4\pi \times 10^{-7}$ T·m/A. The relation is:

$$c = \frac{1}{\sqrt{\varepsilon_0 \mu_0}}$$

The speed of light plays a distinguished role in ALBERT EINSTEIN's special theory of relativity, where it sets an upper limit for the speed of material objects and of the transmission of information. One of the postulates of the theory states that observers in different states of MOTION find the same value for the speed of light. (*See* RELATIVITY, SPECIAL THEORY OF.)

The speed of light is also involved in defining the astronomical distance unit of the light-year. This is the distance light travels in a year, whose value is about 9.5×10^{15} meters (m). In a similar manner, one can refer to a light-second, light-minute, etc. Note that, due to the finite speed of light, astronomical observations are views of the past. The light reaching us from a GALAXY at a distance of, say 1 million light-years left that galaxy 1 million years ago. So the images of the galaxy that we obtain show the state of the galaxy that long ago. In this manner, observations of very distant galaxies give information about the early UNIVERSE.

The speed at which electromagnetic radiation propagates through transparent matter might differ from c. So the term *speed of light* is not to be taken literally but, rather, as a convenient label for the speed of light (and of all electromagnetic radiation) in vacuum. (*See* INDEX OF REFRACTION.)

spin Spin is an inherent property of ELEMENTARY PARTICLEs, whereby each kind of particle possesses a certain quantity of ANGULAR MOMENTUM that is characteristic of the kind of particle. So elementary particles are assigned a spin QUANTUM NUMBER, s. The range of values of s is the half-integers, 0, 1/2, 1, 3/2, . . . The significance of s is that the maximal absolute value of the component of the particle's angular momentum in any direction equals $sh/(2\pi)$, where h denotes the PLANCK CONSTANT, whose value is $6.62606876 \times 10^{-34}$ joule·second (J·s). The full set of allowed values for the component of spin angular momentum in any direction for a particle of spin s is: $sh/(2\pi)$, $(s - 1)h/(2\pi)$, . . ., $-(s - 1)h/(2\pi)$, $-sh/(2\pi)$. For instance, the allowed values of the spin component of an ELECTRON, a particle with a half unit of spin, are $h/(4\pi)$ and $-h/(4\pi)$. Particles of integer spin belong to the family of BOSONs, while those of spin 1/2, 3/2, etc. are FERMIONs. Examples of the former are the PION (spin 0) and the PHOTON (spin 1). The electron, the PROTON (both spin 1/2), and the omega-minus (spin 3/2), for example, are fermions.

Stanford Linear Accelerator Center (SLAC) Founded in 1962 at Menlo Park, California, on the campus of Stanford University, the Stanford Linear Accelerator Center is a U.S. government laboratory under the Department of Energy. The center is devoted to research in the physics of ELEMENTARY PARTICLES, both theoretical and experimental, using its linear particle accelerator. It maintains a database of more than 500,000 articles on elementary particles written since 1974. SLAC also offers a user facility in which high-intensity X-RAY beams serve for research in medical science, biology, chemistry, physics, materials science, and environmental science. SLAC's staff numbers some 1,300 people, and almost 3,000 users, from nearly 400 institutions, perform research there each year. (*See* ACCELERATOR, PARTICLE; EXPERIMENTAL PHYSICS; THEORETICAL PHYSICS.)

The center boasts three Nobel Prize awards (1976, 1990, 1995) for discoveries concerning QUARKs and NEUTRINOs. SLAC is home to the Kavli Institute for Particle Physics and Cosmology, where work is carried out on the interface among elementary-particle physics, ASTROPHYSICS, and COSMOLOGY.

The URL of the Stanford Linear Accelerator Center website is http://www.slac.stanford.edu/. SLAC's mailing address is 2575 Sand Hill Road, Menlo Park, CA 94025.

state of matter A form in which MATTER can exist that is significantly different from other possible forms is termed a state of matter. The term PHASE, in one of its meanings, is synonymous with state of matter, and a PHASE TRANSITION is a change from one state of matter to another. The so-called ordinary states of matter are

Two beam pipes of the PEP-II Storage Ring at Stanford Linear Accelerator Center (SLAC). The upper pipe carries positrons, and the lower pipe carries electrons. When the two beams have reached sufficient density, they are made to meet head-on, and the products of the resulting electron-positron collisions are studied. ***(Photo, Peter Ginter. Courtesy of Stanford Linear Accelerator Center)***

SOLID, LIQUID, and GAS, where the latter two are also taken together as the FLUID state and the former two as condensed matter. The possible phase transitions among these three states of matter include FREEZING, MELTING, BOILING, and CONDENSATION. (*See* CONDENSED-MATTER PHYSICS.)

In addition to the ordinary states of matter, there are other possible states that matter might take. One is the LIQUID CRYSTAL state. Additional ones include the superconducting state and the superfluid state. At sufficiently high TEMPERATURES, the PLASMA state is created. (*See* SUPERCONDUCTIVITY; SUPERFLUIDITY.)

The solid state can be subcategorized into crystalline state, polycrystalline state, and amorphous state. A material might possess more than a single crystalline state, which can differ in their crystal type or unit cell. (*See* CRYSTAL; CRYSTALLOGRAPHY; POLYCRYSTALLINE SOLID.)

States that might be called "extreme" become possible under extreme conditions. One such state is nuclear matter, in which matter takes the form of a compact conglomeration of NUCLEONs. NEUTRON STARs are in this state. As an even more extreme state, the theoretical possibility of quark matter is being considered. In such a state, extreme PRESSURE would crush the nucleons together to such an extent that they lose their individual identity, and the QUARKs that normally constitute the nucleons would be free to move throughout the VOLUME of the body. The BLACK HOLE is thought to hold the position of nature's most extreme state. (*See* NUCLEUS.)

statics The subfield of MECHANICS that deals with EQUILIBRIUM situations, when no MOTION takes place, is called statics. The foundation of statics is ISAAC NEWTON's first law of MOTION, stating that in the absence of FORCEs acting on it, a body will remain at rest or in MOTION at constant SPEED in a straight line. It follows that for a body, such as a building, to be at rest and remain at rest, the resultant of all forces acting on the body must equal zero, and the resultant of the TORQUEs of all those forces, with respect to all axes of ROTATION, must equal zero too. Then the body will neither undergo translational motion nor rotate. (*See* NEWTON'S LAWS.)

statistical mechanics The field of statistical mechanics studies the statistical properties of physical systems containing large numbers of constituents. Examples of such systems are a GAS (which comprises a very large number of ATOMs or MOLECULEs) and a GALAXY (which contains billions, even thousands of billions, of stars). Although it might be possible in principle to follow the detailed evolution of such a system in terms of the behavior of every individual constituent, in practice it is impossible. Even if it were possible, it might not be very useful. Statistical mechanics develops methods of deriving useful information about the system as a whole from some, however limited, knowledge about its constituents. In particular, it explains the properties of bulk MATTER in terms of the properties and INTERACTIONs of its constituent particles (atoms, IONs, molecules). An example of that is KINETIC THEORY, which explains properties of certain bulk matter in terms of the MOTION of its constituents. The kinetic theory of gases derives the properties of PRESSURE, TEMPERATURE, and quantity from the particle nature of gases. More generally, statistical mechanics offers a connection between THERMODYNAMICS, which deals with macroscopic properties of MATTER, and the underlying microscopic structure of matter and its properties.

Statistical mechanics allows testing theories about the microscopic aspects of nature by means of macroscopic measurements. From assumptions about the interatomic FORCEs in a SOLID and the nature of ELECTRONs, for instance, methods of statistical mechanics allow predictions of bulk properties such as HEAT CAPACITY and electric and thermal CONDUCTIVITIES. These properties can be measured and the predictions tested. In this manner, the microscopic assumptions can be evaluated. (*See* ELECTRICITY; HEAT.)

See also BOLTZMANN DISTRIBUTION; BOSE-EINSTEIN STATISTICS; ENTROPY; EQUIPARTITION OF ENERGY; FERMI-DIRAC STATISTICS; FREE ENERGY; KINETIC THEORY; MAXWELL-BOLTZMANN STATISTICS; MAXWELL DISTRIBUTION; TEMPERATURE; THERMODYNAMICS.

steady-state process A steady-state process is a process in a physical system during which, although changes are taking place, there are nevertheless large-scale aspects of the system that do not change in TIME. As an example, consider a steady-state flow of air around an obstacle in a wind tunnel. Air MOLECULEs are continually entering the tunnel at one end, passing the obstacle, and leaving the tunnel at the other end. Yet the flow pattern remains constant; specifically, the

VELOCITY of the air at every location in the pipe does not change in time. The steady-state cosmological model can serve as an additional example. According to this model, the UNIVERSE, although it is observed to be expanding, with the GALAXIES generally flying away from each other, is nevertheless in a steady state and maintains a constant average DENSITY of MATTER over time. That is accomplished by the steady creation of new matter at a sufficient rate to fill in the gaps left by the expansion. (This model is now generally considered invalid.) (*See* BIG BANG; COSMOLOGY; HUBBLE EFFECT.)

Stefan-Boltzmann law Named for the Austrian physicists Josef Stefan (1835–93) and LUDWIG BOLTZMANN, the Stefan-Boltzmann law concerns the electromagnetic ENERGY radiated by a BLACK BODY. The law states that the total POWER (energy per unit TIME) of the RADIATION is proportional to the fourth power of the black body's absolute TEMPERATURE. One mathematical expression of the law is in terms of the radiated power per unit surface AREA, P, of the black body and has the form:

$$P = \sigma T^4$$

Here P is in watts per square meter (W/m²), T denotes the absolute temperature in kelvins (K), and σ is the Stefan-Boltzmann constant, whose value is $\sigma = 5.67051 \times 10^{-8}$ W/(m²·K⁴). The Stefan-Boltzmann law follows from MAX KARL ERNST LUDWIG PLANCK'S formula for the black-body SPECTRUM. (*See* ELECTROMAGNETISM.)

Stern-Gerlach experiment The Stern-Gerlach experiment was performed by the German/American physicist Otto Stern (1888–1969) and the German physicist Walter Gerlach (1889–1979). Its importance lies in its demonstrating that the ELECTRON possesses SPIN. In the experiment, a beam of silver ATOMs was passed through a very inhomogeneous magnetic FIELD. That caused the beam to split into two components, and the two beams were detected when they exited the apparatus. The silver atoms were electrically neutral, so their deflection could be due only to their possessing a magnetic dipole moment, connected with atomic ANGULAR MOMENTUM. The electron configuration of the silver atom is such that the atom possesses no angular momentum that is due to the electrons' orbits. So it was possible to attribute the atom's angular momentum to the inherent angular momentum of an individual electron, the electron's spin. Since the beam split into two, the value of the electron's spin was determined to be one-half. (*See* CHARGE; ELECTRICITY; MAGNETISM; MOMENT, MAGNETIC DIPOLE; QUANTIZATION; QUANTUM NUMBER.)

Stokes's law Named for its discoverer, the Irish/British physicist and mathematician George Gabriel Stokes (1819–1903), Stokes's law gives the magnitude of the viscous FORCE that acts on a SOLID sphere moving at sufficiently low SPEED through a FLUID. If we denote the magnitude of the force by F_v, Stoke's law gives:

$$F_v = 6\pi r \eta v$$

where F_v is in newtons (N), r denotes the radius of the sphere in meters (m), η is the VISCOSITY of the fluid in newton·seconds per square meter (N·s/m²), and v represents the speed of the sphere in the fluid (or the speed of the fluid flowing past the sphere) in meters per second (m/s).

If a sphere whose DENSITY is greater than that of the fluid is released from rest in a fluid, it falls and accelerates until it reaches its TERMINAL SPEED, the speed at which the viscous force balances the net downward force. It then continues to fall at the terminal speed. The magnitude of the net downward force acting on the sphere equals the magnitude of the force of gravity less the magnitude of the buoyant force. For a homogeneous sphere, this quantity, F_d, is:

$$F_d = \frac{4\pi r^3}{3}(\rho - \rho_f)g$$

where $4\pi r^3/3$ is the volume of the sphere; ρ and ρ_f denote the densities of the sphere and the fluid, respectively, in kilograms per cubic meter (kg/m³); and g is the ACCELERATION due to gravity, with nominal value 9.8 meters per second per second (m/s²). The net downward force is independent of the sphere's speed. At terminal speed, $F_v = F_d$. Equating the above two expressions, we obtain the terminal speed, v_t:

$$v_t = \frac{2r^2(\rho - \rho_f)g}{9\eta}$$

Alternatively, the fluid's viscosity, η, can be found by measuring the terminal speed:

$$\eta = \frac{2r^2(\rho - \rho_f)g}{9v_t}$$

(*See* BUOYANCY; GRAVITATION.)

strain Strain is the relative change in some spatial quantity characterizing the state of a body that results from STRESS applied to the body. Strain is the ratio of a change of a quantity to the quantity itself and is thus dimensionless. For linear deformation—stretching or compression—the strain is called tensile strain and is given by the change in length divided by the length. Bulk strain is the ratio of the change of VOLUME to the volume. Angular deformation, called SHEAR, is described by shear strain, which equals the angle of deformation. For elastic materials up to their proportionality limit, the strain is proportional to the applied stress. (*See* DIMENSION; ELASTICITY; HOOKE'S LAW.)

stress A set of balanced FORCEs acting on a body can cause a deformation of the body. The effect is described in terms of stress and STRAIN, where the stress is the ratio of a deforming force to the AREA of the surface over which the force acts. The SI UNIT of stress is the pascal (Pa), equivalent to newton per square meter (N/m^2). In linear deformation—stretching or compression—a pair of equal-magnitude, oppositely directed forces act on a body along the same line of action. The stress for this situation is tensile stress, which equals the magnitude of one of the forces divided by the cross-section area of the body. Bulk stress causes volume change. It is simply the PRESSURE acting on the body. Shear stress is the effect of a pair of equal-magnitude, oppositely directed forces acting on a body along parallel lines of action. It equals the ratio of the magnitude of one of the forces to the area of the surface parallel to the force over which the force acts. Shear stress brings about angular deformation. (*See* ELASTICITY; HOOKE'S LAW; SHEAR.)

string theory String theory is an approach to the unification of the four fundamental INTERACTIONS: the STRONG INTERACTION, the ELECTROMAGNETIC INTERACTION, the WEAK INTERACTION, and GRAVITATION. String theory is based on the OSCILLATION, splitting, and joining of entities at the PLANCK SCALE—of around 10^{-35} meter (m) and 10^{-43} second (s)—called strings. The theory requires additional DIMENSIONs to the four of SPACE and TIME. Actually, there are a number of string theories, which apparently form different limiting cases of a more general theory, called M-theory. The field is presently in active development. Although it seems to afford some insight, it has yet to achieve definitive successes.

strong interaction Also referred to as the color FORCE or gluon force, the strong interaction forms one of the four fundamental INTERACTIONs among the ELEMENTARY PARTICLEs, along with the ELECTROMAGNETIC INTERACTION, the WEAK INTERACTION, and GRAVITATION. Of all the interactions, the strong interaction possesses the shortest range, about 10^{-15} meter (m). At its most fundamental, the strong interaction acts among QUARKs and is mediated by GLUONs, which couple to the color CHARGE of quarks and are exchanged by quarks. The strong interaction is responsible for binding together the quarks that form the HADRONs, which consist of the BARYONs, such as the NUCLEONs (the PROTON and the NEUTRON), and the MESONs, which include the PION. The strong interaction is also responsible for the existence of NUCLEI as bound conglomerations of nucleons. Its manifestation at the nuclear level is described by exchange of mesons among baryons and, in particular, exchange of pions among nucleons. The fundamental mechanism of the strong nuclear force, however, is understood to lie in the exchange of gluons among the quarks composing the nucleons and pions.

See also CONSERVATION LAW; INVARIANCE PRINCIPLE; SYMMETRY.

superconductivity The total lack of electric RESISTANCE that certain SOLID materials exhibit at sufficiently low TEMPERATUREs (i.e., at temperatures below their respective critical temperatures) is termed superconductivity. When an electric CURRENT is caused to flow in a material that is in the superconducting state, the current will continue to flow forever without diminishing, as far as is presently known. Another characteristic of the superconducting state is that it does not suffer the presence of a magnetic FIELD within itself and even expels a preexisting field when the material is cooled to its superconducting state while immersed in a magnetic field. An applied magnetic field that is greater than the material's critical field, even though it cannot penetrate the material in its superconducting state, nevertheless destroys the superconducting state. The value of the critical field depends on temperature, is greatest approaching 0 kelvin (K), and declines to zero at the critical temperature. An effect of the critical field is to limit the current that can be passed through a superconducting material. If the current is too great, the magnetic field it produces will surpass the critical field and suppress superconductivity. (*See* ELECTRICITY; MEISSNER EFFECT; MAGNETISM.)

Type I superconductors are certain pure METALS, including aluminum, mercury, and niobium, for example. They are characterized by critical temperatures below 10 K and possess relatively low critical fields, below about 0.2 tesla (T). Type II superconductors comprise various ALLOYS and metallic COMPOUNDS. Their critical temperatures are more or less in the 14–20-K range and their critical fields higher than 15 T. So type II superconductors are much more suitable than type I for the construction of superconducting ELECTROMAGNETS: they do not have to be cooled to as low temperatures and they can carry higher currents. Whereas in a normal electromagnet energy must be invested both to set up the magnetic field and to maintain it, in a superconducting electromagnet, energy is invested only to set up the field, as long as the field is kept below the critical field. A successful theory of superconductivity for types I and II, called the BCS theory (after the American physicists JOHN BARDEEN, Leon Neil Cooper [1930–], and John Robert Schrieffer [1931–]), explains the effect as conduction by paired ELECTRONS, called COOPER PAIRS, whose existence involves the participation of the material's CRYSTAL structure.

A class of high-temperature superconductors, with critical temperatures ranging to above 130 K, has been discovered and is being expanded. These are metallic oxides in ceramic form and all contain copper. One importance of this class is that many of its members can be cooled to the superconducting state with liquid nitrogen, whose temperature is about 77 K and which is much cheaper than the liquid helium (at about 4 K) required for the cooling of superconductors of types I and II. High-temperature superconductivity holds promise of many useful future applications, including loss-free electricity transmission, MAGNETIC LEVITATION, and liquid-nitrogen-based superconducting electromagnets. The BCS theory does not seem applicable to high-temperature superconductivity, and no other explanation for it is currently available.

See also JOSEPHSON EFFECT; STATE OF MATTER.

superfluidity The term *superfluidity* describes a low-TEMPERATURE state of LIQUID matter that flows with no VISCOSITY. Superfluidity has been discovered in the isotope helium-4, ^{4}He, below about 2.2 K, where it is a form of Bose-Einstein condensate. It has also been detected in helium-3, ^{3}He, at temperatures of thousandths of a kelvin. A superfluid can flow through tiny pores with no FRICTION and is extremely mobile over surfaces it is in contact with. The superfluid state possesses unusual thermal and acoustical properties as well. It flows under the effect of a temperature difference between different locations and transmits temperature WAVES, as examples. (*See* ACOUSTICS; BOSE-EINSTEIN STATISTICS; HEAT; SOUND; STATE OF MATTER.)

superposition principle The superposition principle states that the combined effect of two (or more) effects occurring simultaneously is their algebraic sum. Systems in which the superposition principle holds are called linear systems; those in which it is invalid are termed nonlinear systems. As an example, consider the effect of forces on a spring. One force might stretch the spring by 4.0 centimeters (cm), while another force, acting alone, might stretch it by 3.0 cm. As long as the spring is not stretched beyond its proportionality limit, it forms a linear system, and the combined effect of the two forces is to stretch the spring by 7.0 cm. Beyond the proportionality limit, however, the spring is nonlinear, and the superposition principle is not valid for it. If the spring's proportionality limit is 5.0 cm, say, then the combined effect of the two forces might be to stretch the spring by only 6.8 cm. As an additional example, while the response of a transparent material to the passage of a single LIGHT beam is linear for sufficiently low INTENSITIES, two light beams propagating through it simultaneously might cause the material to exceed its linear limit and produce nonlinear effects. One such effect is that the exiting light contains FREQUENCIES that are not present in the entering beams. If f_1 and f_2 are the frequencies of the two beams, the light leaving the material might contain also the frequencies $f_1 + f_2$ and $|f_1 - f_2|$, for example. (*See* ELASTICITY.)

One expression of the validity of the superposition principle for a physical system is that the system's behavior is describable by linear differential equations, at least for some range of values of the PARAMETERS characterizing the system. In general, if S_1 and S_2 are any two solutions of a set of differential equations, then their linear combination is any expression of the form $aS_1 + bS_2$, where a and b are two numbers. The equations are designated linear when, for every pair of solutions S_1 and S_2 and for every pair of numbers a and b, the linear combination $aS_1 + bS_2$ is also a solution of

a)

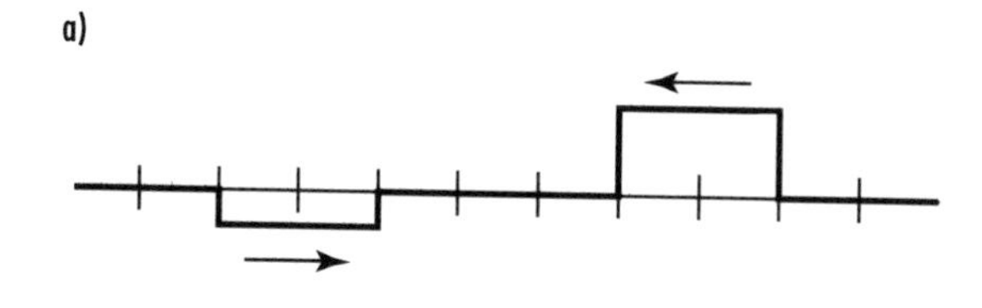

b)

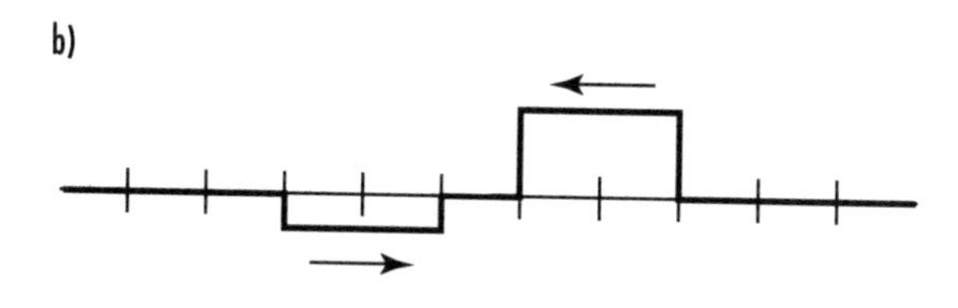

c)

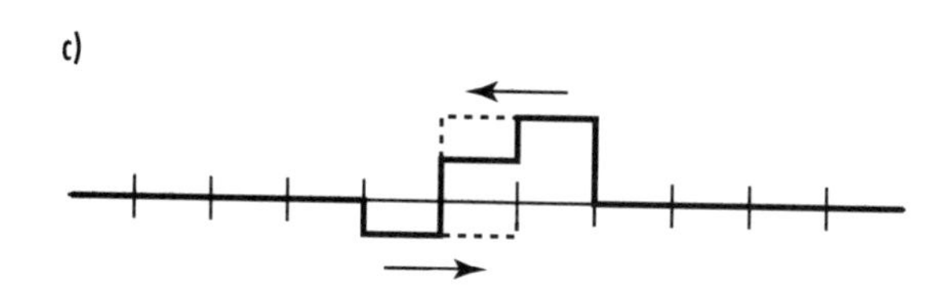

d)

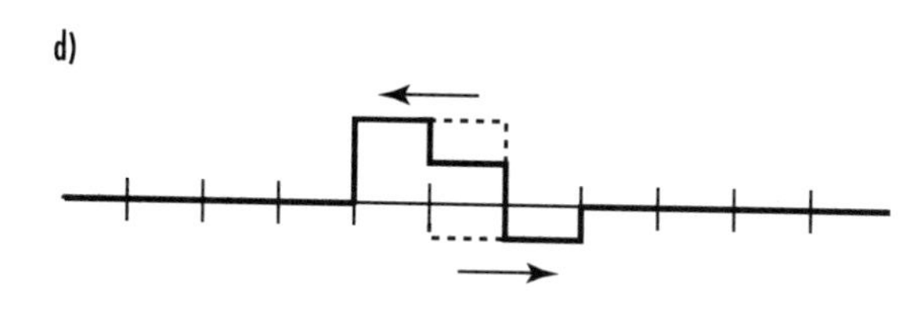

e)

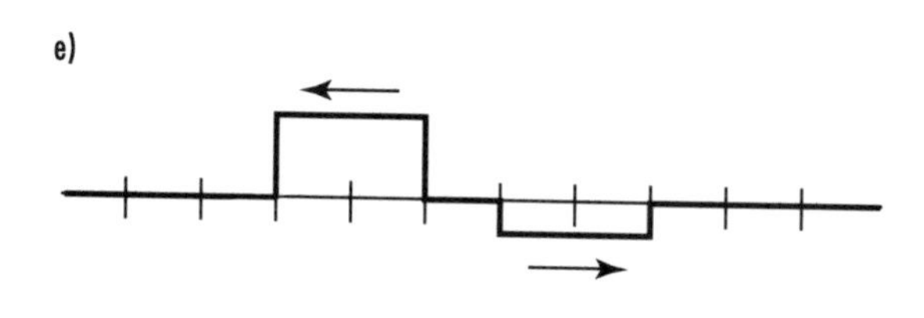

f)

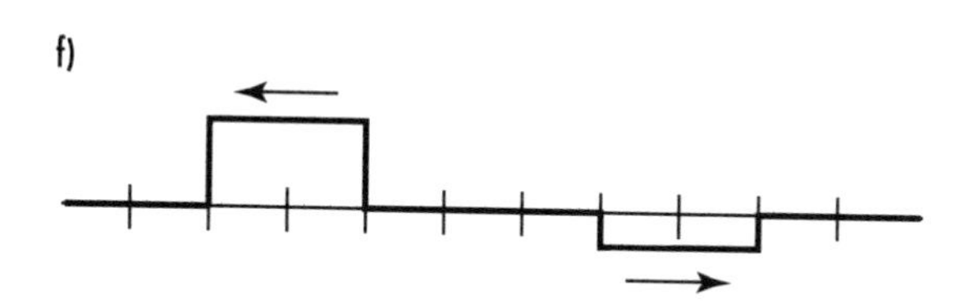

An example of the superposition principle, valid for linear systems, which states that the combined effect of two (or more) effects occurring simultaneously is their algebraic sum. Two pulse waves in a medium, one positive and the other negative, move toward each other at equal speeds of one scale division per unit time. The figures depict consecutive situations at intervals of one time unit. (a), (b) The pulses are approaching each other. (c) The pulses partially overlap, with their net effect shown. (d) The pulses still partially overlap. (e), (f) The pulses are departing from each other, unaffected by their encounter.

them. So if S_1 and S_2 represent two possible behaviors of a system that is describable by linear differential equations, then any linear combination, $aS_1 + bS_2$, also represents a behavior of the system, a superposition of the two behaviors. Putting $a = b = 1$, we obtain the superposition principle as stated above.

The superposition principle is precisely valid in QUANTUM PHYSICS. That is because the SCHRÖDINGER EQUATION, which governs the behavior of quantum systems, is a linear differential equation. The state of any system is described by a wave function, Ψ. If Ψ_1 and Ψ_2 represent two possible states of a system, then $a\Psi_1 + b\Psi_2$, with any complex numbers a and b such that $|a|^2 + |b|^2 = 1$, also represents a possible state, a superposition of the two states. In the quantum case, that can bring about nonclassical results, such as in the following example. When the states Ψ_1 and Ψ_2 are each characterized by a definite, but different, value of ENERGY, say, E_1 and E_2, respectively, their superposition $a\Psi_1 + b\Psi_2$, with nonzero a and b, is a state that does not possess a definite energy. Energy measurements of this state will give a range of results. What can be said, however, is that the average value of a large number of energy measurements will equal $|a|^2E_1 + |b|^2E_2$. (*See* CLASSICAL PHYSICS; QUANTUM MECHANICS.)

supersymmetry Supersymmetry is a type of gauge symmetry that is proposed by some researchers in an attempt to better understand the ELEMENTARY PARTICLES and their INTERACTIONs. The FIELD transformations of supersymmetry mix FERMIONs and BOSONs. Accordingly, according to supersymmetry, every known kind of fermion or boson should possess a bosonic or fermionic partner, respectively, and the partners should have the same MASS. That is clearly not the case; the known elementary particles do not appear to possess such partners. So supersymmetry, if valid, would not be exact. The as-yet undiscovered partner particles would have masses that are higher than can be produced by present-day particle accelerators. The names of the particles predicted by supersymmetry are constructed as follows. For a known fermion say, ELECTRON or QUARK, the name of the corresponding supersymmetric boson is found by adding an "s" before the name, giving selectron or squark. The name of the supersymmetric fermionic partner of a known boson, say, PHOTON or W or GLUON, is obtained by changing the "on" to "ino" or by adding "ino" to the end, resulting in photino or wino or gluino. (*See* ACCELERATOR, PARTICLE; GAUGE THEORY; SYMMETRY.)

surface tension The property of LIQUIDs that WORK is required to increase the AREA of their free surface (i.e., the surface separating the liquid from VACUUM or from GAS) is known as surface tension. Accordingly, a liquid minimizes its POTENTIAL ENERGY by minimizing its free surface, all other effects being equal. Surface tension comes about as follows. Every MOLECULE of a liquid is attracted by its nearby neighbors. For a molecule inside the VOLUME of the liquid, the attractive FORCEs cancel out. A molecule at a free surface has a net force pulling it into the liquid. Increasing the area of a free surface brings more molecules to the surface and requires overcoming the force pulling those molecules back into the liquid. In this way, work is performed when the surface area is increased. The surface of a liquid acts like a membrane under TENSION. But the force required to "stretch the membrane" is independent of the "membrane's" extension, whereas for a real elastic membrane, the force increases with extension. The surface tension of a liquid is quantified and defined as the magnitude of the force required to extend the surface, per unit length—perpendicular to the force—along which the force acts. It is determined by stretching the surface and measuring the required force and the length of the edge of the stretching device, then dividing the former by the latter. Its SI UNIT is newton per meter (N/m). Surface tension is TEMPERATURE dependent, tending to decrease with rising temperature. (*See* ELASTICITY.)

One effect of surface tension is that objects that are not too heavy, but have a DENSITY greater than that of water, can be made to float on the surface of water, whereas ordinarily they sink in water. You might try that with small pieces of aluminum foil. While floating, the object slightly depresses and stretches the water's surface at its points of contact. Some insects exploit this phenomenon to walk on the surface of water. Another effect is the tendency of drops of a liquid to take a spherical shape. For a given volume, the shape of a sphere gives it the smallest surface area. (*See* BUOYANCY.)

See also CAPILLARY FLOW.

symmetry The possibility of making a change in a situation while some aspect of the situation remains unchanged is symmetry. As a common example, many animals, including humans, possess approximate bilateral symmetry. Consider the plane that bisects the animal's body front to back through the middle. With respect to that plane, the external appearance of the body has this balance: for every part on the right there is a similar, matching part on the left (and vice versa). If the body were reflected through the bisecting plane, as if the plane were a two-sided MIRROR, the result would look very much like the original. Here we have the possibility of reflection as a change under which the body's external appearance remains unchanged (more or less). (*See* REFLECTION.)

Spatial symmetry is symmetry under spatial changes. Those include reflection (as in the example), ROTATION, DISPLACEMENT, and scale change. Here are some examples. A starfish possesses approximate fivefold rotation symmetry about the axis through its center and perpendicular to its plane, with respect to external appearance. It can be rotated through angles that are integer multiples of 360°/5 = 72°, and such rotations do not change its appearance. The LAWS OF NATURE are symmetric under displacements, in that the same laws are valid at all locations in SPACE (as far as is presently known). Perform experiments here or perform them there, in both cases you discover the same laws of nature. Symmetry under change of scale is a property of certain natural phenomena and mathematically generated patterns. Such systems are called FRACTALs. The shape of coastlines and the form of mountain ranges are approximately fractals, as they present more or less the same structure at different scales.

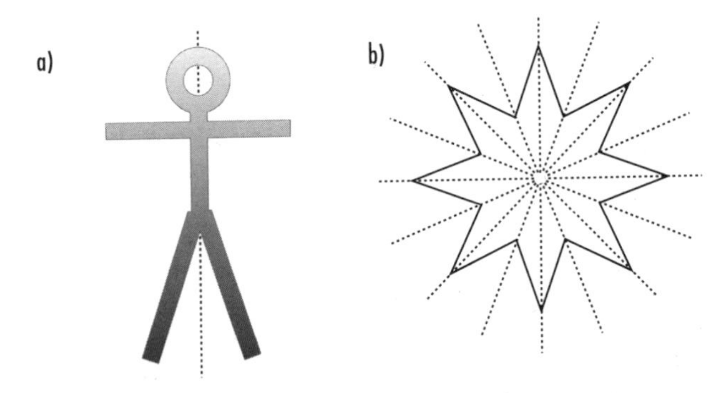

Examples of spatial symmetry. (a) The figure possesses bilateral symmetry (i.e., reflection symmetry) with respect to the dashed line. (b) Eightfold rotation symmetry about the figure's center is accompanied by bilateral symmetry with respect to each of the eight dashed lines.

Besides spatial symmetry, there are many additional types of symmetry. Temporal symmetry involves changes in TIME, time ordering, or time intervals. As an example, the laws of nature also appear to be symmetric under time displacements (again, at least as far as we know): perform experiments now or at any other time, and you will find the same laws of nature. Spatiotemporal symmetry has to do with changes involving both space and time. An example is a change from one REFERENCE FRAME to another moving at constant VELOCITY with respect to the first. These changes are expressed by GALILEI TRANSFORMATIONs, LORENTZ TRANSFORMATIONs, and POINCARÉ TRANSFORMATIONs. The laws of nature are found not to change under the latter two kinds of transformations. (Galilei transformations are low-SPEED approximations of Lorentz transformations.) Observers in relative straight-line MOTION at constant speed perform experiments and discover the same laws of nature. This symmetry forms an essential component of ALBERT EINSTEIN's special theory of relativity. An even more general spatio-temporal symmetry lies at the heart of Einstein's general theory of relativity: observers in relative motion of any kind, including accelerated motion, find the same laws of nature. (*See* ACCELERATION; RELATIVITY, GENERAL THEORY OF; RELATIVITY, SPECIAL THEORY OF.)

Other changes, of more abstract character, are also useful in connection with symmetry. They include rearrangement (permutation) of parts or aspects of a system, changes of PHASE of wave functions, and mixing of FIELDs, among others. (*See* FIELD THEORY; QUANTUM FIELD THEORY; QUANTUM MECHANICS.)

Symmetry of the laws of nature under some change is called an INVARIANCE PRINCIPLE. Invariance principles are related to CONSERVATION LAWs. In addition to the invariance principles that underlie the theories of relativity, other invariance principles play major roles in physics. The theories of the STRONG INTERACTION, ELECTROMAGNETIC INTERACTION, WEAK INTERACTION, and ELECTROWEAK INTERACTION are founded on invariance principles, called gauge symmetries. (*See* GAUGE THEORY.)

Note that symmetry implies equivalence of parts, components, or aspects of a system. The more symmetries a system possesses, or the higher a system's degree of symmetry, the more equivalence it contains. Thus symmetry correlates with uniformity. The higher the degree of symmetry, the higher is the level of uniformity in a system. Compare a square plate and a circular disc with regard to their symmetry under rotations about their axes (through their centers and perpendicular to their planes), for instance. The former maintains its appearance only under rotations through integer multiples of 360°/4 = 90°, while the latter is symmetric under rotation through any angle whatsoever. The disc possesses perfect rotational uniformity, while that of the square is severely limited.

ORDER correlates with distinguishability, inequivalence, nonuniformity. Its opposite, DISORDER, thus correlates with indistinguishability, equivalence, and uniformity. So symmetry correlates with disorder. In this connection, let us compare the same substance in its crystalline state and its gaseous state. The CRYSTAL possesses a relatively high degree of order: the molecules are positioned on the crystal LATTICE and not elsewhere (inhomogeneity), and the physical properties of the crystal are generally different in different directions (anisotropy). The crystal is characterized by a corresponding set of symmetries. The GAS, on the other hand, is highly disordered: the molecules can be anywhere in the container (homogeneity), and all directions are equivalent with regard to the substance's physical properties (isotropy). As a result, the gas possesses a much higher degree of symmetry than does the crystal. Every symmetry of the crystal is also a symmetry of the gas, while the gas has many more additional symmetries. (*See* STATE OF MATTER.)

The symmetry principle, or Curie's symmetry principle, or the Curie principle, is of major importance in the application of symmetry. This principle states that the symmetry of a cause must appear in its effect. An alternative, equivalent statement of the principle is that any asymmetry (lack of symmetry) in an effect must also characterize the cause of the effect. So an effect is at least as symmetric as its cause. Consider this example. The value of the INDEX OF REFRACTION of a substance—as an effect—depends on the INTERACTION of electromagnetic RADIATION, such as LIGHT, with the substance's constituent ATOMs, MOLECULEs, or IONs as the cause. By the symmetry principle, any anisotropy (dependence on direction) of the index of refraction, which is an asymmetry, must be found also in the distribution of the substance's constituents. When a substance is in a LIQUID state, its constituents are randomly and uniformly distributed, with no distinguished directions. Consequently, the index of refraction of liquids is isotropic (independent of direction). In a crystalline state of a substance, however, when the constituents are arrayed on a lattice, not all directions are equivalent. Then the index of refraction might very well have different values in different directions (an effect called BIREFRINGENCE). (*See* CURIE, PIERRE; ELECTROMAGNETISM.)

See also ENTROPY; SUPERSYMMETRY.

T

tachyon Any hypothetical particle that travels faster than the SPEED OF LIGHT is designated a tachyon. Contrary to common belief, ALBERT EINSTEIN's special theory of relativity does not forbid superluminal SPEEDS for particles. What it does forbid is accelerating particles to the speed of light. So no ordinary particle, called a tardyon, can reach light speed or higher speeds. For them the speed of light sets an upper speed limit. The existence of tachyonic MATTER is not precluded. Such matter, if it existed, could be accelerated to infinite speeds, but could not be decelerated to the speed of light or lower speeds. For tachyons, too, the speed of light is a limiting speed, but it serves as a lower limit rather than an upper limit. The possibility of INTERACTION between tachyons and ordinary matter raises issues of cause and effect. That comes about because the temporal order of events as experienced by a tachyon can be the opposite of the order that we ascribe to the same events. As for now, there seems to be no evidence or theoretical need for the existence of tachyons. (*See* ACCELERATION; RELATIVITY, SPECIAL THEORY OF.)

temperature Temperature is a measure of the average random KINETIC ENERGY of a particle of MATTER. One of the most important properties of temperature is that if two systems possess the same temperature, they will be in thermal EQUILIBRIUM when they are brought into contact with each other. In other words, when two systems are brought into contact so that HEAT might flow from one to the other (but with no transfer of MATTER or WORK), their properties do not change if they are at the same temperature. Another of temperature's most important properties is that heat spontaneously flows from a region or body of higher temperature to one of lower temperature, never the opposite. (*See* THERMODYNAMICS.)

Absolute temperature, or thermodynamic temperature, is defined by the behavior of an ideal gas or, equivalently, by using thermodynamic considerations. For the Kelvin absolute scale, a property of water is utilized as well. The lowest conceivable temperature, called absolute zero, is assigned the value 0 kelvins (K) in SI UNITs, and the TRIPLE POINT of water is given the value 273.16 K. (Note that the degree sign "°" is not used with kelvins.) The FREEZING point of water, which is one-hundredth of a kelvin lower than the triple point, is then 273.15 K. In the commonly used Celsius temperature scale, the freezing point of water is taken as 0 degrees Celsius (°C). In this scale, the BOILING point of water is 100°C and absolute zero is –273.15°C. The magnitude of a kelvin is derived from the Celsius scale and equals one Celsius degree, which is one-hundredth of the temperature difference between water's freezing and boiling points. The conversion between the scales is given by:

$$T = T_c + 273.15$$

where T and T_c denote the same temperature in the Kelvin and Celsius scales, respectively. (*See* GAS, IDEAL.)

In the Fahrenheit scale, still used in the United States and only in the United States, the freezing point of water is set at 32 degrees Fahrenheit (°F), and a Fahrenheit degree is defined as 5/9 of a Celsius degree. As a result, the conversion between the Fahrenheit and Celsius scales is given by:

$$T_F = (9/5)T_C + 32°$$
$$T_C = (5/9)(T_F - 32°)$$

where T_F and T_C denote the same temperature in the Fahrenheit and Celsius scales, respectively.

Absolute zero temperature is not attainable, although it can be approached as closely as desired. As it is approached, the random kinetic energy of the particles of matter approaches a minimum. The minimum is not zero, due to a quantum effect involving the HEISENBERG UNCERTAINTY PRINCIPLE, whereby some residual energy, called zero-point energy, remains. For general temperatures, the average random energy of a matter particle is approximated by the value of kT, where T is the absolute temperature in kelvins (K) and k denotes the BOLTZMANN constant, whose value is $1.3806503 \times 10^{-23}$ joule per kelvin (J/K). (*See* QUANTUM PHYSICS; THERMODYNAMICS.)

See also KINETIC THEORY.

tension A stretching FORCE is referred to as a tension. A string under tension, for instance, has a pair of equal-magnitude, oppositely directed longitudinal forces acting on its ends, tending to stretch it.

See also ELASTICITY; STRAIN; STRESS; SURFACE TENSION.

terminal speed The SPEED at which all FORCEs acting on a body falling in a FLUID cancel out is called the terminal speed. There are three such forces, one downward and two upward. The downward force is the force of gravity. Its magnitude is mg, where m denotes the body's MASS in kilograms (kg) and g is the ACCELERATION due to gravity, with nominal value 9.8 m/s^2. One upward force is the buoyant force, whose magnitude equals $\rho_f Vg$. Here ρ_f denotes the DENSITY of the fluid in kilograms per cubic meter (kg/m^3), and V is the body's VOLUME in cubic meters (m^3). The other upward force is the viscous force. Its magnitude, F_v, increases as the body's speed through the fluid increases, and its direction is opposite to the body's VELOCITY, thus upward for a falling body. Neither of the first two forces depends on the body's speed. So the net speed-independent force on the body is the downward force whose magnitude, F_d, equals the difference between the magnitude of the force of gravity and the magnitude of the buoyant force:

$$F_d = (m - \rho_f V)g$$

The magnitude of the total force acting on the falling body, F, is then:

$$F = F_d - F_v = (m - \rho_f V)g - F_v$$

All forces are in newtons (N). (*See* BUOYANCY; GRAVITATION; VISCOSITY.)

When a body is released from rest in a fluid, the viscous force acting on it is momentarily zero. With a density greater than that of the fluid, the body accelerates downward. As its speed increases, so does the viscous force. In this way, the total force acting on the body decreases as the body's speed of fall increases. When the body's speed reaches the terminal speed, v_t, the viscous force completely cancels the speed-independent force, so:

$$F = F_d - F_v = 0$$

With no net force acting on it, the body continues its fall at constant speed, the terminal speed, v_t, according to ISAAC NEWTON's first law of MOTION. This is a stable situation. We just saw that when a body is falling at less that terminal speed, F_d is greater than F_v, and the body is accelerated to terminal speed. When the body is falling at a speed greater than the terminal speed—it might have been pushed to such a speed and then released, for instance—F_v is greater than F_d, and the body is decelerated to terminal speed. (*See* NEWTON'S LAWS.)

STOKES'S LAW gives the magnitude of the viscous force acting on a sphere moving in a fluid:

$$F_v = 6\pi r \eta v$$

Here r denotes the sphere's radius in meters (m), η is the viscosity of the fluid in newton·seconds per square meter (N·s/m^2), and v represents the speed of the sphere with respect to the fluid in meters per second (m/s). For a homogeneous sphere, the magnitude of the net speed–independent force acting on it, due to gravity and buoyancy, equals:

$$F_d = \frac{4\pi r^3}{3}(\rho - \rho_f)g$$

where $4\pi r^3/3$ is the volume of the sphere, and ρ denotes the density of the sphere in kilograms per cubic meter (kg/m^3). Equating F_v and F_d and solving for v, we obtain the terminal speed for a spherical falling body:

$$v_t = \frac{2r^2(\rho - \rho_f)g}{9\eta}$$

theoretical physics Physics is broadly divided into two endeavors: EXPERIMENTAL PHYSICS and theoretical physics. Experimental PHYSICISTs, or experimentalists, are in direct contact with the natural phenomena. They plan, set up, and perform the experiments. They collect data from their experiments, process the data, and publish the results so others can (hopefully) confirm them or (unfortunately) prove them wrong. Confirmed results then become grist for the mill of theoretical physicists, or theoreticians. They try to put the results in broader contexts, relate them to other experimental results, and gain an understanding of them in the sense of being able to explain them.

The domain of theoretical physics is the realm of ideas, concepts, patterns, relations, generalizations, abstractions, and unifications. Mathematics forms the essential tool for expressing all that in coherent ways and obtaining results that can be compared with experiment. One goal of theoretical physics is to discover patterns and order in natural phenomena and experimental results, which allow the prediction of new phenomena and results. These schemes are called laws. The STEFAN-BOLTZMANN LAW can serve as an example. This law states that the total POWER of electromagnetic RADIATION emitted by a BLACK BODY is proportional to the fourth power of the body's absolute TEMPERATURE. The law was discovered and formulated based on a pattern that was found among the results of many measurements of temperature and radiated power in situations that approximate a black body. It allows the prediction of radiated power for all temperatures, such as the temperatures of stars, even if the temperatures are not among those of the set of experimental data that formed the basis for the law. So the law forms a generalization from a particular set of

Theories: Not Mere Speculations!

"It's only a theory!" We have all heard this put-down, or maybe even declared it ourselves on occasion. It means "Don't take the idea too seriously; it's only a speculation," or, "It's merely a hypothesis." To give one example, this is the all-too-common creationist reaction to Charles Darwin's (1809–82) theory of biological evolution. "It's only a theory; it's not fact." However, Darwin's theory, together with such theories as ALBERT EINSTEIN's theories of relativity, are not mere speculations, not off-the-cuff hypotheses. They are well-founded, solidly confirmed explanations of very many facts and phenomena of nature. They offer consistent, unifying, and far-reaching explanations of the way things are and how things happen. They can predict new effects and thus be retested over and over again. (*See* RELATIVITY, GENERAL THEORY OF; RELATIVITY, SPECIAL THEORY OF.)

What, then, is the problem? Simply, the term *theory* is used in different ways in everyday conversation and in science. In the language of scientists, a theory is an explanation, no more and no less. It might indeed be a hypothesis, such as a GRAND UNIFIED THEORY (GUT) of ELEMENTARY PARTICLES and their INTERACTIONS. It might even form a speculation, such as any of various ideas about why the expansion of the UNIVERSE seems to be accelerating. Many theories do start their lives as plausible, or even implausible, hypotheses. But a theory might offer a unifying explanation of a broad range of phenomena and might, furthermore, predict new phenomena and thus be testable. And such a theory might continually pass test after test after test. It then becomes accepted as *the* theory. Such in fact are Darwin's and Einstein's theories, each in its field of application.

But whether a scientific theory is speculative, hypothetical, promising, competing, or reigning, it always forms an attempt to explain. It always proposes a way of understanding what is known. Moreover, to be taken seriously, it should also predict what is not yet known. That makes it falsifiable and thus testable. Correct predictions confirm a theory. Sufficiently many such successes can make it accepted. A single failure, however, causes it to crash. Then a revision might be called for, or possibly a new theory altogether.

Do successful theories live forever? Not ordinarily, at least not so far. And at least not in the field of physics. While the basic ideas of Darwin's theory of evolution, for instance, might possibly maintain their validity for all time, that does not seem to be the case in physics. What has been happening is that as theories get tested under increasingly extreme conditions, they invariably reveal their limitations. They then become subsumed under more general theories of broader

(continues)

Theories: Not Mere Speculations! *(continued)*

applicability. ISAAC NEWTON's laws of MOTION proved inaccurate for SPEEDs that are a large fraction of the SPEED OF LIGHT. Einstein's special theory of relativity fixed that problem. Newton's law of GRAVITATION could not correctly deal with strong gravitational FIELDs and high speeds. Einstein's general theory of relativity gave the correct treatment and, moreover, extended the limits of validity of the special theory. As for limitations on the general theory of relativity, it does seem that the theory is not compatible with quantum theory, itself another very successful physics theory. It is commonly assumed that both Einstein's general theory of relativity and quantum theory will eventually be subsumed under a very general theory, which, it is hoped, will offer deep insight into SPACE, TIME, and the quantum aspect of nature. (*See* NEWTON'S LAWS; QUANTUM PHYSICS.)

Here is an archetypal example of a physics theory and how it develops. JOHANNES KEPLER carefully studied Tycho Brahe's (1546–1601) observations of the motions of the Sun and the planets that were then known—Mercury, Venus, Mars, Jupiter, and Saturn. As a result, he came up with three laws that correctly describe the motions of the planets, including that of planet Earth, around the Sun. KEPLER'S LAWS, however, do not form a theory, since they do not, and were not intended to, explain anything. What they do, though, is show that the motions of the six planets are not independent. The motions present particular cases of certain general laws of planetary motion. Kepler's laws have proven valid for the additional planets that were discovered—Uranus, Neptune, and Pluto—and also for the systems of moons around the multimoon planets.

Being an excellent physicist, Newton did not just accept the laws that Kepler had discovered, but inquired into their cause. *Why does nature behave in just this way?* He looked for an explanation, a theory, for Kepler's laws. And he found a theory, in the form of his own three laws of motion and law of gravitation. This theory not only explains Kepler's laws in the sky, but also a host of other phenomena, including, very importantly and usefully, mechanical effects on Earth. His theory was so successful that it lasted some 200 years until its limitations became apparent. It is still the theory we use for everyday life, such as for designing buildings and vehicles and for athletic activities, including baseball and pole vaulting. But in principle, Newton's theory has been subsumed under Einstein's theories and quantum theory.

Of course, we cannot change the way *theory* is commonly used in ordinary conversation. But we *can* be cognizant of the very different meaning of the term in science. That way we can avoid the absurdity of labeling as mere hypotheses such monumental theories of science as Darwin's and Einstein's.

measured values of power at certain temperatures to the values of power at all temperatures. (*See* ELECTROMAGNETISM; LAWS OF NATURE.)

Beyond discovering laws, however, a more ambitious and fundamental goal of theoreticians is to devise explanations for laws. Such an explanation is termed a theory. A theory provides a fundamental framework from which the validity of a law follows as a logical (i.e., mathematical) consequence. In the case of the Stefan-Boltzmann law, for instance, MAX KARL ERNST LUDWIG PLANCK devised a framework for the exchange of energy between matter and the electromagnetic FIELD that led to a formula for the black-body radiation SPECTRUM. The Stefan-Boltzmann law follows immediately from this formula. Planck's theory contributed to the development of QUANTUM MECHANICS and the concept of the PHOTON and eventually to QUANTUM ELECTRODYNAMICS (QED), which is one of the most successful theories of physics. QED offers a wide range of theoretical results that are exceedingly precise (they include many significant digits) and accurate (they compare well with experimental measurements). One example of such a result is the value of the magnetic moment of the ELECTRON. (*See* MOMENT, MAGNETIC DIPOLE.)

In different fields at different times, the physics might be experiment driven or theory driven. In the former case, the experimentalists of the field are taking the initiative by examining interesting (to them), unexplained phenomena. The theoreticians are, so to speak, running along behind them, trying to make sense of the experimental observations. In theory-driven physics, on the other hand, the theoreticians are in the lead, devising theories of nature that not only explain what is already known, but predict new phenomena. The experimentalists' job is then to check those theories against the real world, to disprove the false ones and to confirm the more successful ones.

In the field of ELEMENTARY PARTICLE physics, there is a specialized intermediate endeavor of phenomenology. Phenomenologists are positioned somewhere between the experimentalists and the theoreticians. They study the experimental results and look for order and regularities. Then the theoreticians take over from there.

In most fields of physics, a physicist is normally either a theoretician or an experimentalist, although there are some exceptions, when people are involved in both endeavors. For their doctoral degree, physicists usually do either a theoretical or an experimental project for their dissertation, and then most often remain either a theoretician or an experimentalist, respectively, for the rest of their professional lives.

thermal capacity *See* HEAT CAPACITY.

thermal efficiency A heat engine is a device that converts ENERGY in the form of HEAT into WORK, from which other forms of energy, such as mechanical or electric energy, might be derived. The thermal efficiency, or simply efficiency, of a heat engine equals the ratio of the net work output during a single cycle of operation to the heat input during the cycle. It is a dimensionless quantity. By the first law of THERMODYNAMICS, which is the WORK-ENERGY THEOREM, or the law of conservation of energy, the efficiency of a heat engine cannot be greater than 100 percent. In other words, you cannot get out more work or energy than you put in. According to the second law of thermodynamics, moreover, the efficiency of a heat engine cannot even reach 100 percent. Not all the invested heat can be converted to work. That happens because a heat engine must discharge some heat during each cycle. Again using conservation of energy, the heat input equals the sum of the work output and the discharged waste heat. So the heat input is always greater than the work output, and the efficiency is always less than 100 percent. (*See* CONSERVATION LAW; DIMENSION; ELECTRICITY; MECHANICS.)

Let us denote, for a single cycle of operation of a heat engine, the heat input by Q_H, the discharged heat by Q_L, and the work performed by W, all in the same units. Since at the end of a cycle the engine returns to its initial state, its internal energy has not changed, and by the work-energy theorem the net work output for the cycle is $W = Q_H - Q_L$, and the thermal efficiency, e, is given by:

$$e = \frac{W}{Q_H} = \frac{Q_H - Q_L}{Q_H} = 1 - \frac{Q_L}{Q_H}$$

The theoretical upper limit for the efficiency of any heat engine that takes in heat at a higher TEMPERATURE,

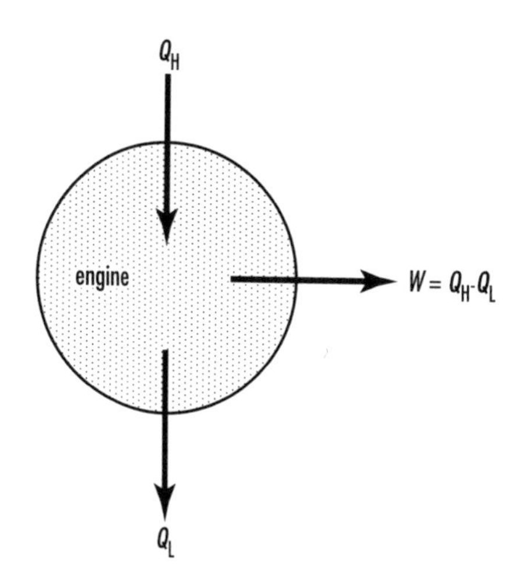

The thermal efficiency of a heat engine equals the ratio of the net work output during a single cycle of operation, W, to the heat input during the cycle, Q_H. The net work equals $W = Q_H - Q_L$, where Q_L, denotes the heat discharged by the engine during a cycle. The thermal efficiency is then $e = W/Q_H = 1 - Q_L/Q_H$.

T_H, and discharges heat at a lower one, T_L, is set by the efficiency of an ideal, reversible engine operating in a CARNOT CYCLE, called a Carnot engine. For absolute temperatures, the efficiency of this engine, e_C, is given by:

$$e_C = 1 - \frac{T_L}{T_H}$$

Efficiency is increased in general by raising the higher temperature and lowering the lower one.

thermal expansion Thermal expansion is the expansion that most materials undergo as their TEMPERATURE rises. Consider an unconfined piece of SOLID material. When its temperature changes by a small amount from T_1 to T_2 and any of its linear DIMENSIONs increases concomitantly from L_1 to L_2, the relative change in length is approximately proportional to the temperature change:

$$\frac{L_2 - L_1}{L_1} = \alpha(T_2 - T_1)$$

Here L_1 and L_2 are both in the same unit of length; T_1 and T_2 are both either in kelvins (K) or in degrees Celsius (°C); and α, in inverse kelvins (K^{-1}), denotes the coefficient of linear expansion of the material. This coefficient is generally temperature dependent. As the solid expands, any AREA on its surface expands also,

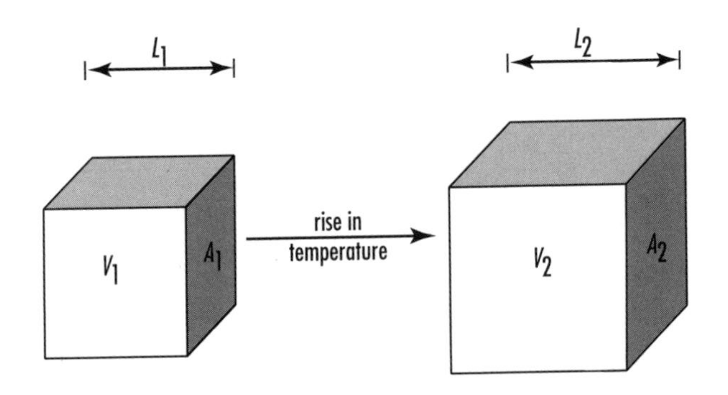

Most materials expand when their temperature rises. For a solid body, every length, area, and volume increases. The figure shows this, with the length of one edge of the body expanding from L_1 to L_2, the area of one of the body's surfaces increasing from A_1 to A_2, and the body's volume expanding from V_1 to V_2. In every case, the relative change is proportional to the change in temperature (for small temperature changes).

from A_1 to A_2 (both in the same unit of area). Then, to a good approximation, the relative change in area is:

$$\frac{A_2 - A_1}{A_1} = 2\alpha(T_2 - T_1)$$

The coefficient 2α is the coefficient of area expansion. Similarly, any VOLUME of the material undergoes expansion as well, from V_1 to V_2 (both in the same unit of volume), where:

$$\frac{V_2 - V_1}{V_1} = 3\alpha(T_2 - T_1)$$

So the coefficient 3α is the coefficient of volume expansion.

For LIQUIDs and GASes, it is only volume expansion that has significance, and the coefficient of volume expansion is denoted by β in the expression:

$$\frac{V_2 - V_1}{V_1} = \beta(T_2 - T_1)$$

thermionic emission The spontaneous emission of ELECTRONs from the surface of a sufficiently hot METAL is called thermionic emission. The emission occurs when a significant fraction of the free electrons in the metal possess sufficient thermal ENERGY to overcome the FORCEs they encounter at the metal's surface, which normally confine them to the VOLUME of the metal. The thermal energy of those electrons must be greater than the work function of the metal. As the temperature of the metal increases, both the rate of emission and the average energy of an emitted electron increase as well. Thermionic emission from a hot cathode serves as the source of electrons for the electron beam in television picture tubes, computer monitor tubes, and oscilloscope tubes, as examples. Such tubes are called cathode ray tubes, or CRTs. Once emitted, the electrons are accelerated by an electric FIELD, associated with an applied VOLTAGE, and collimated to form a beam. The beam is shaped, focused, and directed by magnetic fields. When the beam strikes the surface of the tube that serves as the screen, LIGHT is emitted at the region of incidence through the process of FLUORESCENCE by a suitable material coating the inside surface of the screen. Additional methods for causing electron emission from metals include FIELD EMISSION and the PHOTOELECTRIC EFFECT. (*See* CONDUCTION; CONDUCTION BAND; ELECTRICITY; HEAT; MAGNETISM; TEMPERATURE.)

thermocouple A pair of wires of different metals that have both their ends welded together serves as a thermocouple. When the two junctions are at different TEMPERATUREs, an ELECTROMOTIVE FORCE (emf) is produced in the circuit and an electric CURRENT flows. The emf and current increase with increase of temperature difference. This effect is named the Seebeck effect, for the German physicist Thomas Seebeck (1770–1831). By introducing a GALVANOMETER or other suitable electric or electronic device into the circuit, the current or emf can be measured. When such a device is calibrated, it can serve to measure temperature differences between the two junctions. By maintaining one of the junctions at a known temperature, the device becomes a THERMOMETER. Since the thermocouple wires can be extremely fine and the HEAT CAPACITY of the probe junction consequently very small, a thermocouple can be suitable for

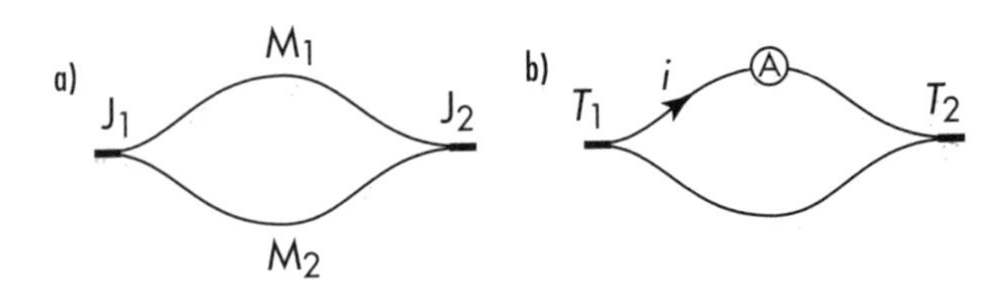

(a) A thermocouple consists of a pair of wires of different metals M_1 and M_2 welded together at their ends, forming two junctions J_1 and J_2. (b) When the junctions are at different temperatures T_1 and T_2, an electric current *i* flows in the circuit and can be measured by an ammeter A.

measuring the TEMPERATURE of very small objects. (*See* ELECTRICITY.)

An inverse effect is the Peltier effect, named for the French physicist Jean Charles Peltier (1785–1845). This is the creation of a temperature difference between the two junctions of a thermocouple as a result of an electric current flowing in the circuit.

thermodynamics Thermodynamics is the field of physics that deals with the conversion of ENERGY from one form to another, while taking TEMPERATURE into account, and derives relationships among macroscopic properties of MATTER. Thermodynamics is founded on four laws, which are based on and form generalizations of our experience with nature.

Zeroth Law

The zeroth law of thermodynamics. This law states that physical systems may be endowed with the property of temperature, which is described by a numerical quantity. When two systems possess the same temperature (i.e., have the same value for this quantity), they will then be in thermal EQUILIBRIUM while in contact with each other. In other words, when two systems are brought into contact so that HEAT might flow from one to the other (but with no transfer of matter or WORK), their properties do not change if they are at the same temperature. In thermodynamics, temperature is usually denoted T and is expressed on an absolute scale in the SI UNIT of the kelvin (K). (The zeroth law is not commonly included in introductory presentations of thermodynamics.)

The notion of temperature is derived from the fact that if systems A and B are in thermal equilibrium when in contact and so are systems B and C, then invariably systems A and C are also found to be in thermal equilibrium when brought into contact. That suggests the existence of a property that is a physical quantity and has the same value for systems that are in thermal equilibrium, hence temperature.

First Law

The first law of thermodynamics. This law is equivalent to the WORK-ENERGY THEOREM, or the law of conservation of energy, while recognizing that systems possess internal energy. The law states that the increase in the internal energy of a system equals the sum of work performed on the system and heat flowing into the system. In symbols:

$$\Delta U = W + Q$$

where ΔU denotes the increase of the system's internal energy, W is the work done on the system, and Q is the HEAT flowing into the system. All quantities are in the same unit of energy, which in the SI is the joule (J). (*See* CONSERVATION LAW).

Note that while internal energy is a property of a system, neither work nor heat flow is a property. The latter two are modes of energy transfer. Work is the transfer of orderly energy. An external FORCE compressing a system is performing work on it. Heat flow is the transfer of disorderly energy, the random motion of the particle constituents of matter. Heat flows as a result of a temperature difference between a system and its surroundings. It follows, then, that for an isolated system, the internal energy must remain constant. That is a constraint on processes occurring in isolated systems.

Second Law

The second law of thermodynamics. This law goes beyond the first law and adds a further constraint. Not all processes that are allowed by the first law will indeed occur. They may occur if they are consistent with the second law and will never occur if they violate it. One formulation of the second law is that heat does not flow spontaneously from a cooler body to a warmer one (i.e., from lower to higher temperature).

Another equivalent formulation is in terms of a heat engine, which is any device that converts thermal energy, heat, into mechanical energy, from which other forms of energy might be derived, such as electric energy. It does so by taking in heat from an external source and performing work on its surroundings. In these terms, the second law can be stated thus: a cyclic heat engine must discharge heat at a lower temperature than it takes heat in. In other words, not all the heat input is convertible to work. The THERMAL EFFICIENCY of a cyclic heat engine is less than 100 percent. (*See* ELECTRICITY; MECHANICS.)

In order to quantify these notions, a physical quantity, called ENTROPY, is introduced as a property of physical systems. The entropy of a nonisolated system can be increased or decreased by the flow of heat into it or from it, respectively. The increase in entropy, ΔS, in joules per kelvin (J/K), due to the amount of heat, ΔQ, in joules (J) flowing reversibly into the system at absolute TEMPERATURE T in kelvins (K), is given by:

$$\Delta S = \Delta Q/T$$

The entropy of a system can change even when the system is isolated, as a result of internal processes. (*See* REVERSIBLE PROCESS.)

The second law of thermodynamics is formulated in terms of entropy as: the entropy of an isolated system cannot decrease over time. Spontaneous, irreversible evolution of an isolated system is always accompanied by an increase of the system's total entropy. That might involve a decrease of entropy in part of the system, but it will be more than compensated for by an increase in the rest of the system.

In connection with the second law, the CARNOT CYCLE describes the operation of an ideal, reversible heat engine that takes in heat at higher temperature T_H and discharges heat at lower temperature T_L. In a single cycle of operation, heat Q_H is absorbed at the higher temperature and heat Q_L discharged at the lower. For absolute temperature, or thermodynamic temperature, the relation:

$$T_L/T_H = Q_L/Q_H$$

holds for a Carnot engine. The thermal efficiency of a Carnot engine, e_C, sets a theoretical upper limit for the efficiency of any heat engine operating between the same two temperatures. It is given by:

$$e_C = 1 - T_L/T_H$$

This shows absolute zero temperature as the discharge temperature of a Carnot engine that takes in heat at a finite temperature and possesses perfect efficiency ($e_C = 1$).

Third Law

The third law of thermodynamics. This law has to do with absolute zero temperature. One statement of it is: it is impossible to reach absolute zero temperature through any process in a finite number of steps. Equivalently, a Carnot engine cannot have perfect efficiency. Another formulation is that at absolute zero temperature, all substances in a pure, perfect crystalline state possess zero entropy. This law sets the zero point for entropy, from which the entropy of any substance can be measured and calculated by using the above formula for entropy change due to reversible heat flow. (*See* CRYSTAL; STATE OF MATTER.)

Perpetual Motion

Various and diverse devices are occasionally proposed by inventors as sources of free or almost-free energy. Such devices seem too good to be true, and indeed, they *are* too good to be true. There are those devices that are claimed to produce more energy than needs to be invested or that produce energy with no need for investment. Their operation is termed perpetual motion of the first kind, which means they violate the first law of thermodynamics (i.e., the law of conservation of energy). Since, at least as far as energy is concerned, you cannot get something for nothing ("there's no free lunch"), such devices cannot work as a matter of principle, and there is no need to further analyze the details of their operation.

Then there are those devices that obey conservation of energy, but violate the second law of thermodynamics. Their operation is called *perpetual motion of the second kind.* They do not create energy—such as mechanical or electric—from nothing. But they are claimed to be able to convert thermal energy, heat, into such orderly forms of energy at 100 percent efficiency, or at least at efficiencies that exceed the allowable according to the second law. As an example, there exists a tremendous amount of internal energy in the oceans. Some of it can be converted to orderly energy, as long as heat is discharged at a lower temperature. But that is neither a simple nor an inexpensive procedure, since warm and cold ocean water do not coexist sufficiently near each other and at a sufficient temperature difference to make the operation possible without considerable investment. A perpetual motion device of the second kind might be purported to dispense altogether with the discharge and simply take thermal energy from the ocean and convert all of it to electricity. Such claims violate the second law of thermodynamics and are thus invalid in principle, so again, it is not necessary to look further into how the devices operate.

Statistical Mechanics

The properties of bulk matter that thermodynamics deals with—temperature, pressure, internal energy, entropy, etc.—are macroscopic manifestations of the situation that exists at the level of constituent particles, at the microscopic level. So thermodynamics should be, and is, translatable into microscopic terms. The field of physics that fulfills this function is STATISTICAL MECHANICS, which studies the statistical properties of physical systems containing large numbers of constituents and relates them to the system's macroscopic properties.

See also KINETIC THEORY.

thermometer Any apparatus for measuring TEMPERATURE is known as a thermometer. Its operation is based on the temperature dependence of some physical quantity. The THERMAL EXPANSION of a substance can be used for that purpose, for instance. The ordinary mercury or alcohol thermometer is based on this effect. Another application of thermal expansion to thermometry is the bimetallic strip, which consists of two strips of different METALS, possessing different linear expansion coefficients, attached together along their length. As the temperature varies, one of the metals contracts or expands more than the other, causing the strip to bend in a direction and by an amount that depend on temperature. The ELECTROMOTIVE FORCE (emf) or electric CURRENT in a THERMOCOUPLE circuit is another example of a temperature-dependent quantity. Here the quantity depends on the temperature difference of the two thermocouple junctions. When one of the junctions is held at a known temperature, the other serves as a thermometer probe. An additional temperature quantity that is utilized for constructing a thermometer is electric RESISTIVITY. For almost all SOLIDs, the resistivity increases with increasing temperature. By what amounts to a measurement of the RESISTANCE of a piece of suitable material, usually a metal, its temperature, and thus the temperature of whatever it is in thermal contact with, can be determined. For very high temperatures, such that any material thermometer would melt or be otherwise damaged, a PYROMETER is used. This device measures the temperature of a system by the SPECTRUM of the electromagnetic RADIATION the system emits. Commonly, it compares the color of the emitted LIGHT with a color scale. Thermometers need to be calibrated in order to determine the temperature that corresponds to each value of the physical quantity that underlies their operation. (*See* ELECTRICITY.)

thin film A thin layer of a substance, which might have a width of as little as a single ATOM, is termed a thin film. The physics of such systems can differ from the usual due to their effectively two-dimensional character. In metallic thin films, for instance, the conduction ELECTRONs are free in only two dimensions rather than in the more usual three. Such thin films are important in the technology of integrated circuits. Transparent thin films produce colorful INTERFERENCE effects in LIGHT that is reflected from them. The interference occurs between the light WAVEs reflected from the front surface and those reflected from the rear. Examples are oil floating on water and soap bubbles. (*See* CONDUCTION; CONDUCTION BAND; DIMENSION; METAL.)

time The concept of time is one of the most difficult concepts in physics, rivaled in the depth of its philosophical ramifications perhaps only by the concept of SPACE. For the purpose of PHYSICS, however, much of that difficulty must be ignored. In order to make progress in physics, PHYSICISTs cannot, in general, allow themselves to become mired in philosophical detail. So they make do with whatever general understanding is sufficient for carrying out the goals of physics.

One of the dominant characteristics of nature, one that we possess powerful intuitive awareness of, is *becoming*. Things do not stay the same; changes occur; situations evolve. That is what is meant by *becoming.* In physics, time can be described, even defined, as the DIMENSION of *becoming,* where *dimension* is used in the sense of the possibility of assigning a measure. Time's measure is the answer to the question "When?" for an event. The answer to this question always takes the form of a single number, composed of, say, the date and "time" (in the sense of clock reading) of the event. Thus, time involves an ordering of events into earlier and later, just as numbers are ordered into smaller and larger. Also, time involves a duration, an interval, between every pair of events, which we call the elapsed time between the earlier event and the later one. Since a single number is required to answer the question "When?" time is said to be one-dimensional.

Avoiding the technicalities, though, time is generally viewed by physicists as the backdrop for becoming, for change and evolution. In physics, time appears as a PARAMETER, or independent variable, usually denoted by t. The SI UNIT of time is the second (s). Physical quantities are represented as functions of t, which give the value of their corresponding quantity for any time. As for becoming, however, as powerful as our intuition of it is, that aspect of nature is *not* comprehended by physics. There is nothing in physics that requires t to take any particular value or to continually increase in value. This does not mean that physics will not eventually achieve a better understanding of the nature of time. But for now, the nature of time remains largely beyond the grasp of physics.

ALBERT EINSTEIN's special and general theories of relativity reveal that time and space are not independent of each other and that fundamentally they are best considered together as four-dimensional SPACE-TIME. (*See* RELATIVITY, GENERAL THEORY OF; RELATIVITY, SPECIAL THEORY OF.)

time constant When the rate of decrease in TIME of any physical quantity, N, is proportional to the value of that quantity, the time dependence of the quantity takes the form of exponential decay:

$$N = N_0 e^{-\gamma t}$$

Here N_0 is the value of N at time zero, t denotes the time, and γ is the decay constant, whose UNIT is the inverse of the unit used for time. Note the time derivative of N:

$$dN/dt = -\gamma N_0 e^{-\gamma t} = -\gamma N$$

So the rate of decrease of N in time, which is $-dN/dt$, is indeed proportional to N, with the decay constant, γ, as the proportionality coefficient. (*See* DECAY.)

Examples of exponential decay include RADIOACTIVITY and the damped HARMONIC OSCILLATOR. For radioactivity, N might represent the amount of undecayed radioactive material or it might represent the INTENSITY of radioactive RADIATION. Alternatively, N might represent the amplitude of a damped harmonic oscillator.

The expression for exponential decay can be given the equivalent form:

$$N = N_0 e^{-t/\tau}$$

where τ is the time constant and is the reciprocal of the decay constant. Its unit is the same as the unit used for time. The time constant is the time interval required for N to decrease to $1/e$ of its value (i.e., for its value to decrease by a factor of approximately 2.718). Note that after a time interval of at least five time constants, N is reduced to less than 1 percent of its initial value. For radioactive decay, it is common, instead of the time constant, to use the HALF-LIFE, which is the time interval in which half the radioactive material decays, or in which the intensity of radiation decreases to half its value. The half-life equals $\tau \ln 2 \approx 0.693\tau$. After at least six half-lives, less than 1 percent of the initial amount of radioactive material is still undecayed.

time dilation The relativistic effect that a moving clock is measured as ticking more slowly than the same clock at rest is called time dilation. More generally, let T_0 denote the TIME interval between two events occurring at the same location with respect to some REFERENCE FRAME. The events might be consecutive ticks of a clock at rest or consecutive heartbeats of a person at rest, as examples. The time interval, T_0, that is measured in this reference frame for the two events is called their proper time interval. Now, consider an observer moving at constant SPEED and in a straight line relative to this reference frame. With respect to the observer's reference frame, the two events do not occur at the same location. The ticking clock or beating heart is moving in the observer's reference frame, so the two ticks or two heartbeats do not happen at the same place. The observer measures the time interval between the two events and finds the value T. This value is always greater than the proper time interval, T_0. The relation between them is:

$$T = \frac{T_0}{\sqrt{1 - v^2/c^2}}$$

Here v denotes the relative speed of the two reference frames, which is the speed of the observer relative to the clock/person as well as the speed of the clock/person relative to the observer, and c is the SPEED OF LIGHT, whose value is 2.99792458×10^8 meters per second (m/s). The speed, v, is in meters per second (m/s), and the time intervals are in the same UNIT of time. (*See* RELATIVITY, SPECIAL THEORY OF.)

Note that at everyday speeds, time dilation is undetectable, while at ten percent of the speed of light, the dilation is about a half percent. In order to achieve 100 percent dilation (increase by a factor of two), the speed must reach around 87 percent of the speed of light. As the speed approaches the limit of the speed of light, the measured time interval dilates toward infinity. However, no object or observer can achieve the speed of light, so no time interval can dilate to infinity. The formula for time dilation follows from the LORENTZ TRANSFORMATION.

The effect of time dilation is seen in the considerably extended lifetimes of rapidly moving unstable ELEMENTARY PARTICLES, as compared with their lifetimes at rest. Such situations occur in connection with COSMIC RAYS and particle accelerators. Time dilation plays a role in the DOPPLER EFFECT for LIGHT or any other electromagnetic RADIATION. It is the cause of the relativistic Doppler effect, or relativistic red shift. (*See*

ACCELERATOR, PARTICLE; ELECTROMAGNETISM; HALF-LIFE; RED SHIFT.)

Time dilation is a mutual effect; each of two observers in relative MOTION measures the time intervals of the other's clock as dilated. That leads to what is known as the TWIN PARADOX.

torque Also called the moment of force, torque is a VECTOR quantity that expresses the rotating, or twisting, capability of a FORCE with respect to some point. Let force **F** act on a particle or body, and denote by **r** the position vector of the point of application of the force with respect to some reference point O. So **r** is the spatial vector extending from reference point O to the point of application of the force. Then the torque of the force with respect to O, $\boldsymbol{\tau}_O$, is given by the vector product:

$$\boldsymbol{\tau}_O = \mathbf{r} \times \mathbf{F}$$

Here **F** is in newtons (N), **r** is in meters (m), and torque is in newton·meters (N·m). Note that the SI UNIT of ENERGY or WORK, the joule (J), is equivalent to newton·meter. But torque and energy are very different quantities, and the unit of torque, the newton·meter, is *not* called a joule. (*See* ROTATION.)

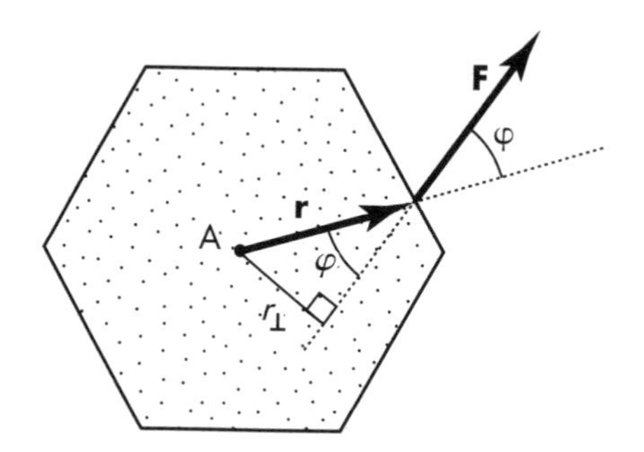

The torque of a force that is perpendicular to a rotation axis. The rotation axis A of the body is perpendicular to the plane of the figure. The force F acting on the body lies in the plane. r denotes the position vector of the force's point of application with respect to the point of intersection of the axis and the figure plane. The smaller angle (less than 180°) between r and F is denoted φ. The force's moment arm, the perpendicular distance between the force's line of action and the axis, is $r\perp$. The magnitude of the force's torque with respect to the axis equals $\tau_0 = r\perp F = rF \sin \varphi$, where r and F denote the magnitudes of r and F, respectively. The torque's direction is perpendicular to the figure plane and, in this example, points out of the page.

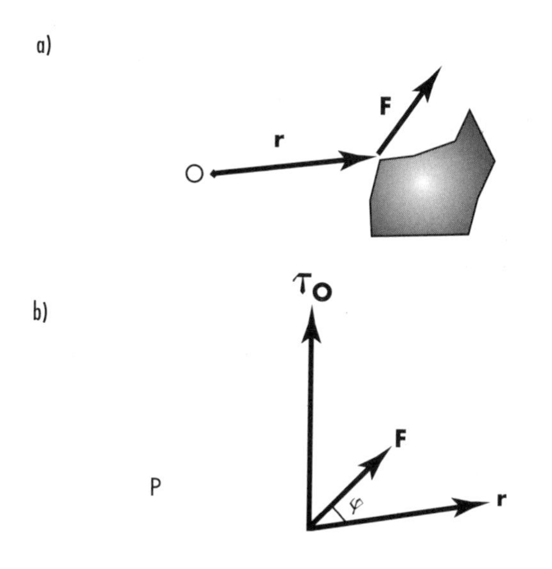

The torque of a force. (a) Force F acts on a body. r denotes the position vector of the force's point of application with respect to reference point O. The torque of the force with respect to O is $\tau_0 = \mathbf{r} \times \mathbf{F}$. (b) The relation among vectors r, F, and τ_0. Plane P contains r and F. Vector τ_0 is perpendicular to P in the sense shown. The smaller angle (less than 180°) between r and F is denoted φ. The magnitude of the torque equals $rF \sin \varphi$, where r and F denote the magnitudes of r and F, respectively.

The magnitude of a force's torque, τ_O, is given by:

$$\tau_O = rF \sin \varphi$$

where r and F denote the magnitudes of **r** and **F**, respectively, and φ is the smaller angle (less than 180°) between **r** and **F**. The direction of the vector τ_O is perpendicular to the plane in which the two vectors **r** and **F** lie. Its sense is such that if the base of **F** is attached to the base of **r** and the base of $\boldsymbol{\tau}_O$ joined to those bases, then from the vantage point of the tip of $\boldsymbol{\tau}_O$, looking back at the plane of **r** and **F**, the rotation from **r** to **F** through the smaller angle between them (φ, the angle less than 180°) is counterclockwise. The torque $\boldsymbol{\tau}_O$ measures the capacity of the force **F** to cause a rotation about an axis passing through point O in the direction of $\boldsymbol{\tau}_O$.

In many practical cases, a body is constrained to rotate about an axis that is fixed in a given direction, and the force is perpendicular to that direction, such as in the application of a wrench to a bolt or a screwdriver to a screw. Then the magnitude of the force's torque for rotation about that axis equals the product of the magnitude of the force and the perpendicular distance (which is also the shortest distance), $r_\perp$, between the force's line of action and the axis, called the moment arm of the force:

$$\tau_O = r_\perp F$$

This can be expressed in terms of the position vector of the force's point of application with respect to the axis, i.e., taking reference point O to lie on the axis, as near to the application point as possible. Then the position vector, **r**, is perpendicular to the axis and its magnitude, r, is the perpendicular distance of the application point from the axis. The torque's magnitude then equals:

$$\tau_O = rF \sin \varphi$$

where, as defined above, φ is the smaller angle (less than 180°) between **r** and **F**. Its direction is parallel to the axis and in the sense indicated by the right-hand thumb when the right hand grasps the axis with the fingers curving in the sense of rotation that the force tends to impart to the body.

See also ANGULAR MOMENTUM; MOMENT OF INERTIA.

transducer Any device that converts one form of signal into another or one form of ENERGY into another is termed a transducer. A microphone, for example, converts an acoustic signal into an electric one. It converts oscillating air PRESSURE into oscillating VOLTAGE or electric CURRENT. A loudspeaker, on the other hand, does just the opposite and transforms an electric signal into a sound WAVE. Transducers that convert a physical quantity into a voltage or current are useful as measuring devices. A THERMOCOUPLE, for instance, converts a TEMPERATURE difference into a current by means of the Seebeck effect and can thus serve as a THERMOMETER. Or, an electric STRAIN gauge converts an object's deformation to a voltage by means of the piezoelectric effect. (*See* ACOUSTICS; ELECTRICITY; OSCILLATION; PIEZOELECTRICITY; SOUND.)

transistor A transistor is an electronic component made of SEMICONDUCTOR. A transistor operates as an electric CURRENT regulator: small changes in one current through the transistor, the base current, bring about large changes in another current, the collector current. Most commonly, transistors serve as amplifiers or switches. The simplest transistors are constructed of three layers of semiconductor, with a conducting lead connected to each layer. The layers are termed emitter, base, and collector, with the very thin base separating the other two layers. The layers are doped so that the base is either p-type or n-type, and the emitter and collector both of the other type, either n-type or p-type respectively. Accordingly, a transistor is either an npn

Three transistors of various kinds. Each possesses three leads: emitter, base, and collector. The metal case of the large transistor serves both as collector lead and to help dissipate heat. To set the scale of the figure, the length of the transistor on the right, including its leads, is 1.8 centimeter (0.75 inch). ***(Photo by Frost-Rosen)***

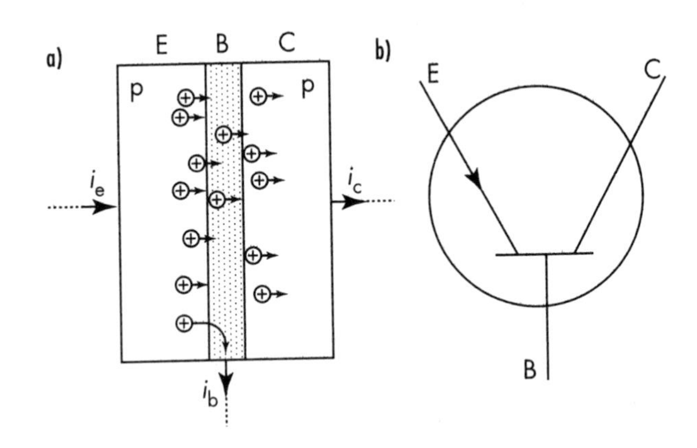

(a) The structure of a pnp transistor, consisting of a very thin layer of n-type semiconductor sandwiched between two layers of p-type semiconductor. A lead is attached to each of the emitter E, the base B, and the collector C. Holes, denoted by ⊕, carry the electric current in a pnp transistor. The emitter current i_e, base current i_b, and collector current i_c are indicated. They obey the relation $i_e = i_b + i_c$. (b) The symbol that is used for a pnp transistor in a circuit diagram.

or a pnp transistor. In an npn transistor, the current carriers are ELECTRONs, while HOLEs carry the current in a pnp transistor. (*See* CONDUCTION; ELECTRICITY.)

triple point The unique set of conditions under which a substance's SOLID, LIQUID, and GAS PHASES are in EQUILIBRIUM and can thus coexist is known as the substance's triple point. The triple point of water, for example, has a TEMPERATURE of 273.16 kelvins (K) and a PRESSURE of 0.610×10^3 pascals (Pa). At that temperature and pressure, ice, liquid water, and water vapor coexist in equilibrium, with no tendency for either of them to transform to another. Water's triple point serves for the definition of the absolute temperature scale.

tunneling Tunneling is a quantum effect in which systems can achieve states that are separated from their initial states by an ENERGY barrier. In particular, particles, such as ELECTRONs and ALPHA PARTICLEs, can pass from one energetically allowed region to another, although, according to CLASSICAL PHYSICS, to do so they would need more energy than they possess. The effect can be likened to a car that does not have sufficient fuel to go over a mountain, yet there is a finite possibility of finding the car on the other side, as if the car passed through a tunnel (hence the name of the effect). Tunneling is important for the understanding of various physical phenomena, such as ALPHA DECAY, FIELD EMISSION, and the JOSEPHSON EFFECT, and for the operation of devices, including scanning electron microscopes and certain SEMICONDUCTOR electronic components. (*See* ELECTRON MICROSCOPY; QUANTUM MECHANICS; QUANTUM PHYSICS.)

turbulence The situation in FLUID flow where the flow VELOCITY and fluid PRESSURE at every point vary wildly in time is called turbulence. This is in contrast to laminar flow, which is a steady flow in which the velocity and pressure remain constant in time, or vary only slowly, throughout the fluid. The REYNOLDS NUMBER is a PARAMETER that characterizes the flow of incompressible, viscous fluids. It is a dimensionless quantity defined as $\rho v l/\eta$, where ρ denotes the fluid's DENSITY in kilograms per cubic meter (kg/m^3), v is the flow SPEED in meters per second (m/s), l is a length in meters (m) that is characteristic of the geometry of the system, and

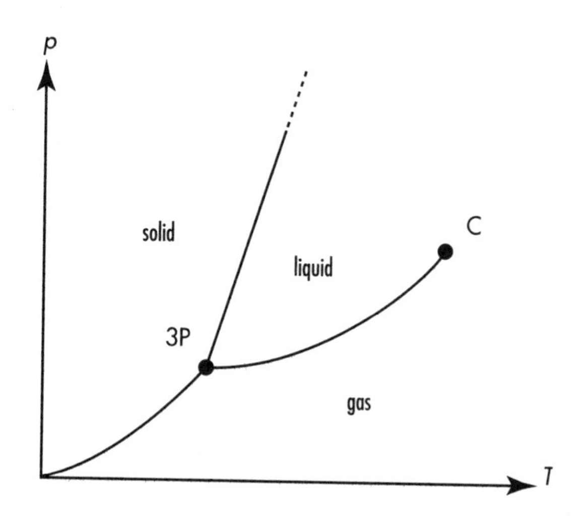

The triple point of a substance is the unique set of conditions under which its solid, liquid, and gas phases are in equilibrium and can thus coexist. It is denoted 3P in this typical *p-T* (pressure-temperature) phase diagram. Regions of the diagram are labeled according to the state of the substance. The critical point is denoted C.

Turbulence in air flow. Smoke carried by the upward flow of warm air from the wick of a freshly extinguished candle shows initial laminar flow (lower left) becoming turbulent (top). *(Photo by Frost-Rosen)*

η represents the fluid's VISCOSITY in newton·seconds per square meter (N·s/m^2). When the value of the Reynolds number is less than about 2,000, the flow is laminar. For values of Reynolds number over around 3,000, the flow is turbulent, with velocity and pressure varying wildly in time. (*See* DIMENSION.)

Note that for a given incompressible fluid, the density, ρ, and viscosity, η, are fixed. The geometry of the situation determines l. Then the character of flow is controlled solely by the flow speed, v. Accordingly, for sufficiently low flow speeds, the flow will be laminar, while turbulent flow develops at high speeds.

twin paradox The twin paradox is related to the relativistic effect of TIME DILATION, whereby a moving CLOCK is measured as running more slowly than when it is at rest. The effect is not only for clocks, but for any process, including the aging of an organism: a moving heart is observed to beat more slowly than when it is at rest. Consider a pair of twins on Earth. One twin, the sister, takes a long round trip in a very fast spaceship, while her brother remains on Earth. During the outward leg of the journey, the brother on Earth observes his traveling sister aging more slowly than he is. Then, on the return leg, the Earthbound twin again finds his sister aging more slowly than he. When the twin sister finally lands and steps off the space ship, she is indeed younger than her twin brother, who has aged more than she has during her absence. (*See* RELATIVITY, SPECIAL THEORY OF.)

The paradox, or actually *apparent* paradox, is as follows. Since time dilation is a mutual effect, it should be expected that after the trip, the twins will somehow find themselves at the same age. It is, after all, their *relative* MOTION that matters. Each twin should have the same right to consider him- or herself to be at rest and the other one moving. Just as the Earthbound brother observes his sister aging more slowly than he, both while she is moving away and while she is returning, the spaceship-bound sister should see her brother aging more slowly than she, both while Earth is moving away from the spaceship and while it is returning. The situation seems to be symmetric. Why, then, does one twin age more slowly than the other?

The paradox is only an apparent paradox because, in fact, the twins are not in equivalent situations. While the Earthbound twin remains in the same inertial reference frame, that of the Earth, during his sister's trip, the traveling twin finds herself in four successive inertial reference frames. She starts in her brother's. While she is on the outward leg she is in her second. During her return she is in the third. And upon landing she returns to her initial reference frame, her fourth. She objectively differs from her brother in that she feels three acceleration FORCEs, while the brother on Earth feels none. First, she is accelerated to the outward speed. Then she is accelerated to turn the spaceship back toward Earth. And finally, she is decelerated for landing. (*See* ACCELERATION; REFERENCE FRAME, INERTIAL.)

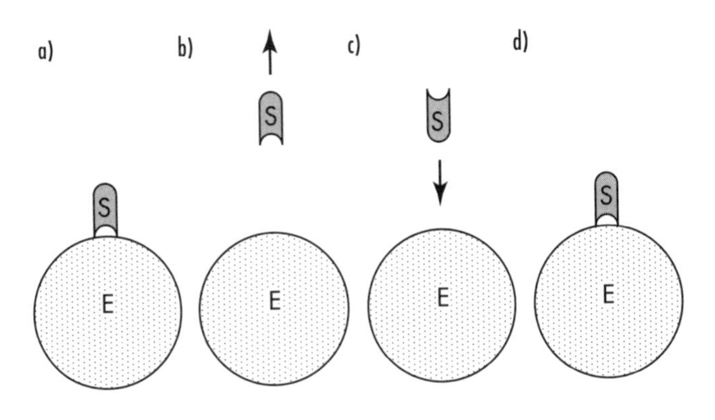

In the twin paradox, when the Earthbound twin greets his twin sister upon her return from a trip into space and back, she is younger than he is. Their different rates of aging can be attributed to the fact that, during the trip, the Earthbound twin remains in a single inertial reference frame, while the traveling twin finds herself in four consecutive inertial reference frames and undergoes an acceleration from each frame to the next. (a) She and her spaceship S are in her first inertial reference frame, that of Earth E. Then she accelerates to (b) her second inertial reference frame, which is in constant-velocity motion away from Earth. Another acceleration turns her around and puts her in (c) her third inertial reference frame, moving toward Earth at constant velocity. Her third acceleration—a deceleration as viewed from Earth—brings her to rest on Earth and into (d) her fourth inertial reference frame.

Under the simplifying assumption that the accelerations, in and of themselves, do not affect aging, the time dilation calculations can be carried out within the special theory of relativity. The result is that the traveler does indeed age more slowly. Since accelerations are involved, however, the general theory of relativity should be invoked. When it is, the result does not change. This effect has been tested and confirmed experimentally by synchronizing two ultraprecise atomic clocks and then taking one on a round-the-Earth trip by airplane, with the other remaining fixed on Earth. (*See* RELATIVITY, GENERAL THEORY OF.)

unit In order to meaningfully specify a value for a physical quantity, it must be made clear just what the number is counting, i.e., what the reference value of the quantity is. If a TIME interval is specified simply as 203, for example, we still do not know the duration of the interval. We need to know also what we have 203 of, i.e., what it is that is being counted. Such a reference amount for a physical quantity is a unit of the quantity. A unit must be well defined in order for it be useful. The unit for time, as an example, in the INTERNATIONAL SYSTEM OF UNITS (SI) is the second (s). The second is defined as the duration of 9,192,631,770 PERIODs of microwave electromagnetic RADIATION corresponding to the transition between two hyperfine levels in the GROUND STATE of the ATOM of the ISOTOPE cesium-133. Given that, it is clear what a time interval of 203 seconds (s) is. The specification of the value of any physical quantity must include both a number and a unit. The only exceptions are dimensionless quantities, such as the FINE STRUCTURE CONSTANT or STRAIN, whose values are independent of the system of units. (*See* DIMENSION; ELECTROMAGNETISM.)

See also BASE QUANTITY.

universe The universe is usually understood as comprising all of nature, everything of a physical character. COSMOLOGY is the field of physics that deals with the universe as a whole. The range of sizes in the universe runs from around 15 billion light-years down to perhaps about 10^{-35} meter (m). According the BIG-BANG picture, the age of the universe is around 15 billion years. The future of the universe is presently unknown. (*See* PLANCK SCALE; SPEED OF LIGHT.)

The contents of the universe appear organized in a hierarchy. The largest structures seem to be superclusters of GALAXIES, which consist of galaxy clusters, themselves composed of galaxies. Galaxies contain many ingredients, the most noticeable of which are stars, dust, and, apparently in most galaxies, a gigantic BLACK HOLE at the center. They also seem to contain additional, nonradiating matter, whose nature is presently unknown. At smaller scales, there are planets, asteroids, comets, and such. And at even smaller scales, there are MOLECULES, ATOMS, and ELEMENTARY PARTICLES.

The universe appears to be expanding, in that the galaxies seem to be moving away from each other at speeds that increase with distance. The big-bang model of the evolution of the universe has everything coming into being some 15 billion years ago in a primordial explosion, from which the universe has been expanding and cooling ever since. Although it might well be expected that gravitational attraction would bring about a slowing down of the expansion, recent observations appear to indicate that the rate of expansion of the universe is actually increasing. Clearly, there is still much to be learned and understood about the universe, not the least of which is the nature of the VACUUM. (*See* COSMIC MICROWAVE BACKGROUND; GRAVITATION; HUBBLE EFFECT.)

Unification: Finding Oneness in Diversity

In the operation of PHYSICS, PHYSICISTS search for pattern and order among the most fundamental phenomena of nature; attempt to discover laws for those phenomena, based on their pattern and order, which allow the prediction of new phenomena; and try to explain the laws by means of theories. (Note that in the language of science, *theory* does not mean speculation or hypothesis, as it does in everyday language, but explanation.) Consistently throughout this procedure, what appear to be different phenomena are found to be only different aspects of the same phenomenon; different laws are revealed as particular manifestations of the same law; and different theories turn out to be subsumed under a single, broader, more fundamental one. The coming together and merging of diverse facts or concepts to form a more general, encompassing framework is unification. Unification is the recognition and demonstration that an apparent diversity is, in fact, a unity, whose various elements are but different aspects of a common core.

Unification continually takes place throughout physics as an essential ingredient of physicists' striving to gain understanding: when apparently different things are recognized as manifestations of the same thing, they become better understood. Here are some noteworthy examples of unification that have occurred at various times and at various levels in the operation of physics.

Let us start with a unification of phenomena, and let us take the phenomena to be the motions of the Sun and planets in the sky, as observed from Earth. Judged by appearance alone, those motions seem to have almost nothing to do with each other. That was the situation until the 16th and 17th centuries, when the only known planets were those that can be seen with the naked eye: Mercury, Venus, Mars, Jupiter, and Saturn. JOHNANNES KEPLER studied Tycho Brahe's (1546–1601) earlier observations of the motions of the Sun and those planets and came to the recognition that the diversity of motion is only apparent. He showed that all the seemingly unrelated motions fit into the SOLAR SYSTEM model we are familiar with, together with three laws of planetary motion that are obeyed by all the known planets, including the Earth as one of the planets. When the other planets—Uranus, Neptune, and Pluto—were discovered, they, too, were found to obey KEPLER'S LAWS.

That was indeed unification on an astronomical scale! Rather than planets moving in any which way along just any orbits, the picture became one of nine particular cases of a single set of laws of motion. Perhaps, one might even say, it revealed nine manifestations of a single generic planetary orbit.

Consider another example of unification of phenomena. Let these phenomena comprise the various and diverse chemical ELEMENTS: hydrogen, helium, lithium, . . ., etc. In the 19th century, Dmitri Mendeleev (1834–1907) discovered a pattern among the elements, which led to his PERIODIC TABLE. In the 20th century, QUANTUM MECHANICS explained Mendeleev's periodic table of the elements. What became clear was that the fundamental unit of a chemical element is the ATOM, which consists of a positively charged NUCLEUS and a number of (negatively charged) ELECTRONS. Each element is characterized by the number of electrons in a neutral atom of the element, its ATOMIC NUMBER, which serves to order the elements in the periodic table. The chemical and physical properties of an element depend on the states that are allowed to its atomic system of nucleus and electrons, in particular its GROUND STATE, its state of lowest ENERGY. Quantum mechanics determines those states. (*See* CHARGE.)

Here was unification on the atomic scale. The chemical elements, as widely differing as they are in so many ways, became understood as manifestations of the same basic phenomenon: a number of electrons in the electric FIELD of a positively charged nucleus. (*See* ELECTRICITY.)

Turn now to an example of unification of laws. Until the 19th century, the understanding of electric and magnetic effects and of relations between the two had been gradually developing. Various patterns that had been discovered were expressed as a number of laws, including such as COULOMB'S LAW (for CHARLES AUGUSTIN DE COULOMB), AMPÈRE'S LAW (for ANDRÉ-MARIE AMPÈRE), the BIOT-SAVART LAW (for Jean-Baptiste Biot [1774–1862] and Félix Savart [1791–1841]), and GAUSS'S LAW (for Karl Friedrich Gauss [1777–1855]). In the 19th century, JAMES CLERK MAXWELL proposed a theory of ELECTROMAGNETISM in the form of a set of equations, called MAXWELL'S EQUATIONS. These equations unified all the previously known laws, in the sense that all the latter can be derived from Maxwell's equations for particular cases. All the various laws became understood as different aspects of the laws of electromagnetism as expressed by Maxwell's

What is clear, however, is that our understanding of the very large cannot be separated from our understanding of the very small. The better we understand the elementary particles and their INTERACTIONs, the better we understand the earliest stages of the universe, and vice versa. Recent attempts at better understanding include the possibility of extra DIMENSIONs, beyond the four dimensions of SPACE-TIME. They also suggest that the equations. Furthermore, Maxwell's theory showed that electricity and MAGNETISM themselves must be viewed as two aspects of the more general phenomenon of electromagnetism. In this way, electricity and magnetism were unified within electromagnetism.

Unification takes place also at the level of theories. Here is an example. The modern, quantum treatment of the ELECTROMAGNETIC INTERACTION (electromagnetism), the WEAK INTERACTION, and the STRONG INTERACTION, is carried out in the framework of QUANTUM FIELD THEORY, where the three interactions are described by means of GAUGE THEORIES. During the 20th century, it became apparent that the electromagnetic and weak interactions have much in common, although they appear to be quite different. For instance, the electromagnetic interaction is mediated by the SPIN-1, massless, neutral PHOTON, while the weak interaction has three kinds of mediating particle, all with spin 1, all possessing MASS, and two carrying electric charge.

Yet, at sufficiently high energy, meaning high TEMPERATURE, when the MASS ENERGY of the weak intermediaries becomes negligible, the neutral one becomes very similar to the photon, and the two interactions merge into one, the ELECTROWEAK INTERACTION. The electroweak interaction, then, forms a unification of the electromagnetic and weak interactions that is valid at high energy. It is mediated by four particles. Each of the electromagnetic and weak interactions, with its corresponding intermediating particle(s), is an aspect of the electroweak interaction. Presumably, that was the situation during an early stage in the evolution of the universe, when temperatures, and thus energies, were considerably higher than they are now. As the UNIVERSE expanded and cooled, the two interactions eventually "parted ways," so to speak, each developing its own individual identity. In this way, the theory of the electroweak interaction unifies the theories of the electromagnetic and weak interactions. (*See* COSMOLOGY.)

Even beyond the electroweak unification, there are indications that a further unification should exist. This one would unify the electroweak interaction with the strong interaction, giving what is called a GRAND UNIFIED THEORY (GUT). If true, this unification should be valid only at even higher energies than are required for electroweak unification. And in the life of the universe, the GUT should have held sway for a shorter duration than did the electroweak theory.

The ultimate goal of unifying all the interactions dazzles physicists. It would unify the GUT with GRAVITATION and encompass all the known interactions within a single theory, grandly labeled a "theory of everything" (TOE). Its reign in the evolution of the universe would have to have been only for the briefest of times, when the universe was extremely hot. As the expansion of the universe cooled it, the gravitational and GUT interactions would become distinct from each other. Later, at a lower temperature, the strong and electroweak interactions would separate from each other and gain their identities. Then the weak and electromagnetic interactions would come into separate existence. The unification of electricity and magnetism within electromagnetism remains valid yet today.

Physicists are actively working on GUTs, but so far with no definitive result. Work on a TOE must first overcome the hurdle of making ALBERT EINSTEIN's general theory of relativity, which is the current theory of gravitation, compatible with QUANTUM PHYSICS. The result, if and when achieved, would be termed a theory of QUANTUM GRAVITY. This theory would, it is hoped, then be unified with the GUT—assuming one is found—to produce a TOE. It is difficult to imagine what, if anything, might lie beyond a theory of everything. The unification scenario just described—leading to a TOE—is merely conjecture, only an extrapolation from the present situation into the future. It is very likely that, as has so often happened in physics, new discoveries will change the picture dramatically, and the path of unification will lead in very unexpected directions. (*See* RELATIVITY, GENERAL THEORY OF.)

Whatever might occur in that regard, unification has so far served physics very well, in the ways described above as well as in very many other ways. Unification makes for simplicity. Rather than two or more disparate entities needing to be understood (think of electricity and magnetism, or of all the chemical elements), each requiring its own laws and theory, only a single, broader, and deeper entity takes the investigational stage (e.g., electromagnetism, or electrons in the electric field of a nucleus). The many laws and theories are replaced by fewer laws and by a single, encompassing, more fundamental theory (Maxwell's equations, or quantum mechanics, in the examples), and physics takes a further step toward a deeper understanding of nature.

universe, *our* universe, might be but one of many universes that compose what could be called a multiverse.

With telescopes of various kinds looking out to increasingly greater distances (i.e., farther and farther into the past), astronomical data of great cosmological importance are coming in and allowing an increasingly clearer picture of the behavior of the universe. These are exciting times for cosmology.

The Universe

The term *universe* is often used in PHYSICS. What it refers to differs somewhat from its meaning in ordinary conversation. In the latter, *universe* seems to equate to *all,* where *all* comprises the totality of whatever one is considering. One speaks of a "universe of discourse." On the cosmic scale, the everyday implication of *universe* is the totality of being, existence, etc., the complete collection of all there is. That is commonly understood to include nature, the supernatural, the spiritual, the mental, the emotional, and so on and so on.

PHYSICISTS restrict the term to mean only the material universe, or physical universe (i.e., those aspects of the UNIVERSE that involve only MATTER and ENERGY in their various forms, including FIELDS and WAVES). And although physics deals only with the material universe, and not with any other aspect of the universe, it does not consider even the whole material universe. Physics is confined to studying only those components of the material universe with which we can, or can conceivably, interact. The notion of INTERACTION comprises affecting and being affected. What physics really needs in this regard is to be able to perform observations and measurements and to receive data, or to conceivably be able to do so. The term *conceivably* here means that, although we might not be able to interact at present, interaction is not precluded by any principle known to us and is considered attainable through further technological research and development.

One might then ask whether there exist components of the material universe with which we cannot interact, not even conceivably. This question will forever remain a question. Because, how could we ever know? One might speculate, assume, imagine, or believe, but that does not make it so. With no possibility of interaction, any such hypothetical component might as well not exist at all.

Take the concept of BLACK HOLE. No LIGHT, other RADIATION, or matter can leave from a black hole. Does that fact then relegate black holes to effective nonexistence? No, it does not. Black holes can indeed be observed and measured. The observation is indirect, but it is valid nevertheless. Black holes have very observable effects. They affect the orbits of their binary partners, if they make up part of a binary (i.e., a two-star) system, and they emit characteristic radiation as a result of matter falling into them. The gigantic black holes located at the centers of GALAXIES affect the orbits of all the stars of the galaxy. So there is no problem with the existence of black holes.

The "universe of discourse" of physics is the material universe that we can, or can conceivably, observe, measure, and receive data from. That is also called nature. Often the idea of interaction is simply taken for granted, and "universe" is equated with "nature."

So physics studies that component of the universe called nature. But not all aspects of nature lie within the strict domain of physics and present themselves as candidates for being understood through physics. Not all natural phenomena form grist for the mill of physics. One requirement, which is applicable as well for the rest of the natural sciences (i.e., also for chemistry and biology), is that a phenomenon be *reproducible.* This means that the effect is observed and measured consistently again and again by the same investigator and by others. Nature might well possess irreproducible phenomena, but the natural sciences would not be able to deal with them. Effects such as extrasensory perception (ESP), if they existed, would belong to this category.

For a phenomenon to belong to the domain of the natural sciences, it must also be *predictable.* It must exhibit sufficient pattern and ORDER to allow prediction of the phenomenon's behavior in as-yet-untested situations. That is done through the formulation of laws. It is not precluded that nature possesses unpredictable phenomena, but if so, the natural sciences would not be able to study them. ESP, if it existed, would be such a phenomenon. (*See* LAWS OF NATURE.)

So the natural sciences, including physics, are capable of dealing only with the reproducible and predictable aspects of nature. What is unique to physics is that it studies only the most *fundamental* of those aspects, leaving the less fundamental ones for chemistry and biology to attempt to make sense of. What is meant by *fundamental* here is that physics requires no explanations from other sciences. When chemists need to better understand what underlies phenomena in their field, they turn to physics. And biologists similarly turn to chemistry and physics. But physicists turn only to themselves. Nothing underlies physics but more physics.

Having considered various aspects of the universe and how the natural sciences, and physics in particular, relate to them, let us turn to the universe as a whole. What is its mode of existence? Until the advent of ALBERT EINSTEIN's general theory of relativity in the early part of the 20th century, it was generally taken for granted that the material universe exists within an independent, absolute framework of SPACE and TIME. In the hypothetical absence of matter, the framework would still be there. Einstein's theory changed that. It shows that matter/energy and SPACE-TIME are intimately related, with each affecting the other. Now it is generally thought that, rather than the universe existing within space-time, they are both tightly bound to each other. (*See* RELATIVITY, GENERAL THEORY OF.)

In the big-bang picture of the evolution of the universe, the universe started its life in an extremely dense, hot state, from which it has been expanding and cooling ever since. Prerelativistic thinking would have the universe expanding into preexisting space. The present commonly held conception is that space came into being together with the universe and is expanding with the universe. At the BIG BANG, space was "small." Since then, it has been growing "larger." Time, too, is understood to have meaning only within the universe. It makes no sense to consider the situation "before" the big bang.

COSMOLOGY is the field of physics that studies the universe as a whole: its present state, its past and future evolution, and its origin. A big conceptual problem with cosmology is that the subject of its investigation is not reproducible. The universe we can interact with and study is unique. It is not a member of a class of universes that we can examine. We cannot run experiments on universe evolution. There is no pattern or order to be discovered in the evolution of the universe, which is, therefore, unpredictable in principle. Any proposed evolution scenario, such as the big-bang scenario, can indeed be compared with data from the universe's past. (Astronomical observations show us the past, due to the finite SPEED OF LIGHT: the more distant the observed object, the farther in its past we are seeing it.) But, due to the time scales involved, there is no way of testing a scenario's predictions for the future of the universe.

As for explaining why the universe is as it is, that too cannot be done in a manner that follows the normal procedure of physics. Some nonscientific reasoning must be introduced. This unsatisfactory state of explanatory affairs is overcome if the universe is considered to be one of a class of universes. Patterns, order, and laws have meaning for a class of universes, whereas they are meaningless for a unique universe. There are those cosmologists who are considering just such ideas.

One type of proposal is that of a cyclic universe, with the universe being born, expanding, collapsing, being born again, and so on forever. Each cycle is in effect a new universe. So the class of universes, according to those proposals, is not a class of simultaneously existing universes, but rather is a class of sequentially existing ones. Our universe is but one in this temporal sequence of universes. Other proposals have our universe existing simultaneously with an infinite number of other universes in what is termed a multiverse. Clearly, these ideas are reverting to a universe existing within space or time. Another type of proposal has universes multiplying by giving birth to "baby universes" within themselves. Others even invoke extra DIMENSIONS, beyond the four of space and time, in their quest for additional universes.

We have seen how physics relates to various aspects of the universe, including nature, and how physics tries to make sense of the universe as a whole. The operation of physics within the universe is quite well defined: physics studies the most fundamental reproducible and predictable aspects of the material universe with which we can, or can conceivably, interact. On the other hand, cosmology, the field of physics that deals with the universe as a whole, is operating at or beyond the edge of the normal domain of physics. Since, at least to the extent of our observational reach, the universe is a unique phenomenon, making scientific sense of it is not a straightforward matter. That will require some stretching of the domain's edge.

Humankind's image of the universe has continuously developed over the millennia. At each stage, people surely looked back, as we do today, at how their predecessors pictured the universe and thought, as we think today: "How naive and primitive!" With the recent and present acceleration in the progress of science, one cannot help wondering how our own image of the universe, as sophisticated as it might seem to us, will appear in as little as 20 years from now.

vacuum In CLASSICAL PHYSICS, vacuum is the condition of an absence of MATTER. Vacuum pumps can produce very good vacuums by removing GAS MOLECULES from an enclosed VOLUME until a very low DENSITY is achieved. Intergalactic SPACE presents an even better vacuum. But there is no perfect macroscopic vacuum. Intergalactic space, as good a vacuum as it is, might have an ATOM in every cubic meter or so, on the average. At the microscopic level, however, the space between atoms, or between molecules, or even most of the volume of atoms themselves, forms what is considered to be a perfect classical vacuum: complete nothingness. (*See* GALAXY.)

Yet, according to QUANTUM PHYSICS, the situation is not so simple. The HEISENBERG UNCERTAINTY PRINCIPLE does not allow the certainty that is posed by the classical picture of the vacuum. During sufficiently short TIME intervals and over small enough distances, ENERGY and LINEAR MOMENTUM are "borrowed" for the spontaneous materialization and subsequent annihilation of particle-antiparticle pairs, PHOTONs, and other particles. These are called virtual particles, since they almost never achieve the status of real particles, which maintain their existence. In this manner, the quantum vacuum is teeming with activity and presents a very different picture from the classical one. (*See* ANTIMATTER; ELEMENTARY PARTICLE.)

Particles can be extracted from the vacuum through a process in which virtual particles are made real. This is done naturally by BLACK HOLEs and forms the mechanism for black-hole "evaporation." In this case, it is the gravitational FIELD that separates the members of a virtual particle-antiparticle pair and causes them to materialize. A very strong electric field can similarly separate a virtual electrically charged pair, such as an ELECTRON-POSITRON pair. Then the oppositely charged particles are pulled apart by the field and become real, with the required energy being supplied by the field. (*See* CHARGE; ELECTRICITY; GRAVITATION; MASS ENERGY.)

The recently discovered ACCELERATION in the rate of expansion of the UNIVERSE raises the possibility that the vacuum harbors energy, called dark energy, that causes an effect of negative PRESSURE (i.e., gravitational repulsion throughout the universe). The general theory of relativity allows such an effect. The nature of dark energy, if indeed there is such a thing, is far from clear at present. (*See* RELATIVITY, GENERAL THEORY OF.)

valence band When many ATOMs are in close proximity and are interacting with each other, such as in a SOLID, the ENERGY levels allowed to the ELECTRONs in the outer atomic shells become very much more numerous than for individual atoms. The energy levels are so numerous that they can be considered to be continuous, rather than discrete, and are described in terms of energy "bands" (i.e., ranges of energy). Electrons in the GROUND STATE are said to be in the valence band. Electrons with sufficient energy to move freely among the relatively fixed atoms—such electrons are called free

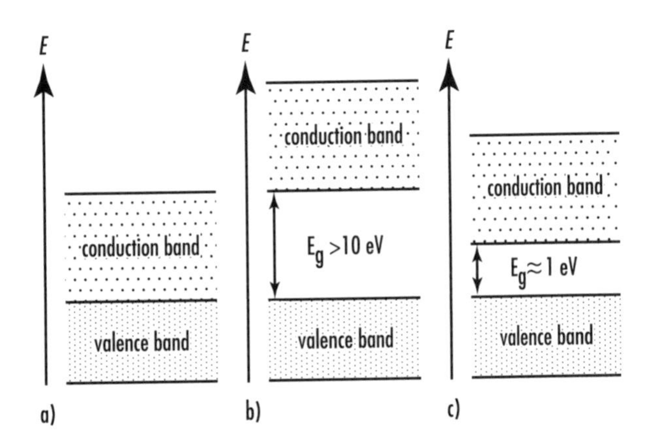

The valence band of a solid is the range of energies of electrons in the ground state. The conduction band of a solid represents the range of energies at which electrons can move about freely. The energy axis is labeled *E*. (a) When the bands overlap or almost overlap, the material is a conductor. (b) When the energy gap E_g between the bands is large, greater than about 10 eV, the material is an insulator. (c) Semiconductors have intermediate energy gaps, on the order of 1 eV.

electrons—are said to be in the CONDUCTION BAND. If the two bands overlap or almost overlap, then the material is endowed with naturally occurring free electrons and has low electric RESISTIVITY. Such a material is called a CONDUCTOR. If there is a large energy gap between the valence and conduction bands, greater than, say, around 10 electron volts (eV), very few electrons will occupy the conduction band, and the material's resistivity will be high. It is called an INSULATOR. With intermediate energy gaps of about 1 eV, electrons can be boosted into the conduction band with relatively small investments of energy. Such materials are called SEMICONDUCTORS. (*See* ELECTRICITY; INTERACTION.)

See also CONDUCTION.

van der Waals equation Named for the Dutch physicist Johannes Diderik van der Waals (1837–1923), this equation forms a modification of the ideal GAS law that takes into account the finite VOLUME of the gas MOLECULES and the FORCES among them. The idea behind the van der Waals equation starts with the ideal gas law:

$$pV = nRT$$

where p denotes the PRESSURE in pascals (Pa), equivalent to newtons per square meter (N/m^2); V is the VOLUME in cubic meters (m^3); T is the absolute TEMPERATURE in kelvins (K); n is the amount of gas in MOLES (mol); and R is the gas constant, whose value is 8.314472 joules per mole per kelvin ([J/mol·K]). Now, two effects are taken into account. (1) The volume available to the molecules of the gas is less than the volume of the container. Letting V represent the volume of the container, the volume that a gas molecule sees is less than V by the total volume of all the gas molecules, which is proportional to n. So in the ideal gas law, V is replaced by $(V - nb)$, where b is a constant for each gas and equals the effective volume of a single molecule of the gas multiplied by AVOGADRO'S NUMBER (the number of molecules in a mole). The SI UNIT of b is cubic meters per mole (m^3/mol). (2) The pressure within the volume of the gas is greater than the pressure measured on the container's walls, because of the intermolecular attraction. The effect is proportional to the square of the number of molecules per unit

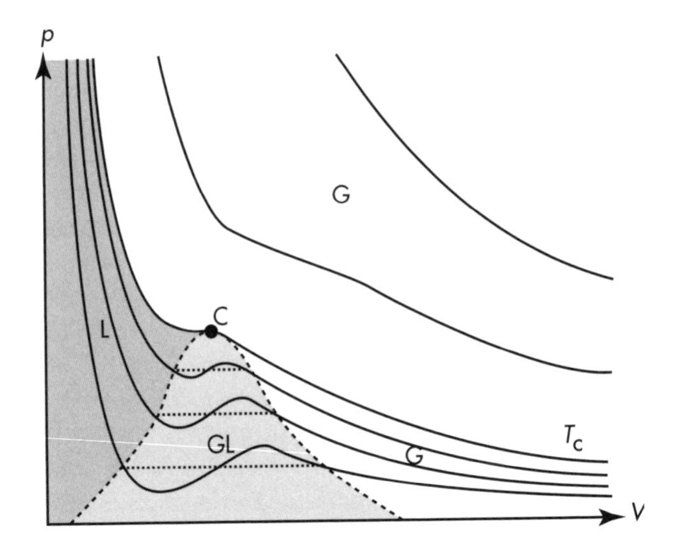

A number of isotherms are shown for a fluid obeying the van der Waals equation in a typical pressure-volume *(p-V)* phase diagram. The critical point and critical-temperature isotherm are labeled C and T_c, respectively, where T_c denotes the critical temperature, the temperature at the critical point. (Above the critical temperature, no amount of pressure will bring about a gas-liquid phase transition.) Higher curves represent greater temperatures. Regions of the diagram are labeled according to whether under those conditions only a gas can exist (G), only a liquid can exist (L), or gas and liquid can exist in equilibrium (GL). In the GL region, the fluid's isothermal behavior is described by the dotted horizontal straight lines (isobars), rather than by the van der Waals isotherms, as the fluid undergoes a gas-liquid phase transition (evaporation or condensation) under change of volume at its vapor pressure for each temperature.

volume, and thus proportional to $(n/V)^2$. So, with p denoting the pressure on the walls, in the ideal gas law, p is replaced by $(p + n^2a/V^2)$, with a a constant for each gas. The SI unit of a is meter^6·pascal per mole^2 ($m^6 \cdot Pa/mol^2$). (*See* GAS, IDEAL.)

In this way, the ideal gas law takes the form of the van der Waals equation:

$$\left(p + \frac{n^2 a}{V^2}\right)(V - nb) = nRT$$

Note that for sufficiently low DENSITY (i.e., for small n/V), we have $V - nb \approx V$ and $p + n^2 a/V^2 \approx p$, and the ideal gas law is recovered. Otherwise, the van der Waals equation better represents the behavior of real gases than does the ideal gas law.

vapor pressure The PRESSURE of the GAS state, or gas PHASE, of a substance when it is in EQUILIBRIUM with the LIQUID phase is termed vapor pressure. (When the gas and liquid phases of a substance are coexisting, the gas is often referred to as a vapor.) A liquid in an open container will eventually evaporate completely. In a closed container, however, assuming a sufficient amount of liquid, the evaporated molecules accumulate. The gaseous-phase pressure increases until the rate of molecules reentering the liquid phase equals the rate that the molecules leave it. The pressure of the gas phase in that steady-state equilibrium situation is defined as the vapor pressure of the substance at that temperature. Vapor pressure increases with temperature. When the vapor pressure equals or exceeds the pressure of the liquid, BOILING occurs. (*See* CONDENSATION; EVAPORATION; STATE OF MATTER; STEADY-STATE PROCESS.)

vector A physical quantity that is characterized by both a magnitude, which is a nonnegative number, and a direction is designated a vector. Contrast this with a SCALAR, which is a nondirectional physical quantity. Examples of vectors are VELOCITY, ACCELERATION, FORCE, and the electric FIELD. More specifically, the velocity of an aircraft can be specified by its magnitude, say 50 meters per second (m/s), and its direction, perhaps north by northwest and parallel to the surface of the Earth. (*See* ELECTRICITY.)

A vector is often denoted by a bold symbol, such as **v**. That is the notation used in this book. Other notations, commonly used in handwriting, are $\vec{v}$ and $\underline{v}$. The magnitude of a vector is indicated by $|\mathbf{v}|$, $|\vec{v}|$, $|\underline{v}|$, or simply v. A vector's direction can be specified by giving the three angles between its direction and the positive x-, y-, and z-axes of a Cartesian COORDINATE SYSTEM, its direction angles, (α, β, γ), respectively. Alternatively, its direction cosines can be given, $(\cos \alpha, \cos \beta, \cos \gamma)$. A vector's components are the values of its projections along the three axes. Thus, the x-, y-, and z-components of vector **v**, denoted (v_x, v_y, v_z), equal $(v \cos \alpha, v \cos \beta, v \cos \gamma)$, where v represents the magnitude of **v**.

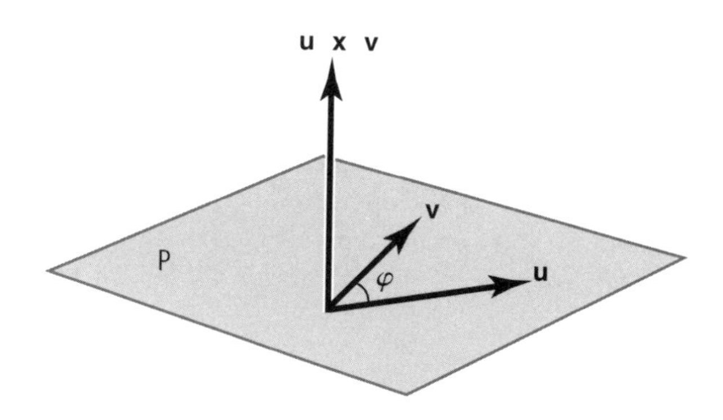

The vector product of vectors u and v. Plane P contains u and v. Their vector product, u × v, is a vector perpendicular to P in the sense shown. The smaller angle between u and v is denoted φ. The magnitude of u × v equals *uv* sin φ, where *u* and *v* denote the magnitudes of u and v, respectively.

Vectors are represented graphically by arrows that point in the vectors' direction. The length of the arrow is put proportional to the magnitude of the represented vector. So it is common to speak of a vector as "pointing" and to refer to the "head," or "point," of a vector and to its "tail," or "base."

Taking again the above example of velocity, let the x-axis point east, the y-axis north, and the z-axis directly upward. For the direction north-by-northwest and parallel to the Earth's surface, the direction angles are (112.5°, 22.5°, 90°). (You might find it useful to draw this.) The corresponding direction cosines equal (–0.383, 0.924, 0). The x-, y-, and z-components of the velocity are (–19.1 m/s, 46.2 m/s, 0 m/s).

A number of operations are defined for vectors.

Product of Vector and Scaler

One operation is multiplication of a vector, **v**, by a scalar, s, denoted $s\mathbf{v}$. The result of this operation, $s\mathbf{v}$, is a vector whose magnitude equals $|s|$ times the magnitude of **v**:

$$|s\mathbf{v}| = |s| v$$

The direction of $s\mathbf{v}$ is the same direction as that of $\mathbf{v}$, if s is positive, and the opposite direction of $\mathbf{v}$, if s is negative. The components of $s\mathbf{v}$ are (sv_x, sv_y, sv_z).

Using the velocity example, an aircraft flying at twice the velocity of the one given is traveling at 100 m/s, north by northwest and parallel to the Earth's surface. The components of its velocity are (–38.2 m/s, 92.4 m/s, 0 m/s).

Scalar Product

Another operation is the scalar product of two vectors, **u** and **v**, denoted $\mathbf{u}\cdot\mathbf{v}$. The scalar product is a scalar whose value equals the product of the vectors' magnitudes and the cosine of the smaller angle (≤ 180°) between their directions:

$$\mathbf{u}\cdot\mathbf{v} = uv \cos \varphi$$

Here φ denotes the angle between the vectors' directions. This can be understood as the product of the magnitude of one of the vectors and the projection of the other vector along the direction of the first:

$$\mathbf{u}\cdot\mathbf{v} = u(v \cos \varphi) = v(u \cos \varphi)$$

Note that for two vectors of fixed magnitudes, the value of their scalar product is greatest when they are parallel, with $\varphi = 0$ and $\cos \varphi = 1$. Then $\mathbf{u}\cdot\mathbf{v} = uv$. When the vectors are antiparallel, with $\varphi = 180°$ and $\cos \varphi = -1$, their scalar product is maximally negative, $\mathbf{u}\cdot\mathbf{v} = -uv$. The scalar product of a pair of perpendicular vectors, with $\varphi = 90°$ and $\cos \varphi = 0$, equals zero.

WORK forms an important example of a scalar product, since it involves the scalar product of force and DISPLACEMENT. When a constant force, **F**, acts along the straight-line displacement, **d**, the work that the force performs equals $\mathbf{F}\cdot\mathbf{d}$.

The scalar product can conveniently be expressed also in terms of the vectors' components, as:

$$\mathbf{u}\cdot\mathbf{v} = u_x v_x + u_y v_y + u_z v_z$$

The magnitude of a vector equals the square root of its scalar product with itself:

$$|\mathbf{v}| = \sqrt{\mathbf{v}\cdot\mathbf{v}} = \sqrt{v_x^2 + v_y^2 + v_z^2}$$

We can confirm this for the first velocity example. The sum of the squares of the components is $(-19.1 \text{ m/s})^2 + (46.2 \text{ m/s})^2 + (0 \text{ m/s})^2 = 2499.25 \text{ m}^2/\text{s}^2$, and $\sqrt{2499.25 \text{ m}^2/\text{s}^2} = 50.0$ m/s (to the three-significant-digit precision in which we are working), which is indeed the magnitude of the velocity.

Vector Product

Yet a third operation is the vector product of two vectors, denoted $\mathbf{u} \times \mathbf{v}$. This is a vector. Its magnitude equals the product of the two vectors' magnitudes and the sine of the smaller angle between their directions:

$$|\mathbf{u} \times \mathbf{v}| = uv \sin \varphi$$

Note here that for two vectors of fixed magnitudes, the magnitude of their vector product is greatest when they are perpendicular, with $\varphi = 90°$ and $\sin \varphi = 1$. Then $|\mathbf{u} \times \mathbf{v}| = uv$. When two vectors are parallel or antiparallel, with $\varphi = 0$ or $\varphi = 180°$ and $\sin \varphi = 0$, their vector product vanishes.

The direction of $\mathbf{u} \times \mathbf{v}$ is perpendicular to the plane defined by the directions of **u** and **v**. The sense of this vector is such that, if we imagine it pointing from one side of the **u**-**v** plane to the other, it is pointing toward the side from which the angle φ is seen as involving a counterclockwise rotation from the direction of **u** to the direction of **v** in the **u**-**v** plane. The vector product obeys the relation:

$$\mathbf{u} \times \mathbf{v} = -\mathbf{v} \times \mathbf{u}$$

The x-, y-, and z-components of $\mathbf{u} \times \mathbf{v}$ can be expressed in terms of the components of **u** and **v**, as:

$$(\mathbf{u} \times \mathbf{v})_{x,\, y,\, z} = (u_y v_z - u_z v_y,\ u_z v_x - u_x v_z,\ u_x v_y - u_y v_x)$$

TORQUE can serve as an example of the vector product. Let force **F** act on a particle or body, and denote by **r** the position vector of the point of application of the force with respect to reference point O. So **r** is the spatial vector extending from reference point O to the point of application of the force. Then the torque of the force with respect to O, $\boldsymbol{\tau}_O$, is given by the vector product:

$$\boldsymbol{\tau}_O = \mathbf{r} \times \mathbf{F}$$

Fields

Now, turn to FIELDs, which are physical quantities that depend on location. The divergence of a vector field, $\mathbf{v}(x, y, z)$, is denoted div **v**, or $\nabla\cdot\mathbf{v}$. It is a scalar field, given in terms of the components of **v** by:

$$\text{div } \mathbf{v}(x, y, z) = \partial v_x(x, y, z)/\partial x + \partial v_y(x, y, z)/\partial y + \partial v_z(x, y, z)/\partial z$$

For an interpretation of the divergence, let the field, $\mathbf{v}(x, y, z)$, represent the velocity at every location in an incompressible FLUID in flow. Then a nonzero value of div **v** at a location indicates that there is a source or sink of fluid at that location. If div **v** > 0, fluid is entering the flow from outside the system (a source) there. For div **v** < 0, fluid is disappearing from the flow (a sink).

On the other hand, every scalar field, $s(x, y, z)$, defines the vector field that is its gradient, denoted grad s, or ∇s. The x-, y-, and z-components of grad s are given by:

$$(\text{grad } s)_{x,y,z} = (\partial s(x, y, z)/\partial x, \partial s(x, y, z)/\partial y, \partial s(x, y, z)/\partial z)$$

Take a TEMPERATURE field, as an example, where $s(x, y, z)$ represents the temperature at every location. The direction of grad s at any location L indicates the direction at L in which the temperature has the greatest rate of increase. In other words, if the temperatures are compared at all locations on the surface of a small sphere centered on L, the temperature at the location in the direction of grad s from L will be the highest. The magnitude of grad s equals the rate of temperature increase in that direction.

From a vector field, $\mathbf{v}(x, y, z)$, it is also possible to obtain another vector field, called the curl of **v** and denoted curl **v**, or $\nabla \times \mathbf{v}$, or sometimes rot **v**. The x-, y-, and z-components of curl **v** are given by:

$$(\text{curl } \mathbf{v})_{x,y,z} = (\partial v_z/\partial y - \partial v_y/\partial z, \partial v_x/\partial z - \partial v_z/\partial x, \partial v_y/\partial x - \partial v_x/\partial y)$$

The interpretation of the curl is more complicated than for the divergence and the gradient, and we forgo it here.

velocity The TIME rate of DISPLACEMENT (i.e., the change of position per unit time) is called velocity, or more precisely, instantaneous velocity. Like displacement, velocity is a VECTOR quantity, possessing both magnitude and direction. In a formula:

$$\mathbf{v} = d\mathbf{r}/dt$$

where **v** denotes the velocity in meters per second (m/s), and t is the time in seconds (s). The symbol **r** represents the position vector in meters (m). The position vector associated with a point in space is the vector whose magnitude equals the distance from the origin of the COORDINATE SYSTEM to the point under consideration and whose direction is the direction of the point as viewed from the origin. Then $d\mathbf{r}$ denotes the infinitesimal displacement vector, the infinitesimal change in **r**. Instantaneous speed is the magnitude of the instantaneous velocity. (*See* SPEED.)

The average velocity, $\mathbf{v}_{av}$, over the time interval from t_1 to t_2 equals the net displacement during that time interval divided by the time interval. So:

$$\mathbf{v}_{av} = \frac{\mathbf{r}(t_2) - \mathbf{r}(t_1)}{t_2 - t_1}$$

where $\mathbf{r}(t_1)$ and $\mathbf{r}(t_2)$ denote the position vectors at times t_1 and t_2, respectively. The instantaneous velocity is obtained from the average velocity by taking the limit of t_2 approaching t_1.

In a simple situation, the motion of a body might be constrained to one DIMENSION, say, along the x direction. In that case, we can drop the vector notation and use the formula:

$$v = dx/dt$$

where v and x denote, respectively, the velocity and position in the x direction. If, moreover, the velocity is constant (i.e., the position changes [increases or decreases] by equal increments during equal time intervals), the formula can be further simplified to:

$$v = \Delta x/\Delta t$$

where Δx is the (positive or negative) change in position during time interval Δt.

Relative velocity is the velocity of a moving object as observed in a REFERENCE FRAME that is itself moving. The velocity of aircraft B as viewed from moving aircraft

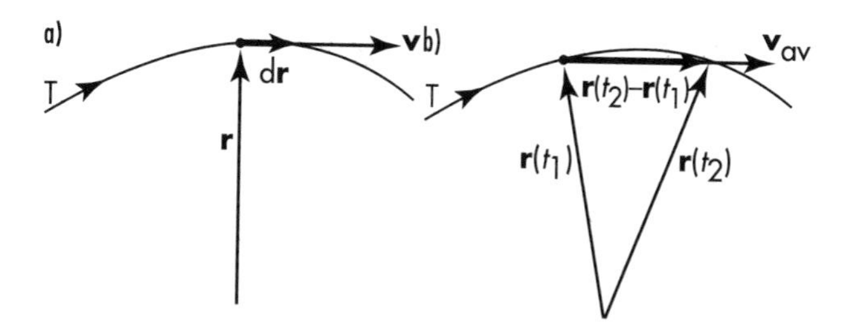

Instantaneous and average velocities for a particle moving along a trajectory T. (a) The instantaneous velocity of a particle with position vector r is v = *dr/dt,* where *dr* is the infinitesimal change in the particle's position vector during infinitesimal time *dt*. *dr* is in the direction of the trajectory at the particle's location, so v is tangent to the trajectory there. (b) The average velocity between times t_1 and t_2 is given by $\mathbf{v}_{av} = [\mathbf{r}(t_2) - \mathbf{r}(t_1)]/(t_2 - t_1)$, where r(*t*) denotes the particle's position vector at time *t*. $\mathbf{v}_{av}$ is parallel to $\mathbf{r}(t_2) - \mathbf{r}(t_1)$, the displacement that the particle undergoes during the time interval.

A, for instance, is the relative velocity of B with respect to A and is different from the velocity of B with respect to the Earth. Let $\mathbf{v}_0$ denote the velocity of the reference frame and $\mathbf{v}$ the velocity of an object. Then the relative velocity of the object with respect to the reference frame, $\mathbf{v}_{rel}$, is given by:

$$\mathbf{v}_{rel} = \mathbf{v} - \mathbf{v}_0$$

The angular velocity, $\boldsymbol{\omega}$, of a point particle with velocity $\mathbf{v}$ is defined as:

$$\boldsymbol{\omega} = \frac{\mathbf{r} \times \mathbf{v}}{r^2}$$

where r denotes the magnitude of the position vector, $\mathbf{r}$, of the particle. Note that angular velocity depends on the origin chosen, while linear velocity does not. So angular velocity is always with respect to some point. The SI UNIT of angular velocity is radian per second (rad/s). For rotational MOTION, the direction of the angular velocity vector is taken parallel with the rotation axis and pointing in the direction that a corotating right-hand screw would advance. The magnitude of angular velocity in this case equals the angular speed of ROTATION.

See also ANGULAR MOMENTUM; KINEMATICS.

vibration Oscillatory MOTION is referred to as vibration. The motion of the diaphragm of an operating loudspeaker is an example of vibration. (*See* OSCILLATION.)

viscosity The internal FRICTION of FLUIDs, as manifested by their resistance to flow, is termed viscosity. As a physical quantity, the viscosity of a fluid is a measure of the shear stress resulting from layers of fluid flowing past each other. Its SI UNIT is newton·second per square meter (N·s/m^2). Note the difference in the flow of maple syrup or molasses, for example, and water. The former possess considerably higher viscosities than does water. Viscosity plays a major role in various effects, including those described by POISEUILLE'S LAW and STOKES'S LAW. The latter is related to TERMINAL SPEED. (*See* SHEAR; STRESS.)

See also REYNOLDS NUMBER; TURBULENCE.

voltage This is an equivalent term for electric potential difference. (*See* POTENTIAL, ELECTRIC.)

volume The amount of SPACE is termed volume. Its SI UNIT is the cubic meter (m^3). The volume of a brick whose sides have lengths a, b, and c, for example, equals abc. Or, the volume of a sphere of radius r is $(4/3)\pi r^3$. If all lengths characterizing a system are changed by multiplication by the same factor, then all volumes of the system will be correspondingly multiplied by the cube of that factor. If the sides of a brick are each doubled, for instance, then the volume grows by a factor of eight. Or, if the radius of a sphere is halved, then its volume is reduced to an eighth of its former value.

wave A propagating disturbance is called a wave. A wave might propagate in a material medium or in a FIELD. In the former case, called a mechanical wave, a disturbance is the condition of the matter locally being out of EQUILIBRIUM, such as the result of giving the lower end of a hanging rope a horizontal jerk. In regaining equilibrium at one location, the medium causes adjacent locations to depart from equilibrium. Then these, too, return to equilibrium, which causes *their* adjacent locations to depart from equilibrium, and so on. In this way, the disturbance travels through the medium by means of self-sustaining propagation. It is the elastic character of a medium (or, for fields, a property analogous to elasticity) that causes the medium to tend to regain equilibrium when disturbed. Note that the medium itself does not propagate, but only undergoes local displacement about equilibrium at every point. Waves carry and transmit ENERGY. Examples of mechanical waves are: (1) Waves in a stretched rope or string. They might be brought about by hitting the rope or by plucking the string or rubbing it with a bow, as in playing a guitar or violin. The string or rope is disturbed by locally being displaced laterally from its equilibrium position. (2) SOUND in air. This is a PRESSURE wave. The sound source alternately compresses and decompresses the air, locally raising and lowering its pressure and DENSITY from their equilibrium values. (3) Surface waves on a LIQUID, such as ocean waves. The disturbance here is a local displacement of the surface from its equilibrium level. (*See* ELASTICITY.)

An important case of waves in fields is electromagnetic RADIATION, which propagates as waves in the electromagnetic field. The propagating disturbance in this case is a deviation of the electric and magnetic fields from their equilibrium values. The electric and magnetic fields are coupled to each other in such a way that a change over time in one field generates the other. That brings about self-sustaining propagation of disturbances. Electromagnetic radiation includes such as visible LIGHT, ultraviolet and infrared radiation, X-RAYS, GAMMA RAYS, radio and television waves, and microwaves. Electromagnetic waves in VACUUM all travel at the same SPEED, called the SPEED OF LIGHT and conventionally denoted c, which has the value 2.99792458×10^8 meters per second (m/s). (*See* ELECTRICITY; ELECTROMAGNETISM; MAGNETISM.)

Types of Wave

Waves are categorized according to the character of the propagating disturbance. In a SCALAR wave, the disturbance is in a scalar quantity (i.e., a quantity that does not possess a spatial direction), such as TEMPERATURE, pressure, and density. An example is a sound wave in air, which can be described as a pressure wave. A longitudinal wave is one in which the disturbance is a local displacement along the direction of wave propagation, such as is produced in a rod by striking the end of the rod longitudinally. For another example, take a stretched-out Slinky and give the end a push or pull in the direction of the Slinky. A sound wave in air can

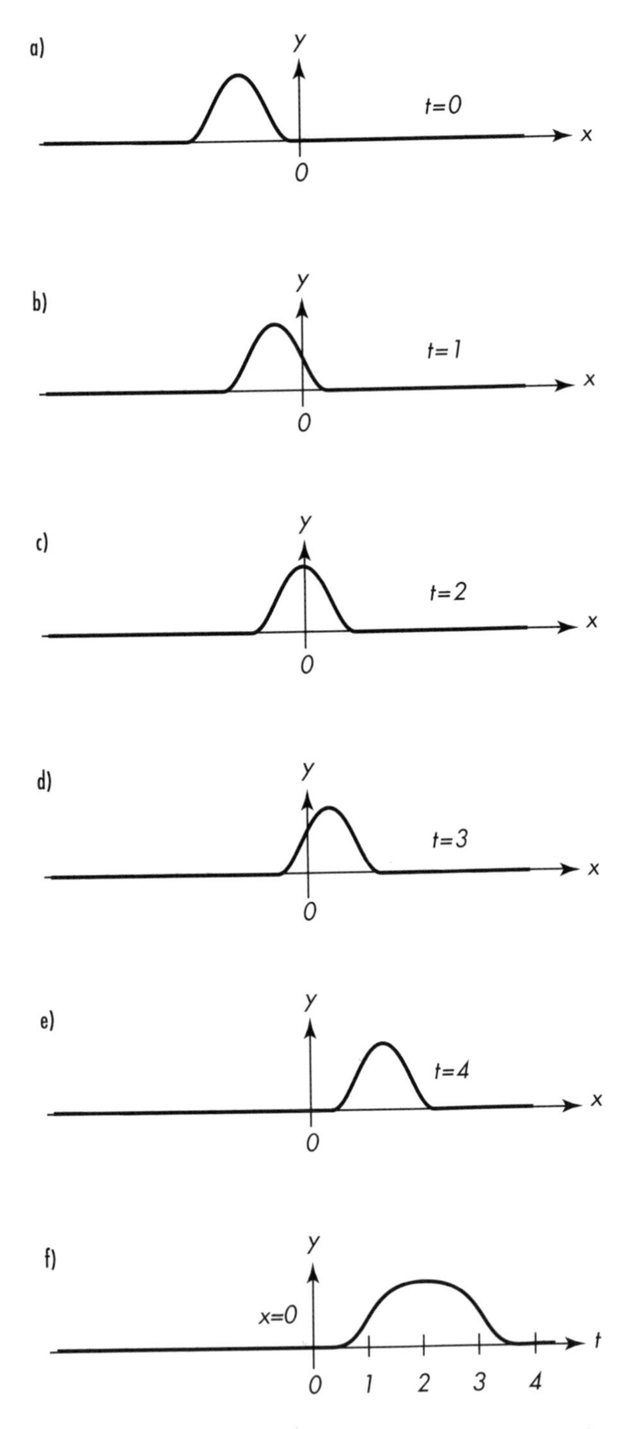

A vertically polarized transverse pulse wave propagating in the positive *x* direction. (a)–(e) Snapshots of the waveform for a sequence of times *t*. (f) The corresponding displacement at location *x* = 0 as a function of time.

also be described as a longitudinal wave, with the air undergoing longitudinal disturbance as a result of the pressure disturbance.

When the disturbance is a local displacement perpendicular to the direction of propagation, the wave is transverse. Such are waves in stretched strings. In the Slinky example, shake the end perpendicularly to the Slinky. Also, electromagnetic waves are of this type. Besides those, there are additional types of waves. A torsion wave, for instance, propagates a twist displacement and is a version of a transverse wave. Waves need not be of any pure type. Waves through the Earth, such as are produced by earthquakes, might consist of a mixture of types. Surface waves on a liquid involve a combination of longitudinal and transverse disturbances.

A transverse wave may possess the property of POLARIZATION. Put inversely, a transverse wave is said to be unpolarized when the propagating disturbance occurs in all directions perpendicular to the direction of propagation, with no particular relation among the disturbances in the various directions. In a polarized wave, on the other hand, there are constraints and relations among the disturbances in the various directions. In linear, or plane, polarization, for instance, the disturbance is confined solely to a single direction. An example is a horizontal stretched string in which a disturbance involving purely vertical local displacement (i.e., with no horizontal component) is propagated. There are other kinds of polarization, such as circular and elliptical.

Wave Propagation

As a wave propagates, the medium might absorb energy from it, thus reducing its INTENSITY. The wave is then said to be attenuated as it propagates. (*See* ABSORPTION.)

When the transmission properties of a medium change rather abruptly along the path of propagation of a wave, such as when a wave is incident on an interface between two different media, the wave undergoes (generally partial) REFLECTION. This is a breakup of the wave into two waves, one of which, the reflected wave, propagates back into the region of initial properties (into the initial medium), and the other, the transmitted wave, proceeds into the new-property region (through the interface into the second medium). If the direction of the transmitted wave is different from that of the original wave, the wave is said to be refracted. In the case of a well-defined interface, which we take to be planar for our discussion, the directions of propagation of the reflected and refracted waves relate to that of the

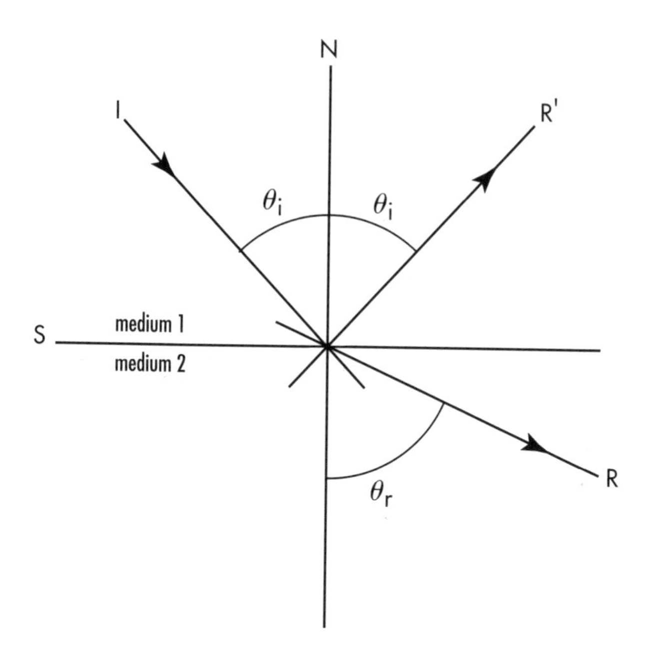

Reflection and refraction of a wave at the interface surface S between two media. The line of incidence I (in the direction of the incident wave), the normal N (perpendicular to the interface surface), the line of reflection R′ (in the direction of the reflected wave), and the line of refraction (in the direction of the refracted wave) R all lie in the plane of incidence, which is the plane of the figure. The angles of incidence and refraction are denoted θ_i and θ_r, respectively. They are related through Snell's law. The angle of reflection equals the angle of incidence.

incident wave in the following manner. Consider a straight line perpendicular to the interface surface, called the normal, and another line that passes through the point where the normal penetrates the interface and is in the direction of propagation of the incident wave, the line of incidence. Denote the smaller angle between the two lines, the angle of incidence, by θ_i. The two lines define a plane, the plane of incidence. The direction of reflection is indicated by the direction of a line lying in the plane of incidence, passing through the penetration point, on the opposite side of the normal from the line of incidence, and making the same angle with the normal as does the line of incidence, θ_i. This angle is the angle of reflection.

The direction of REFRACTION is similarly indicated by the direction of a line lying in the plane of incidence, passing through the penetration point, on the same side of the normal as the line of incidence, and making angle θ_r with the normal, the angle of refraction. According to SNELL'S LAW:

$$\frac{\sin\theta_i}{v_i} = \frac{\sin\theta_r}{v_r}$$

where v_i and v_r denote the propagation speeds of the waves in the medium of incidence (and reflection) and the medium of transmission (refraction), respectively. Note that if $v_r > v_i$ (i.e., the wave is incident on a medium of higher propagation speed), there is a range of incident angles θ_i—from a minimal angle, the critical angle, up to 90°—for which no refraction angle θ_r satisfies the above relation. In that case, there is no transmission at all, a situation called total internal reflection. Such is the case, for example, when light waves in glass or water impinge on the interface with air. The critical angle, θ_c, is given by:

$$\sin\theta_c = v_i/v_r$$

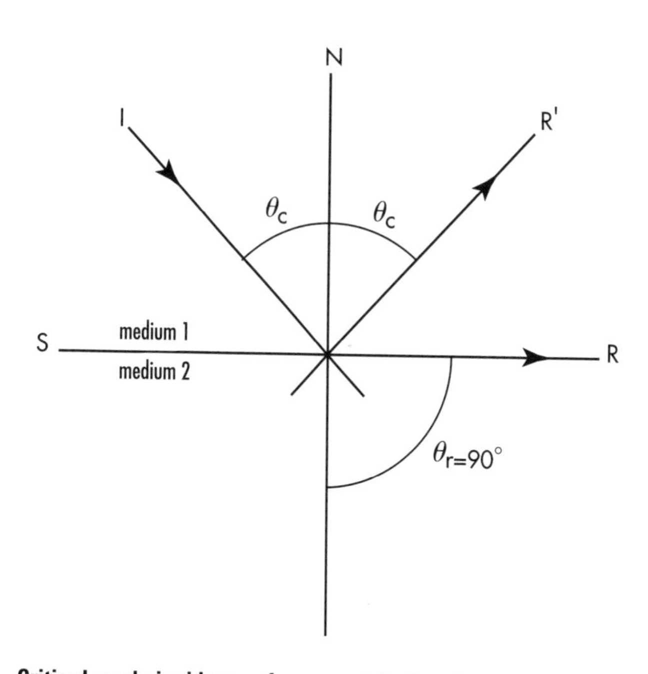

Critical-angle incidence of a wave at the interface surface S between two media. The line of incidence I (in the direction of the incident wave), the normal N (perpendicular to the interface surface), the line of reflection R′ (in the direction of the reflected wave), and the line of refraction R′ (in the direction of the refracted wave) all lie in the plane of incidence, which is the plane of the figure. The critical angle θ_c is that angle of incidence for which the angle of refraction θ_r equals 90°, which is the largest value it can have. At angles of incidence greater than the critical angle, there is no refracted wave and the incident wave is totally reflected, an effect called total internal reflection. The angle of reflection always equals the angle of incidence, both indicated by θ_c in the figure.

Periodic Wave

An important particular kind of wave is a periodic wave. It is produced when a medium (or the electromagnetic field) is disturbed periodically (i.e., is made to oscillate in a repetitive manner), which generates a moving repetitive spatial pattern, or repetitive waveform, along the direction of propagation. At every point of the medium along the line of propagation, the medium undergoes local OSCILLATION. At any instant, a snapshot of the whole medium shows a spatially repetitive waveform. As the waveform moves along, at all points the medium undergoes local, but well-correlated, oscillations. The oscillations at every point are characterized by their FREQUENCY, f, the number of cycles of oscillation per unit time in hertz (Hz), and their period, T, which is the time of a single cycle of oscillation in seconds (s) and equals $1/f$. The repetitive waveform, as revealed by a snapshot, has a characteristic repeat distance, the WAVELENGTH, λ. After a time interval of one period, every point of the medium completes a single full cycle of oscillation and returns to its initial state. During that time, the waveform moves such that it becomes indistinguishable from its initial appearance, which means it moves the repeat distance, a distance of a single wavelength. So during time T, the wave propagates through distance λ. Its propagation speed, v, in meters per second (m/s), is then:

$$v = \frac{\lambda}{T} = \frac{\lambda}{1/f} = f\lambda$$

This is a fundamental relation for all periodic waves. The speed of propagation is determined by the properties of the medium, while the frequency is the same as that of whatever is exciting the periodic wave. Together they determine the wavelength through this relation.

A particularly important kind of periodic wave is a sinusoidal wave. At every point along the line of propagation, the medium's local oscillation has a sinusoidal dependence on time, while at any instant a snapshot of the whole medium shows a spatially sinusoidal waveform. For a wave traveling in the positive x direction, that can be expressed as:

$$\text{Disturbance} = A \cos 2\pi(x/\lambda - ft)$$

where A denotes the wave's amplitude, x is the position in meters (m), and t is the time in seconds (s). The disturbance is a spatial and temporal sinusoidal oscillation between the limits of $+A$ and $-A$. This can be seen

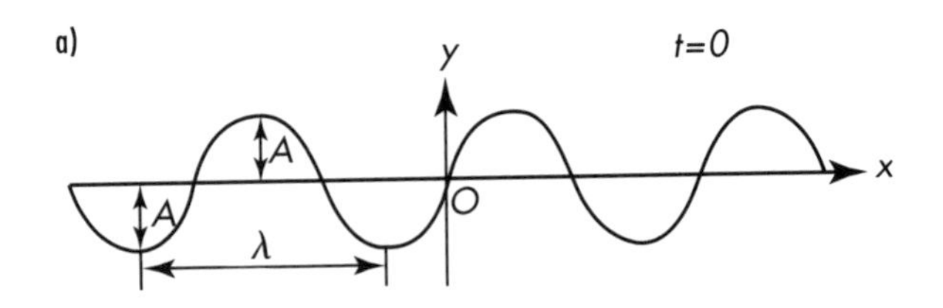

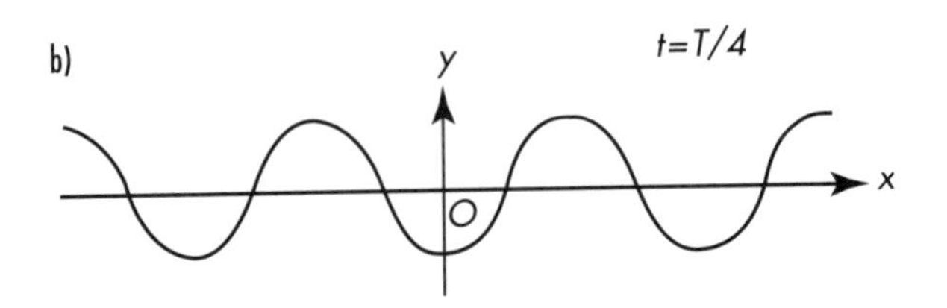

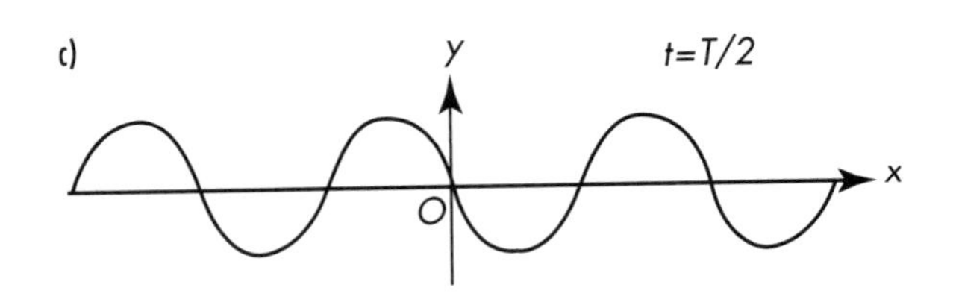

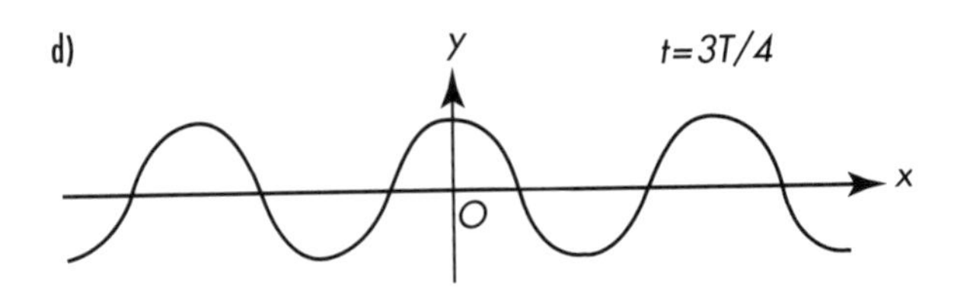

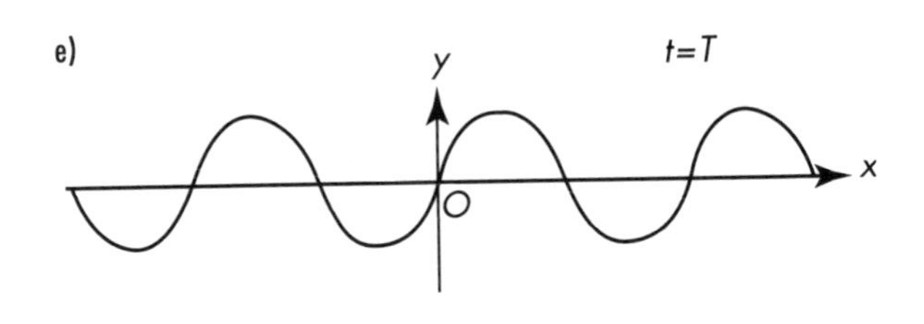

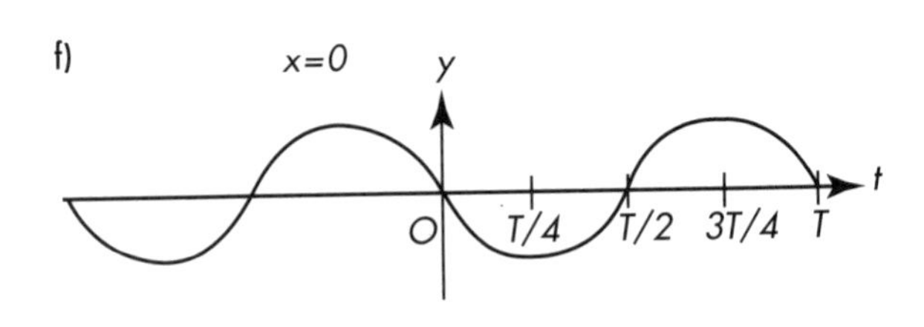

A vertically polarized transverse sinusoidal wave propagating in the positive *x* direction. (a) Wavelength λ and amplitude *A* are indicated. (a)–(e) Snapshots of the waveform for a sequence of times *t* at quarter-period intervals, where *T* denotes the period. (f) The corresponding displacement at location *x* = 0 as a function of time.

by keeping x constant, which results in sinusoidal time dependence of the disturbance at a fixed location, or by keeping t fixed, giving a sinusoidal waveform at a particular time. Another, equivalent form of this relation, one that shows the propagation speed, v, explicitly, is:

$$\text{Disturbance} = A \cos 2\pi(x - vt)/\lambda$$

(*See* HARMONIC MOTION.)

The effect of two or more waves passing through a medium and affecting it simultaneously is INTERFERENCE. The bending of waves around obstacles is DIFFRACTION.

A wave in a system can reflect from the system's boundaries, interfere with itself, and bring about a situation called a standing wave, in which the system oscillates while no wave propagation is apparent. The oscillation of a plucked stretched string (such as on a guitar), for instance, is in reality a standing wave, in which transverse waves are traveling along the string in both directions, reflecting from the ends, and interfering with each other. Those locations along a standing wave where the medium is at rest are termed NODEs. Positions of maximal oscillation amplitude are called antinodes. Examples are readily observed by shaking the end of a stretched rope transversely and finding those frequencies at which the shaking generates standing waves. The nodes and antinodes will be obvious. Their number is greater for higher standing-wave frequency.

See also DISPERSION; NATURAL FREQUENCY SHOCK WAVE; WAVEFRONT.

wavefront A surface in space that a WAVE has reached after propagating some time from its source is referred to as a wavefront. For example, in an isotropic medium (i.e., a medium in which the propagation SPEED is the same in all directions), the wavefronts of a wave emanating from a point source are concentric spherical surfaces. For a periodic wave. the OSCILLATIONs produced by the wave on a wavefront are all in PHASE (i.e., are all at the same phase at the same time).

wavelength The repeat distance of the repetitive spatial pattern of a periodic WAVE in the medium in which the wave is propagating is called the wavelength of the wave. A periodic wave is produced by a periodic disturbance imposed on the medium (i.e., when the medium is made to oscillate). The wave propagates away from the disturbance in the form of a moving, spatially repetitive waveform. The minimal distance a spatially repetitive pattern needs to be shifted in order to coincide with itself is its repeat distance. In the case of a periodic wave, the repeat distance is termed the wavelength. (By generalization, wavelength is sometimes used for repeat distance in the description of a repetitive spatial pattern that is not related to a wave in any way.) The SI UNIT of wavelength is the meter (m). (*See* OSCILLATION.)

wave-particle duality The term *wave-particle duality* refers to the dual nature of WAVE and particle phenomena, such that each possesses characteristics of the other. Wave-particle duality is a quantum effect. A particle is a discrete, localized entity, described by its position at any time and by its MASS, ENERGY, and LINEAR MOMENTUM. A wave is a continuous, nonlocalized effect, spread out over space, for which such quantities as position, momentum, mass, and energy are not defined. On the other hand, a periodic wave is characterized by its FREQUENCY and WAVELENGTH, which are irrelevant to a particle. In spite of the apparent incompatibility of wavelike and particlelike properties, they do coexist in wave-particle duality. An electromagnetic wave such as LIGHT, for example, in addition to its wavelike character—exhibited by DIFFRACTION, for instance—can also behave as a flow of particles, called PHOTONs, as demonstrated by the localized, grainy effect on a photographic film. In addition, ELECTRONs—particles, which can be counted individually by detectors—can exhibit wavelike behavior, which is the foundation of electron microscopy. The particles associated with sound waves are called PHONONs. (*See* ELECTROMAGNETISM; QUANTUM PHYSICS.)

Whether a wave-particle phenomenon exhibits its wave or particle aspect depends on how it is investigated. If diffraction is measured, the phenomenon will show it, thus revealing its wave nature. On the other hand, if interaction with a localized entity, such as an atom, is investigated, particle behavior will become manifest. In each instance, the other aspect is "suppressed." In the first, free particles do not undergo diffraction, but travel in straight lines. In the second, waves are not localized, but are spread out over space.

QUANTUM MECHANICS gives relations between the frequency and wavelength of waves, on the one hand, and the energy and momentum of the particles associated with the waves, on the other. For a wave of frequency f in hertz (Hz), the particles associated with it individually possess energy E in joules (J), such that:

$$E = hf$$

where h is the PLANCK CONSTANT and has the value $6.62606876 \times 10^{-34}$ joule·second (J·s). In addition, the wavelength of a wave, λ, in meters (m) and the magnitude of the momentum of each of its associated particles, p, in kilogram·meters per second (kg·m/s) are related by:

$$p = h/\lambda$$

This relation was proposed by the French physicist LOUIS-VICTOR PIERRE RAYMOND DE BROGLIE to better understand the wave nature of matter. The wavelength that corresponds to a particle according to this relation is known as the particle's de Broglie wavelength.

Let us consider YOUNG'S EXPERIMENT in the spirit of this discussion. In the experiment, light from a single source is directed through a pair of parallel slits to a screen or a photographic plate. As a result, an INTERFERENCE pattern appears. The pattern is completely understood in terms of coherent waves emanating from the slits and undergoing interference at the photographic plate. That is the wave picture of the experiment. The interference pattern on the plate is formed of many silver grains that result from the interaction of light with silver halide MOLECULES. This interaction is completely understood in terms of a PHOTON interacting with each molecule. So we are led to the particle picture, in which two streams of photons flow from the slits to the plate. But a photon, as a particle, can pass through only one slit or the other, not through both at the same time. Let us reduce the light INTENSITY to, say, one photon per minute and use a photon detector to find out which slit each photon passes through. We discover that each photon passes through this slit or that on its way to the plate, but the interference pattern never forms. Instead, we find a random scatter of silver grains. On the other hand, if we do not use the detector and wait long enough—at one photon per minute—the interference pattern does gradually take form. (*See* COHERENCE.)

Young's experiment is basically asking a wave question: how does the light from the two slits interfere (interference is a typical wave phenomenon), and how does the interference pattern depend on the light's wavelength (wavelength is also a typical wave phenomenon)? As long as that question is in force, there is no meaning to which slit the light passes through; it passes through both. That remains true even at such a low intensity that only one photon would pass through the system per minute if the particle picture were valid. But a single photon does not interfere with itself, and there is no other photon around for it to interact with. Still, the interference pattern forms. So the particle picture is not valid. By asking a wave question, we make the wave picture valid and invalidate the particle picture (i.e., we demonstrate the wave aspect and suppress the particle aspect).

When we place a photon detector at the slits to discover which slit the photon passes through, we are changing the question to a particle question: which path does the photon take? When that question is answered, the particle picture is valid: a photon passes through this or that slit. Then the interference pattern disappears, since particles do not undergo interference. In this manner, we make the particle picture valid and invalidate the wave picture; we demonstrate the particle aspect and suppress the wave aspect.

See also COMPLEMENTARITY; MATTER WAVE.

weak interaction This is one of the four fundamental INTERACTIONs among the ELEMENTARY PARTICLEs, along with the STRONG INTERACTION (also called the color FORCE or the GLUON force), the ELECTROMAGNETIC INTERACTION, and GRAVITATION. Like the strong interaction, the weak interaction has a short range, which is about 10^{-17} meter (m). The weak interaction affects all the MATTER particles (i.e., the QUARKs [and thus the HADRONS] and the LEPTONs). It is mediated by the intermediate vector BOSONS, which are the W^+ (possessing one elementary unit of positive electric CHARGE), the W^- (similarly charged negatively), and the Z^0 (electrically neutral). All three possess MASS. (*See* ELECTRICITY.)

According to present understanding, at sufficiently high ENERGIES (or, equivalently, high TEMPERATURES), high enough that the masses of the intermediate vector bosons are negligible, the weak and the electromagnetic interactions become unified to a single interaction, called the ELECTROWEAK INTERACTION. Then the massless PHOTON joins the effectively massless intermediate vector bosons to form a quartet of mediating bosons for the unified interaction.

See also CONSERVATION LAW; INVARIANCE PRINCIPLE; SYMMETRY.

weight The term *weight* refers to the FORCE on an object due to the gravitational attraction of the Earth or of whatever astronomical body the object is near. The direction of weight is toward the center of Earth (or of the attracting body). Its magnitude, W, is given by the product of the object's MASS, m, and the free-fall ACCELERATION, also called acceleration due to gravity or gravitational acceleration, g, at the object's location:

$$W = mg$$

Here W is in newtons (N), m in kilograms (kg), and g in meters per second per second (m/s^2), with nominal value of 9.8 meters per second per second m/s^2 at the surface of the Earth. (*See* FREE FALL; GRAVITATION.)

Care should be taken to distinguish between weight and mass, which are very different but are often confused with each other. Whereas mass is a SCALAR property of an object that does not depend on the object's location, an object's weight is a VECTOR quantity (a force) and varies from place to place in the universe. On the Moon, for example, the acceleration due to gravity, g, equals about 1.7 m/s^2, compared with around 9.8 m/s^2 on the surface of the Earth. So an object's weight on the Moon is about a sixth of its weight on Earth, or as commonly expressed, it weighs about a sixth as much as it does on Earth. The UNITS of mass and weight are different: the SI unit of mass is the kilogram (kg), while that of weight, as force, is the newton (N). However, due to the above relation between mass and weight, for fixed g, the weight of an object is proportional to its mass and can be used to measure its mass. So mass units are commonly used for weight. One's weight might be 65 kg (143 pounds), for instance, using mass units to express a force of some 637 N in magnitude. One's mass is then indeed 65 kg. On the Moon, one's mass would still be 65 kg, but one's weight there would be only about 111 N, which is approximately the weight of an 11-kg object on Earth.

There is often confusion also over the idea of weightlessness. True weightlessness simply means that an object's weight is zero. It follows from the above relation that this occurs only when the acceleration due to gravity is zero. This means that an object is weightless when it is far from any astronomical body or when the forces on it due to the gravitational attraction of nearby bodies cancel out. As an example of the latter, there exists a point between the Moon and Earth where the attraction of a body by the Moon exactly cancels its attraction by the Earth. At that point, an object is weightless.

Perceived, or apparent, weightlessness is the situation when an object is not supported against gravitational attraction and, as a result, is falling freely. It has weight, since it is attracted to the Earth (or to a nearby astronomical body). But with support lacking, a person in that situation feels the same as if she were indeed weightless, and moreover, she is not able to detect by any measurement performed in her immediate vicinity whether she is truly weightless or in free fall. Such is the case for astronauts in an orbiting space laboratory, as an example. This effect was among ALBERT EINSTEIN's considerations in developing the general theory of relativity. (*See* RELATIVITY, GENERAL THEORY OF.)

Weinberg, Steven (1932–) American *Physicist* A theoretical physicist, Steven Weinberg is best known for his work on the ELECTROWEAK INTERACTION. He received his Ph.D. in 1957 from Princeton University at Princeton, New Jersey. Weinberg was at Columbia University in New York, New York, during 1957–59, at the University of California at Berkeley during 1959–69, and at the Massachusetts Institute of Technology in Cambridge during 1969–73. Then he joined the physics department of Harvard University, also in Cambridge, and in 1986 moved to the University of Texas at Austin, where he is currently serving. Weinberg's research has been mostly in the field of ELEMENTARY PARTICLES. Notably, he worked on unifying the WEAK INTERACTION and the ELECTROMAGNETIC INTERACTION as GAUGE THEORIES to the electroweak interaction. Weinberg has also been working on the theory of the STRONG INTERACTION (quantum chromodynamics) and its possible unification with the electroweak interaction to a GRAND UNIFIED THEORY (GUT). In addition, he has performed some work in ASTROPHYSICS and COSMOLOGY. Weinberg shared the 1979 Nobel Prize in physics with Sheldon L. Glashow and Abdus Salam "for their contributions to the theory of the unified weak and electromagnetic interaction between elementary particles, including inter alia the prediction of the weak neutral current." (*See* THEORETICAL PHYSICS.)

With the help of Feynman diagrams, Steven Weinberg explains the electroweak theory of elementary particles, which he played a major role in developing. He shared the 1979 Nobel Prize in physics. ***(AIP Emilio Segrè Visual Archives, Weber Collection)***

Weyl, Claus Hugo Hermann (1885–1955) German/American *Mathematician, Physicist* Known among physicists as a mathematical physicist, Hermann Weyl contributed both to QUANTUM MECHANICS and, as a colleague of ALBERT EINSTEIN, to the general theory of relativity. He received his doctorate in mathematics from the University of Göttingen, Germany, in 1908. After teaching at Göttingen a few years, Weyl served as professor of mathematics at the Zürich Technische Hochschule, Switzerland, during 1913–30. He then returned to the University of Göttingen for three years, after which he joined the Institute for Advanced Study at Princeton, New Jersey, from 1933 until his retirement in 1952. Weyl worked on differential geometry in connection with the general theory of relativity and attempted to unify ELECTROMAGNETISM and GRAVITATION, whereby they would both be manifestations of geometric properties of SPACE-TIME. He also applied SYMMETRY considerations to quantum mechanics through the mathematical tool of group theory, which is especially suitable to quantum mechanics.

See also RELATIVITY, GENERAL THEORY OF.

Wigner, Eugene Paul (1902–1995) Hungarian/American *Physicist* A theoretical physicist, Eugene Wigner is considered to be one of the most outstanding of 20th-century physicists. His higher education was in chemical engineering, and he received his doctorate in that field from the Technische Hochschule in Berlin, Germany, in 1925. During 1933–36 Wigner was at Princeton University at Princeton, New Jersey, and then at the University of Wisconsin at Madison for 1936–38. In 1938, he returned to Princeton, where he remained as professor of mathematical physics until his retirement in 1971. During 1942–45 Wigner worked on the Manhattan Project, which was the U.S. government's

The 20th-century mathematical physicist Hermann Weyl contributed both to quantum mechanics, where he applied symmetry considerations, and to the general theory of relativity. *(AIP Emilio Segrè Visual Archives, Nina Courant Collection)*

World War II effort to develop a nuclear-FISSION ("atomic") bomb and nuclear ENERGY. Wigner possessed a wide range of interests in physics. He is best known for his application of SYMMETRY considerations to QUANTUM MECHANICS by means of the mathematical tool of group theory. Fields to which he contributed include atomic structure, CRYSTAL structure, and NUCLEAR PHYSICS. In 1963, Wigner was awarded one-half of the Nobel Prize in physics "for his contributions to the theory of the atomic nucleus and the elementary particles, particularly through the discovery and application of fundamental symmetry principles."

See also ATOM; ELEMENTARY PARTICLE; NUCLEUS; THEORETICAL PHYSICS.

work A SCALAR quantity, work is defined as a result of a FORCE acting over a distance in the direction of the force. Quantitatively, when a force **F** acts over an infinitesimal directed distance $d\mathbf{s}$, the infinitesimal amount of work performed, dW, is given by the scalar product:

$$dW = \mathbf{F} \cdot d\mathbf{s}$$

This can also be expressed as:

$$dW = F\, ds \cos \theta$$

where F and ds are the magnitudes of **F** and $d\mathbf{s}$, respectively, and θ is the angle between the two VECTORS. Work is in joules (J), force in newtons (N), and distance in meters (m).

When the force is in the direction of motion (i.e., when $\theta = 0$ and $\cos \theta = 1$), the work done equals simply $F\, ds$. When the force is opposite to the direction of

Eugene Wigner was a theoretical physicist who worked in a variety of fields, including nuclear physics, and is known for his application of symmetry considerations to quantum mechanics. He shared the 1963 Nobel Prize in physics. *(AIP Emilio Segrè Visual Archives, Wigner Collection)*

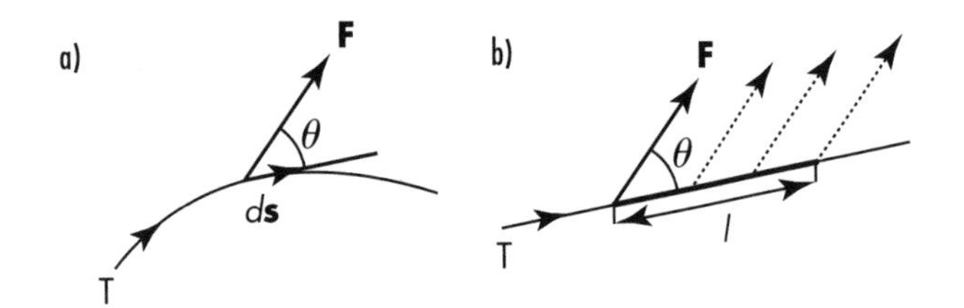

(a) The work done when a force F acts over an infinitesimal directed distance *ds* is given by $dW = \mathbf{F} \cdot d\mathbf{s}$. This can also be expressed as $dW = F\,ds \cos\theta$, where *F* and *ds* denote the magnitudes of the vectors F and *ds*, respectively, and θ is the angle between them. In the figure, the particle upon which the force acts is moving along a trajectory T. (b) In the special case when the force is constant and the trajectory straight, the work done by the force acting along path length *l* is $W = Fl \cos\theta$, where θ is the angle between F and the direction of motion.

motion (such as a braking force), $\theta = 180°$ and $\cos\theta = -1$, the work is negative and equals $-F\,ds$. When the force is perpendicular to the direction of motion (such as in the case of a CENTRIPETAL FORCE or a magnetic force), then $\theta = 90°$ and $\cos\theta = 0$, and no work is performed. (*See* MAGNETISM.)

The total work performed when a (possibly varying) force acts along a path is found by summing up (i.e., by integrating) dW along the path. A constant force of magnitude F, acting along a straight line segment of length l at constant angle θ from the direction of motion, performs work W such that:

$$W = Fl \cos\theta$$

Work is intimately related to ENERGY through the WORK-ENERGY THEOREM, which states that the work done on an otherwise isolated system equals the increase of total energy content (in all forms) of the system. Conversely, when an otherwise isolated system performs work on its surroundings, the work equals the decrease of the system's total energy content. In this manner, work can be viewed as a "currency" through which energy of one system is converted to energy of another system.

From the basic definition for work, presented above, it is possible to derive expressions for work performed in various additional specific cases. When a constant torque of magnitude τ acts through angle φ around an axis of rotation, for instance, the amount of work performed is:

$$W = \tau\varphi$$

work-energy theorem The work-energy theorem states that the WORK done on an otherwise isolated system equals the increase of total ENERGY content (in all forms) of the system. Conversely, when an otherwise isolated system performs work on its surroundings, the work done by the system equals the decrease of the system's total energy content. In this manner, work can be viewed as a "currency" through which energy of one system is converted to energy of another system. It follows that when a system is completely isolated, so no work is done on it and it does no work on its surroundings, the system's total energy content remains constant. That is the law of conservation of energy. The energy of such a system can change form over time. It might convert from KINETIC ENERGY to POTENTIAL ENERGY and back again, for instance. But the total of all forms of energy of an isolated system remains constant. (*See* CONSERVATION LAW.)

X-ray Electromagnetic RADIATION in the region of the electromagnetic SPECTRUM between ultraviolet and GAMMA RAYs is called X-rays. The range of WAVELENGTHs of X-rays spans from around 10^{-8} meter (m) down to about 10^{-12} m, with the longer wavelengths termed soft X-rays and the shorter called hard X-rays. X-rays are produced artificially by bombarding targets, normally made of METAL, with high-energy ELECTRONs. One source of such X-rays is BREMSSTRAHLUNG, which is the radiation emitted by decelerating electrons. The other source is the X-rays emitted by the ATOMs of the target after they are excited to high-ENERGY states by the electrons impacting on them. (*See* ELECTROMAGNETISM.)

X-rays are very useful for imaging, such as in medicine, dentistry, and industry. A modern development in X-ray use is computed tomography, or CT (also called CAT), whereby a computer analyzes the results of passing X-rays through the scanned object in many different directions throughout 360° around the object and generates an image of a slice through the object. X-rays are used also in X-ray CRYSTALLOGRAPHY, which is the determination of the structure of crystals and of their components from the DIFFRACTION pattern of scattered X-rays that results when an X-ray beam of known wavelength is directed at the crystal.

Natural X-rays emitted by celestial objects and phenomena, which are detected by X-ray telescopes, are of great interest to astronomers. One source of such X-rays is the hot matter in orbit around NEUTRON STARs and BLACK HOLEs. Such matter can be excited to high-energy states, causing it to emit X-rays. It can also become so hot that X-ray-emitting nuclear reactions are initiated. The matter comes from a companion star, from which it is pulled by the strong gravitational attraction of the neutron star or black hole. It collects around the latter in what is called an accretion disk. (*See* BRAGG'S LAW; GRAVITATION; NUCLEAR PHYSICS; NUCLEUS.)

Young's experiment Named for the British scientist Thomas Young (1773–1829), Young's experiment was carried out in the early 19th century and was pivotal in determining the WAVE nature of LIGHT. In that experiment and in others like it, light from a single source illuminates a screen, but only after the light is allowed to pass through a close pair of holes or parallel slits in a barrier. As a result, patterns of variously colored and illuminated areas are found on the screen. They are very different from those produced by illumination through a single opening. Wave behavior explains the effect. The light passing through the openings spreads out due to DIFFRACTION, and so the screen has an area where there is an overlapping of illumination from the two openings. At each point in the area of overlap, there is INTERFERENCE between the rays from the two sources, depending on the WAVELENGTH (i.e., the color) of the light and on the difference of distances between the point and each of the openings. In the extreme, constructive interference produces maximal INTENSITY, while destructive interference gives minimal intensity and even cancellation. Intermediate degrees of interference correspondingly result in intermediate intensity. Thus a pattern of colors and of brightness and darkness appears on the screen.

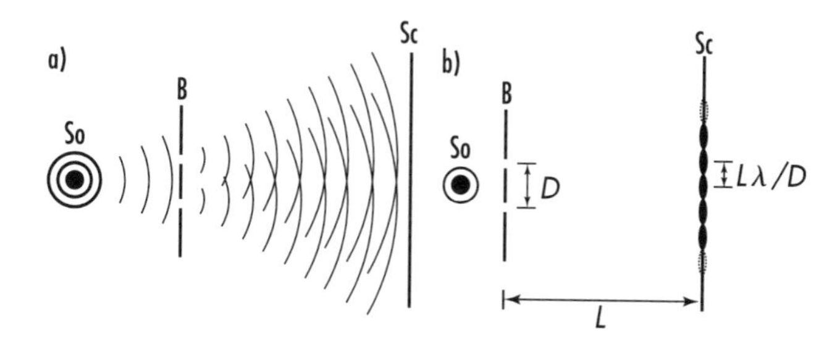

Young's experiment demonstrates interference of light and thus light's wave nature. (a) Monochromatic light from a single source, So, falls on two narrow, close, parallel slits in a barrier B. The light passing through the slits spreads out, due to diffraction, and illuminates the screen, Sc. (b) As a result of interference of light from the two slits, there appears on the screen a pattern of light bands, called fringes, separated by dark bands. The distance between adjacent fringes near the center of the screen is $L\lambda/D$, where λ denotes the wavelength of the light, L is the distance from the barrier to the screen, and D is the distance between the centers of the slits.

In the most common way the experiment is performed today, light of a single wavelength, called monochromatic light, is passed through two parallel slits. The resulting pattern on the screen is a series of light bands, called fringes, separated by dark bands (with the single-slit diffraction pattern superposed on it). In the usual case, when the distance D between the centers of the slits is much smaller than the distance L from the slits to the screen, the spacing between adjacent fringes near the center of the interference pattern is given in meters (m) by $L\lambda/D$, where λ denotes the wavelength of the monochromatic light, and L, λ, and D are all in meters (m).

Zeeman effect Named for the Dutch physicist Pieter Zeeman (1865–1943), the Zeeman effect is the splitting of atomic spectral lines when the atoms are put in a magnetic FIELD. It is a result of the splitting of atomic ENERGY levels in the presence of a magnetic field. In the absence of a magnetic field, ELECTRON configurations that differ only in their directions of orbital ANGULAR MOMENTUM and of electron SPIN all possess the same energy, a situation termed degeneracy. A magnetic field breaks the degeneracy by causing such configurations to have different energies, since the MAGNETIC DIPOLE moments associated with electron angular momentum and spin are affected differently when they are oriented differently to the direction of the magnetic field. According to QUANTUM MECHANICS, the allowed orientations are finite in number for any degenerate atomic energy level and are characterized by the QUANTUM NUMBERS called orbital magnetic quantum number and spin magnetic quantum number. (*See* ATOM; BOHR THEORY; MAGNETISM; MOMENT, MAGNETIC DIPOLE; SPECTROSCOPY; SPECTRUM.)

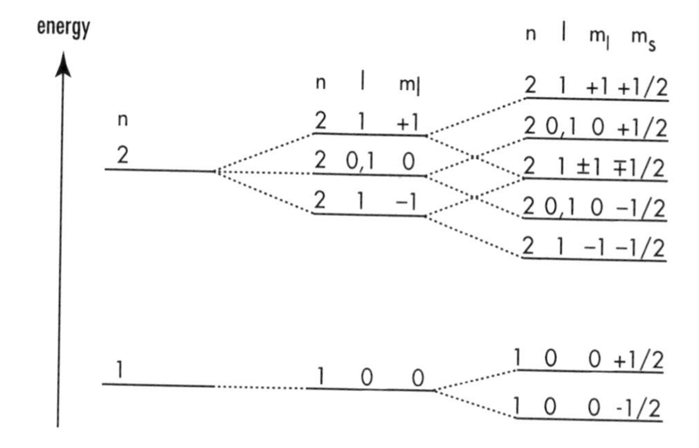

The Zeeman effect—the splitting of atomic spectral lines when the atoms are put in a magnetic field—is caused by the splitting of atomic energy levels in the presence of a magnetic field. This splitting is shown for the two lowest energy levels of hydrogen. *Left:* The energy levels with no magnetic field. *Center:* The splitting brought about by a magnetic field, while ignoring the effect due to electron spin. *Right:* The full array of split energy levels, taking electron spin into account. In all cases, the relevant quantum numbers of the energy levels are shown: principal quantum number n, orbital angular momentum quantum number l, magnetic quantum number m_l, and spin quantum number m_s. The diagram is not drawn to scale.

APPENDIX I

BIBLIOGRAPHY AND WEB RESOURCES

For a general source on physics—to fill in details and obtain further information—one can use any of the many college and university introductory and modern-physics textbooks. Some are listed here, but others can serve equally well:

Beiser, Arthur, and Isabel Berg. *Concepts of Modern Physics.* 6th ed. New York: McGraw-Hill, 2002.

Feynman, Richard P., Robert B. Leighton, and Matthew L. Sands. *The Feynman Lectures on Physics.* San Francisco: Addison-Wesley, 1989.

Halliday, David, Robert Resnick, and Jearl Walker. *Fundamentals of Physics.* 6th ed. New York: Wiley, 2002.

Hecht, Eugene. *Physics: Calculus.* Pacific Grove, Calif.: Brooks/Cole, 1996.

———. *Physics: Algebra and Trigonometry.* 3d ed. Pacific Grove, Calif.: Brooks/Cole, 2002.

Krane, Kenneth S. *Modern Physics.* 2d ed. New York: Wiley, 1995.

Moore, T. A. *Six Ideas That Shaped Physics.* 2d ed. New York: McGraw-Hill, 2003.

Serway, Raymond A., and John W. Jewett, Jr. *Principles of Physics.* 3d ed. Pacific Grove, Calif.: Brooks/Cole, 2002.

Serway, Raymond A., Clement J. Moses, and Curt A. Moyer. *Modern Physics.* 2d ed. Pacific Grove, Calif.: Brooks/Cole, 1997.

A Classic Popularization of Ideas and Concepts of Modern Physics

Gamow, George. *Mr Tompkins in Paperback* (comprising *Mr Tompkins in Wonderland* and *Mr Tompkins Explores the Atom*). New York: Cambridge University Press, 1993.

One Source for Chemistry (A University Textbook); Other Such Textbooks Can Be of Similar Service

Moore, John W., Conrad L. Stanitski, and Peter C. Jurs. *Chemistry: The Molecular Science.* Fort Worth, Tex.: Harcourt, 2002.

Two Reference Books for Physics

Daintith, John, and John O.E. Clark. *The Facts On File Dictionary of Physics.* 3d ed. New York: Facts On File, 1999.

The Diagram Group. *The Facts On File Physics Handbook.* New York: Facts On File, 2001.

For the History of Cosmology

Harrison, Edward R. *Masks of the Universe.* 2d ed. New York: Cambridge University Press, 2003.

ON COSMOLOGY ITSELF AND ASTRONOMY, A SAMPLING FROM AMONG MANY AVAILABLE BOOKS

Bennett, Jeffrey, Megan Donahue, Nicholas Schneider, and Mark Voit. *The Cosmic Perspective.* 2d ed. San Francisco: Addison-Wesley, 2002.

Freedman, Roger A., and William J. Kaufmann, III. *Universe.* 6th ed. New York: Freeman, 2001.

Greene, Brian. *The Elegant Universe: Superstrings, Hidden Dimensions, and the Quest for the Ultimate Theory.* New York: Norton, 1999.

Hawking, Stephen. *A Brief History of Time.* New York: Bantam Doubleday Dell, 1998.

Thorne, Kip S. *Black Holes and Time Warps: Einstein's Outrageous Legacy.* New York: Norton, 1995.

AN APPROACHABLE INTRODUCTION TO ELEMENTARY PARTICLES

Ne'eman, Yuval, and Yoram Kirsh. *The Particle Hunters.* 2d ed. Cambridge, U.K.: Cambridge University Press, 1996.

A SOURCE FOR SYMMETRY

Rosen, Joe. *Symmetry in Science: An Introduction to the General Theory.* New York: Springer, 1995.

PERIODICALS

Scientific American (a monthly) and *Science News* (a weekly) are excellent sources for the latest developments and recent thinking in physics.

INTERNET SOURCES OF INFORMATION ON PHYSICS

The American Physical Society (APS) offers a wealth of possibilities on its website. URL: www.aps.org/resources/. The home page of the APS is at URL: www.aps.org/. Downloaded: May 28, 2003.

On-line information about Nobel Prize laureates in physics is available on the Nobel Foundation website. URL: www.nobel.se/physics/. Downloaded: May 28, 2003.

For other Nobel Prize-related information, the Nobel Foundation's home page is at URL: www.nobel.se/. Downloaded: May 28, 2003.

Appendix II

NOBEL LAUREATES IN PHYSICS

This is a list, in chronological order, of all the Nobel Prize laureates in physics since the inception of the prize. For each year, the name of the laureate (or laureates) is given, with dates of birth and death, nationality (at the time of the award), and the Nobel Committee's description of the accomplishment(s) for which the prize was awarded.

1901 Wilhelm Conrad Röntgen (1845–1923), Germany

"In recognition of the extraordinary services he has rendered by the discovery of the remarkable rays subsequently named after him."

1902 The prize was awarded jointly to:

Hendrik Antoon Lorentz (1853–1928), the Netherlands
Pieter Zeeman (1865–1943), the Netherlands

"In recognition of the extraordinary service they rendered by their researches into the influence of magnetism upon radiation phenomena."

1903 The prize was divided, with one half awarded to:

Antoine Henri Becquerel (1852–1908), France

"In recognition of the extraordinary services he has rendered by his discovery of spontaneous radioactivity."

The other half was jointly awarded to:

Pierre Curie (1859–1906), France
Marie (Sklodowska) Curie (1867–1934), France

"In recognition of the extraordinary services they have rendered by their joint researches on the radiation phenomena discovered by Professor Henri Becquerel."

1904 Lord Rayleigh (John William Strutt) (1842–1919), Great Britain

"For his investigations of the densities of the most important gases and for his discovery of argon in connection with these studies."

1905 Philipp Eduard Anton von Lenard (1862–1947), Germany

"For his work on cathode rays."

1906 Joseph John Thomson (1856–1940), Great Britain

"In recognition of the great merits of his theoretical and experimental investigations on the conduction of electricity by gases."

1907 Albert Abraham Michelson (1852–1931), United States

"For his optical precision instruments and the spectroscopic and metrological investigations carried out with their aid."

1908 Gabriel Jonas Lippmann (1845–1921), France

"For his method of reproducing colors photographically based on the phenomenon of interference."

1909 The prize was awarded jointly to:

Guglielmo Marconi (1874–1937), Italy
Karl Ferdinand Braun (1850–1918), Germany

"In recognition of their contributions to the development of wireless telegraphy."

1910 Johannes Diderik van der Waals (1837–1923), the Netherlands

"For his work on the equation of state for gases and liquids."

1911 Wilhelm Carl Werner Otto Fritz Franz Wien (1864–1928), Germany

"For his discoveries regarding the laws governing the radiation of heat."

1912 Nils Gustaf Dalen (1869–1937), Sweden

"For his invention of automatic regulators for use in conjunction with gas accumulators for illuminating lighthouses and buoys."

1913 Heike Kamerlingh Onnes (1853–1926), the Netherlands

"For his investigations on the properties of matter at low temperatures which led, inter alia, to the production of liquid helium."

1914 Max Theodor Felix von Laue (1879–1960), Germany

"For his discovery of the diffraction of x-rays by crystals."

1915 The prize was awarded jointly to:

William Henry Bragg (1862–1942), Great Britain

and his son

William Lawrence Bragg (1890–1971), Great Britain

"For their services in the analysis of crystal structure by means of x-rays."

1916 The prize money was withheld and not awarded for this year.

1917 The prize was reserved and awarded in 1918 to:

Charles Glover Barkla (1877–1944), Great Britain

"For his discovery of the characteristic Roentgen radiation of the elements."

1918 The prize was reserved and awarded in 1919 to:

Max Karl Ernst Ludwig Planck (1858–1947), Germany

"In recognition of the services he rendered to the advancement of physics by his discovery of energy quanta."

1919 Johannes Stark (1874–1957), Germany

"For his discovery of the Doppler effect in canal rays and the splitting of spectral lines in electric fields."

1920 Charles Edouard Guillaume (1861–1938), Switzerland

"In recognition of the service he has rendered to precision measurements in physics by his discovery of anomalies in nickel steel alloys."

1921 The prize was reserved and awarded in 1922 to:

Albert Einstein (1879–1955), Germany

"For his services to theoretical physics, and especially for his discovery of the law of the photoelectric effect."

1922 Niels Henrik David Bohr (1885–1962), Denmark

"For his services in the investigation of the structure of atoms and of the radiation emanating from them."

1923 Robert Andrews Millikan (1868–1953), United States

"For his work on the elementary charge of electricity and on the photoelectric effect."

1924 The prize was reserved and awarded in 1925 to:

Karl Manne Georg Siegbahn (1886–1978), Sweden

"For his discoveries and research in the field of X-ray spectroscopy."

1925 The prize was reserved and awarded jointly in 1926 to:

James Franck (1882–1964), Germany
Gustav Ludwig Hertz (1887–1975), Germany

"For their discovery of the laws governing the impact of an electron upon an atom."

1926 Jean Baptiste Perrin (1870–1942), France

"For his work on the discontinuous structure of matter, and especially for his discovery of sedimentation equilibrium."

1927 The prize was divided, with one-half awarded to:

Arthur Holly Compton (1892–1962), United States

"For his discovery of the effect named after him."

The other half was awarded to:

Charles Thomson Rees Wilson (1869–1959), Great Britain

"For his method of making the paths of electrically charged particles visible by condensation of vapour."

1928 The prize was reserved and awarded in 1929 to:

Owen Willans Richardson (1879–1959), Great Britain

"For his work on the thermionic phenomenon and especially for the discovery of the law named after him."

1929 The prize was reserved and awarded in 1930 to:

Louis-Victor Pierre Raymond de Broglie (1892–1987), France

"For his discovery of the wave nature of electrons."

1930 Chandrasekhara Venkata Raman (1888–1970), India

"For his work on the scattering of light and for the discovery of the effect named after him."

1931 The prize money was withheld for this year.

1932 The prize was reserved and awarded in 1933 to:

Werner Heisenberg (1901–1976), Germany

"For the creation of quantum mechanics, the application of which has, inter alia, led to the discovery of the allotropic forms of hydrogen."

1933 The prize was awarded jointly to:

Erwin Schrödinger (1887–1961), Austria
Paul Adrien Maurice Dirac (1902–1984), Great Britain

"For the discovery of new productive forms of atomic theory."

1934 The prize money was withheld for this year.

1935 James Chadwick (1891–1974), Great Britain

"For the discovery of the neutron."

1936 The prize was divided, with one-half awarded to:

Victor Franz Hess (1883–1964), Austria

"For his discovery of cosmic radiation."

The other half was awarded to:

Carl David Anderson (1905–1991), United States

"For his discovery of the positron."

1937 The prize was awarded jointly to:

Clinton Joseph Davisson (1881–1958), United States
George Paget Thomson (1892–1975), Great Britain

"For their experimental discovery of the diffraction of electrons by crystals."

1938 Enrico Fermi (1901–1954), Italy

"For his demonstrations of the existence of new radioactive elements produced by neutron irradiation, and for his related discovery of nuclear reactions brought about by slow neutrons."

1939 Ernest Orlando Lawrence (1901–1958), United States

"For the invention and development of the cyclotron and for results obtained with it, especially with regard to artificial radioactive elements."

1940 The prize money was withheld for this year.

1941 The prize money was withheld for this year.

1942 The prize money was withheld for this year.

1943 The prize was reserved and awarded in 1944 to:

Otto Stern (1888–1969), United States

"For his contribution to the development of the molecular ray method and his discovery of the magnetic moment of the proton."

1944 Isidor Isaac Rabi (1898–1988), United States

"For his resonance method for recording the magnetic properties of atomic nuclei."

1945 Wolfgang Pauli (1900–1958), Austria

"For the discovery of the Exclusion Principle, also called the Pauli Principle."

1946 Percy W. Bridgman (1882–1961), United States

"For the invention of an apparatus to produce extremely high pressures, and for the discoveries he made therewith in the field of high pressure physics."

1947 Edward V. Appleton (1892–1965), Great Britain

"For his investigations of the physics of the upper atmosphere especially for the discovery of the so-called Appleton layer."

1948 Patrick M. S. Blackett (1897–1974), Great Britain

"For his development of the Wilson cloud chamber method, and his discoveries therewith in the fields of nuclear physics and cosmic radiation."

1949 Hideki Yukawa (1907–1981), Japan

"For his prediction of the existence of mesons on the basis of theoretical work on nuclear forces."

1950 Cecil F. Powell (1903–1969), Great Britain

"For his development of the photographic method of studying nuclear processes and his discoveries regarding mesons made with this method."

1951 The prize was awarded jointly to:

John D. Cockcroft (1897–1967), Great Britain
Ernest T. S. Walton (1903–1995), Ireland

"For their pioneer work on the transmutation of atomic nuclei by artificially accelerated atomic particles."

1952 The prize was awarded jointly to:

Felix Bloch (1905–1983), United States
Edward Mills Purcell (1912–1997), United States

"For their development of new methods for nuclear magnetic precision measurements and discoveries in connection therewith."

1953 Frits (Frederik) Zernike (1888–1966), The Netherlands

"For his demonstration of the phase contrast method, especially for his invention of the phase contrast microscope."

1954 The prize was divided, with one-half awarded to:

Max Born (1882–1970), Great Britain

"For his fundamental research in quantum mechanics, especially for his statistical interpretation of the wavefunction."

The other half was awarded to:

Walther W. G. Bothe (1891–1957), Germany

"For the coincidence method and his discoveries made therewith."

1955 The prize was divided, with one-half awarded to:

Willis Eugene Lamb (1913–), United States

"For his discoveries concerning the fine structure of the hydrogen spectrum."

The other half awarded to:

Polykarp Kusch (1911–1993), United States

"For his precision determination of the magnetic moment of the electron."

1956 The prize was awarded jointly to:

William Shockley (1910–1989), United States
John Bardeen (1908–1991), United States
Walter Houser Brattain (1902–1987), United States

"For their researches on semiconductors and their discovery of the transistor effect."

1957 The prize was awarded jointly to:

Chen Ning Yang (1922–), China and United States
Tsung-Dao Lee (1926–), China and United States

"For their penetrating investigation of the so-called parity laws which has led to important discoveries regarding the elementary particles."

1958 The prize was awarded jointly to:

Pavel Alekseyevich Cherenkov (1904–1990), USSR
Ilya Mikhailovich Frank (1908–1990), USSR
Igor Yevgenyevich Tamm (1885–1971), USSR

"For the discovery and the interpretation of the Cherenkov effect."

1959 The prize was awarded jointly to:

Emilio Gino Segrè (1905–1989), United States
Owen Chamberlain (1920–), United States

"For their discovery of the antiproton."

1960 Donald A. Glaser (1926–), United States

"For the invention of the bubble chamber."

1961 The prize was divided, with one-half awarded to:

Robert Hofstadter (1915–1990), United States

"For his pioneering studies of electron scattering in atomic nuclei and for his thereby achieved discoveries concerning the structure of the nucleons."

The other half was awarded to

Rudolf L. Mössbauer (1929–), Germany

"For his researches concerning the resonance absorption of gamma radiation and his discovery in this connection of the effect which bears his name."

1962 Lev Davidovich Landau (1908–1968), USSR

"For his pioneering theories for condensed matter, especially liquid helium."

1963 The prize was divided, with one-half awarded to:

Eugene P. Wigner (1902–1995), United States

"For his contributions to the theory of the atomic nucleus and the elementary particles, particularly through the discovery and application of fundamental symmetry principles."

The other half was awarded jointly to:

Maria Goeppert-Mayer (1906–1972), United States
Johannes Hans D. Jensen (1907–1973), Germany

"For their discoveries concerning nuclear shell structure."

1964 The prize was divided, with one-half awarded to:

Charles H. Townes (1915–), United States

"For fundamental work in the field of quantum electronics, which has led to the construction of oscillators and amplifiers based on the maser-laser principle."

The other half was awarded jointly to:

Nikolai Gennadievich Basov (1922–2001), USSR
Alexander Mikhailovich Prokhorov (1916–2002), USSR

"For basic researches in the field of experimental physics, which led to the discovery of the maser and the laser."

1965 The prize was awarded jointly to:

Sin-Itiro Tomonaga (1906–1979), Japan,
Julian S. Schwinger (1918–1994), United States
Richard P. Feynman (1918–1988), United States

"For their fundamental work in quantum electrodynamics, with profound consequences for the physics of elementary particles."

1966 Alfred Kastler (1902–1984), France

"For the discovery and development of optical methods for studying hertzian resonances in atoms."

1967 Hans Albrecht Bethe (1906–), United States

"For his contributions to the theory of nuclear reactions, especially his discoveries concerning the energy production in stars."

1968 Luis W. Alvarez (1911–1988), United States

"For his decisive contributions to elementary particle physics, in particular the discovery of a large number of resonance states, made possible through his development of the technique of using hydrogen bubble chamber and data analysis."

1969 Murray Gell-Mann (1929–), United States

"For his contributions and discoveries concerning the classification of elementary particles and their interactions."

1970 The prize was divided, with one-half awarded to:

Hannes Olof Gosta Alfven (1908–1995), Sweden

"For fundamental work and discoveries in magneto-hydrodynamics with fruitful applications in different parts of plasma physics."

The other half was awarded to:

Louis Eugene Felix Neel (1904–2000), France

"For fundamental work and discoveries concerning antiferromagnetism and ferrimagnetism which have led to important applications in solid state physics."

1971 Dennis Gabor (1900–1979), Great Britain

"For his invention and development of the holographic method."

1972 The prize was awarded jointly to:

John Bardeen (1908–1991), United States
Leon N. Cooper (1930–), United States
J. Robert Schrieffer (1931–), United States

"For their jointly developed theory of superconductivity, usually called the BCS-theory."

1973 The prize was divided, with one-half awarded jointly to:

Leo Esaki (1925–), Japan
Ivar Giaever (1929–), United States

"For their experimental discoveries regarding tunneling phenomena in semiconductors and superconductors, respectively."

The other half was awarded to:

Brian D. Josephson (1940–), Great Britain

"For his theoretical predictions of the properties of a supercurrent through a tunnel barrier, in particular those phenomena which are generally known as the Josephson effects."

1974 The prize was awarded jointly to:

Martin Ryle (1918–1984), Great Britain
Antony Hewish (1924–), Great Britain

"For their pioneering research in radio astrophysics: Ryle for his observations and inventions, in particular of the aperture synthesis technique; Hewish for his decisive role in the discovery of pulsars."

1975 The prize was awarded jointly to:

Aage Niels Bohr (1922–), Denmark
Benjamin R. Mottelson (1926–), Denmark
Leo James Rainwater (1917–1986), United States

"For the discovery of the connection between collective motion and particle motion in atomic nuclei and the development of the theory of the structure of the atomic nucleus based on this connection."

1976 The prize was awarded jointly to:

Burton Richter (1931–), United States
Samuel C. C. Ting (1936–), United States

"For their pioneering work in the discovery of a heavy elementary particle of a new kind."

1977 The prize was awarded jointly to:

Philip Warren Anderson (1923–), United States
Nevill Francis Mott (1905–1996), Great Britain
John H. Van Vleck (1899–1980), United States

"For their fundamental theoretical investigations of the electronic structure of magnetic and disordered systems."

1978 The prize was divided, with one-half awarded to:

Pyotr Leonidovich Kapitsa (1894–1984), USSR

"For his basic inventions and discoveries in the area of low-temperature physics."

The other half was awarded jointly to:

Arno A. Penzias (1933–), United States
Robert W. Wilson (1936–), United States

"For their discovery of cosmic microwave background radiation."

1979 The prize was awarded jointly to:

Sheldon L. Glashow (1932–), United States
Abdus Salam (1926–1996), Pakistan
Steven Weinberg (1933–), United States

"For their contributions to the theory of the unified weak and electromagnetic interaction between elementary particles, including inter alia the prediction of the weak neutral current."

1980 The prize was awarded jointly to:

James W. Cronin (1931–), United States
Val Logsdon Fitch (1923–), United States

"For the discovery of violations of fundamental symmetry principles in the decay of neutral K-mesons."

1981 The prize was divided, with one-half awarded jointly to:

Nicolaas Bloembergen (1920–), United States
Arthur L. Schawlow (1921–1999), United States

"For their contribution to the development of laser spectroscopy."

The other half was awarded to:

Kai M. B. Siegbahn (1918–), Sweden

"For his contribution to the development of high-resolution electron spectroscopy."

1982 Kenneth G. Wilson (1936–), United States

"For his theory for critical phenomena in connection with phase transitions."

1983 The prize was divided, with one-half awarded to:

Subrahmanyan Chandrasekhar (1910–1995), United States

"For his theoretical studies of the physical processes of importance to the structure and evolution of the stars."

The other half was awarded to:

William Alfred Fowler (1911–1995), United States

"For his theoretical and experimental studies of the nuclear reactions of importance in the formation of the chemical elements in the universe."

1984 The prize was awarded jointly to:

Carlo Rubbia (1934–), Italy
Simon Van Der Meer (1925–), the Netherlands

"For their decisive contributions to the large project, which led to the discovery of the field particles W and Z, communicators of weak interaction."

1985 Klaus Von Klitzing (1943–), Germany

"For the discovery of the quantized Hall effect."

1986 The prize was divided, with one-half awarded to:

Ernst Ruska (1906–1988), Germany

"For his fundamental work in electron optics, and for the design of the first electron microscope."

The other half awarded jointly to:

Gerd Binning (1947–), Germany
Heinrich Rohrer (1933–), Switzerland

"For their design of the scanning tunneling microscope."

1987 The prize was awarded jointly to:

J. Georg Bednorz (1950–), Germany
Karl Alexander Muller (1927–), Switzerland

"For their important breakthrough in the discovery of superconductivity in ceramic materials."

1988 The prize was awarded jointly to:

Leon M. Lederman (1922–), United States
Melvin Schwartz (1932–), United States
Jack Steinberger (1921–), United States

"For the neutrino beam method and the demonstration of the doublet structure of the leptons through the discovery of the muon-neutrino."

1989 The prize was divided, with one-half awarded to:

Norman F. Ramsey (1915–), United States

"For the invention of the separated oscillatory fields method and its use in the hydrogen maser and other atomic clocks."

The other half was awarded jointly to:

Hans G. Dehmelt (1922–), United States
Wolfgang Paul (1913–1993), Germany

"For the development of the ion trap technique."

1990 The prize was awarded jointly to:

Jerome I. Friedman (1930–), United States
Henry W. Kendall (1926–1999), United States
Richard E. Taylor (1929–), Canada

"For their pioneering investigations concerning deep inelastic scattering of electrons on protons and bound neutrons, which have been of essential importance for the development of the quark model in particle physics."

1991 Pierre-Gilles de Gennes (1932–), France

"For discovering that methods developed for studying order phenomena in simple systems can be generalized to more complex forms of matter, in particular to liquid crystals and polymers."

1992 Georges Charpak (1924–), France

"For his invention and development of particle detectors, in particular the multiwire proportional chamber."

1993 The prize was awarded jointly to:

Russell A. Hulse (1950–), United States
Joseph H. Taylor, Jr. (1941–), United States

"For the discovery of a new type of pulsar, a discovery that has opened up new possibilities for the study of gravitation."

1994 The prize was awarded jointly to:

Bertramin N. Brockhouse (1918–), Canada

"For pioneering contributions to the development of neutron scattering techniques for studies of condensed matter, specifically for the development of neutron spectroscopy."

Clifford G. Shull (1915–), United States

"For pioneering contributions to the development of neutron scattering techniques for studies of condensed matter, specifically for the development of the neutron diffraction technique."

1995 The prize was awarded jointly to:

Martin L. Perl (1927–), United States

"For pioneering experimental contributions to lepton physics, specifically for the discovery of the tau lepton."

Frederick Reines (1918–), United States

"For pioneering experimental contributions to lepton physics, specifically for the detection of the neutrino."

1996 The prize was awarded jointly to:

David M. Lee (1931–), United States
Douglas D. Osheroff (1945–), United States
Robert C. Richardson (1937–), United States

"For their discovery of superfluidity in helium-3."

1997 The prize was awarded jointly to:

Steven Chu (1948–), United States
Claude Cohen-Tannoudji (1933–), France
William D. Phillips (1948–), United States

"For development of methods to cool and trap atoms with laser light."

1998 The prize was awarded jointly to:

Robert B. Laughlin (1950–), United States
Horst L. Stormer (1949–), Germany
Daniel C. Tsui (1939–), United States

"For their discovery of a new form of quantum fluid with fractionally charged excitations."

1999 The prize was awarded jointly to:

Gerardus 't Hooft (1946–), the Netherlands

Martinus J. G. Veltman (1931–), the Netherlands

"For elucidating the quantum structure of electroweak interactions in physics."

2000 The prize "for the researchers' work [which] has laid the foundations of modern information technology (IT), particularly through their invention of rapid transistors, laser diodes, and integrated circuits (chips)" was divided, with one half awarded jointly to:

Zhores I. Alferov (1930–), Russia
Herbert Kroemer (1928–), Germany

"For developing semiconductor heterostructures used in high-speed and opto-electronics."

The other half was awarded to:

Jack S. Kilby (1923–), United States

"For his part in the invention of the integrated circuit."

2001 The prize was awarded jointly to:

Eric A. Cornell (1961–), United States
Wolfgang Ketterle (1957–), Germany
Carl E. Wieman (1951–), United States

"For the achievement (in 1995) of Bose-Einstein condensation in dilute gases of alkali atoms, and for early fundamental studies of the properties of the condensates."

2002 The prize for pioneering contributions in astrophysics was divided, with one-half awarded jointly to:

Raymond Davis, Jr. (1914–), United States
Masatoshi Koshiba (1926–), Japan

"For pioneering contributions to astrophysics, in particular for the detection of cosmic neutrinos."

The other half was awarded to:

Riccardo Giacconi (1931–), United States

For "pioneering contributions to astrophysics, which have led to the discovery of cosmic X-ray sources."

2003 The prize was awarded jointly to:

Alexei A. Abrikosov (1928–), United States
Vitaly L. Ginzburg (1916–), Russia
Anthony J. Leggett (1938–), United States

"For pioneering contributions to the theory of superconductors and superfluids."

2004 The prize was awarded jointly to:

David J. Gross (1941–), United States
H. David Politzer (1949–), United States
Frank Wilczek (1951–), United States

"For the discovery of asymptotic freedom in the theory of the strong interaction."

APPENDIX III

Periodic Table of Elements

1 — atomic number
H — symbol
1.008 — atomic weight

Numbers in parentheses are the atomic mass numbers of radioactive isotopes.

1 H 1.008																	2 He 4.003
3 Li 6.941	4 Be 9.012											5 B 10.81	6 C 12.01	7 N 14.01	8 O 16.00	9 F 19.00	10 Ne 20.18
11 Na 22.99	12 Mg 24.31											13 Al 26.98	14 Si 28.09	15 P 30.97	16 S 32.07	17 Cl 35.45	18 Ar 39.95
19 K 39.10	20 Ca 40.08	21 Sc 44.96	22 Ti 47.88	23 V 50.94	24 Cr 52.00	25 Mn 54.94	26 Fe 55.85	27 Co 58.93	28 Ni 58.69	29 Cu 63.55	30 Zn 65.39	31 Ga 69.72	32 Ge 72.59	33 As 74.92	34 Se 78.96	35 Br 79.90	36 Kr 83.80
37 Rb 85.47	38 Sr 87.62	39 Y 88.91	40 Zr 91.22	41 Nb 92.91	42 Mo 95.94	43 Tc (98)	44 Ru 101.1	45 Rh 102.9	46 Pd 106.4	47 Ag 107.9	48 Cd 112.4	49 In 114.8	50 Sn 118.7	51 Sb 121.8	52 Te 127.6	53 I 126.9	54 Xe 131.3
55 Cs 132.9	56 Ba 137.3	57-71*	72 Hf 178.5	73 Ta 180.9	74 W 183.9	75 Re 186.2	76 Os 190.2	77 Ir 192.2	78 Pt 195.1	79 Au 197.0	80 Hg 200.6	81 Tl 204.4	82 Pb 207.2	83 Bi 209.0	84 Po (210)	85 At (210)	86 Rn (222)
87 Fr (223)	88 Ra (226)	89-103‡	104 Rf (261)	105 Db (262)	106 Sg (263)	107 Bh (262)	108 Hs (265)	109 Mt (266)	110 Ds (271)	111 Uuu (272)	112 Uub (285)	113 Uut (284)	114 Uuq (289)	115 Uup (288)			

*lanthanide series	57 La 138.9	58 Ce 140.1	59 Pr 140.9	60 Nd 144.2	61 Pm (145)	62 Sm 150.4	63 Eu 152.0	64 Gd 157.3	65 Tb 158.9	66 Dy 162.5	67 Ho 164.9	68 Er 167.3	69 Tm 168.9	70 Yb 173.0	71 Lu 175.0
‡actinide series	89 Ac (227)	90 Th 232.0	91 Pa 231.0	92 U 238.0	93 Np (237)	94 Pu (244)	95 Am (243)	96 Cm (247)	97 Bk (247)	98 Cf (251)	99 Es (252)	100 Fm (257)	101 Md (258)	102 No (259)	103 Lr (260)

The Chemical Elements

element	symbol	a.n.
actinium	Ac	89
aluminum	Al	13
americium	Am	95
antimony	Sb	51
argon	Ar	18
arsenic	As	33
astatine	At	85
barium	Ba	56
berkelium	Bk	97
beryllium	Be	4
bismuth	Bi	83
bohrium	Bh	107
boron	B	5
bromine	Br	35
cadmium	Cd	48
calcium	Ca	20
californium	Cf	98
carbon	C	6
cerium	Ce	58
cesium	Cs	55
chlorine	Cl	17
chromium	Cr	24
cobalt	Co	27
copper	Cu	29
curium	Cm	96
darmstadtium	Ds	110
dubnium	Db	105
dysprosium	Dy	66
einsteinium	Es	99
erbium	Er	68
europium	Eu	63
fermium	Fm	100
fluorine	F	9
francium	Fr	87
gadolinium	Gd	64
gallium	Ga	31
germanium	Ge	32
gold	Au	79
hafnium	Hf	72
hassium	Hs	108
helium	He	2
holmium	Ho	67
hydrogen	H	1
indium	In	49
iodine	I	53
iridium	Ir	77
iron	Fe	26
krypton	Kr	36
lanthanum	La	57
lawrencium	Lr	103
lead	Pb	82
lithium	Li	3
lutetium	Lu	71
magnesium	Mg	12
manganese	Mn	25
meitnerium	Mt	109
mendelevium	Md	101
mercury	Hg	80
molybdenum	Mo	42
neodymium	Nd	60
neon	Ne	10
neptunium	Np	93
nickel	Ni	28
niobium	Nb	41
nitrogen	N	7
nobelium	No	102
osmium	Os	76
oxygen	O	8
palladium	Pd	46
phosphorus	P	15
platinum	Pt	78
plutonium	Pu	94
polonium	Po	84
potassium	K	19
praseodymium	Pr	59
promethium	Pm	61
protactinium	Pa	91
radium	Ra	88
radon	Rn	86
rhenium	Re	75
rhodium	Rh	45
rubidium	Rb	37
ruthenium	Ru	44
rutherfordium	Rf	104
samarium	Sm	62
scandium	Sc	21
seaborgium	Sg	106
selenium	Se	34
silicon	Si	14
silver	Ag	47
sodium	Na	11
strontium	Sr	38
sulfur	S	16
tantalum	Ta	73
technetium	Tc	43
tellurium	Te	52
terbium	Tb	65
thallium	Tl	81
thorium	Th	90
thulium	Tm	69
tin	Sn	50
titanium	Ti	22
tungsten	W	74
ununbium	Uub	112
ununpentium	Uup	115
ununquadium	Uuq	114
ununtrium	Uut	113
unununium	Uuu	111
uranium	U	92
vanadium	V	23
xenon	Xe	54
ytterbium	Yb	70
yttrium	Y	39
zinc	Zn	30
zirconium	Zr	40

a.n. = atomic number

INDEX

Note: Page numbers in **boldface** indicate main entries; *italic* page numbers indicate photographs and illustrations.

G

N